CHEMOTAXIS

Michael Eisenbach
The Weizmann Institute of Science, Israel

With contributions from:

Joseph W. Lengeler
University of Osnabrück, Germany

Mazal Varon, David Gutnick
Tel-Aviv University, Israel

Ruedi Meili, Richard A. Firtel, Jeffrey E. Segall
Albert Einstein College of Medicine, USA

Geneva M. Omann
University of Michigan, Ann Arbor, USA

Atsushi Tamada, Fujio Murakami
Osaka University, Japan

CHEMOTAXIS

Imperial College Press

Published by

Imperial College Press
57 Shelton Street
Covent Garden
London WC2H 9HE

Distributed by

World Scientific Publishing Co. Pte. Ltd.
5 Toh Tuck Link, Singapore 596224
USA office: Suite 202, 1060 Main Street, River Edge, NJ 07661
UK office: 57 Shelton Street, Covent Garden, London WC2H 9HE

British Library Cataloguing-in-Publication Data
A catalogue record for this book is available from the British Library.

CHEMOTAXIS

ISBN 1-86094-413-2

Printed in Singapore by World Scientific Printers (S) Pte Ltd

This book is dedicated to the memory of Dr Robert M. Macnab—a good friend, a great human being, and an excellent scientist, being among the first to reveal the secrets of bacterial swimming and flagellar function and assembly—who passed away prematurely in September 2003.

Contents

Introduction

How does a sperm cell find the ovum? How do white blood cells direct themselves to the site of injury or inflammation? How do unicellular microorganisms find their food and avoid hostile environments? The common answer to all these questions is: By chemotaxis (Figure 1). Similarly to a butterfly that is attracted to a flower by odor, or a male insect that is attracted to the female by pheromones, many organisms (primarily, but not solely, unicellular) and cells of multicellular organisms are attracted to their targets or repelled from certain chemicals by chemotaxis.

What is chemotaxis? Today this term is used to denote cell movement towards or away from a chemical source, defined as positive and negative chemotaxis, respectively. The chemical is defined as chemoattractant or chemorepellent, respectively. According to the common, broad definition of chemotaxis, any cell motion that is affected by a chemical gradient in a way that results in net propagation up a chemoattractant gradient or down a chemorepellent gradient is defined as chemotaxis. This definition includes three narrower definitions that were used in the past to distinguish between different behavioral mechanisms by which cells approach chemoattractants and avoid chemorepellents:

(a) The original, narrow definition of chemotaxis, also termed topotaxis—a change in the *direction* of movement resulting from active alignment of the cell's axis according to a chemical gradient.

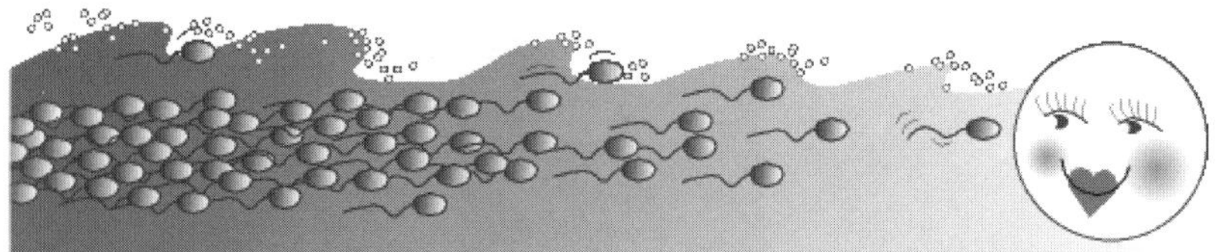

Figure 1. Chemotaxis of sperm cell to the egg: "the origin of life."

Chemotaxis of cells with amoeboid movement (e.g., white blood cells) usually follows this mechanism (Chapters 5 and 6).

(b) A phobic response—a decreased linear velocity in response to a chemical stimulus (normally a stop response) followed by a change of direction. Some bacteria (e.g., *Rhodobacter sphaeroides*) follow this mechanism (Chapter 3).

(c) Klinokinesis—a change in the frequency of spontaneous directional changes in response to a chemical stimulus [7–9, 11, 14, 30, 31]. Some bacteria (e.g., *E. coli*) employ this behavioral mechanism (Chapter 3).

The common denominator of all these mechanisms is that their end result is a directional change up a chemoattractant gradient or down a chemorepellent gradient. These mechanisms should be distinguished from chemokinesis (also termed orthokinesis [7]), which is a mechanism of response that does not involve directional changes and is, therefore, not a part of the broad definition of chemotaxis. In chemokinesis, the *linear* velocity of the cell or organism is altered by the stimulus [7]. Chemotaxis and chemokinesis may occur in parallel, as in the case of the response of sperm cells to substances secreted from the egg (Chapter 7).

The outcome of positive chemotaxis would usually be accumulation of cells or organisms in a region with higher concentration of the chemoattractant. The outcome of negative chemotaxis would usually be dispersal of cells or organisms from a region with higher concentration of the chemorepellent. However, accumulation and dispersal may also be caused by chemokinesis and trapping, for which reason they alone cannot be used as a criterion for chemotaxis (Chapter 7).

Chemicals are not the only stimuli sensed by cells and organisms. Other stimuli include light, temperature, touch, etc. In all cases, the name of the response includes a prefix that describes the stimulus (chemo-, photo-, thermo-, etc.) and the suffix *taxis*, meaning moving towards or away from the stimulus. For example, movement directed by

Table 1. Nomenclature of directed movements in response to various stimuli.

Term	Stimulus	Examples of responsive species	References
Chemotaxis	Chemical	Bacteria, archaea, amoebae, white blood cells, sperm cells	This book
Elasticotaxis	Elastic force	Some gliding bacteria (*Myxococcus xanthus*)	[10]
Electrotaxis	Electrical field	Amoebae	[40]
Galvanotaxis	Electrical current	Bacteria, spermatozoa	[1, 39]
Geotaxis or gravitaxis	Gravity	Bacteria, ciliates (*Paramecium*), flagellates	[3, 12, 15, 33]
Magnetotaxis	Magnetic field	Bacteria	[5, 25, 38]
Phototaxis	Light	Bacteria, archaea, amoebae, flagellates	[2, 6, 13, 17, 18, 22, 26, 29, 32, 35, 36]
Thermotaxis	Temperature	Bacteria, ciliates, amoebae, nematodes, spermatozoa, trophoblastic cells, leukocytes	[4, 16, 19, 20, 27, 28, 34, 37]
Thigmotaxis or mechanotaxis	Touch, mechanical force	Ciliates, flagellates, endothelial cells	[21, 23, 24]

a light or a temperature gradient is termed phototaxis or thermotaxis, respectively. Table 1 includes the nomenclature of the known responses to various stimuli. When the response to the stimulus is directed growth rather than directed movement, the suffix is *tropism*, meaning growing towards or away from the stimulus: chemotropism, phototropism, etc. As will be discussed later in this book, some of these responses, e.g., bacterial chemotaxis and thermotaxis, or archaeal chemotaxis and phototaxis, share a common molecular mechanism. Others, e.g., galvanotaxis and magnetotaxis, are more passive processes.

This book reviews some of the best-characterized chemotaxis systems, from bacteria to human cells. In so doing, the book demonstrates how basic chemotaxis is to life, how widespread it is, and how versatile its physiological functions are. The book attempts to present the state of the art of a number of representative molecular mechanisms of chemotaxis, to indicate unanswered questions surrounding each mechanism, and to suggest future directions for research. In some systems, the implications for health conditions are discussed. Thus, in the next chapter (Chapter 2), Joseph Lengeler surveys the systems and phenomena in which chemotaxis appears to have a role. Some issues raised in Chapter 2 are dealt with again, usually in more detail, in subsequent chapters. In Chapter 3, I review in depth the behavioral and molecular mechanisms of bacterial chemotaxis—the best-understood behavioral system. In Chapter 4, Mazal Varon and David Gutnick discuss how chemotaxis may also be involved in prokaryotic cell–cell communication. In Chapter 5, Jeffrey E. Segall and colleagues do the same for eukaryotes, describing chemotaxis of amoebae and other cells of multicellular eukaryotes with amoeboid movement. In Chapter 6, Geneva M. Omann describes the physiology and the molecular mechanisms underlying chemotaxis of another fascinating system of eukaryotic cells with amoeboid movement—white blood cells. In Chapter 7, I describe what is known about chemotaxis of sperm cells. In Chapter 8, Atsushi Tamada and Fujio Murakami review how growing axons are guided in the nervous system and find their targets by multiple processes of chemotropism. Finally, in Chapter 9, I indicate points of universality and individuality, or commonality and diversity, among the mechanisms and processes discussed in this book.

This publication is aimed to serve as both a textbook for beginners and a reference book for professionals. For easier reading, each chapter that reviews a certain chemotaxis system follows a standard scheme: it starts with a description of the motility mechanism specific to the system (motility is a prerequisite for chemotaxis) and ends with a description of how this motility is modulated by chemotaxis.

References

1. Adler, J. and Shi, W. (1988). Galvanotaxis in bacteria. *Cold Spring Harbor Symp. Quant. Biol.* **53**, 23–25.
2. Armitage, J.P. and Evans, M.C.W. (1981). The reaction centre in the phototactic and chemotactic response of photosynthetic bacteria. *FEMS Microbiol. Lett.* **11**, 89–92.
3. Aswad, D. and Koshland, D.E., Jr. (1975). Isolation, characterization and complementation of *Salmonella typhimurium* chemotaxis mutants. *J. Mol. Biol.* **97**, 225–235.
4. Bahat, A., Tur-Kaspa, I., Gakamsky, A., Giojalas, L.C., Breitbart, H. and Eisenbach, M. (2003). Thermotaxis of mammalian sperm cells: a potential navigation mechanism in the female genital tract. *Nature Med.* **9**, 149–150.
5. Blakemore, R.P., Frankel, R.B. and Kalmijn, A.J. (1980). South-seeking magnetotactic bacteria in the southern hemisphere. *Nature* **286**, 384–385.
6. Choi, J.-S., Chung, Y.-H., Moon, Y.-J., Kim, C., Watanabe, M., Song, P.-S., Joe, C.-O., Bogorad, L. and Park, Y.M. (1999). Photomovement of the gliding cyanobacterium *Synechocystis* sp. PCC 6803. *Photochem. Photobiol.* **70**, 95–102.
7. Diehn, B., Feinleib, M., Haupt, W., Hildebrand, E., Lenci, F. and Nultsch, W. (1977). Terminology of behavioral responses of motile microorganisms. *Photochem. Photobiol.* **26**, 559–560.

8. Dunn, G.A. (1981). Chemotaxis as a form of directed cell behaviour: some theoretical considerations. In *Soc. Exp. Biol. Seminar Ser. (Biology of the Chemotactic Response)* (J.M. Lackie and P.C. Wilkinson, eds.), pp. 1–26. Cambridge University Press, Cambridge.

9. Ewer, D.W. and Bursell, E. (1950). A note on the classification of elementary behaviour patterns. *Behaviour* **3**, 40–47.

10. Fontes, M. and Kaiser, D. (1999). Myxococcus cells respond to elastic forces in their substrate. *Proc. Natl. Acad. Sci. U.S.A.* **96**, 8052–8057.

11. Fraenkel, G.S. and Gunn, D.L. (1940). *The Orientation of Animals: Kineses, Taxes and Compass Reactions.* Clarendon Press, Oxford.

12. Fukui, K. and Asai, H. (1985). Negative geotactic behavior of *Paramecium caudatum* is completely described by the mechanism of buoyancy-oriented upward swimming. *Biophys. J.* **47**, 479–482.

13. Gest, H. (1995). Phototaxis and other sensory phenomena in purple photosynthetic bacteria. *FEMS Microbiol. Rev.* **16**, 287–294.

14. Häder, D.-P. (1979). Photomovement. In *Encyclopedia of Plant Physiology: Physiology of Movement* (W. Haupt and M.E. Feinleib, eds.), pp. 268–309. New Series, Berlin, Heidelberg, N.Y.

15. Häder, D.-P. (1997). Gravitaxis in flagellates. *Biol. Bull.* **192**, 131–133.

16. Higazi, A.A., Kniss, D., Manuppello, J., Barnathan, E.S. and Cines, D.B. (1996). Thermotaxis of human trophoblastic cells. *Placenta* **17**, 683–687.

17. Hildebrand, E. and Dencher, N. (1975). Two photosystems controlling behavioral responses of *Halobacterium halobium*. *Nature (London)* **257**, 46–48.

18. Hoff, W.D., Jung, K.-H. and Spudich, J.L. (1997). Molecular mechanism of photosignaling by archaeal sensory rhodopsins. *Annu. Rev. Biophy. Biomol. Struct.* **26**, 223–258.

19. Hong, C.B., Fontana, D.R. and Poff, K.L. (1983). Thermotaxis of *Dictyostelium discoideum* amoebae and its possible role in pseudoplasmodial thermotaxis. *Proc. Natl. Acad. Sci. U.S.A.* **80**, 5646–5649.

20. Imae, Y. (1985). Molecular mechanism of thermosensing in bacteria. In *Sensing and Response in Microorganisms* (M. Eisenbach and M. Balaban, eds.), pp. 73–81. Elsevier, Amsterdam.

21. Iwatsuki, K. and Hirano, T. (1996). An increase in the influx of calcium ions into cilia induces thigmotaxis in *Paramecium caudatum*. *Experientia* **52**, 831–833.

22. Kreimer, G. (1994). Cell biology of phototaxis in flagellate algae. *Int. Rev. Cytol.* **148**, 229–310.

23. Kreimer, G. and Witman, G.B. (1994). Novel touch-induced, Ca^{2+}-dependent phobic response in a flagellate green alga. *Cell Motil. Cytoskeleton* **29**, 97–109.

24. Li, S., Butler, P., Wang, Y., Hu, Y., Han, D.C., Usami, S., Guan, J.L. and Chien, S. (2002). The role of the dynamics of focal adhesion kinase in the mechanotaxis of endothelial cells. *Proc. Natl. Acad. Sci. U.S.A.* **99**, 3546–3551.

25. Matsunaga, T. and Sakaguchi, T. (2000). Molecular mechanism of magnet formation in bacteria. *J. Biosci. Bioeng.* **90**, 1–13.

26. Melkonian, M., Meinicke-Liebelt, M. and Häder, D.-P. (1986). Photokinesis and photophobic responses in the gliding flagellate, *Euglena mutabilis*. *Plant Cell Physiol.* **27**, 505–513.

27. Metzner, P. (1919). Die Bewegung und Reizbeantwortung der bipolar gegeisselten Spirillen. *Jahrb. Wiss. Botan.* **59**, 325–412.

28. Mizuno, T., Kawasaki, K. and Miyamoto, H. (1992). Construction of a thermotaxis chamber providing spatial or temporal thermal gradients monitored by an infrared video camera system. *Analyt. Biochem.* **207**, 208–213.
29. Nultsch, W. (1970). Photomotion of microorganisms and its interactions with photosynthesis. In *Photobiology of Microorganisms* (P. Halldal, ed.), pp. 213–251. Wiley-Interscience, London.
30. Pfeffer, W. (1883). Lokomotorische Richtungsbewegungen durch chemische Reize. Ber. Deutsche Botan. *Gesellschaft* **1**, 524–533.
31. Pfeffer, W. (1884). Lokomotorische richtungsbewegungen durch chemische reize. *Untersuch. aus d. Botan. Inst. Tübingen* **1**, 363–482.
32. Ragatz, L., Jiang, Z.Y., Bauer, C.E. and Gest, H. (1995). Macroscopic phototactic behavior of the purple photosynthetic bacterium *Rhodospirillum centenum*. *Arch. Microbiol.* **163**, 1–6.
33. Roberts, A.M. (1970). Geotaxis in motile microorganisms. *J. Exp. Biol.* **53**, 687–699.
34. Ryu, W.S. and Samuel, A.D.T. (2002). Thermotaxis in *Caenorhabditis elegans* analyzed by measuring responses to defined thermal stimuli. *J. Neurosci.* **22**, 5727–5733.
35. Schlenkrich, T., Fleischmann, P. and Häder, D.-P. (1995). Biochemical and spectroscopic characterization of the putative photoreceptor for phototaxis in amoebae of the cellular slime mould *Dictyostelium discoideum*. *J. Photochem. Photobiol.* **B 30**, 139–143.
36. Spudich, J.L. and Stoeckenius, W. (1979). Photosensory and chemosensory behavior of *Halobacterium halobium*. *J. Photobiochem. Photobiophys.* **1**, 43–53.
37. Tawada, K. and Miyamoto, H. (1973). Sensitivity of *Paramecium* thermotaxis to temperature change. *J. Protozool.* **20**, 209–292.
38. Vainshtein, M.B., Suzina, N.E., Kudryashova, E.B., Ariskina, E.V. and Sorokin, V.V. (1998). On the diversity of magnetotactic bacteria. *Microbiology* (English translation) **67**, 670–676.
39. Zhang, X., Jin, L. and Takenaka, I. (2000). Galvanotactic response of mouse epididymal sperm: *in vitro* effects of zinc and diethyldithiocarbamate. *Arch. Andrology* **45**, 105–110.
40. Zhao, M., Jin, T., McCaig, C.D., Forrester, J.V. and Devreotes, P.N. (2002). Genetic analysis of the role of G-protein-coupled receptor signaling in electrotaxis. *J. Cell Biol.* **157**, 921–927.

Chemotaxis—A Basic and Universal Phenomenon Among Microorganisms and Eukaryotic Cells*

1. Introduction

> *"Leben ist Bewegung!"*
>
> —Wilhelm Pfeffer [64]

To a layperson, the most obvious difference between living and non-living systems is whether they are motile or not. The tendency of living organisms, from single cells to entire populations, to move actively from less to more favorable environments is a basic and widespread, though very complex, behavior. It rests upon the perception and integration of sometimes conflicting environmental stimuli and a coordinated response. This response consists surprisingly often in changing the speed and direction of their locomotion. Motility is easily visible provided all reactions are fast, as in most animals and in freely motile plants, fungi, and microorganisms, while slow growth movements are often not recognized a priori as motility phenomena. Traditionally, growth

---

* Written by Joseph W. Lengeler, Fachbereich Biologie/Chemie, Universität Osnabrück, D-49069 Osnabrück, Germany.

7

movements and differentiation processes which involve cell migration were treated separately from fast motility reactions, and considered as central in sensory biology, a distinction that obviously relied more on practical and historical reasons than on logical arguments.

All cells have the ability to not only survey their environment through the detection of external stimuli, but also perceive their own physiological states, which act as internal stimuli. For most unicellular organisms, the outside world corresponds directly to the physical world and to the other members of the *biocenosis*, i.e., the community of all living organisms of their ecosystem. The corresponding organisms are usually clustered in clones or populations of one to many species. Even unicellular microorganisms, although they are autonomous cells and complete organisms, cannot be fully understood if viewed only as single cells [82]. Life in temporary biocenoses that comprise mixed populations of different organisms with complementary metabolic and morphological capacities is thus the microorganismic equivalent of multicellular life in higher organisms [48, 57]. In contrast, for cells of a multicellular organism, the outside world is mostly other cells of this organism, and only specialized sensory cells have direct contact with the external physical world and other organisms. Chemical stimuli, in particular, hold information about food availability and quality; about other attractants; about the presence of noxious substances and repellents; and about the density and location of other members of the same species and biocenosis. Hence their added importance for multicellular organisms and for life in social groups, populations, and biocenoses.

Across the entire prokaryotic and eukaryotic world, chemosensory systems for intracellular and extracellular stimuli start by means of receptor proteins with a more or less pronounced stimulus specificity. For external stimuli, receptors are normally located at the cell surface, and stimulation corresponds almost invariably to binding or dissociation of a stimulatory molecule to or from the receptor. This information is transduced over the membrane to intracellular sensors and converted into a signal which eventually elicits a cellular response. Signal transduction systems usually involve protein kinases and phosphatases. These control directly or indirectly the activity of linear and rotary motors when cells respond to stimulation with changes in locomotion. Very often, specific indicator molecules such as alarmones, second-messengers, neuropeptides, hormones, and pheromones, as well as targeting protein subunits that can carry information over longer distances couple sensors and targets to form complex signal transduction networks.

Chemosensory systems were among the first to appear during evolution and the components of signal transduction systems and of motility devices are among the most highly conserved ones [1, 14, 52, 53, 63, 89, 91]. Consequently, the logic behind these systems is remarkably universal. The most characteristic examples of chemosensing originate from three areas:

1. Search for optimal surroundings and partners;
2. Colonization of new biotopes and other social activities;
3. Cell movements during growth, ontogeny, and differentiation.

The former two classes of examples are found in animals, in many plants and fungi, at least during some stages of their life cycle, and in about half the prokaryotes. The third class is normal for all multicellular eukaryotes, and is also found among specialized microalgae, primitive fungi, and even social prokaryotes.

In this chapter, the examples have been chosen to show the generality of chemosensory phenomena. They also point out the often unsuspected common elements found in the behavior of freely motile microorganisms, and within the unicellular stages and multicellular tissues of eukaryotes. In the original definition of W. Pfeffer, any process that causes the oriented movement of a cell or an organism relative to a chemical stimulus was called *chemotaxis* (pl. *chemotaxes*) if the organism was freely motile, and *chemotropism* if only parts of the organism reacted, e.g., when only the roots or shoots of a plant or the hyphal tip of a fungus answered to a stimulus by oriented growth [64, 65]. Chemical substances triggering such a directed and active movement towards the source, hence a positive chemotaxis, were called *attractants* (*"Lock-stoffe"*), the ones causing a negative chemotaxis were called *repellents* (*"Schreckstoffe"*). We will touch only briefly on tropisms, mainly to demonstrate that taxis and tropism share broadly similar chemosensory mechanisms. We will exclude passive movements, undirected motility phenomena, and taxes in more organized animals, and concentrate instead on the oriented movements of whole cells or unicellular organisms from one location to another. No molecular and other details will be given here. They are treated in depth in the subsequent chapters using a few model organisms as paradigms. Wherever needed, we will refer to these chapters with their extensive reference lists for more details. On the other hand, the references of this chapter will concentrate on overview and review articles that allow one to find the published literature, and on books that treat and compare pro- and eukaryotic sensory systems. The list also includes a few systematic textbooks that describe most organisms mentioned [3, 7, 11, 13, 24, 25, 85, 94]. Finally, to facilitate reading of the sections relating to eukaryotes, their systematic classification and the names of relevant model organisms mentioned throughout the chapter are given in Table 1 (see Section 3.2.2).

2. Cell Motility Is a Basic and Universal Phenomenon Among Living Organisms

Although rooted firmly in the ground, plants can colonize new biotopes at a rate of up to 200 m per year, and they can transfer their spores, pollen, and seeds with the help of water, winds, and animals over very large distances; the rabbits from Europe also imported their small parasites to faraway Australia; the same bacteria colonized the so-called black-smokers in the middle of the Atlantic and the Pacific ocean, and the nearest coast thousands of kilometers away. These three examples illustrate the various means of rapid and efficient movement of organisms. Their locomotion depends on structures such as attachment devices and on endurance forms such as spores and seeds, which evolved through the generations to ensure mobility. Although the corresponding locomotory components must be actively synthesized by the organisms, this locomotion is still considered as passive movement. It contrasts *motility* that involves active movements performed by motile organisms for the purpose of locomotion.

2.1. *Active swimming by means of flagella and cilia*

Active swimming by means of cell appendages is among the phylogenetically oldest forms of motility. However, there are two different types of flagella depending on whether we are considering the prokaryotes or the eukaryotes. *Prokaryotes* (Monera), i.e., organisms without a true cell nucleus, comprise the *eubacteria* or ordinary bacteria, and the *archaea*, once considered the "old" or "*archaebacteria*" in Greek. The prokaryotes use rotating flagella to swim through liquid media of low to medium viscosity. The corresponding rotary flagella motors depend on the proton (or sodium) motive force of the cell as an energy source. These flagella and their derivatives, used for swimming in more viscous environments such as soil, will be described in detail elsewhere (see Section 2.2 in Chapter 3). In contrast, *eukaryotic* ("with a true nucleus") microorganisms, as well as *zoospores*, *microgametes*, *spermatozoids*, and *sperms* (see Section 4 for definitions) use a different type of flagella and cilia for swimming. These rest upon microtubule-associated, i.e., linear motors. Despite their different mechanics of movement, flagella and cilia both contain a core of precisely arranged microtubules, the axonema, in a characteristic $(9 + 2)$ arrangement that comprises an outer ring of nine doublet microtubules surrounding two inner singlet microtubules (see Chapter 7 and [34]). It is generally agreed that an ATP-driven sliding of the microtubules generates the force for the periodic flagella beating and cilia bending that determine the overall cell speed [21, 28, 32, 45, 71, 78, 92].

Eukaryotic *flagella* are whiplike motile appendages of the body surface. Their number per cell is usually small although it may increase, e.g., in the Hypermastigida, to considerable numbers [3, 7, 11, 94]. Rather typical eukaryotes, e.g., most flagellated protista and most spermatozoids of algae, mosses, and ferns possess two types of flagella per cell, one causing a net cell movement, the other actively steering the cell [3, 13, 25, 85]. The flagellum may be fused to the cell body by means of an undulating membrane that also allows steering, e.g., in the parasitic Trichomonada and Trypanosomida. *Steering*, i.e., the active orientation of the cell's axis relative to a stimulating gradient, allows a more efficient form of taxis, originally termed *topotaxis* by W. Pfeffer (see Chapter 1, and Section 2.5). In contrast, *cilia* are shorter motile organelles that either cover the entire cell surface or are locally concentrated in bands, e.g., around the peristom ("mouth area") into which they swirl food. While flagella tend to beat, cilia tend to bend in a highly coordinated way, thus creating waves that move over the cell body, e.g., in the Holotrichia like *Paramecium*. By reverting the cilia movements, these organisms can swim forwards and backwards. Cilia may be transformed by fusing many together into an undulating "membrane", and into *cirri* (s. *cirrus*). Specialized cirri, e.g., at the front and end of *Paramecium*, may have sensory functions, while those at the ventral side of, e.g., *Stylonychia* are converted into primitive *crawling* devices allowing these unicellular ciliophora to move along solid surfaces [3, 5, 94].

2.2. *Swarming movements*

Swarming migration is a specific form of bacterial locomotion across solid surfaces that is observed in a number of flagellated conventional bacteria, now renamed eubacteria (see Chapters 3 and 4). In liquid media, these bacteria exist as short (1–5 μm) *swimmer* cells which are propelled by a single polar flagellum or by a flagellar bundle. If swimmer cells encounter media with increased viscosity, e.g., agar surfaces, they begin to form long, nonseptated "snakes" of up to 100 μm length. Most conspicuously, however, their entire surface becomes profusely covered by hundreds of new or *lateral flagella*. The result of this process is a so-called *swarmer* cell that has the unique ability to swarm over solid surfaces, though only as a member of a multicellular group of swarmers [10, 26]. Swarming periods are part of a typical life cycle in which the cells swim and divide as swimmers, differentiate and swarm across solid surfaces or within biofilms (see Section 5.2), stop swarming, and dedifferentiate back to vegetative swimmers.

Based on our understanding of swarming in eubacteria such as *Proteus, Serratia, Vibrio, E. coli,* and *S. enterica,* differentiation processes involve widely conserved pathways controling flagellum biosynthesis, cell motility, and septation rather than evolutionary new programs. Thus in *E. coli,* the flagellar components are identical in swarmer and swimmer cells and the corresponding global regulatory networks are closely related to those detecting carbon and nitrogen starvation, and to those involved in motility control including chemotaxis, osmo-control, cell division and virulence (see Chapter 4). Similar generation or life cycles in which sessile and motile stages alternate periodically are also found among the eukaryotes, e.g., when multicellular algae and fungi release unicellular motile spores and gametes (see Section 3.2.2).

2.3. *Gliding and twitching movements*

To move on surfaces covered with little water, in soils, biofilms, and microbial mats, some microorganisms use gliding and twitching motility (see Chapter 3). *Gliding* is defined as the movement of a nonflagellated or nonciliated cell in the direction of its long axis on a solid surface [86]. *Twitching,* however, is described as intermittent and jerky movements that can occur in all directions and are not coordinated among different cells [53]. Very often, gliding bacteria show a complex lifecycle that alternates between vegetative and sporulating periods (see Chapters 4 and 5). The cells move under *vegetative,* (i.e., normal, well-fed) conditions individually or in small swarms of a few thousands of cells. These behave as multicellular "wolf-packs," which search for food and feed in a coordinated way, i.e., they show social behavior. When cells become starved for nutrients, they stop dividing, and tens of thousands begin to aggregate to form an aerial fruiting-body or *sporangium*. Simultaneously and similar to eukaryotic organisms, the cells begin to specialize and to differentiate into nondividing stalk cells and into myxospores that, similar to other *spores*, represent an endurance or survival form. The developing sporangium heaps up and the myxospores are lifted slightly above surface. Fruiting bodies probably allow easier passive spreading of spores and other endurance forms into the environment [5, 10, 13, 45].

In *M. xanthus*, the *pil* genes (for *pilus* or hair production) that are involved in gliding and in spore formation (see Chapter 4), and their products are related to other systems with a role in cell–cell contact, e.g., between donor and recipient cells during conjugation, or host infection by a parasite in pathogenesis, thus indicating common evolutionary roots [22, 41, 84]. During gliding, some pili are exported, attach to solid surfaces or other cells, are retracted and drag the cell grapnel-like along the surface, or pull two cells together. Motility is coordinated furthermore by a set of proteins (products of the "frizzy" or *frz* genes), which correspond to the *che* genes and their products in enteric bacteria with a central function in chemotaxis (see Section 7 in Chapter 3). Similar observations to those on *M. xanthus* have also been made in other social and multicellular prokaryotes, e.g., *Cytophaga spp.*, *Flexibacter spp.*, and many cyanobacteria [31, 54].

Interestingly, gliding movements are also found among the sporozoa, i.e., mostly parasitic eukaryotic protozoa and a few algae (Desmidiaceae, Diatomeae) (see Table 1). These eukaryotic organisms, too, show a complex life cycle alternating between motile and immobile stages (see Section 3.2.2 and [3, 85, 94]). Their gliding is even less understood than in the bacteria.

2.4. *Crawling and amoeboid movements*

A last form of motility, which is characteristic of all eukaryotic cells, is an apparent streaming or rather *crawling* over solid surfaces and within multicellular tissue layers. The cells move by pushing out temporary cytoplasmic projections called *pseudopodia*, or in specialized forms *filopodia*, *lamellipodia*, *lobopodia*, and *rhizopodia*. Crawling is most conspicuous among the amoebae ("amoeboid movement") and their immediate relatives, and also among the slime molds, e.g., *Dictyostelium* (see Chapters 5 and 6). Central are actin-associated linear motors that involve actin-binding proteins and myosins related to those underlying the contraction of muscles. Actin-associated and microtubule-associated motors with kinesins are also responsible for many intracellular motility phenomena, in particular cytoplasmic streaming, chromatid movement during mitosis and meiosis, transport of vesicles between cell organelles and redistribution of organelles, phagocytosis, cell division, etc. In the context of this chapter, however, only cell spreading and migration in a directed way during growth under the influence of, e.g., chemical gradients is central. But obviously, these various phenomena rest on similar and highly conserved motor sytems that also evolved from chemotactic into chemotropic movements (see Section 6).

As for gliding and twitching, crawling requires that the motile cell forms a temporary, but tight anchorage to the substratum or, when moving in multicellular layers, to neighboring cells. By exerting a dragging force the cells can then move relative to the substratum or to each other. The front is either defined by the largest pseudopodium, the geometry of the substratum, or the position of the neighboring cells. The best-understood examples for crawling are the free-living and the social amoebae (see Chapter 5 and [3, 5, 13, 45]), motile tissue cells of the eukaryotes, e.g., leukocytes, neurons, and fibroblasts (see Chapter 6 and [29, 32, 61, 79]), and cell aggregation during, e.g., regeneration processes (see Section 6).

2.5. *Problems related to nomenclature*

In the absence of external stimuli, movement of freely motile cells is generally a random walk in the free dimensions of space. As stated before, Pfeffer called any process that causes a net cell movement towards or away from a stimulus a *taxis*. Depending on the stimulus, he defined, e.g., chemotaxis, as the process triggered by chemoattractants and chemorepellents, phototaxis as the process triggered by light, etc. (see Table 1 in Chapter 1). In addition, Pfeffer distinguished between the different forms of locomotion by which various organisms reacted in taxes. These were *topotaxis* ("directed reaction") that involves the active orientation of the cell's axis relative to a stimulating gradient, and phob-otaxis ("avoiding reaction") [64, 65]. In *phobotaxis*, the cell does not find the optimal area by a steering mechanism. Instead, it favors the right compared to the wrong directions, i.e., the chemical stimuli change random locomotion into a nonrandom swimming. Phobotaxis is achieved either by altering the cells' frequency of turning, or by altering the cells' swimming speed. This process was later called *kinesis*. It is the favored strategy among animals and hence became almost a synonym for taxis among zoologists. In their literature, taxis triggered by a process which affects the speed is called *orthokinesis,* while processes modulating the swimming turns are called *klinokinesis* (references in [5]). In the literature related to prokaryotes, plants, and fungi, however, the terms introduced by Pfeffer are preferred (see also Section 2 in Chapter 9).

Chemical stimuli are perceived by the larger eukaryotic cells and organisms through comparison of the stimulus concentrations measured at the front and the end of the body, i.e., as spatial gradients. For the smaller prokaryotes, these differences are close to statistical noise, therefore, they must compare concentrations sequentially in time as they move along, a procedure which requires a memory. For pragmatic reasons, definitions such as phobotaxis and topotaxis, or aerotaxis, chemotaxis, osmotaxis, phototaxis, and thermotaxis (see Table 1 in Chapter 1), may be helpful, but they tend to hide common mechanisms or evolutionary roots among the different systems. Thus, chemotaxis involves chemicals and, similar to osmotaxis, ions as stimuli; aerotaxis involves the chemical oxygen but shares many elements with taxes to redox and proton motive force changes that are related themselves to phototactic stimuli (see also Section 3.1.2), etc. Finally, locomotion rests upon chemosensory mechanisms which have most elements in common as will be discussed in Chapters 3 to 8. Not considered in this chapter are forms of locomotion which serve to swirl food towards an animal, cause abrupt flight movements in normally sessile organisms, allow floating by changes in buoyancy, as well as running, swimming, flying, gliding, crawling, and other movements based on muscles in higher animals. These are named here only to complete the picture of locomotion being one of the truly basic and universal phenomena among all living organisms.

3. The Physiological Role of Chemotaxis: Sensory Aspects

As with any organism, the growth of microorganisms in their natural habitat depends on various physicochemical parameters, e.g., pH, temperature, light, the availability of water, oxygen, and, above all, nutrients [3, 5, 48]. In general, when organisms are faced with nutrient limitation, the few nutrients are concentrated locally, and can only be detected at short distance. Microorganisms use in essence the tactics of a hunting dog to find nutrients, food, and prey, and to find their mating partners by means of chemotaxis. After sensing a chemical gradient with one of their multiple sensors, they change their locomotion from a nonbiased to a biased movement. Attractants usually constitute or indicate the presence of food, of prey organisms, and of fellow partners, while most repellents are noxious substances, which indicate unfavorable conditions [36, 98].

Microorganism originally meant any organism not visible to the naked eye, i.e., in the μm range. Today we distinguish between the prokaryotes which lack a true nucleus and the eukaryotic microorganisms represented, e.g., by the Protista (see Table 1). Because microorganisms are so small and in general short-lived, their sensory performances must be viewed in relation to their size, the size of their immediate surroundings, and their lifespan. Thus, *E. coli* (3 μm length) travels the distance of 300 μm in about 10 sec, the equivalent of 300 m for a human being; its memory lasts from 1 to 15 min, corresponding to 4–60 man-years; the spermatozoids (approx. 12 μm length) of the liverwort *Sphaerocarpos* sense the corresponding egg cells from about 5 mm or 80 m for us, a distance they cover in about 40 sec. Chemotaxis in microorganisms is thus equivalent to the reactions triggered by chemosensory stimulation in animals, in particular, taste for close-up chemicals and smell for more distant chemicals. The latter often constitute social signals and are used to recognize and attract mates or individuals of the same population, or to raise alarm and initiate flight movements (see Sections 4 and 5).

3.1. *Chemotaxis among the prokaryotes*

All organisms are surrounded in their environment by tens of thousands of chemicals. However, they survey their immediate environment by sensing only a selected cocktail of substances. These are normally restricted to common nutrient components and to molecules representing their natural habitat in the most faithful way.

3.1.1. *Chemosensing and metabolic activities are coupled*

The principle of this clever strategy is best illustrated using enteric bacteria as an example. Thus, *E. coli* synthesizes permanently only six sensors for external stimuli, i.e., the methyl-accepting proteins or MCPs named Tar, Tsr, Trg, Tap, Tcp, and Aer (see Section 6.1.1 in Chapter 3). These MCPs sense specifically five amino acids, the intermediate γ-amino-isobutyrate, and a few small peptides representing the peptidic world; they also sense citrate, the aeration level, and a few repellents. This may seem a rudimentary and inefficient system until we realize that many more sensors can be synthesized on demand and within minutes. Their synthesis is tightly coupled to the chemicals which are present in the medium or which the cells have contacted in the immediate past; to the current physiological state of the cells; to their metabolism, in particular, that related to carbon catabolism, nitrogen supply, and energetization; and even to the physiological state of the entire bacterial population (see also Chapter 4).

When growing under feast conditions, e.g., in complex media supplemented with 1% glucose, cells of *E. coli* swim slowly and many cells are immobile. On the average, such well-fed cells have less than one flagellum per cell. They contain low levels of the *alarmone* (or indicator molecule called "second messenger" in eukaryotes) cyclic adenosine 3'-5'-monophosphate (cAMP), and express low amounts of all enzymes and proteins which are members of the *crpA* modulon [48, 49]. This *modulon* comprises a group of genes which encode all carbohydrate transporters and catabolic enzymes. Included are also the regulatory proteins FlhD and FlhC, which are required for expression of all flagellar genes and hence flagella synthesis (see Section 2.2. in Chapter 3). Expression of the genes of the *crpA* modulon depends on the global regulatory protein CrpA, which must be complexed with cAMP to activate the corresponding promoters. In the absence of cAMP, e.g., in well-fed cells, gene activation does not occur and the genes remain repressed. Starved cells, in contrast, contain high levels of the alarmone cAMP and swim vigorously by means of 6 to 8 flagella per cell. In addition, all carbohydrate transporters and all catabolic enzymes are now synthesized permanently at a low level. This uninduced level can be increased (induced) within minutes in the presence of their substrates in the medium to high levels.

The proteins and enzymes of the *crpA* modulon include a set of periplasmic binding proteins specific to maltose and maltodextrins (MBP), galactose and glucose (GBP), and ribose (RBP) (see Section 6.2.1 in Chapter 3). These binding proteins form a complex, first with three ABC-transporters for these carbohydrates, and then with the MCPs mentioned above. The MCPs complexed with their binding proteins thus constitute inducible sensors for the carbohydrates mentioned. Under starving conditions, a further class of chemosensors, whose genes are also members of the *crpA* modulon, will also be expressed in the presence of their substrates. There are about 20 enzymes II of the

phosphoenolpyruvate (PEP)-dependent carbohydrate:phosphotrans-
ferase systems or *PTS*s that are specific for a large number of aldoses,
ketoses, amino sugars and polyhydric alcohols ([49] and Section 8.2.7 in
Chapter 3). Because all 20 enzymes II of a cell depend on the unique or
general protein kinase EI for their phosphorylation, the phosphorylation
level of this protein kinase reflects in an integrated way, first the sum of
their activities, i.e., the presence or absence of PTS-carbohydrates in the
medium and their flux into the cell, followed by the ratio of PEP to
pyruvate in the cell, i.e., feast or famine conditions. In *E. coli* the
enzyme *adenylate cyclase*, which converts ATP into cAMP (hence
"cyclase"), is activated by one of the phosphorylated PTS-proteins called
enzyme IIA. Under starvation, where the ratio of PEP to pyruvate, and
hence of phospho-EI to EI, is high, the cyclase is activated and synthesizes high levels of cAMP. As a direct consequence, expression of the
genes included in the *crpA* modulon is high. Increased uptake rates of
any PTS-carbohydrate, however, cause a decrease of the phospho-EI
levels in msec, a transient accumulation of free EI in the cells, and
through inhibition of CheA involved in tumble generation, prolonged
chemotactic runs (see Section 8.2.7 in Chapter 3) as well as an immediate decrease in intracellular cAMP levels. This is the first remarkable
example of how the physiological state and, in particular, feast and
famine conditions, modulate flagella synthesis, cell motility, and the
chemosensory capacities of a cell. And it becomes apparent that motility and chemotaxis are part of a larger cell program, which may be called
the functional unit "quest for food" [48, 49].

3.1.2. *Cellular energy levels are central in chemosensing*

A second, equally remarkable example among microorganisms is
related to the sensing of oxygen, of the cellular energy levels, and even,
for phototrophic organisms, of light [5, 18, 45]. There is increasing evidence that the metabolic effects of oxygen, of the proton motive force,
of the redox potential, and, where relevant, of light are closely interre-
lated at the level of the flow of reducing equivalents through the electron transport systems. As in the previous example, it is thus possible
and sufficient for the individual cell to monitor its total energy level by
sensing one of these parameters. Not surprisingly, in various organismic
groups different parameters are sensed. Surprisingly, however, the
underlying *PAS domains* are highly conserved from the prokaryotes
to the eukaryotes. PAS is an acronym formed from the names of the

proteins in which motives involved in sensing oxygen, light, redox potential, and energy levels were first recognized (see [2, 89] and Section 8.2.7 in Chapter 3). In *E. coli*, MCP VI or Aer is the *"aerotaxis"* transducer that was originally thought to be directly involved in the sensing of oxygen. It contains a PAS-domain with a noncovalently bound flavine adenine dinucleotide (FAD) cofactor whose redox state reflects redox changes in the electron transport systems. Thus, Aer seems to be the ultimate sensor for a variety of stimuli, in particular, changes in the concentration of external oxygen; of electron acceptors such as nitrate, fumarate, and trimethylamine oxide; of carbon sources like succinate and glycerol, whose metabolism generates electron acceptors; and of uncouplers which, like lipophilic and weak organic acids, affect the intracellular pH, and hence the proton motive force [48]. Because redox changes coupled to carbon catabolism, ATP levels, and ion fluxes over membranes are so intimately linked, many more substances which apparently trigger chemotaxis may in fact be triggering what should be more appropriately called *"energotaxis"*. Such candidates are the ions K^+, Na^+, phosphate, and arsenate in enteric bacteria, which affect the ATP and the proton motive force levels; carbon sources which trigger a metabolism-dependent chemotaxis in members of the α-subgroup of the purple eubacteria, e.g., *Rhodobacter spp.* and *Rhizobium spp.*; and the intracellular pH sensed by the MCP I or Tsr in *E. coli*. Similarly, light is apparently sensed in some purple eubacteria, e.g., *Rhodospirillum spp.* as changes in the redox state of cytochrome C molecules, which are cofactors of the photosynthetic electron transport system [4, 72].

Excessive light intensities and high oxygen concentrations are deleterious to all living organisms, probably because they generate highly toxic oxygen forms. In motile organisms both conditions cause a negative chemotactic reaction and trigger stress-resistance programs. Thus it is perhaps not surprising that the global regulators FlhD and FlhC, products of the flagellar master operon *flhDC* (see Section 2.2 in Chapter 3), and the sensor Aer have more complex physiological roles than surmised before [70]. FlhD alone is involved in the regulation of cell division, while the heterotetramer FlhD/FlhC represses enzymes involved in aerobic respiration, e.g., for NADH, glycerol, pyruvate, and succinate metabolism, and certain catabolic enzymes, transporters, and sensors involved in chemotaxis. Among them is Aer, which besides its role as a sensor, also regulates a number of genes involved in anaerobic respiration, e.g., nitrate and nitrite, fumarate, and dimethylsulfoxide reductases, glycerol 3-phosphate and formate dehydrogenases, hydrogenase 2, citrate synthase, cytochrome O-oxidase, and enzymes of the Entner–Doudoroff pathway, a major supplier of redox equivalents. When we add the global regulator cAMP.CrpA, to this regulatory and sensory network which, as discussed above (see Section 3.1.1), links the expression of the flagellar master operon *flhDC* to the functional unit carbon catabolism, it becomes clear again that chemosensing involves more than chemotaxis alone [49]. The emerging picture, even for the enteric bacteria, is obviously much more complex than anticipated before.

3.1.3. *Cellular sensors are variable in structure*

One of the hallmarks of the prokaryotes is their astonishing diversity in metabolic pathways coupled with their fast adaptability to changes in the environment. We have only begun to realize that a similar variety exists among their chemosensing systems (see Section 6 in Chapter 3). Thus, besides membrane-integral and membrane-associated sensors, there are soluble sensors which are located in the cytoplasm and sense intracellular indicator molecules [10, 57]. Furthermore, besides sensors with a broad specificity, which detect more general parameters such as energy levels, or feast and famine stress conditions, there exist other sensors which react to specific stimuli. These sense intracellular concentrations of oxygen molecules directly by means of a heme B-containing transducer [2, 29]. Often, the general and specific sensor activities are not redundant. This increases the flexibility of an organism to adapt, especially when the cell integrates the information gathered through both systems, and hence responds in a coordinated way to, e.g., chemical and phototactic stimuli. Thus, the archeon *Halobacterium salinarum* senses UV and blue light directly through two sensory rhodopsins, SRI and SRII, which contain retinal as light-sensitive cofactors ([5, 76] and Section 4.2 in Chapter 9). The rhodopsins are complexed to an MCP-like protein, called HtrI and HtrII, respectively, and to CheA. At the same time, a proton motive-force-sensitive MCP and MCP-like chemosensors are also present and also modulate the CheA kinase activity. We will see later how the *phototrophic* eubacteria and the cyanobacteria, that is bacteria that depend mostly on photosynthesis as energy source, use similar arrangements to coordinate their response to light and nutrient conditions within the natural habitat (see Section 5.3).

3.1.4. *The role of protein synthesis in chemosensing and learning*

A final aspect, largely neglected in bacterial chemotaxis up to now at the expense of activity control, relates to the synthesis of sensors and its control. The reaction of an organism to a given chemical or any other stimulus cannot be predicted solely based on the knowledge of its genes, its sensory components, and its regulatory networks. Its immediate prehistory and the actual environmental conditions must also be known, in particular, whether the cells were starved or not, and whether the sensors, metabolic enzymes, and the transporters are induced or were repressed. Cells of the wild-type strain *E. coli* K-12, grown in minimal media and on glucose as sole carbon source, show a low chemotactic response to most PTS carbohydrates, and to maltose, galactose, and ribose. As described before, the corresponding enzymes II and periplasmic binding proteins, i.e., (parts of) the sensors, are not induced under these conditions. When such cells are inoculated into soft-agar swarm plates containing $100\,\mu\text{M}$ D-mannitol and $100\,\mu\text{M}$ D-glucitol (also called D-sorbitol), the cells multiply and swarm out in two sharp concentric rings separated by about $5\,\text{mm}$ [48]. Cells from the outer ring grow and follow exclusively mannitol. As a class A substrate, mannitol strongly represses the synthesis of the glucitol-specific enzymes II and the corresponding catabolic enzymes. Cells from the inner ring consume the class B substrate glucitol after a delay of about $45\,\text{min}$ during which these enzymes are induced, and then follow its gradient. When reinoculated into a similar plate, cells of the outer ring form two sequential rings again, while cells of the inner ring move for about three to five generations in one concentric ring and consume both polyhydric alcohols simultaneously. The latter cells have "learned" to react to glucitol in the presence of mannitol. Under appropriate conditions, this "memory or maintenance effect" can be perpetuated for generations until the sensors and metabolic enzymes are repressed and lost from the cells through degradation and dilution at each cell division. The two populations present in the swarm plates are isogenic and constitute a *clone*, i.e., all cells have the same genes and all are descendents of a single cell. On the first plate, these clonal cells were divided into two stable, but physiologically different, subpopulations. Obviously, their sensory reactions are largely determined by the prehistory and other epigenetic parameters (see Section 6.3), e.g., the presence or absence of inducible sensors. "Learning," however, corresponds to the synthesis of such sensors. Cells of *R. sphaeroides* similarly show a strong response, but only to that nutrient which is currently limiting growth, a phenomenon that is almost certainly based on related memory effects [4, 5, 40].

Specific repressors and activators that control operons and regulons normally react to specific molecules such as substrates and metabolic end products. In contrast, global regulators respond to physiological states, e.g., feast and famine, anoxia, and other stress conditions. Different concentrations of indicator molecules such as cAMP, or the phosphorylation levels of signal transduction systems and of two-component systems (see Section 7 in Chapter 3) often mirror these conditions in the cell. Instead of controlling the switch of the flagellar motor, as in the chemotaxis of enteric bacteria, global regulatory systems can also regulate gene expression of entire modulons and other large groups of genes controlling extended metabolic networks in response to the changing physiological conditions [48, 49]. Examples are the FlhD/FlhC and Aer controlled networks discussed above (see Section 3.1.2 and [70]), and the ArcB sensor of *E. coli*, which also contains a PAS domain and senses redox changes [89]. Together with the response regulator ArcA, this sensor forms a two-component system which controls the synthesis of several terminal oxidases. Other sensors control in various bacteria the synthesis of photosynthetic components, of dicarboxylate transporters, of anaerobic respiratory pathways, and of global regulators involved in numerous bacteria in nitrogen fixation and metabolism (Ntr, Nif, Fix, etc.) [4, 5, 40, 89]. Some of these are even involved in controlling the activity of alternative sigma-subunits of the RNA polymerase. Thus, RpoF and RpoS in the enteric bacteria control cell differentiation processes and eventually cell surface rearrangements, including flagella synthesis [10, 48, 57]. This tight coupling between chemosensor sytems, cell locomotion, and differentiation processes has been conserved among all living organisms (see Sections 3.2 and 5). Similarly, preconditioning, adaptation, learning and related memory effects in higher organisms are all based to a large extend on the synthesis of new proteins and the degradation or loss of old proteins (see Chapter 8 and [33, 59, 79]).

3.2. *Chemotaxis among eukaryotic organisms*

Research in the 19th century was largely devoted to the phenomenological description of the types and forms of movements among the larger eukaryotic flagellated or ciliated protista, the zoospores and the motile gametes of otherwise sessile plants, fungi, and animals; of the crawling of amoebae and leucocytes; of the various motility devices; and of the classification of motile organisms [21, 25, 34, 39, 77, 98]. It had also been noted that organisms including the bacteria were attracted by some, and repelled by other chemicals. However, only a few scientists tackled the problem in a systematic and analytical way, and even fewer recognized it as part of sophisticated sensory systems. Among the first to do so was W. Pfeffer [64–67], who was so deeply convinced of the universality of all sensory mechanisms that he even tested the effect of the latest anaesthetics, chloroform and ether, on his pet organism *Bacterium commune*, perhaps a strain of *E. coli* [75].

3.2.1. *The importance of model organisms*

Pfeffer and others realized quickly that the new type of fundamental research should be restricted to a few well-chosen organisms, each best suited for a specific problem. Thus, *B. commune* represented a model organism not only for chemotaxis, but also for medically important bacteria. Several cyanobacteria, e.g., of the genus *Anabaena, Nostoc, Oscillatoria*, and *Phormidium*, were chosen as the most primitive *autotrophic* organisms, i.e., able to obtain most of their cellular carbon by carbon dioxide fixation and photosynthesis, to study phototaxis and gliding motility [31]. Today, cyanobacteria are considered as highly evolved and social prokaryotic algae. *Chlamydomonas, Volvox,* and their relatives represented the green algae with well-developed flagella, and the eukaryotic type of photosynthesis [12, 34, 66]. Hence the use of these primitive plants or *protophyta* in studying flagella structure and movements, phototaxis in plants, and the differentiation from unicellular to pluricellular algae. *Dictyostelium, Physarum,* and other slime molds represented the motile amoebae within the fungal world [13, 38], while freely motile zoospores and gametes of other fungi; of many sessile algae; and of liverworts, hornworts, mosses, and ferns were a curiosity [9, 25, 35, 44, 73, 95]. Although especially well suited for, e.g., chemotaxis tests with specific sex attractants, it was not clear how accurately they represented the sensory capacities of true multicellular organisms, and, in particular, of animals. Therefore, the next models had to be *heterotrophic* organisms that, similar to other animals (hence *protozoa*), needed to obtain their cell carbon assimilation of organic compounds; had to contain a "nervous system" even if it was primitive; and preferentially had to be accessible to electrophysiological methods. These were *Paramecium, Euplotes, Blepharisma*, other ciliophora, some free-living amoebae, and animal sperms [39, 68].

From about 1900 to 1950, interest in biological research shifted to physiological and genetical problems, in particular photosynthesis, differentiation, the role of chromosomal and other genes, and sensory problems which could be tackled with electrophysiological methods. Interest in sensory problems in microorganisms was limited to photo-taxis, gravitaxis, galvanotaxis, and the related tropisms, and eventually ceased almost completely. Since about 1960 the revival of interest has been linked intimately to the development of modern biochemical and molecular genetics methods [1, 34, 36, 77, 93, 98]. This approach allowed not only the identification of genes, gene products, and cellular components, but also their analysis in purified form, i.e., *in vitro*. The combination of biochemical methods with specifically selected mutants also became a powerful alternative to studying regulators and signal transduction networks *in vivo*. Originally, the new approach was restricted to a few model organisms from genetics, e.g., *E. coli, S. enterica, Chlamydomonas reinhardtii* [28, 32, 51, 71], *Paramecium aurelia* [78, 92], and, interestingly, *Drosophila melanogaster*. After the recent establishment of genetical methods for many organisms, the number of model organisms increased very rapidly. Besides the older models it includes today a multitude of single cell and social eubacteria, the archeon *Halobacterium salinarum* (references in Chapters 3 and 4), eukaryotic plants, e.g., *Arabidopsis thaliana*, some molds and fungi, and even the yeast *Saccharomyces cerevisiae* ([5, 27, 38, 52] and Table 1). The importance of immobile organisms like *Saccharomyces* [16, 17, 23] and, *Arabidopsis* [56], of the nematode *Caenorhabditis elegans* [59], and of *D. melanogaster* [30] in the studies of chemotaxis rests upon the universal biochemistry of sensory networks which, as predicted by W. Pfeffer [67], are highly conserved in all living organisms. The results obtained in recent times through this combined approach will be described in depth in Chapters 3–8 for several of such model organisms.

3.2.2. *Primitive eukaryotic groups showing true chemotaxis*

Many eukaryotic organisms have been said to show chemotaxis towards a huge number of chemicals [9, 20, 27, 39, 46, 66, 68, 90, 93]. However, knowledge based on direct tests has rarely progressed beyond identification of the molecules triggering a reaction. Furthermore, for most organisms a physiological role of chemotaxis in their natural habitat is assumed, but has rarely been proven through rigorous tests. Such direct tests would require, e.g., a comparison of the ecological fitness between a wild-type population and populations of isogenic mutants lacking either specific chemosensors, chemosensing in general, or motility [97]. Long lists exist on chemicals which have ever been shown to trigger a reaction within any lower eukaryote (references in [12, 13, 34, 67, 74, 83, 98]). The tests, however, were often inadequate because they did not differentiate properly between metabolic and sensory effects, or between real attraction and simple entrapping ("fly-paper effect"). In the remaining part of this section, rather than giving such lists, we will give a brief overview of the organismic groups for which chemotaxis seems relevant and refer to later chapters for details of the various model organisms representing specific groups (Table 1).

Table 1. Systematic groups of the eukaryotes with predominantly motile organisms.

Official name	Trivial name	Representative classes (In parentheses model organisms)
KINGDOM (REGNUM)	**EUKARYA** (eukaryotes)	
SUBKINGDOM	**PROTISTA** (unicellular eukaryotes)	
DIVISION (PHYLUM)		
Archamoebaea		
Tetramastigota	Flagellates	Hypermastigida (*Giardia, Trichomonas*)
Kinetoplasta	Flagellates	(*Trypanosoma, Leishmania*)
Euglenophyta	Euglenids	(*Astasia, Euglena*)
Gymnomycota	Slime molds	Acrasiomycota (*Dictyostelium*) Myxomycota (*Physarum*) (*Plasmodiophora*)
Dinophyta	Dinoflagellates	(*Noctiluca*)
Apicomplexa	Sporozoa	Gregarinida (*Gregarina*) Coccidia (*Eimeria*) Haemosporidia (*Plasmodium*)
Ciliophora	Ciliates	Holotrichia (*Paramecium*) Spirotrichia (*Blepharisma, Euplotes, Stylonychia*)
Mastigomycota	Oomycetes (water molds)	Oomycota (*Achlya, Peronospora, Phytophtora, Saprolegnia*)
SUBKINGDOM	**EUMYCOTA** (true fungi)	
Chitridiomycota	Chytrids	(*Allomyces, Blastocladiella, Olpidium*)
Zygomycota		(*Phycomyces*)
Ascomycota	Sac fungi	(*Neurospora, Saccharomyces*)
Basidiomycota	Club fungi	(*Ustilago*)

(Continued)

Table 1. *Continued*

Official name	Trivial name	Representative classes (In parentheses model organisms)
SUBKINGDOM	**PROTISTA** (algae)	
Cryptophyta		Cryptomonadales
Heterokontophyta	Yellow–green, Golden algae	Chrysophyceae (*Vaucheria*) Diatomeae
	Brown algae	Phaeophyceae (*Cutleria, Ectocarpus, Fucus*)
Acanthopodida	Amoebae	Sarcodina (*Amoeba, Entamoeba*)
Neomonada		Mesomycetozoa Choanoflagellates
SUBKINGDOM	**METAZOA** (animals)	
Parazoa	Sponges	Porifera
Eumetazoa	True animals	(*Hydra, Caenorhabditis, Drosophila, Mus*)
SUBKINGDOM	**CORMOBIONTA** (algae, plants)	
Rhodophyta	Red algae	
Chlorophyta	Green algae	Chlorophyceae (*Chlamydomonas, Volvox*)
	Mosses	Bryophyta (*Marchantia, Sphaerocarpos*)
	Ferns	Pteridophyta (*Lycopodium, Isoetes, Selaginella, Dryopteris, Pteridium*)
	Gymnosperms	Palm trees (*Zamia*)
	Angiosperms	Flowering plants (*Arabidopsis*)

Originally, the eukaryotes were classified into five *kingdoms* (*regnum*): protozoa, fungi, algae (originally "simple aquatic plants"), green plants, and animals. Among these groups, motile (monadale) forms were restricted to the unicellular protozoa which included the flagellates, the ciliates, and the amoebae; to some primitive fungi and algae; and to the animals. Modern classification systems, which mostly rest upon the 18s-rRNA method, i.e., on DNA similarities, deviate substantially from the traditional view [15, 55, 85, 94]. Thus, the algae are now assigned to different kingdoms, e.g., the "blue–green algae" or cyanobacteria that were reclassified as prokaryotes as described above.

Other organisms have been reclassified as autonomous *divisions* (*phylum*, pl. *phyla*) within the eukaryotic kingdom of protista as listed roughly according to their 18s-rRNA similarities in Table 1. The table also includes those higher evolved systematic divisions that contain permanently or transiently motile forms, and a few model animals mentioned in this chapter. The new classification clarifies many inconsistencies created by the previous accidental lumping together of all simple unicellular organisms into one group and indicates new links. Thus, the divisions Tetramastigota (previously Mastigophora or "whip-bearing") and Kinetoplasta contain heterotrophic flagellated organisms (protozoa) that often show a parasitic (Trypanosomes, Leishmania, Trichomonas) or symbiotic (Hypermastigida) life style. Their closest relatives are the Euglenophyta with *Euglena* whose chlorophyl-free variant *Astasia* has long been classified as an "animal" [7, 11]. The autotrophic (protophyta) flagellated organisms are further subdivided into the divisions Dinophyta, Cryptophyta, Heterokontophyta with the well known brown algae, Rhodophyta (red algae), and Chlorophyta. This last and important division contains all green algae, the mosses, ferns, and all land plants.

The unicellular and freely motile slime molds (Gymnomycota) and oomycetes (Mastigomycota) were long considered as a curiosity among the hyphal fungi. Their reclassification as autonomous divisions close to the other protista now helps to explain why some (e.g., *Dictyostelium*) generate amoeboid forms during their life cycle; others (e.g., *Saprolegnia*, *Achlya*, *Phytophtora*, and *Peronospora*), flagellated zoospores; and still others (e.g., *Physarum* and the parasitic *Plasmodiophora*), both forms. Within the primitive division Archamoebaea, the original ability to form flagella and pseudopods is also preserved. Other phylogenetically based continuous transitions between various groups are the flagellated *Gregarina* and *Eimeria*, among the otherwise immobile sporozoa (Apicomplexa); the common elements found between flagella and cilia; and finally between the amoeboid movements of the slime molds and of the true amoebae or "pseudopod-carriers" (Acanthopodida). The latter now form an independent division closely related to the true fungi (Eumycota), the red algae, and the Heterokontophyta. A decisive event during evolution was the transition from the unicellular to multicellular forms. Transition apparently occurred several times and in different divisions of the protista, in particular, among the Chrysophyceae to the brown algae, among the slime molds, among the amoebae and the Neomonada to the animals, among the motile Chitridiomycota to the fungi (and perhaps to the red algae), and among the Chlorophyceae to the green algae. The transition from unicellular and motile, i.e., *monadale*, to multicellular forms passed the filamental stage to end ultimately either in the siphonal and hyphal organisms, e.g., the fungi in which no septa separate individual cells, or in elaborated thalli represented, e.g., by the familiar seaweeds. All these transition forms can still be found among the extant protista, algae, fungi, and simple animals [3, 7, 11, 24, 44, 85, 94].

An interesting example is the potential ancestors of the sponges. The division Neomonada includes the *choanoflagellates*, which apparently are the ancestors of the *choanocytes*. These specialized "collar cells" of the sponges (Porifera) cover the wall of the interior cavity where they trap bacteria and other small food particles, just like the choanoflagellates. However, when necessary, e.g., during differentiation or regeneration from an injury, the choanocytes relocate by an amoeboid movement within the sponge body and start as totipotent cells (*archaocyta*) the regeneration of a new organism (see Section 6.3 and [55]). In parallel with the increased specialization during evolution, many cells lost, at least transiently, their motility. However, wherever cell motility has been conserved, interestingly, the cells have conserved the same chemosensory capacities as the protista. The description of this trend will be the topic of the next section in which the essential role of chemotaxis in fertilization and reproduction will be described.

4. The Role of Chemotaxis in Fertilization and Reproduction

The continuous reduction in free cell motility is characteristic of the evolution of the eukaryotes. This trend started within the prokaryotes, in particular, among autotrophic and sessile forms which inhabit solid surfaces, e.g., the filamentous cyanobacteria, social eubacteria, and prosthecate-budding bacteria [58]. The animals gained relatively early the ability to move their entire body, and restricted cell motility to differentiation processes and to the microgametes or sperms (see Section 6 and Chapters 7 and 8). The algae, in contrast, had already evolved at the monadale level "mother cells" of increased size which, through a few sequential mitoses, produced several daughter cells or *sporozoids* [25, 41, 85]. These remain together until the maternal envelope is destroyed, then dissipate first by active swimming, and later by passive diffusion. During evolution, the daughter cells grow progressively longer, and in the end permanently together, thus forming, first filaments, then complex thalli and other forms still found among the extant brown, red, and green algae. In a parallel evolution, primitive and motile fungi developed into immobile hyphal forms and into the true fungi (Zygomycota, Ascomycota, Basidiomycota), in which millions of wall-less cells are fused to a single organism [13, 24]. However, even immobile organisms must propagate periodically and invade new biotopes. As a rule and to this end, they produce and release specialized unicellular forms which swim actively by means of one or several flagella. These cells rarely propagate through amoeboid movements. They are called *zoospores* when produced for asexual reproduction, or *gametes* when involved directly in fertilization and sexual reproduction. Gametes are often, but not always, the direct product of a meiosis (meiospores). In any case, they contain a single or *haploid* set of chromosomes. For obvious reasons, the appearance of meiosis, of haploid gametes, and of fertilization, which generates a *diploid* zygote, were coupled during evolution in an obligatory way.

4.1. *The role of chemotaxis in fertilization*

The first gametes had already appeared among the Protista, the Oomycota, the Chrysophyceae and the Chlorophyceae (Table 1), e.g., within the genus *Chlamydomonas* [28, 32, 37, 71]. Under normal conditions, their sporozoids reproduce asexually in a *vegetative* or v-cycle. Under stress, in particular, nitrogen limitation, they switch to the *generative* or G-cycle and sexual reproduction. When the sporozoids are converted into morphologically indistinguishable gametes, they are called *isogametes*. Despite their appearance, isogametes are genetically different and express complementary *mating types* (mt^+ and mt^-). This becomes visible during conjugation that involves in principle two cells of opposing mating type. Sporozoids and gametes are propelled by two flagella at a rate of about $100\,\mu$m/sec. During the G-cycle the flagella display mating-type specific agglutinins. These are rod-like and high molecular weight glycoproteins. Surprisingly, no evidence exists for a chemotactic attraction between gametes before mating [28, 45]. Instead, random collision between the flagella from mt^+ and mt^- gametes causes adhesion and cell clumping similar to the accumulation of flies on a fly-paper. Direct contacts between mt^+ and mt^- flagellar agglutinins cause an immediate adhesion. On first contact, a surface motility system transports further agglutinins to the flagellar tip, thus reinforcing adhesion and initiating cell–cell contact. At the same time, a cAMP-dependent signal cascade triggers several cellular responses. These cause degradation of the cell walls, fusion of the two naked gametes, and ultimately conjugation. After a meiotic cell division, the diploid and immobile zygote releases two mt^+ and two mt^- vegetative cells. Although the flagellar apparatus, the swimming behavior, the conjugation, and especially the rhodopsin-dependent phototaxis of *Chlamydomonas* have been analyzed in great detail [71], our knowledge of its chemotaxis remains comparatively scarce. In view of the results with related organisms, in particular those with anisogamy and oogamy, this knowledge is probably very incomplete.

In many extant algae, fungi, and other primitive organisms, the gametes of opposite mating type differ. When one is more active and smaller (*microgamete*), and the other more sluggish and larger (*macrogamete*), the phenomenon is called *anisogamy*. Later in evolution, microgametes evolved into motile male gametes, i.e., *spermatozoids* in higher plants, and *sperms* in animals, while macrogametes became immobile egg cells or *oocytes*, hence *oogamy*. In all algae, and in most primitive fungi and animals, the gametes are released into surrounding waters. Although swimming actively, they are transported over larger distances by water currents, raindrops, and passive diffusion. Fertilization occurs when the gametes are attracted at short distance through chemotaxis to each other, i.e., from mm to cm. Luring the motile microgametes becomes imperative for the immobile egg cells at the latest when the eggs remain within a sessile maternal organism (see [6, 13, 25, 44] and Chapter 7 for sperms).

4.2. *The role of gamones in the brown algae and in fungi*

Substances excreted by the gametes or their producer organisms to attract gametes of opposite mating type or sex are called *gamones* ("*gam*etic horm*ones*") in plants, and *pheromones* in animals [12, 23, 27, 43, 44, 64, 65, 73, 74, 77, 85, 98]. Gamones have best been analyzed in some brown algae and primitive fungi. All transitions between isogamy, anisogamy and oogamy are found among the brown algae. These alternate periodically between an asexual and a sexual reproduction cycle, that comprises first the *gametophyte* or gamete-forming organism [25, 34, 85]. As a rule, their motile gametes carry a long flagellum, which drives the cell, and a short flagellum serving to steer the cell. In the absence of a chemotactic gradient, the male microgametes swim in long, non-biased circles. In contrast, the "female" isogametes and the macrogametes become immobile upon maturation. Similar to egg cells they stick to solid surfaces, and begin to excrete a cocktail of gamones [37, 43]. As soon as male gametes sense the gamones, they accelerate flagella beating ("chemokinesis"), the swimming circles become narrower, and they orient their body axis towards the female gamete ("chemotopotaxis"). Usually, many male gametes adhere by means of their long flagellum to the female and clump. One gamete fertilizes the female, which stops gamone excretion. The unsuccessful gametes dissipate, and a diploid zygote is formed, which sooner or later develops into a sporophyte (see below). It produces haploid meiospores, from which new gametophytes are generated to complete the life cycle [85]. Many gamones have been isolated from various brown algae and shown to attract male gametes as efficiently as the female gametes do. These gamones are highly unsaturated alicyclic hydrocarbons, which are derived from unsaturated fatty acids and form a new class of natural substances [34, 37, 43]. As expected, the gamone cocktail produced by an organism is sensed best by the gametes of his species, less well by gametes of related genera, and not at all by gametes from more distantly related brown and other algae. Virtually no details are known about the underlying signal transduction or motility control systems.

The chitridiomycete *Allomyces macrogynus* produces from gametophyte mycelia bright orange male and colorless female gametangia, and from these haploid gametes of similar coloration [13, 34, 37, 43]. Both isogametes are propelled by a single polar flagellum. Exposure to water triggers their release and this process is accelerated for male gametes in the presence of female gametangia. These and the female gametes excrete the gamone sirenin, a bicyclic sesquiterpene named after the sirens of Greek mythology. Sirenin attracts male gametes with a threshold affinity constant of about 100 pM, and is highly specific for *Allomyces*. Sirenin seems to be inactivated during the process by the male gametes. Some evidence indicates the existence of a complementary gamone ("Parisin") excreted by the male to lure the female. After fertilization and fusion of the gametes, the zygote retains both flagella until encystment and germination to a new diploid sporophyte mycelium. The biflagellated zygote is insensitive to sirenin, but attracted by amino acids. Thus, it resembles the monoflagellated zoospores which are produced by the sporophyte to propagate the organism through an additional vegetative reproduction cycle.

4.3. *The role of gamones in the archegoniata*

Among the land plants, two large groups with more than 30 000 species depend on water for fertilization [32, 35, 73, 85, 98]. These are the *Bryophyta* or liverworts, hornworts, and mosses, and the *Pteridophyta* or lycopods, horsetails, and ferns. In both groups, two generations alternate between the gametophyte and the sporophyte plant. Within the bryophyta, the haploid *gametophyte* ("gamete-forming plant") is an autonomous plant, the familiar moss. After fertilization, the now diploid zygote develops into a *sporophyte* ("spore-forming plant") that still depends on the gametophyte for nutrients, water, etc. At the end of its life cycle, the sporophyte produces after a meiotic cell division haploid spores that dissipate, and after germination produce new gametophytes. The gametophyte of the pteridophyta, in contrast, is an inconspicuous plant the size of a pea, while their sporophytes (the familiar horsetails and ferns) are complex plants which may reach 25 m in size. Gametophytes produce in the *antheridia* specialized male organs, motile and mostly biflagellated spermatozoids in the bryophyta, but multiflagellated spermatozoids in the pteridophyta, and the egg cells in female organs called *archegonia*. The archegonium is characteristic of the bryophyta and the pteridophyta, hence their common name *archegoniata*. Each gametophyte produces antheridia, archegonia, or both, such that the spermatozoids must swim through (rain or dew) water to reach the archegonia on the same or a close-by plant. Archegonia are bottle-like structures with one egg cell at their bottom. All other interior cells lyse upon maturation and produce a slime which swells in contact with water and extrudes in part from the bottleneck. The egg cell or its immediate neighbors produce powerful gamones that diffuse into the slime and a short distance away from the orifice of the archegonium. Thus, passing spermatozoids of the liverwort *Sphaerocarpos donnellii* were attracted from a distance of about 2 mm, swarmed from 10 to 15 min around the opening of the neck before they dissipated again [47]. Usually, a few spermatozoids enter the slime and, following obviously a gradient of the gamone, reach the egg cell, which is eventually fertilized by a single spermatozoid [35].

Most modern textbooks report that, based on the results of Pfeffer and his contemporaries, "proteins" are the gamones for (some) liverworts, sucrose for many mosses, citrate for some lycopods, and malate for several ferns. In fact, this is an oversimplification of the facts and of the literature [6, 9, 12, 34, 35, 44, 83, 95, 98]. Pfeffer believed that spermatozoids hit randomly the archegonial slime and stick to it before they are attracted by gamones to the egg cells, and that the egg cells excrete the gamones until fertilization. He showed that L-malate in the medium inhibited the attraction of fern spermatozoids to their archegonia, and that malate attracted these spermatozoids in his famous capillary test with a threshold concentration of 0.0001% (75 μM). From this he concluded that the fern gamone was "a substance acting analogous to

L-malate" [64, 65, 67]. No gamone of any archegoniate has been identified, although they clearly exist. Thus, when a mature archegonium of *Sp. donnellii* was ripped of its thallus, about 10 times more spermatozoids were attracted to its bottom than to its neck opening [47]. Neither immature archegonia and other parts of the thallus, nor mature archegonia from distantly related species attracted its spermatozoids. Furthermore, isolated pieces of its slime and agar droplets, into which mature archegonia had been included for a while, but removed before testing, also caused a temporary and specific response. Brought into water, mature antheridia released within 2–5 min their spermatogonia cells, and from these, motile spermatozoids. As in many algae, spermatozoids lying closest to an archegonium were released first, i.e., the gamones accelerated their release. Unstimulated spermatozoids swam in long smooth spirals and filled the medium uniformly. When they passed within 5 mm an agar droplet with gamone or a ripped off archegonium they began to swim within seconds in shortened spirals toward the gamone source ("topotaxis"). At the bottom of such an archegonium, was a "bee-swarm"-like accumulation reaching between 0.25 and 0.5 mm diameter in 2–5 min, the swarm being separated from unstimulated spermatozoids by a comparatively free swimming zone. After about 15 min, the majority dissipated rapidly but a clump of immobile spermatozoids remained at the center of attraction. At this stage, even newly hatched spermatozoids were no longer attracted. Immobilization was apparently due to paralysis of the long swinging flagellum while the ventral steering flagellum kept beating vigorously. Clumping was not seen after attraction to gamone-containing agar droplets.

Based on these results, *Sp. donnellii* produces gamones which are concentrated near the egg cell in the archegonium and trigger chemotaxis in spermatozoids. Either the same or a second gamone accelerates release of the spermatozoids from mature spermatogonia, and agglutinin-like factors seem involved in the attachment of spermatozoids to the archegonial slime. Spermatozoids from *Sp. donnellii* can easily be distinguished from those of the closely related liverwort *Marchantia polymorpha*. When mixed and exposed simultaneously to archegonia from both species, freshly hatched spermatozoids were only attracted to their archegonia, and none swam to archegonia from *Riella spp.*, another close relative [47]. This argues for a high specificity of the liverwort gamones and against amino acids, sucrose, malate, and other trivial molecules being the gamones for, e.g., all liverworts, mosses, and ferns. From entire thalli of *Sp. donnellii*, a fraction containing small peptides (about 20–40 amino acid residues) has been isolated. It resembles the natural gamone(s) in heat stability, hydrophilicity, pH optimum, etc. and attracts *in vitro* spermatozoids with high efficiency [80, 81]. Neither peptone nor other protein hydrolysates attracted spermatozoids, nor did they inhibit attraction to archegonia, and similarly for sucrose and L-malate [47]. In view of these data from archegoniata, and those from many algae and fungi, it seems likely that specific, but as yet unidentified gamones exist among the archegoniata which trigger chemotaxis. In addition, the various spermatozoids react to ions (preferentially K^+ and Ca^{2+}), carbohydrates, amino acids, peptides, dicarboxylic acids, alkaloids, and aliphatic amines as reported in the literature [98]. This variety of chemical stimuli and the fact that the spermatozoids are haploid forms of otherwise diploid organisms make them excellent model organisms for the study of chemotaxis and even chemotropism in higher plants.

4.4. *Embryophyta use chemotropism in fertilization*

Among the highest evolved plants, i.e., the Gymnosperms and the Angiosperms or flowering plants, a continuous evolution is visible [35, 85]. It begins within the primitive palm trees which still release from tubelike gametophytes their motile spermatozoids close to the egg cells. In all other plants, the male gametophyte is reduced essentially to the *pollen tubes*. These grow in a directed way towards the egg cells, obviously attracted by gamones, i.e., they show chemotropism, and after fusion release a nucleus into the egg cell [63]. This conjugation is reminiscent of some primitive green algae (Conjugata) in which filaments of opposite mating-type pair, and a cell from one filament, stripped of its cell wall crawls to a naked cell of the other filament for conjugation and cell fusion [25]. In the true fungi, hyphae of opposite sex also grow in an oriented or chemotropic movement towards each other, and fuse for mating in a process called *gametangiogamy* [13, 24]. Clearly, many elements involved in the chemotaxis of freely motile organisms must be closely related to those underlying chemotropism in plants and fungi, as well as fertilization and differentiation in animals (see Chapters 7 and 8).

4.5. *Relations to other eukaryotic chemosensory systems*

There are good reasons to consider gamones and the chemotactic or chemotropic reactions they trigger as a subset of the larger class of processes triggered and controlled by pheromones. These molecules are central in complex differentiation processes, in particular those involved in sexuality and life cycles [16, 17, 36]. The basic and conserved components upon which pheromone sensing rests include across systems and phyla receptors that are coupled to GTP-binding or G-proteins, second messengers, nucleotide- and voltage-gated channels, and a plethora of protein kinases and phosphatases as described in Chapters 5–7, and summarized in Section 3 in Chapter 9. For example, the budding yeast and many related yeasts begin to mate when starved for nitrogen [23, 27, 45, 50]. A G-protein is involved in this nutrient sensing, but interestingly the corresponding effector is the adenine cyclase, and hence, as in bacteria, the second messenger cAMP. The cyclase is normally controlled in animals by heterotrimeric G-proteins. This is an important example showing how conserved components of various signal transduction systems can be combined in different ways and in different organisms to systems with new properties. As part of the new program started by nitrogen starvation, cells of opposite mating types begin to excrete soluble pheromones, in the budding yeast small peptides [23, 27]. These bind cross-wise to highly specific receptors which trigger a G-protein-dependent signal transduction system, and which ultimately control mating, nucleus fusion, meiosis, and ascospore formation.

As for many other complex processes in which chemosensory mechanisms play an essential role, here again, motile unicellular model organisms have been very helpful. These are for the mating pheromones *Chlamydomonas* as discussed before [28, 32, 71], and several ciliophora, e.g., *Blepharisma japonicum* [37] and *Paramecium* [90, 92]. The ciliophora share with sensory neurons the ability to modulate electrical properties. In this way they can depolarize and hyperpolarize in the presence of chemical stimuli and integrate, similar to many olfactory receptor neurons, this information with signals emanating from other sensory pathways. In contrast to neurons, *Paramecium* and other motile unicellular organisms can be used to analyze—through relatively simple, and yet precise and sensitive assays—differences between, e.g., wild-type and mutant cells in their swimming behavior, mating, or differentiation, and to isolate the corresponding gene products in larger amounts [5, 45, 78]. A similar approach is used increasingly to analyze very complex behavioral systems, e.g., chemotaxis in the nematode *C. elegans* [59], or the courtship of *D. melanogaster* [30].

5. The Role of Chemotaxis in Colonizing New Biotopes: Social Aspects

As mentioned before, motility in one or the other form is almost ubiquitous among the animals and aquatic organisms. Even sessile and terrestric plants and fungi either produce, if only transiently, motile forms during their life cycle, or have developed devices for passive propagation with the help of water, wind, animals, etc. Considering the wide distribution of motility, and the impressive variety of chemicals sensed by most organisms, we can safely conclude that chemosensory capacities, and hence chemotaxis and chemotropism are heritable traits with great adaptive or selective value.

5.1. *Problems related to tests in natural habitats*

Despite the obviousness of this conclusion, not much is known about the actual contribution of chemotaxis in a particular situation. Rather typical is the question about the roles that motility and chemotaxis play within the genera *Agrobacterium*, and *Sinorhizobium* in the early events of plant–pathogen interactions. These bacteria infect a large variety of plants, but only at wounded sites, and then form either plant tumors like *A. tumefaciens*, or nodules in which, e.g., leguminous plant roots and *S. meliloti* live in symbiosis. The bacteria react in laboratory tests to amino acids, sugars, flavones, root exudates, and to excised cells [97]. From *Agrobacterium spp.*, strain A348, four mutants with defects in chemotaxis were isolated and tested for the ability to form tumors. When used to infect plants directly or to inoculate sand, the mutants were almost as virulent as the wild-type, but in predried soil they were avirulent. Furthermore, the reports on their chemotactic responses are contradictory (references in [97]). Thus, strain A348 seems to be less motile than strain C58, normally used in virulence tests. Some groups report that acetosyringone, the natural inducer of the *vir* genes for virulence, and related phenolic compounds are attractants while others see no reaction; some claim that the Ti-plasmid which carries the *vir* genes must be present for these reactions, others say no. A similar controversy exists concerning the attraction of *S. meliloti* by root exudates and luteolin, the inducer of its *nod* genes for nodulation. In these and related studies, the tests are often not reliable, the test conditions not well defined, and general attractants like carbohydrates, amino acids, and organic acids are rarely distinguished from specific chemoattractants originating from the wound or root extracts.

5.2. *Cooperative consortia, biofilms, and other associations*

The more it becomes clear that unicellular organisms can only be fully understood as members of a (clonal) population and of a biocenosis, the more it becomes apparent that in natural habitats cooperation between different organisms is as common as competition. Pheromones, taste, olfaction, and "chemotaxis" are central in social contacts of higher organisms, and similarly in symbiosis and related phenomena of microorganisms. Such cooperations are often facultative, i.e., the partners leave and have to find each other periodically.

The natural habitats of microorganisms are remarkably diverse [87]. For the prokaryotes, their ability to colonize even the most extreme biotopes rests on the unequaled metabolic versatility, coupled with their capacity to react almost instantaneously to any change in the environment. Another key element of the high adaptability of microorganisms is the ability to position themselves rapidly in new ecological niches. At first, this seems unlikely in view of the fact that so many prokaryotes can only move passively, e.g., through spores and other endurance forms, and that the distances covered by even the fastest protista do not exceed mm/sec. Consequently, biotopes like marine beaches that are only accessible to organisms able to move rapidly hence and forth and in short intervals, should not be accessible to slowly motile and immobile organims. Thus, growth of sulfide-oxidizing phototrophic bacteria in marine environments is determined by opposing gradients of light intensity and sulfide concentrations in the surrounding water. Green sulfide-oxidizers (Chlorobiaceae) are always found below purple bacteria (Chromatiaceae) due to their better light exploitation efficiency and higher sulfide tolerance [4,72]. And both accumulate below the sulfide/oxygen interface where light is available in sufficient quality and intensity, and where sulfide is protected from oxidation. It is clear that, due to the daily changes in light intensity, this interface and productive zone changes its position dramatically twice per day, definitely more than immobile organisms could possibly follow. In such situations, transient or long-lasting cooperations and associations between motile and immobile partners should be very helpful.

The term *consortium* describes such a temporary symbiotic association of two or more different bacteria in a physical and structured way. Consortia comprise a colorless central bacterium, and attached to its surface (hence *epibionts*) are up to 20 phototrophic bacteria [48, 62].

The chemotrophic partner reduces sulfate and sulfur to H_2S and seems to supply the phototrophic epibionts with reducing equivalents. In some consortia, e.g., those called "Chlorochromatium aggregatum" and "Pelochromatium roseum", the immobile epibionts "hitch-hike" at the expense of the flagellated central bacterium, all cells divide in synchrony, and the consortium even shows a phototactic response. This behaviour not only suggests an exchange of compounds and signals between the partners, but also that consortia are formed for the purpose of keeping their cells within optimal surroundings for photosynthesis and sulfur reduction.

In line with these observations it has been found that many bacteria grow preferentially on solid surfaces rather than in the surrounding aqueous phases [10, 58]. Furthermore, they have a strong tendency to colonize surfaces in cooperation with many prokaryotic and eukaryotic organisms, and to settle in highly structured *biofilms* [8, 19, 87, 88]. The composition of biofilms is variable, depending mostly on the species available, the particular surface, and the environmental conditions. Biofilms develop in several stages: from the initial adhesion of the first bacteria by means of pili, fimbriae, exopolysaccharides, and other adhesive molecules through several maturation stages, to the dispersion of motile cells or microcolonies from the biofilm. Initially, adhesion sites are located randomly over the surface and the composition of the population is relatively uniform. Surface growth, various metabolic activities, cell movements, and secondary invasion by new organisms rapidly change the homogeneous surface into a heterogeneous and complex architecture populated in the end by hundreds of different species. During biofilm maturation, specialized bacterial forms and organisms assemble at the surface. They produce exopolysaccharides, alginate, slime and other protective macromolecules; they tolerate aerobic conditions; and together they can attack insoluble nutrients such as cellulose, chitin, lignin, and other hard-to-destroy materials. Obligate anaerobes, however, must colonize the interior of the biofilm. This assembling and reassembling during the formation of the biofilm requires lateral motility, very often gliding, swarming, and crawling (see Sections 2.2–2.4). Similar to other social activities, these movements are normally controlled and coordinated through signal transduction and global regulatory networks which react to quorum sensing signals.

When the environmental resources are exhausted or as the consequence of seasonal changes, swarmer cells begin to convert to swimmer cells that leave the biofilm individually or in the form of microcolonies. This detachment, release and subsequent spreading of motile forms is a physologically regulated process. It requires that the organisms change their cell surface components, in particular those related to cell adhesion. Furthermore they must synthesize motility devices that allow active locomotion, e.g., flagella, for swimming and drifting to new habitats. As described before (see Section 3.1.2), in *E. coli* and other enteric bacteria the synthesis of the flagella and "energotaxis" are tightly controlled in relation to the energy level of the cells by means of the MCP-transducer Aer, the global regulators CrpA and FlhD/FlhC, and by the alternative subunit RpoF of the RNA-polymerase. During stationary, starvation, and anoxic periods, most bacteria develop stress-response resistance and, linked to this, a characteristic gene expression and protein profile (references in [48]). Bacteria from biofilms in the late maturation and detachment phase resemble in their physiological state stressed and stationary-phase bacteria. These phases often correspond to high-cell density conditions during which quorum sensing (see Chapter 4) integrates the physiology and the behavior of microbial communities. Biofilm formation thus resembles in many aspects other developmental processes as found in the life cycles of many organisms [6, 7, 10, 13, 41, 54, 58, 84].

5.3. *Microbial associations in natural habitats*

Studies with populations of a single organism grown in sterilized media and tested under laboratory conditions are a prerequisite to analyzing at the molecular level the signal transduction and regulatory networks involved in chemotaxis. In contrast, biofilms reproduce more accurately the environmental conditions as they prevail in natural habitats because they include natural cooperations and competitions between the members of this complex biocenosis. The corresponding studies transgressed recently from a purely academic interest to application-oriented aspects when it was found that biofilm formation causes severe problems for medicine, the environment, and industry [19, 88]. Suffice it to mention that for medicine, the problems are related to accelerated dental decay; to the formation of reservoirs of potential pathogens on catheters, metal-containing artificial joints and other medical implants; and to increased resistance of biofilm members to the immune defense and antimicrobial agents. Water fouling in ponds, and even in larger lakes and seas; the consequences of acid-rain damage to the rhizosphere of trees; metal corrosion in water and oil pipelines or in ship hulls; and "wall growth" in fermenters and sewage plants are further examples of problems caused by biofilm formation.

Beneficial associations between bacteria and eukaryotes, in which motility seems essential, are numerous ([42, 48, 54] and Chapter 4). Very often, heterotrophic and motile eukaryotic organisms host chemotrophic or phototrophic bacteria and organisms, either on their body surface or in specialized organs, and then become secondary autotrophs. The apparent gain for the microorganismic epibiont is that the larger and mobile host provides a more continuous supply of nutrients, redox equivalents, or light, and furnishes the other needs of the epibiont or symbiont. Thus, sulfide-oxidizing chemoautotrophic bacteria related to those described above in bacteria-bacteria consortia, are also found as epibionts which grow on the surface of ciliophora, nematodes or roundworms, and shrimps [69]. These associations are found in habitats containing unstable sulfide/oxygen gradients, e.g., in marine sediments, hydrothermal vents in the volcanic deep sea, or in regularly flooded mangrove forests. The epibionts typically grow in specialized local areas where they can reach very high densities, e.g., bacteria standing upright in regular monolayers, or forming long nonseptated filaments. The host organism, which after a while often begins to feed on his epibionts, serves as a transport vehicle and seems to actively seek locations in which his epibionts reach high growth rates. Thus, nematodes migrate up and down in sediments, which allows their epibiontic bacteria to collect and ingest reduced sulfur derivatives that they metabolize after the host has moved to oxic sediment regions. Shrimp and ciliophora hosts, on the other hand, swim in and out of the volcanic vent plumes, or move up and down sulfide/oxygen/light-gradients for similar reasons. The signals exchanged between these symbionts, which allow coordination of their movements with their metabolic requirements, remain to be identified.

In the rumen, a specialized part of the gut of ruminant animals, and in the hind-gut of termites, two strictly anaerobic ecosystems, which rest upon cellulose degradation, have evolved through the generations [48, 87, 96]. Both systems comprise on the whole ciliophora with extensive ciliation, copiously flagellated Hypermastigida, fungi producing multiflagellated zoospores, and highly motile bacteria specialized to swim in viscous media, e.g., of the genus *Treponema*. Using phototaxis—whose ecological value in seeking optimal light conditions is reasonably well understood for many phototrophic bacteria [4, 31, 36], and algae like

Chlamydomonas, *Volvox*, and *Euglena* [41, 45, 51, 57, 71]—as an analogy, we can assume a similar role for chemotaxis in these motile organisms [18, 77]. Thus, many cyanobacteria grow in long filaments which are basically immobile [10]. The filaments consist of dividing vegetative cells and of heterocysts, i.e., specialized cells able to fix nitrogen. Under nutrient or light limitations, the filaments begin to produce shorter, but motile filament fragments called *hormogonia*. The corresponding cells transiently stop cell division, but begin to spread by means of gliding motility. Other cyanobacteria form symbioses with a large variety of plants and fungi, e.g., the well-known lichens. The host organisms seem to send out pheromones which control hormogonium differentiation and dedifferentiation, and which act as chemoattractants. Thus, hormogonia of the genus *Nostoc* normally show a positive phototaxis [31]. If necessary, they move into gland cells in the interior of their host plant *Gunnera*, obviously attracted by a strong chemical stimulus emanating from the host that must overcome the positive phototactic stimulus [54]. Each of these examples corroborates the assumption that chemotaxis, like phototaxis, has an eminent ecological value. This assumption is corroborated further by the observation that for each organism studied in some detail, the attractants and repellents which show the lowest threshold in stimulation, which are sensed by the greatest variety in receptors, and which represent the strongest stimuli, invariably indicate the presence of prey, food, or food components; of sexual and social partners; and of noxious substances found regularly in the natural habitats of that organism.

In summary we can conclude that chemosensing, and consequently chemotaxis and the closely related chemotropism are important for all motile cells and organisms in at least two areas: first, long-ranging ("vertical") locomotion mostly involved in finding rapidly new biotopes and new partners, and second, short-ranging ("horizontal") locomotion, traditionally often considered as slow cell growth events, and mostly related to differentiation processes. This will be the topic of the last section.

6. The Role of Chemotaxis in Differentiation Processes of Multicellular Organisms

By instinct we do not associate movement and motility with slow growth phenomena as seen during various differentiation processes. These are, in particular, the development of a complex plant or animal from a unicellular zygote during ontogeny, and from spores within sexual and asexual reproduction cycles, and also extensive regeneration processes after wounding or cloning of an organism. Despite appearances, the capacity to move is conserved in most cells and the corresponding genes are conserved in all cells of multicellular organisms. This can be best seen in the regular occurence of motile unicellular forms during the life cycles of otherwise sessile plants, fungi, and primitive animals, e.g., spores and gametes, and of motile cells in higher evolved animals, e.g., leucocytes, granulocytes, and fibroblasts. Embryonic and larval stages which move actively by means of flagella and cilia emanating from cells covering their surface, also reveal this capacity. While many cells, especially those located in multilayered and fully differentiated tissues, remain static, other cells are capable of considerable movement. Cell spreading and migration under the influence of chemotactic gradients will be discussed in this section.

6.1. *The cytoskeleton is central in amoeboid crawling movements*

As indicated in Section 2.4 in this chapter, amoeboid movements within tissues and crawling over solid surfaces are characteristic of motile eukaryotic cells. Their migration rests on the plasticity of the cytoskeleton, which is capable of rapid and sustained reorganization. Effective cell movement requires the establishment of a leading edge where the various forms of pseudopodia, e.g., filopodia and lamellipodia or microspikes, are pushed temporarily out of the cell while the rest of the plasmalemma is quiescent. In the process, microtubules help to transport membrane vesicles to the leading edge whereas actin filaments help to drag the cell body. A forward movement requires, furthermore, that transient adhesions be formed with the substratum or with neighboring cells, that can support the dragging forces exerted by the cell motor (see Chapters 5, 6, and 8). This form of movement in which only a part of the body moves during each single step, and net cell movement requires several steps again shows how closely related taxis and tropism are, at least when amoeboid movements are involved.

Chemotaxis, i.e., the oriented movement of freely motile and free living amoebae, and in particular of the cellular slime molds, has been analyzed in great detail (see Chapter 5). Like other organisms, *Dictyostelium* and other Acrasiales react preferentially to compounds which signal the presence of prey bacteria, or of other cells of their species.

The latter chemoattractants, called acrasins, are derivatives of pterins, folic acid, cAMP, and dipeptides [5, 27, 45]. Apparently, many sensory transduction pathways found among the vertebrates are already present among the slime molds, e.g., G-protein coupled receptors for extracellular cAMP, and the corresponding heterotrimeric G-proteins, adenylate and guanylate cyclases, and components of the inositolphosphate pathway. In contrast to bacterial chemotaxis but similar to other eukaryotic systems, control of amoeboid motility seems to involve multiple sensory outputs as generated by the binding of chemical stimuli to a receptor protein [38, 52]. Furthermore, these sensory systems measure both temporal and spatial changes. The cellular response, however, rests upon actins, e.g., in pseudopodia extension, and probably myosin in cytoskeleton association and dissociation. As described above for swarmer and other motile cells, e.g., zoospores and gametes, motility phases preclude as a rule cell division. Both processes require dramatic changes in the cytoskeleton that have opposing effects on cell fluidity and rigidity [16, 17, 46, 60]. Such close connections hint again at very old evolutionary roots common to differentiation and sensory processes.

6.2. *Physical guidance must not be confused with chemotaxis*

While it is reasonably easy to distinguish whether an isolated organism reacts to chemicals or other tactic stimuli, i.e., whether it shows chemotaxis or thigmotaxis and stereotaxis, such a distinction can be very difficult to establish for cells moving in packs and within multicellular layers. Conditions which generate "fly-trapping paper" effects or which cause physical guidance of such cells and thus mimic chemotaxis, aggravate the problem even more. In the social bacteria and other gliding microorganisms, the first or pioneer cell moving over a solid surface already generates a path covered with slime molecules ([10] and Chapter 5). Other cells which hit such a path have a strong tendency to follow it, e.g., because the slime molecules have a slightly antiadhesive effect that facilitates gliding. Theoretically, other molecules with strong antiadhesive effects could inhibit adhesion of the pseudopods to the substratum and hence net crawling movements. A structured deposition of both types of antiadhesive molecules would then provide a clearly delineated path for migrating cells. Cells following such a path would seem to move in an oriented way, perhaps guided by chemoattractants which have been excreted by the pioneer cell or by cells lining the path, when in fact orientation is caused exclusively by physical guidance. As will be shown in detail in Chapters 5, 6, and 8, substrate adhesion, physical guidance, and chemoattraction act in concert to guide eukaryotic leucocytes and fibroblasts by chemotaxis, and neurons by chemotropism [79]. Another important determinant of these movements is the direct interaction of the moving cells with other cells. Thus, in contact inhibition, cells stop moving after collision with another cell and will attempt to move in another direction. When surrounded completely by other cells, movement ceases such that cells in culture form monolayers rather than pile up. To invade, e.g., other tissues, the cells must overcome this contact inhibition.

6.3. *Model systems to analyze motility in differentiation processes*

During the development and growth of plants, fungi, and animals, the clonal cells originating from the zygote differentiate, i.e., they begin to perform different tasks and to organize themselves in space and time to generate the species-specific morphology or body, physiology, and behavior. As stated above, no cell or organism can be described completely solely based on the knowledge of its genes. At any given moment, the state of each cell depends as much on its genome as on the past and present environmental conditions. Such mechanisms that can control gene expression without themselves being under the direct control of genes are called *epigenetic* ("behind the genes"). They decide ultimately which of the numerous genes encoded in the genome will actually be expressed, e.g., in specific tissues or in the different members of a clonal population (see Section 3.1.4), and start the corresponding program of protein synthesis, cell proliferation, differentiation, and eventually cell death. These environmental conditions include, especially for cell clones developing from a zygote to a mature multicellular organism, the various cells of this clone. Realization of such a controlled differentiation requires intensive cell–cell interaction and communication. This communication, however, rests heavily on chemosensing and information transfer, often over long distances using chemicals as information carriers. The signaling molecules are excreted by the cells for long-range effects which must reach distant cells, or they remain attached to the producer cell for local effects [33, 45, 61, 79]. The molecules range from nucleotides (second messengers) and steroids, to peptides (neurotransmitters, hormones), and polypeptides (growth factors). The by now classic example of a differentiation from an apparently homogeneous initial stage, the zygote, is the embryogenic development of *Drosophila* [45]. During the egg formation or oogenesis, the maternal organism imposes on the zygote a network of gene regulators. These define at the earliest developmental stages the frontal, caudal, dorsal, ventral, and segmental outline of the body. At this stage, and in particular during the *gastrulation* phase in which the various tissue layers (ectoderm, endoderm, mesoderm) are generated and positioned in the growing embryo, oriented nucleus and cell movements become extensive. Transfer of cells to a new location will bring the genes under the control of new regulators that trigger new cell translocations, and so on, until a mature fly hatches after finishing the larval and pupal development. Obviously, cell movements and chemotaxis are central to these epigenetic effects during differentiation and concomitant cell migration. During gastrulation, major parts of the blastoderm are translocated over the embryo and begin to invade its interior to form the endoderm. At about the same time, in *Drosophila* so-called pole cells evade the syncytium and move together with the developing endoderm into the embryo. Here they will form at a later stage the germ line, migrate through the mid-gut wall, and eventually into the gonads. Amoeboid movements are especially suited for such processes in multicellular layers. Thus, in the absence of a solid surface, one cell can adhere to a neighbor cell as substratum to pile up, or stretch, for tissue invasion or evasion, and for cell adhesion or cell sorting.

The species-specific reaggregation of macerated sponges whose cells were filtered through a fine mesh has provided another model system for cell–cell adhesion and cell movements during regeneration processes of primitive animals (Parazoa), *Hydra* being a further example for higher evolved or true animals (Metazoa). Sponges lack real structured tissues, but contain a series of motile cells with a variety of functions in stabilizing the body and in transporting food through the animal [7, 11, 94]. Central are so-called *archaocyta* ("Urzellen") which constitute a population of totipotent cells, i.e., all other cells can be generated from archaocyta including the germ cells (egg and motile sperm cells) and the choanocytes mentioned above (see Section 3.2.2). Upon germination, the zygote generates from the first cell divisions unequal blastomer cells, one class being flagellated. The resulting embryonic stage called *amphiblastula* is an autonomous organism that swims actively by means of these flagella. During gastrulation, the flagellated cells are invaginated and move into the interior where they line as choanocytes ("collar cells") the interior cavity of the animal. In *Hydra* with its true tissues, so-called *interstitial cells* also constitute a class of totipotent cells which remain permanently able to crawl through the body [7, 11, 94]. In this respect, they resemble leucocytes, granulocytes, fibroblasts, and other motile cells of the vertebrates (see Chapter 6). In both groups of animals, extensive regeneration processes after wounding, and the development of a new organism from the zygote require that these mobile cells remain intact. After killing specifically these cells, e.g., by irradiation or inhibiting their activity with colchicine, tissue and body regeneration is no longer possible.

Except for embryonic development, neuronal guidance and axon growth has become recently one of the best studied systems for analyzing the role of chemotaxis in differentiation processes (see Chapter 8). During development, learning, and regeneration, neurons project axons that move long distances along stereotypic pathways to find their targets [79]. The trajectory of an axon is selected by its motile tip or *growth cone*. The cone constitutes a sensory and motor structure that recognizes and responds to a combination of short- and long-range cues [14, 33, 61]. The cone resembles free-living amoebae in many respects. Thus, the axon projects from its growth cone long extensions called *filopodia*, and shorter ones called *lamellipodia*. Both contain actin-based motors and allow crawling. The filopodia are highly motile and they carry essential parts of the sensory system, e.g., membrane-bound receptors that detect the directional cues. The length, mobility and flexibility of the filopodia allows them to advance, retract, or turn, and hence to navigate the growing axon around obstacles and toward their ultimate target. The directional cues include cell surface molecules, adhesives and antiadhesives, as well as extracellular matrix proteins that provide short-range and physical guidance. Others are excreted molecules that diffuse from their source to establish chemical gradients, and give long-range guidance information. As a rule, specialized ("guide-post") cells are positioned along the presumptive growth path, and the cone navigates in many small steps from guide-post to guide-post until it reaches eventually the faraway target. The axon grows consequently in an oriented way guided in the long-range by chemical cues, i.e., it shows chemotropism. The mechanisms involved, however, are virtually the same as those active in the chemotaxis of their free living relatives (see Chapters 5, 6, and 8). Some of these chemoattractants are members of large gene families, e.g., the netrins and semaphorins, or the receptors for the immunoglobin adhesion molecules together with the corresponding protein tyrosine kinases and phosphatases, and are highly conserved among the species from worm (*C. elegans*) and fly (*D. melanogaster*) to all vertebrates, further corroborating the hypothesis that chemotaxis and chemotropism are closely related. In contrast to taxes, tropisms are usually accompanied by an increase in cell and body mass as documented by axon growth, and often by cell division as can be seen in a tree growing towards light.

7. Conclusions

As documented by the numerous examples given above, movement and especially motility are basic and universal phenomena in the living universe. Their major advantage is to enlarge considerably the environment surveyed by an organism and to increase its fitness by making accessible more and larger biotopes than could possibly be colonized by immobile organisms. Oriented movements guided by environmental cues increase fitness further because they allow partners to be found over longer distances and the detection of food and prey through chemoattractants that diffuse away and produce gradients which then guide the sensing organisms.

Molecular biology and a series of technical improvements, however, have revealed dramatically new insights into the underlying and complex mechanisms. They showed first, that the mechanisms and strategies used by all organisms are very old, going back in many cases to the unicellular and social prokaryotes, and that the components are highly conserved, i.e., universal [1, 91]. Second, it became apparent that no living organism, in particular no unicellular prokaryotic and eukaryotic microorganism, can be described fully when viewed solely as living in splendid isolation. Their capabilities in cell–cell communication and in cell specialization are remarkably sophisticated. Third, we have only begun to understand the importance of protein synthesis and of cell growth in sensory phenomena, especially when it comes to learning, adaptating, preconditioning and other long-term effects considered traditionally as epigenetic, and as characteristic for slow growth and differentiation processes. Finally, many phenomena which were once considered as unrelated now show a plethora of common elements, e.g., chemotaxis and chemotropism; cytoskeleton, cell division, and amoeboid motility; or chemotaxis, cell–cell communication, and differentiation of the prokaryotes and of the most highly evolved eukaryotes. Thus it seems clear by now why even the most sessile of all organisms, i.e., the large terrestric plants, never lost from their genome the genes necessary to return, if only transiently, back to motile stages. Because Pfeffer was right in stating: "Life is motility!"

References

1. Adler, J. (1987). How motile bacteria are attracted and repelled by chemicals: an approach to neurobiology. *Biological Chemistry Hoppe-Seyler* **368**, 163–173.
2. Alexandre, G. and Zhulin, I.B. (2001). More than one way to sense chemicals. *J. Bacteriol.* **183**, 4681–4686.
3. Anderson, O.R. (1987). *Comparative Protozoology. Ecology, Physiology, Life History.* Springer-Verlag, Berlin.
4. Armitage, J. (1997). Behavioural responses of bacteria to light and oxygen. *Arch. Microbiol.* **168**, 249–261.
5. Armitage, J.P. and Lackie, J.M. (eds.) (1990). *Biology of the Chemotactic Response.* Cambridge University Press, Cambridge, UK.
6. Banks, J.A. (1999). Gametophyte development in ferns. *Annu. Rev. Plant Physiol. Plant Mol. Biol.* **50**, 163–186.
7. Barnes, R. (1988). *Invertebrate Zoology*, 5th ed. W.B. Saunders and Co., Philadelphia, USA.
8. Ben-Jacob, E., Cohen, I. and Gutnick, D.L. (1998). Cooperative organization of bacterial colonies: from genotype to morphotype. *Annu. Rev. Microbiol.* **52**, 779–806.
9. Brokaw, C.J. (1974). Calcium and flagellar response during the chemotaxis of bracken spermatozoids. *J. Cell. Physiol.* **83**, 151–158.
10. Brun, Y.V. and Shimkets, L.J. (eds.) (2000). *Prokaryotic Development.* ASM Press, Washington, D.C., USA.
11. Buchsbaum, R., Buchsbaum, M., Pearse, J. and Pearse, V. (1987). *Animals Without Backbones*, 3rd ed. University Chicago Press, Chicago, USA.
12. Bünning, E. (1953). *Entwicklungs- und Bewegungsphysiologie der Pflanzen.* 3. Aufl. Berlin
13. Carlile, M.J. and Watkinson, S.C. (1994). *The Fungi.* Academic Press London, UK.
14. Carr, W.E.S. (1990). Chemical signaling systems in lower organisms: a prelude to the evolution of chemical communication in the nervous system, in *Evolution of the First Nervous Systems* (P.A.V. Anderson, ed.). Plenum Press, New York, USA.
15. Cavalier-Smith, T. (1998). A revised six-kingdom system of life. *Biol. Rev.* **73**, 203–266.
16. Chang, F. (2001). Establishment of a cellular axis in fission yeast. *Trends in Genet.* **17**, 273–278.
17. Chung, C.Y., Funamoto, S. and Firtel, R.A. (2001). Signaling pathways controlling cell polarity and chemotaxis. *Trends in Biochem. Science* **26**, 557–566.
18. Colombetti, G. and Lenci, F. (eds.) (1984). *Membranes and Sensory Transduction.* Plenum Press, New York and London.
19. Davey, M.E. and O'Toole, G.A. (2000). Microbial biofilms: from ecology to molecular genetics. *Microbiol Mol. Biol. Rev.* **64**, 847–867.

20. Doetsch, R.N. and Sjoblad, R.D. (1980). Flagellar structure and function in eubacteria. *Annu. Rev. Microbiol.* **34**, 69–108.

21. Drews, G. and Nultsch, W. (1962). Spezielle Bewegungsmechanismen von Einzellern (Bakterien, Algen), in *Encyclopedia of Plant Physiology: Physiology of Movements* (W. Ruhland, ed.), pp. 876–920. Springer-Verlag, Berlin.

22. Dworking, M. (1996). Recent advances in the social and developmental biology of the myxobacteria. *Microbiol. Mol. Biol. Rev.* **60**, 70–102.

23. Elion, E.A. (2000). Pheromone response, mating and cell biology. *Curr. Opinion Microbiol.* **3**, 573–581.

24. Esser, K. and Lemke, P.A. (eds.) (1994–1997). *The Mycota*, Vols. 1–6. Springer-Verlag, Berlin.

25. Fritsch, F.E. (1977). *The Structure and Reproduction of the Algae*, Vols. 1, 2. Cambridge University Press, Cambridge, UK.

26. Fuqua, C., Parsek, M.R. and Greenberg, E.P. (2001). Regulation of gene expression by cell-to-cell communication: Acyl-homoserine lactone quorum sensing. *Annu. Rev. Genet.* **35**, 439–468.

27. Gibbs (1989). Chemotaxis in lower eukaryotes. *Microbiol. Mol. Biol. Rev.* **53**, 171–185.

28. Goodenough, U.W., Armbrust, E.V., Campbell, A.M. and Ferris, P.J. (1995). Molecular genetics of sexuality in *Chlamydomonas*. *Annu. Rev. Plant Physiol. Plant Mol. Biol.* **46**, 21–44.

29. Goodman, C.S. (1996). Mechanisms and molecules that control growth cone guidance. *Annu. Rev. Neurosci.* **19**, 341–377.

30. Greenspan, R.J. and Ferveur, J-F. (2000). Courtship in *Drosophila*. *Annu. Rev. Genet.* **34**, 205–232.

31. Häder, D-P. (1987). Photosensory behavior in procaryotes. *Microbiol. Rev.* **51**, 1–21.

32. Harris, E.H. (2001). *Chlamydomonas* as a model organism. *Annu. Rev. Plant Physiol. Plant Mol. Biol.* **52**, 363–406.

33. Hatten, M.E. (1999). Central nervous system neuronal migration. *Annu. Rev. Neurosci.* **22**, 27–41.

34. Haupt, W. and Feinlieb, M.E. (eds.) (1979). *Encyclopedia of Plant Physiology: Physiology of Movements.* New Series, Vol. 7. Springer-Verlag, Berlin.

35. Haustein, E. (1967). Befruchtung der Archegoniaten und Blütenpflanzen, in *Encyclopedia of Plant Physiology: Sexuality, Reproduction, and Alternation of Generations* (W. Ruhland, ed.), pp. 407–447. Springer-Verlag, Berlin.

36. Hazelbauer, G.L. (ed.) (1978). *Taxis and Behavior. Elementary Systems in Biology.* Chapman and Hall, London, UK.

37. Jaenicke, L. (1993). *Protisten—Modellorganismen der Zellbiologie. Biol. In unserer Zeit* **23**, 36–47.

38. Janssens, P.M.W. and Van Haastert, P.J.M. (1987). Molecular basis of transmembrane signal transduction in *Dictyostelium discoideum*. *Microbiol. Rev.* **51**, 396–418.

39. Jennings, H.S. (1906). *Behavior of Lower Organisms*. Columbia University Press, New York.

40. Jeziore-Sassoon, Y., Hamblin, P.A., Bootle-Wilbraham, C.A., Poole, P.S. and Armitage, J.P. (1998). Metabolism is required for chemotaxis to sugars in *Rhodobacter sphaeroides*. *Microbiology* **144**, 229–239.

41. Kaiser, D. (2001). Building a multicellular organism. *Annu. Rev. Genet.* **35**, 103–123.

42. Kent, A.D. and Triplett, E.W. (2002). Microbial communities and their interactions in soil and rhizosphere ecosystems. *Annu. Rev. Microbiol.* **56**, 211–236.

43. Kochert, G. (1978). Sexual pheromones in algae and fungi. *Annu. Rev. Plant Physiol.* **29**, 461–486.

44. Köhler, K. (1967). *Die chemischen Grundlagen der Befruchtung (Gamone)*, in *Encyclopedia of Plant Physiology: Sexuality, Reproduction, and Alternation of Generations* (W. Ruhland, ed.), pp. 282–320. Springer-Verlag, Berlin.

45. Kurjan, J. and Taylor, B.L. (eds.) (1993). *Signal Transduction: Prokaryotic and Simple Eukaryotic Systems*. Academic Press, San Diego, USA.

46. Leick, V. and Hellung-Larsen, P. (1992). Chemosensory behaviour of *Tetrahymena*. *BioEssays* **14**, 61–66.

47. Lengeler, J.W. (1961). *Chemotaxis-Versuche an Spermatozoiden von Archegoniaten*. Diploma Thesis. Universität Köln.

48. Lengeler, J.W., Drews, G. and Schlegel, H.G. (eds.) (1999). *Biology of the Prokaryotes*. Blackwell Science, Oxford, UK.

49. Lengeler, J.W. and Jahreis, K. (1996). Phosphotransferase systems or PTSs as carbohydrate transport and as signal transduction systems, in *Transport Processes in Eukaryotic and Prokaryotic Organisms: Handbook of Biological Physics* (W.N. Konings, H.R. Kaback and J.S. Lolkema, (eds.), pp. 573–598. Elsevier Science, Amsterdam, The Netherlands.

50. Lengeler, K.B., Davidson, R.C., D'Souza, C., Harashima, T., Shen, W., Wang, P., Pan, X., Wuagh, M. and Heitman, J. (2000). Signal transduction cascades regulating fungal development and virulence. *Microbiol. Mol. Biol. Rev.* **64**, 746–785.

51. Levandowsky, M. (2001). Sensory responses in *Euglena*. *ASM News* **67**, 299–303.

52. Loomis, W.F., Kuspa, A. and Shaulsky, G. (1998). Two-component signal transduction systems in eukaryotic microorganisms. *Curr. Opin. Microbiol.* **1**, 643–648.

53. Mattick, J.S. (2002). Type IV pili and twitching motility. *Annu. Rev. Microbiol.* **56**, 289–314.

54. Meeks, J.C. and Elhai, J. (2002). Regulation of cellular differentiation in filamentous cyanobacteria in free-living and plant-associated symbiotic growth states. *Microbiol. Mol. Biol. Rev.* **66**, 94–121.

55. Mendoza, L., Taylor, J.W. and Ajello, L. (2002). The class *Mesomycetozoa*: a heterogeneous group of microorganisms at the animal-fungal boundary. *Annu. Rev. Microbiol.* **56**, 315–344.

56. Meyerowitz, E.M. (1987). *Arabidopsis thaliana. Annu. Rev. Genet.* **21**, 93–111.

57. Mohan, S., Dow, C. and Cole, J.A. (eds.) (1992). *Prokaryotic Structure and Function. A new perspective*. Cambridge University Press, Cambridge, UK.

58. Moore, R.L. (1981). The biology of *Hyphomicrobium* and other prosthecate, budding bacteria. *Annu. Rev. Microbiol.* **35**, 567–594.

59. Mori, I. (1999). Genetics of chemotaxis and thermotaxis in the nematode *Caenorhabditis elegans. Annu. Rev. Genet.* **33**, 399–422.

60. Morrissette, N.S. and Sibley, L.D. (2002). Cytoskeleton of apicomplexan parasites. *Microbiol. Mol. Biol. Rev.* **66**, 21–38.

61. Mueller, B.K. (1999). Growth cone guidance: first steps towards a deeper understanding. *Annu. Rev. Neurosci.* **22**, 351–388.

62. Overmann, J. and Schubert, K. (2002). Phototrophic consortia: model systems for symbiotic interrelations between prokaryotes. *Arch. Microbiol.* **177**, 201–208.

63. Palanivelu, R. and Preuss, D. (2000). Pollen tube targeting and axon guidance: parallels in tip growth mechanisms. *Trends Cell Biol.* **10**, 517–524.

64. Pfeffer, W. (1883). *Locomotorische Richtungsbewegungen durch chemische Reize. Ber. Deutsche Botan. Gesellschaft* **1**, 524–533.

65. Pfeffer, W. (1884). Lokomotorische Richtungsbewegungen durch chemische Reize. *Untersuch. aus d. Botan. Institut Tübingen.* **1**, 363–482.

66. Pfeffer, W. (1888). *Über chemotaktische Bewegungen von Bacterien, Flagellaten und Volvocineen. Untersuch. aus d. Botan. Institut Tübingen.* **2**, 582–661.

67. Pfeffer, W. (1904). *Pflanzenphysiologie.* Bd. II, 2.Aufl. Engelmann, Leipzig.

68. Perez-Miravete, A. (ed.) (1973). *Behavior of Microorganisms.* Plenum Press, New York.

69. Polz, M.F., Ott, J.A., Bright, M. and Cavanaugh, C.M. (2000). When bacteria hitch a ride. *ASM News* **66**, 531–539.

70. Prüß, B.M., Campbell, J.W., Van Dyk, T.K., Zhu, C., Kogan, Y. and Matsumura, P. (2003). FlhD/FlhC is a regulator of anaerobic respiration and the Entner-Doudoroff pathway through induction of the methyl-accepting chemotaxis protein Aer. *J. Bacteriol.* **185**, 534–543.

71. Rochaix, J-D. (1995). *Chlamydomonas reinhardtii* as the photosynthetic yeast. *Annu. Rev. Genet.* **29**, 209–230.

72. Romagnoli, S., Packer, H.L. and Armitage, J.P. (2002). Tactic responses to oxygen in the phototrophic bacterium *Rhodobacter sphaeroides* WS8N. *J. Bacteriol.* **184**, 5590–5598.

73. Rothschild Lord (1956). *Fertilization.* Methuen and Co., London, U.K., and Wiley, New York, USA.

74. Rothert, W. (1901). Beobachtungen und Betrachtungen über taktische Reizerscheinungen. *Flora* **88**, 371–420.

75. Rothert, W. (1904). Über die Wirkung des Äthers und Chloroforms auf die Reizbewegungen der Mikroorganismen. *Jahrb. Wiss. Botanik* **39**, 1–70.

76. Rudolph, J. and Oesterhelt, D. (1995). Chemo- and phototaxis require a CheA histidine kinase in the archeon *Halobacterium salinarium. EMBO J.* **14**, 667–673.

77. Ruhland, W. (ed.) (1959–1967). *Encyclopedia of Plant Physiology*, Vol. 17, *Physiology of Movements*, and Vol. 18, *Sexuality, Reproduction, and Alternation of Generations.* Springer-Verlag, Berlin.

78. Saimi, Y. and Kung, C. (1987). Behavioral genetics of *Paramecium. Annu. Rev. Genet.* **21**, 47–65.

79. Sanes, J.R. and Jessel, T.M. (2000). The guidance of axons to their targets, in *Principles of Neural Science*, 4th ed. (E.R. Kandel, J.H. Schwartz and T.M. Jessel, eds.), pp. 1062–1086. McGraw Hill, USA.

80. Schieder, O. (1968). Untersuchungen über das chemotaktisch wirksame Gamon von *Sphaerocarpos donnellii Austr. Z. Planzenphysiol.* **59**, 258–273.

81. Schieder, O. (1971). *Über das Chemotaktikum des Lebermooses* Sphaerocarpos donnellii. *Austr. Naturwissenschaften* **58**, 568.

82. Shapiro, J.A. (1998). Thinking about bacterial populations as multicellular organisms. *Annu. Rev. Microbiol.* **52**, 81–104.

83. Shibata, K. (1911). *Über die Chemotaxis der Pteridophyten Spermien. Jahrb. Wiss. Botanik.* **49**, 1–60.

84. Shimkets, L.J. (1990). Social and developmental biology of the myxobacteria. *Microbiol. Rev.* **54**, 473–501.

85. Sitte, P., Weiler, E.W., Kadereit, J.W., Bresinsky, A. and Körner, C. (2002). *Strasburger—Lehrbuch der Botanik für Hochschulen.* 35. Aufl. Gustav Fischer Verlag, Stuttgart.

86. Spormann, A.M. (1999). Gliding motility in bacteria: insights from studies of *Myxococcus xanthus. Microbiol. Mol. Biol. Rev.* **63**, 621–641.

87. Stolp, H. (1988). *Microbial Ecology: Organisms, Habitats, Activities.* Cambridge University Press, Cambridge, UK.

88. Stoodley, P., Sauer, K., Davies, D.G. and Costerton, J.W. (2002) Biofilms as complex differentiated communities. *Annu. Rev. Microbiol.* **56**, 187–209.

89. Taylor, B.L. and Zhulin, I.B. (1999). PAS domains: internal sensors of oxygen, redox potential, and light. *Microbiol. Mol. Biol. Rev.* **63**, 479–506.

90. Van Houten, J.L. (1990). Chemosensory transduction in unicellular eukaryotes, in *Evolution of the First Nervous Systems* (P.A.V. Anderson, ed.), pp. 343–356. Plenum Press, New York, USA.

91. Van Houten, J.L. (1994). Chemosensory transduction in microorganisms: trends for neuroscience? *Trends Neurosci.* **17**, 62–71.

92. Van Houten, J.L. (1998). Chemosensory transduction in *Paramecium. Eur. J. Protistol.* 34, 301–307.

93. Weibull, C. (1960). Movement, in *The Bacteria: A Treatise in Structure and Function*, Vol. 1, Structure (I.C. Gunsalus and R.Y. Stanier, eds.), pp. 153–205. Academic Press, San Diego, USA.

94. Westheide, W. and Rieger, R. (eds.) (1996). *Spezielle Zoologie, Teil 1.* Einzeller und Wirbellose. G. Fischer Verlag, Jena.

95. Wilkie, D. (1954) The movements of spermatozoa of bracken (*Pteridium aquilinum*). *Exp. Cell Res.* **6**, 384–391.

96. Williams, A.G. (1986). Rumen holotrich ciliate protozoa. *Microbiol. Rev.* **50**, 25–49.

97. Winans, S.T. (1992). Two-way chemical signaling in *Agrobacterium*—plant interactions. *Microbiol. Rev.* **56**, 12–31.

98. Ziegler, H. (1962). Chemotaxis, in *Encyclopedia of Plant Physiology: Physiology of Movements* (W. Ruhland, ed.), pp. 484–526. Springer-Verlag, Berlin.

Bacterial Chemotaxis

1. Introduction

Bacterial populations may encounter a large spectrum of environments during their life cycles. Due to their small size and relative simplicity, their ability to adjust the environment to their needs is very limited. Instead, they apparently adopted a strategy of moving from one environment to another [457]. Chemotaxis as well as other types of taxis (e.g., thermotaxis and phototaxis) thus enable bacteria to approach (and remain in) beneficial environments and escape from hostile ones. It is, therefore, not surprising that a very large number of bacterial species are motile and chemotactic. (Another strategy of some bacteria for coping with environmental changes is forming spores and waiting for environmental improvement. Bacteria can also form biofilms and thereby change, to some extent, their microenvironment [769].)

The fact that bacterial movement is not random but is rather responsive to light, oxygen, and certain chemicals was already discovered in the 1880s. Engelmann found that *Bacterium thermo* accumulates in oxygen-rich regions [218]. Pfeffer demonstrated that if a thin capillary tube, containing a chemoattractant (e.g., meat extract), is immersed in a suspension of bacteria, the bacteria accumulate near the mouth of the capillary and, with time, within the capillary [576]. In this finding Pfeffer provided the basis for more recent chemotaxis assays. Subsequently, other investigators demonstrated the phenomenon of chemotaxis in other bacterial species, but then, for unclear reasons, the research of chemotaxis was essentially neglected. Renewed interest in the phenomenon was generated in 1959 by Baracchini and Sherris [52], demonstrating chemotaxis of *Pseudomonas viscosa* to oxygen, and in the nineteen sixties by Adler [4–6], who shifted the research from phenomenological to quantitative and who initiated studies to reveal the molecular mechanism of bacterial chemotaxis. Since then, the number of groups studying bacterial chemotaxis has been continuously on the rise.

Bacterial motility and chemotaxis have been studied most intensively in *Escherichia coli* and its close relative *Salmonella enterica* serovar Typhimurium (earlier called *Salmonella typhimurium*; in this chapter it will be called "*Salmonella*" in short). The focus of this chapter is, therefore, on these species. Other bacterial species are discussed in brief.

2. Bacterial Motility

2.1. *Motility types*

Most rod-shaped bacteria are motile, independently of their classification (e.g., Gram positive or negative, aerobes or anaerobes, spore-formers or not, etc.). In contrast, most round bacteria, Cocci, are nonmotile. Bacteria have acquired a number of motility strategies, as follows.

2.1.1. *Flagellar motility*

The most common strategy for motility is movement driven by flagella—external organelles that serve as "propellers." This strategy involves swimming and swarming.

Swimming
In swimming, the flagella, which are helical, rotate and thereby exert thrust that drives the bacteria. While the flagella rotate in one direction, the cell body rotates more slowly in the other direction [78]. Bacteria swim relatively quickly. For example, the swimming speed of a rod-shaped cell (usually 1–5 μm in length) of enteric bacteria like *E. coli* and *Salmonella* is 10–35 μm/s [74, 451, 752], and that of rod-shaped soil bacteria like *Pseudomonas aeruginosa* is even 2–3-fold higher [275, 752]. Marine bacteria swim much faster, up to ~200 μm/s [510].

There are various types of flagellar motility, which depend on the number and location of the flagella in the cell and on the species (Table 1). In some species (e.g., *Pseudomonas* species, *Spirillum* species, *Chromatium* species and *Halobacteria*), the cells swim forward and backward, and

Table 1. Varieties of flagellar motility in bacteria.[a]

Flagellation	Schematic drawing	Species for example	Description of motility	References[b]
A single flagellum at (or near) one of the cell poles.		*Pseudomonas* Spp.	The flagellum—depending on its direction of rotation—pushes or pulls the cell. Consequently, the cell goes back and forth.	[455]
A single flagellum roughly in the middle between the poles.		*Rhodobacter sphaeroides*	The flagellum either rotates clockwise or pauses. Consequently the cell swims in a rather straight line and occasionally stops for reorientation. During a pause the flagellum relaxes to a coil-like form. This form then rotates slowly and reorients the cell.	[32–34]
A bundle of flagella at one of the poles.		*Chromatium okenii*, some cells of *Halobacterium salinarium*	The bundle—depending on its direction of rotation—pushes or pulls the cell. Consequently, the cell goes back and forth.	[16, 77]
A bundle of flagella at each of the two poles.		Some cells of *H. salinarium*	The bundles—depending on their direction of rotation—push or pull the cell. Consequently, the cell goes back and forth or stops.	[16]
A bundle of flagella at each of the two poles in *Spirillum* Spp.		*Spirillum volutans*	Forward and backward swimming is carried out in the same manner, only that the bundles flip over when the cell reverses. The helical cell body rotates in reaction to the rotation of the flagella and this rotation produces the thrust for motility.	[384, 500]
5–10 flagella randomly distributed around the cell.		*Escherichia coli*, *Salmonella*, *Bacillus subtilis*	Most of the time the flagella rotate in one direction (counterclockwise in the case of *E. coli* and *Salmonella*, clockwise in the case of *B. subtilis*) and the cell swims in a rather straight line (a run). Intermittently, the flagella rotate in the opposite direction or pause, as a result of which the cell turns or tumbles (depending on the number of flagella that do so).	[77, 209, 460, 750]

(Continued)

Table 1. *Continued*

Flagellation	Schematic drawing	Species for example	Description of motility	References[b]
5–10 flagella randomly distributed around the cell.		*Sinorhizobium meliloti*	The flagella rotate incessantly counterclockwise and the cell swims most of the time in a rather straight line. Occasional changes in the speed of flagellar rotation cause the cell to turn (without tumbles).	[33]
A polar tuft of 2 flagella + 2–4 lateral flagella.		*Agrobacterium tumefaciens*	Flagella rotate clockwise or pause; consequently the cell swims in a rather straight line or turns.	[645]
Excessive flagellation around the cell.		*E. coli*, *Salmonella*, *Serratia marcescens*, *Proteus mirabilis*	Swarming—surface motility in a colony	[280]
One flagellum at one end, one or more flagella sub-terminally at each end. All the flagella are contained within the periplasmic space.		Spirochetes	The periplasmic flagella cause the cell to bend and gyrate. The cells exhibit smooth swimming, reversals, flexing, and pausing. When the flagellar bundles at both cell poles rotate in opposite directions (one pulls and one pushes), the cell swims in a rather straight line. When the two bundles switch synchronously, the cell reverses. When the two bundles rotate in the same direction, the cell flexes.	[151, 232]

[a] Taken with permission from Eisenbach [214].
[b] Whenever possible, the references are to reviews.

reorientation appears to be passive by Brownian motion. In other species (e.g., *E. coli*, *Salmonella*, *Sinorhizobium meliloti*, *Rhodobacter sphaeroides*, and *Agrobacterium tumefaciens*), the cells move in a rather straight line and, occasionally, actively reorient themselves. Combinations of the varieties mentioned in Table 1 are also possible. For example, cells of marine *Vibrio* (e.g., *Vibrio alginolyticus* and *Vibrio parahaemolyticus*) have each a single flagellum, located at one of the poles and sheathed by what appears to be an extension of the outer membrane [485]. This flagellum pushes the cell forward or pulls it backward. When on the surface of a solid medium or when in a viscous medium, the cells produce lateral flagella *in addition to* the polar ones. The lateral flagella enable swarming [47, 485] (see below) and movement in a viscous medium [48, 624]. A similar situation of a single polar flagellum synthesized in liquid medium and additional lateral flagella synthesized during growth on solid medium exists in *Azospirillum brasilense* [827]. Spirochetes may seem as an exception due to their invisible, periplasmic flagella and their unique mode of swimming (Table 1). However, their swimming speed, 26 µm/s [269], is quite similar to that of *E. coli* and *Salmonella*.

Swarming
Flagella are not only swimming tools; they also serve as tools for swarming (Table 1). Swarming is an organized surface motility of cells in a colony, which depends on massive flagellation and cell-to-cell communication [202, 280]. This organized surface translocation has been demonstrated in both Gram-negative (primarily) and Gram-positive species. Even enteric bacteria with well-characterized swimming motility (e.g., *E. coli* and *Salmonella*), when on a hard surface, are able to differentiate into filamentous (up to 50 µm long), multinucleated, hyperflagellated cells that translocate together as a colony on the surface. Similar to swarming bees, the differentiated bacteria in the colony are organized in such a way that the outer layer of the colony moves like a swirl and expands outwardly, and the evacuated space inside the colony is filled with newly grown bacteria [280]. The result is fast colony expansion (up to ~3 µm/s). Apparently, under conditions that induce swarming, the bacteria themselves produce slime that improves wettability and facilitates expansion of the swarming colony [739]. Quorum sensing (cell density-dependent regulation of gene expression [242]) appears to be essential for swarming.

2.1.2. *"Swimming" without flagella*

Other strategies of motility, which do not depend on flagella, have also been recognized.

Cyanobacteria
A fascinating and apparently unique kind of rapid motility through liquid—"swimming" without flagella or other visible appendages—was

discovered in strains of the marine cyanobacteria *Synechococcus* [767]. The cells of these strains are rod-shaped and they move, without any apparent shape change, as fast as 25 μm/s. The mechanism underlying this movement is obscure [577]. It is known, however, that this "swimming" requires extracellular Ca^{2+} [578]. Recently, Samuel *et al.* [619] identified spicules extending from the cytoplasmic membrane and protruding from the cell surface up to 0.15 μm outwardly, suggesting that these spicules might comprise part of the motility apparatus of *Synechococcus*. They further found that extracellular Ca^{2+} is required for maintaining the structure of the bacterial surface layer. Based on these findings, Samuel *et al.* speculated that oscillations, which may occur within the cytoplasmic or outer membrane, might be transmitted into a rowing motion of the spicules.

Spiroplasma
Another type of flagella-less swimming occurs in *Spiroplasma*, small helical bacteria lacking cell walls and flagella. These bacteria have an internal cytoskeleton in the form of a flat, monolayered ribbon, which is constructed from seven contractile fibrils connected to the inner side of the cell membrane [743]. This internal cytoskeleton functions as a linear motor: the fibrils change their length differentially in a coordinated manner, resulting in cell coiling and uncoiling. It is not yet clear how this cell coiling and uncoiling brings about swimming: by reciprocating helical extension and compression, by propagation of a deformation along the helical path, by propagation of a reversal of the helical sense along the cell body, by irregular flexing and twitching, or by any combination thereof [259, 744].

2.1.3. *Gliding motility*

The most abundant flagella-less motility is gliding, i.e., the movement of an organism on a solid surface with no visible external organelles for the movement and no shape change [271]. Unlike swarming motility, gliding requires a solid surface covered with a liquid film [457]. Gliding bacteria can be divided into three or more classes according to their speed and possibly their motility mechanism:

Fast gliders
Gliding bacteria that belong to this class (e.g., cyanobacteria [blue-green algae] and *Cytophaga*) move as fast as 1–10 μm/s [271]. (Some of the

fast gliders, e.g., *Deleya marina*, have, depending on the conditions, both flagellar and gliding motility [646].) Although gliding organelles have not been found, latex beads, artificially attached to *Cytophaga* cells, were seen to rotate or move back and forth in the speed of gliding [404]. Recently, a carbohydrate-secreting organelle was identified in cyanobacteria, *Cytophaga* and *Flexibacter*, suggesting that steady secretion of slime through this organelle generates the thrust required for gliding motility in these species [301]. In addition, an ordered array of parallel fibrils between the peptidoglycan layer and the outer membrane was discovered in cyanobacteria, suggesting that the fibrillar array in the cell wall may also be involved in gliding motility of this species [3]. (For a review on gliding motility in cyanobacteria, see [302].)

Mycobacteria, as well, slide on wet surfaces (e.g., 0.3% agar) even though they do not possess pili or any other extracellular structures. The mechanism of this translocation is not known, but the movement appears to depend on the presence of glycopeptidolipids, a mycobacterium-specific class of amphiphilic molecules located in the outermost layer of the cell envelope [475].

Recently, an undefined surface migration, independent of flagella, was discovered in *E. coli* and *Vibrio* [141]. This movement, at a speed of about 4 μm/s, can be observed on 0.3% agarose but not on agar and is dependent on lipopolysaccharide biosynthesis.

Slow gliders
The class of slow gliders includes myxobacteria (e.g., *Myxococcus xanthus*)—Gram-negative bacteria that live in soil. They glide very slowly (1–20 μm/min, 4–5 μm/min on average) [683]. The speed of gliding within this range depends on the distance between the moving cell and its nearest neighbor—the greater the distance; the lower the speed. Myxobacteria have two independent motility systems, "adventurous" and "social," which are genetically and functionally distinct [299, 300, 648]. The adventurous motility is the motility of cells located more than a cell's length from any neighboring cell. It is effective mainly on relatively hard and dry surfaces (such as 1.5% agar). Social motility is movement in groups involving continuous reorientation and reassociation of the cells in the group. It is mostly effective on softer and wetter surfaces (such as 0.3% agar). Chainlike aggregates, termed strands and found within the bacterial cell wall, are thought to be one part of the myxobacterial gliding machinery [239]. Structures that resemble the slime-secreting nozzles of cyanobacteria were recently discovered at both cell

poles of *M. xanthus*. They, too, were suggested to have a role in gliding of these bacteria [786]. Social motility (but not adventurous motility) involves pili, present in tufts at one or both poles of the cell [328, 608, 791] and energized by ATP (for a review, see reference [329]). By tethering cells via their pili to a surface, Sun *et al.* [710] nicely demonstrated that these pili are essentially the motility apparatus. The observations were consistent with a model in which the pili are extruded from one cell pole, adhere to a surface, and then retract, pulling the cell in the direction of the adhering pili [710]. Social motility also involves production of extracellular slime fibrils [150, 806]. On the basis of the finding that ADP ribosylation activity is associated with the fibrils [294], it was proposed that the ADP-ribosyl transferase might be an enzymatic activity that is activated upon contact of the fibrils with an appropriate receptor on an adjacent cell [201]. According to this proposal, which still awaits a proof, the fibrils function as tactile antennae that transmit a signal back to the cell indicating the proximity of another cell.

Mycoplasma

Mycoplasmas are wall-less bacteria, widespread in nature. Many of them can glide on hard surfaces such as glass, plastic, or eukaryotic cell surface, but the mechanism of gliding is obscure [511]. Several mycoplasma species have a head-like structure—a protruding membrane extension at one cell pole, which drags the cell body [514]. The gliding appears to have a role, yet unknown, in the parasitic life cycle of mycoplasmas [512]. The uniqueness and the gliding speed of the mycoplasmas, 0.1–5 μm/s (depending on the species and temperature) [365, 511, 513], distinguish them as a separate class of gliders.

2.1.4. *Twitching motility*

Twitching is another kind of surface motility, involving intermittent and jerky movement of single bacterial cells or group of cells in a colony, not necessarily along the long axis of the cell [289]. Due to the lengthy intermissions without movement, the progressive velocity, averaged over time, is very low (2–10 μm/min). It appears that, as in the case of slow gliders, twitching motility is powered by retraction of polar pili [497, 666]. As a matter of fact, it has been proposed, on the basis of morphological and genetic data, that twitching motility and social gliding motility of slow gliders are essentially the same process [639].

2.1.5. *Propulsion by actin filaments*

A unique mode of motility, first described in 1989, is the movement of bacteria such as *Listeria*, *Shigella*, and *Rickettsia* in host eukaryotic cells [408, 579]. The bacteria use a continuous actin filament assembly for propulsion in the cytoplasm of the infected host cell. The actin assembly at the bacterial surface is asymmetrical, with the filaments growing like a comet tail at one end of the bacterial cell and pushing the cell in the other direction. The shape of the actin filaments appears to affect the motility. For example, in *Listeria* and *Shigella* tails, the filaments form a branching network and the bacteria progress relatively fast (e.g., ~24 μm/min within cells of a Vero cell line), whereas *Rickettsia* tails display longer and not cross-linked filaments and the bacteria progress relatively slow (e.g., ~8 μm/min within Vero cells) [264]. These differences suggest that there are different mechanisms of actin polymerization in these species [264]. Although factors involved in this motility are gradually being revealed (e.g., references [145, 203, 394, 409, 435, 567]), the mechanism is essentially still obscure.

2.2. *Bacterial flagella*

As mentioned above, flagella are organelles that enable bacteria to swim in an aqueous solution or swarm on a hard surface (Figure 1). But these are not the only roles of flagella. They are involved in bacterial colonization, they contribute in many cases to the bacterial virulence, and, being strong antigens, they are often targets for antibody response [517, 562]. Flagella are multifunctional: even though a flagellum extends far beyond the cell itself, it receives sensory information from within the cytoplasm and generates a behavior (chemotaxis) that is critical for the cell's survival; it rotates at high speed and switches between directions of rotation in a controlled manner; at the same time, it exports its own component proteins through itself and assembles them at its distant tip [461]. The term "flagellum" ("flagella" in plural) comes from Latin, meaning a little whip. Although this term is adequate for the eukaryotic flagellum (e.g., sperm flagellum), which acts like a whip, it is not adequate (and essentially misleading) for the bacterial flagellum, which acts by rotation. As a matter of fact, bacterial flagella and eukaryotic flagella are totally different organelles. Table 2 indicates the main differences between them, with *E. coli* and human spermatozoa representing bacterial and eukaryotic flagella, respectively.

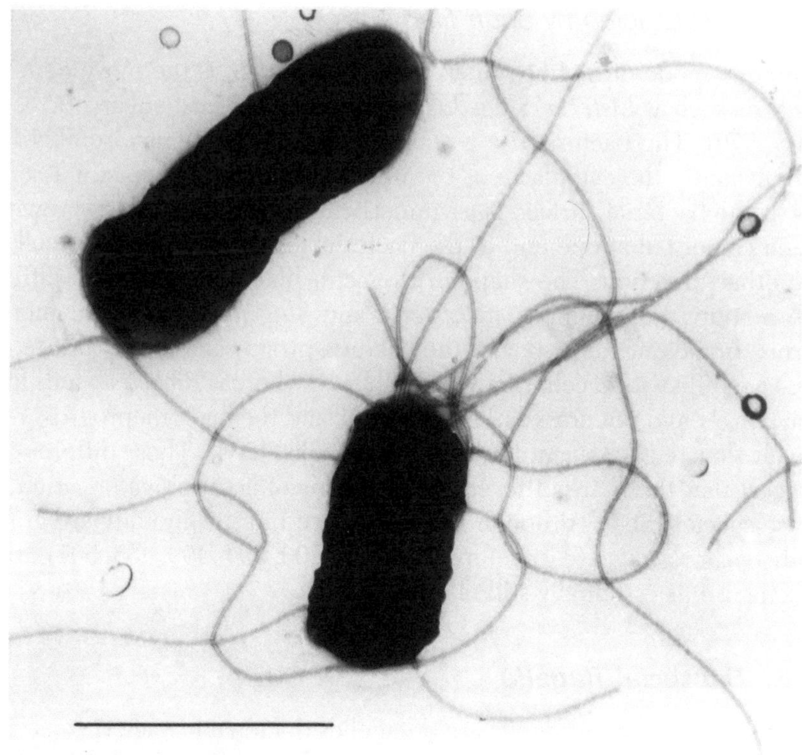

Figure 1. Flagella of *E. coli* observed in the transmission electron microscope. Bar = 1 μm. (Taken with permission from Eisenbach [214].)

Table 2. Comparison between eukaryotic and bacterial flagella.[a]

Property	Sperm flagellum (human)[b]	Bacterial flagellum (*E. coli*)
Filament diameter	0.3–1 μm	0.023 μm
Filament length	60 μm	10–15 μm
Filament structure	Complex structure of tubules surrounded by an extension of the cytoplasmic membrane; the flagellum consists of ~250 proteins	Naked filament consisting of subunits of a single protein
Function	Active beating	Passive rotation, driven by a rotary motor embedded in the cytoplasmic membrane
Energy source	ATP	Proton-motive force across the cytoplasmic membrane

[a] Taken with permission from Eisenbach [214].
[b] More information about the structure and function of sperm flagella can be found in Chapter 7.

2.2.1. *Structure of flagella*

Bacterial flagella consist of three major parts (Figure 2): a basal body, a hook, and a filament. Although the structure of bacterial flagella may vary in some respects between species and families (e.g., Gram-positive

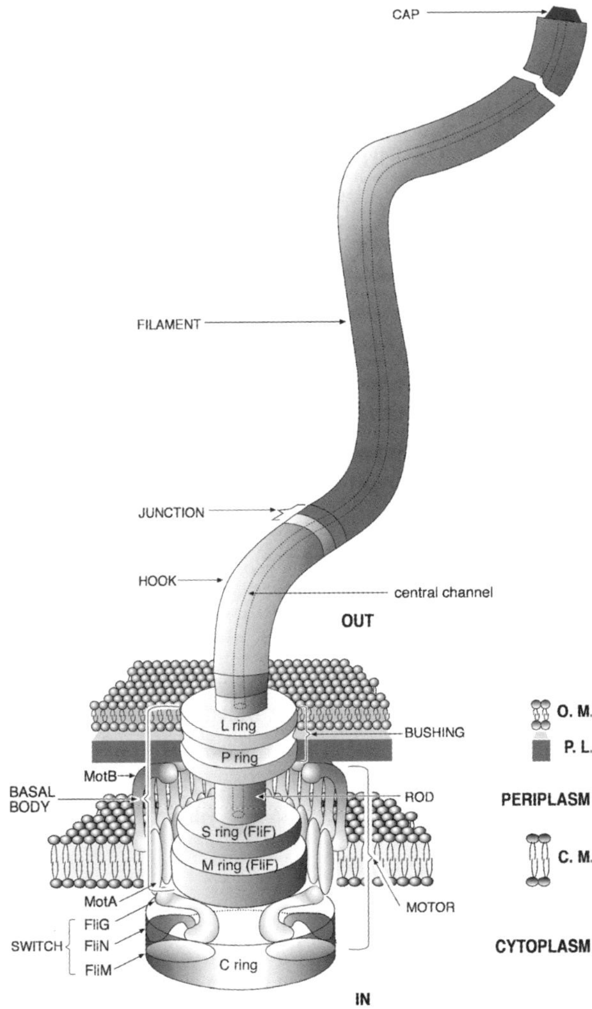

Figure 2. Scheme of *E. coli* or *Salmonella* flagellum. The scheme is not drawn to scale. The actual diameters of the rod, L, P, M, S, and C rings are ~15, 33, 26, 29, 27 and 41–47 nm, respectively [234, 337, 353, 691]. Abbreviations: C.M., cytoplasmic membrane; O.M., outer membrane; P.L., peptidoglycan layer. (Taken with permission from Eisenbach [214].)

and -negative bacteria), the main structural aspects are common to all (for general reviews of bacterial flagella, see [14, 460]).

Basal body
The basal body of *E. coli* and *Salmonella* is composed of a central rod surrounded by four rings: an M ring (M for membrane, as this ring is located in the cytoplasmic membrane), an S ring (S for supramembrane, as this ring is located above the cytoplasmic membrane), a P (for peptidoglycan) ring, and an L (for lipopolysaccharide) ring [13, 182, 184, 460, 535] (Figure 2). The M and S rings are essentially a double ring, composed of a single protein, FliF [751]. The MS ring is the structure on which the functional components of the motor are mounted. The P ring is built from the FlgI protein, and it is linked by a cylindrical wall to the L ring, built from the FlgH protein. (The L and P rings are apparently missing in Gram-positive bacteria [183, 188].) Another ring, the C ring (C for cytoplasm), which contains the proteins FliM and FliN, is attached via the FliG protein to the MS ring from beneath, on the cytoplasmic side. The C ring is involved in switching the direction of flagellar rotation. (The switching system will be discussed in Section 2.2.6) Recently, the protein FliL, as well, was found to be associated with the basal body, but its function is yet unknown [631].

Hook
The hook of *E. coli* and *Salmonella*, built from a single protein, FlgE, is a short (~55 nm long, only 130 FlgE subunits), curved structure that connects the basal body to the flagellar filament (Figure 2). The structure of the hook is similar to that of the filament (see below), with 11 fibrils of FlgE units, forming a helix [460]. The hook is believed to serve as a flexible joint that converts the torque, generated by the flagellar motor in the plane of the cell surface, into a force having both vertical and horizontal components [460].

Filament
The filament of *E. coli* and *Salmonella* is a highly rigid, helical structure, 10–15 μm long, 23 nm in diameter [184, 460, 535]. It is connected to the hook via a short junction composed of two hook-associated proteins, FlgK (also termed HAP1 for hook-associated protein 1) and FlgL (HAP3). At the other end of the filament there is a caplike structure, composed of the protein FliD (HAP2). The filament is built from ~20 000 subunits of a ~55 kDa single protein, flagellin [460]. In the

case of *E. coli*, there is only one flagellin, FliC. In the case of the *Salmonella* species, there are two similar genes (*fliC* and *fljB*), at different locations on the chromosome, that code for flagellin. Only one of them is expressed at any given time. In a stochastic manner, every 10^3–10^5 generations, the currently expressed gene stops being expressed, and the other gene starts being expressed [460]. While the mechanism of this phase variation is understood [460, 660], the reasons for it are not known; presumably they are related to defense against host defense systems [460]. The flagellin subunits are arranged at points on a tubular lattice, forming 11 fibrils, almost (but not exactly) parallel to the filament axis [460, 535, 742] (Figure 3). The fibrils differ from each other,

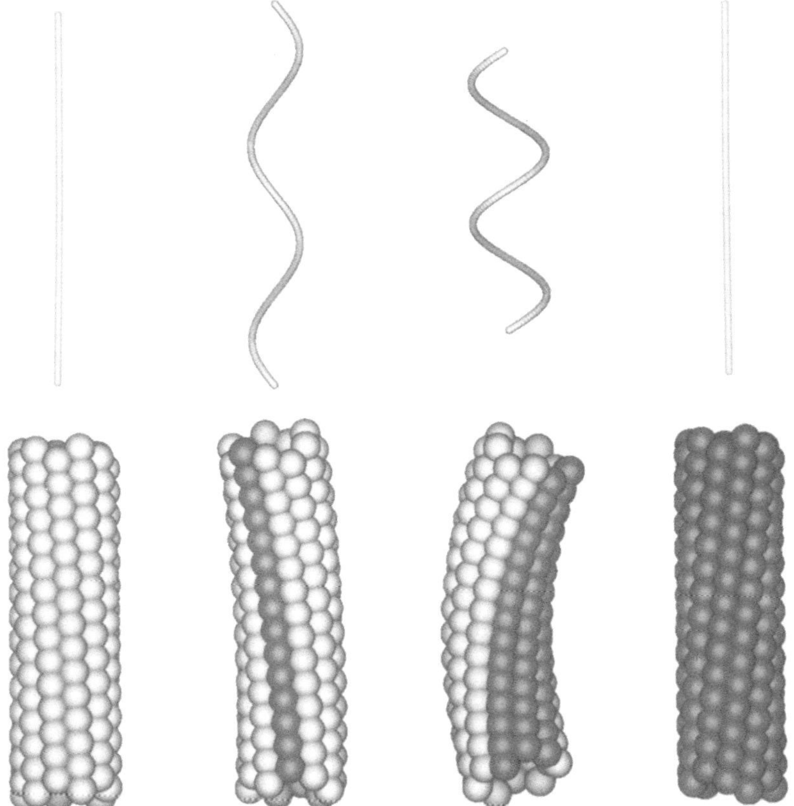

Figure 3. Some flagellar filament morphologies (*top*) and corresponding schemes of the packing arrangements of flagellin molecules in them (*bottom*). The filaments are named (*from left to right*) L-type straight, left-handed helix (normal), right-handed helix (curly), and R-type straight. (Taken with permission from Namba and Vonderviszt [535].)

with a shorter intersubunit distance on one side of the filament than on the other. Because the fibrils are not quite parallel to the filament axis, the length variation around the circumference introduces macroscopic helicity to the filament [460]. The filament can be in a number of helical forms, depending on the conditions [460, 535] (Figure 3). Nine such forms have been observed experimentally. The default physiological form is a left-handed helix. It can be converted to one of the other forms by a mechanical force (for example, when the direction of flagellar rotation is changed—Section 2.3.1) or by changing the pH or the ionic strength of the suspending medium. The filament is passive and its rotation is totally dependent on the flagellar motor (Figure 2).

Not all bacterial species contain filaments made of a single flagellin. For example, the flagellar filament of *Caulobacter crescentus* is made of six different flagellins with distinct molecular sizes [216, 789], and, likewise, archaeal filaments are composed of 1–5 flagellins of different sizes, depending on the species [319]. Other varieties include posttranslational glycosylation of flagellin in *Azospirillum brasilense* [516] as well as in some Archaea (including *Halobacterium* species and *Methanothermus* species) [319] and possibly *Spirochaeta aurantia* [129], presence of short signal sequences on archaeal flagellins [319], a sequence similarity of archaeal flagellins to type IV pilins (proteins constituting the bacterial pilus) rather than to bacterial flagellins [70, 319], and presence of a static, distinct cap structure at the cell pole of *H. salinarium* that carries the filaments of an entire bundle [395].

2.2.2. *Genes involved in motility*

Genes involved in motility were identified by isolation of non-motile mutants and identifying the defective genes. By the assays described in Section 3 below, three classes of motility mutants were isolated: non-motile mutants lacking flagella or having incomplete flagella (Fla⁻ phenotype; this was the largest class), nonmotile mutants with paralyzed flagella (Mot⁻ phenotype; the smallest class), and mutants with aberrant motility (Che⁻ phenotype).

The Fla⁻ class of mutants identified over 40 genes for flagellar assembly, structure, and function (Table 3). In the case of *E. coli*, they are clustered in three regions on the chromosome [657]. Region I (at 24.5 min) contains primarily structural genes. Genes belonging to this region are termed *flg* genes [312] (Table 3). Region II (at 42 min) contains some flagellar genes (termed *flh* genes), involved in regulation and

Table 3. Motility genes of *E. coli*.[a]

Gene[b]	Operon	Hierarchy level	Gene product			
			Size (kDa)	Location	Copies per flagellum[c]	Function
Genes involved in regulation of gene expression						
flhC	*flhDC*	1	22	Cytoplasm		Positive regulator of class II gene expression
flhD	*flhDC*	1	14	Cytoplasm		Positive regulator of class II gene expression
fliA	*fliAZY*	2	27	Cytoplasm		Sigma factor for class III gene expression
flgM	*flgMN*	3	11	Cytoplasm & exterior		Antisigma factor
flk	*flk*	1	35	Cytoplasmic membrane		Senses P- and L-rings completion and couples it to *flgM* translation
Genes involved in flagellar assembly and structure						
flgA	*flgA*	2	24	Periplasm		P-ring assembly; a chaperone?
flgB	*flgBCDEFGHIJ*	2	16	Periplasm	7 ± 1	Proximal rod
flgC	*flgBCDEFGHIJ*	2	14	Periplasm	6 ± 1	Proximal rod
flgD	*flgBCDEFGHIJ*	2	24	Exterior		Hook cap
flgE	*flgBCDEFGHIJ*	2	42	Exterior	132 ± 21	Hook
flgF	*flgBCDEFGHIJ*	2	26	Periplasm	6 ± 1	Proximal rod
flgG	*flgBCDEFGHIJ*	2	28	Periplasm	26 ± 4	Distal rod
flgH	*flgBCDEFGHIJ*	2	22	Outer membrane	28 ± 5	L ring
flgI	*flgBCDEFGHIJ*	2	36	Periplasm	24 ± 4	P ring
flgJ	*flgBCDEFGHIJ*	2	34	Periplasm		By its muramidase activity it possibly makes a hole in the peptidoglycan layer for the rod

(Continued)

Table 3. *Continued*

Gene[b]	Operon	Hierarchy level	Gene product			
			Size (kDa)	Location	Copies per flagellum[c]	Function
flgK	*flgKL*	3	59	Exterior	13±3	HAP1—hook-filament junction
flgL	*flgKL*	3	34	Exterior		HAP3—hook-filament junction
flgN	*flgMN*	3	16	Cytoplasm		Possibly an export chaperone for FlgK and FlgL that prevents oligomerization of HAP1 and HAP3 by binding to their helical domains before export
flhA	*flhBAE*	2	75	Cytoplasmic membrane		Export apparatus
flhB	*flhBAE*	2	39	Cytoplasmic membrane		Export apparatus; hook length control
fliC	*fliC*	3	55	Exterior		Flagellin
fliD	*fliDST*	3	50	Exterior		HAP2—filament cap
fliE	*fliE*	2	11	Presumably periplasm	9	MS ring-rod junction
fliF	*fliFGHIJK*	2	61	Cytoplasmic membrane	27±4	MS ring
fliH	*fliFGHIJK*	2	26	Cytoplasm		Export apparatus; negative regulator of FliI
fliI	*fliFGHIJK*	2	49	Cytoplasm		Export apparatus; ATPase
fliJ	*fliFGHIJK*	2	17	Cytoplasm		Export of flagellar components; a chaperone
fliK	*fliFGHIJK*	2	39	Cytoplasm		Hook length control
fliO	*fliLMNOPQ*	2	11	Cytoplasmic membrane		Export apparatus
fliP	*fliLMNOPQ*	2	23	Cytoplasmic membrane	4−5	Export apparatus
fliQ	*fliLMNOPQ*	2	10	Cytoplasmic membrane		Export apparatus
fliR	*fliR*	2	26	Cytoplasmic membrane	1	Export apparatus
fliS	*fliDST*	3	15	Cytoplasm		Filament elongation; possibly a chaperone, preventing premature interaction of newly synthesized flagellin subunits in the cytoplasm

fliT	*fliDST*	3	14	Cytoplasm		Filament elongation; possibly an export chaperone for FliD, which prevents oligomerization of HAP2 by binding to its helical domain before export. A negative regulatory factor for class 2 expression?

Genes involved in driving the flagellum and controlling its direction of rotation

fliG	*fliFGHIJK*	2	37	Cytoplasm	$\sim$35	Switch-motor complex
fliM	*fliLMNOPQ*	2	38	Cytoplasm	$\sim$35	Switch-motor complex
fliN	*fliLMNOPQ*	2	15	Cytoplasm	$\sim$100	Switch-motor complex
motA	*motAB-cheAW*	3	32	Cytoplasmic membrane		Stator of the motor; H^+-conducting channel
motB	*motAB-cheAW*	3	34	Cytoplasmic membrane		Stator of the motor; anchor of the H^+ channel to the cell wall

Genes whose products are with unknown function

flhE	*flhBAE*	2	12			
fliL	*fliLMNOPQ*	2	17	Cytoplasmic membrane		Associated with the basal body
fliY	*fliAZY*	2	$\sim$27			
fliZ	*fliAZY*	2	$\sim$18			A positive regulatory factor for class 2 expression?
ycgR						Presumably involved in motor function together with *yhjH* and H–NS
yhjH						Presumably involved in motor function together with *ycgR* and H–NS

[a]Based on [50, 71, 224, 225, 237, 317, 326, 335, 336, 353, 370, 400, 460, 506–509, 534, 536, 537, 631, 674, 802] and references cited therein.

[b]According to unified nomenclature for *E. coli* and *Salmonella*. The correlation between this nomenclature and the prior one can be found in [312]. Genes named *flg* are located in region I of the chromosome (at 24.5 min), *flh* in region II (at 42 min), and *fli* in region III (at 43 min). The gene *flk* has a distinct location—see text for details.

[c]±SEM.

assembly. It includes, in addition, genes involved in driving the flagella (*mot* genes) and chemotaxis (*che* genes). Region III (at 43 min) contains genes involved in flagellar structure, assembly, and function, termed *fli* genes. It is divided to two subregions, IIIa and IIIb, due to a disruption consisting of DNA unrelated to motility and chemotaxis [460]. *Salmonella* has a fourth region of genes, *flj* genes, containing genes involved in phase variation. (Phase variation was discussed in Section 2.2.1) The *Salmonella* regions are located at 23, 40, 40, and 56 min (the *Salmonella* regions II and III at 40 min are separated from each other by 23 kb of sequence) [460]. A single-gene operon (at 52 min in *Salmonella*), which does not belong to any of the previously mentioned motility and chemotaxis gene clusters, was recently found in *E. coli* and *Salmonella* and termed *flk* [335]. In view of this distinct location, it is likely that this gene, which regulates flagellin gene expression in response to P- and L-ring assembly, is also involved in cellular processes other than flagellar assembly [335]. In each region, the flagellar genes are further organized in operons, usually according to function (Table 3).

The other two classes of mutants, the Mot⁻ class and the Che⁻ class, identified three types of genes with respect to the phenotypes that result from mutations in them [105, 460]:

(a) Genes that only result in Mot⁻ phenotype when mutated or absent (*motA* and *motB*). The fact that *motA* and *motB* mutants are non-motile in spite of having flagella, indicates that the products of the genes *motA* and *motB* are involved in flagellar rotation.

(b) Genes that only result in defects in chemotaxis, i.e., in Che⁻ phenotype (*che* genes). The *che* genes and their products will be discussed in Sections 5 and 7, respectively.

(c) Genes that, when mutated, can be in one of three phenotypes: absence of flagella, paralysis, or abnormal switching from one direction of rotation to the other (*fliG*, *fliM*, and *fliN*). Paralysis or abnormal switching results from missense mutations or in-frame deletions of these genes. Complete deletion of any of these genes prevents flagellar assembly. Because of the multiphenotype character of *fliG*, *fliM*, and *fliN*, their products were identified as being involved in regulating the direction of flagellar rotation and they were, therefore, termed switch proteins (to be discussed in Section 2.2.6).

Expression of the flagellar genes is controlled by a three-level regulatory hierarchy [377, 397]. Level 1 genes (*flhC* and *flhD*) are required for the expression of level 2 genes, which are, in turn, required for the expression of level 3 genes [105]. Because the *flhDC* operon controls the expression of all other flagellar (and chemotaxis) genes, it is called the "master operon." It is involved in coupling flagellation to the cell cycle [542, 587]. In *E. coli*, the level 1 genes are under the control of cAMP [105, 460]. In addition, they are affected by a number of global regulatory signals. These include, among others, temperature control, heat shock proteins, the carbon storage regulatory system (Csr), DNA supercoiling, growth phase, and high concentrations of either inorganic salts, carbohydrates, or alcohols [160, 771]. The hierarchy levels of the motility genes are indicated in Table 3.

2.2.3. *Function of flagella*

Unlike eukaryotic flagella, bacterial flagella act by rotation. The flagellar motor (to be discussed in Section 2.2.5) rotates the rigid, passive helical filament either counterclockwise or clockwise. The main evidence for a rotary function includes direct visualization of functional flagella [390, 452], rotation of cells tethered by a flagellum to a glass surface [658], and mutual rotation of cells, linked to each other by antiflagellar antibody, of a mutant that cannot swim because of having straight rather than helical flagella [455, 658]. (A relevant early review and a recent historical review of flagellar function can be found in [79, 461], respectively.)

2.2.4. *The energy source of flagellar rotation*

About 0.1% of the cell's total energy consumption under growth conditions is used for flagellar rotation [460]. Larsen *et al.* [406] demonstrated in 1974 that the proton-motive force (Figure 4), not ATP, is the energy source for flagellar rotation of *E. coli* and *Salmonella*. (In their paper, Larsen *et al.* used the term "intermediate in oxidative phosphorylation" instead of proton-motive force. The first who said explicitly that proton-motive force drives bacterial flagella were Manson *et al.* [468], who studied *Streptococcus* cells.) Larsen *et al.* showed that H^+-ATPase mutants (*uncA*), which cannot convert ATP to proton-motive force and *vice versa*, swim normally under aerobic conditions with oxidizable substrates that produce proton motive force, but not with glycolytic substrates that produce ATP. Reduction of the ATP concentration to below the detection level (<0.3% of the pretreatment concentration) by arsenate did not inhibit the motility. (Arsenate is structurally similar to phosphate; therefore, by causing futile cycles of synthesis and spontaneous hydrolysis of

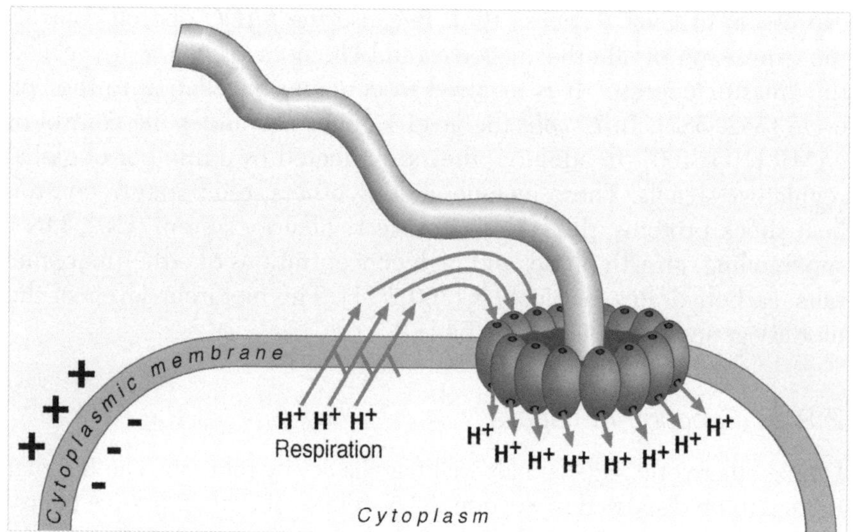

Figure 4. A simplified scheme of the proton-motive force production by respiration and its use for flagellar rotation.

ADP-arsenate, it depletes the cells of ATP.) Abolishing the proton-motive force by the uncoupler carbonylcyanide m-chlorophenylhydrazone (CCCP) abolished the motility. These early studies were followed by many other studies, all reaching the same conclusions [83, 260, 347, 468, 469, 483, 651, 726, 775]. More direct evidence was obtained by demonstrating (a) that flagellar rotation can be powered with an external voltage source [241, 331], and (b) that flagellar rotation in cell envelopes, which contain just buffer instead of cytoplasm, can be powered with an artificially imposed chemical gradient of protons across the membrane (ΔpH), even when the membrane potential is kept at zero [593]. This latter observation indicated that protons (or hydroxyl ions) *per se* drive the motor [593] (Figure 4). Meister *et al.* [495] managed to measure the H^+ flux through the motor of *Streptococcus* by comparing the total H^+ flux through the cell membrane between two states: during flagellar rotation and during lack of rotation. They found that ~ 1200 protons are required for each revolution of the motor.

Some bacterial species such as alkalophilic *Bacilli* grow optimally at alkaline pH values (e.g., pH 11), at which the proton-motive force is almost negligible. By approaches similar to those described above, it was demonstrated that the flagella of these species, as well as the polar flagella of some marine species that live at close to neutral pH values but at high salt concentrations (e.g., *Vibrio* species), are driven by Na^+-motive force instead of proton-motive force [47, 297, 298, 373]. Here a flux of Na^+ ions through Na^+ channels powers the flagella (Section 2.2.5). Interestingly, under certain conditions, *Vibrio* can possess two types of flagella, each driven by a different ion: lateral flagella (used for swarming and movement in viscous media) driven by a flux of protons, and polar flagella (used for swimming in nonviscous media) driven by a flux of Na^+ ions [39, 48, 340]. Both types of flagella are directed by a common chemotaxis signal transduction system [624].

2.2.5. *The flagellar motor*

Structure
Like any other rotary electric motor, the flagellar motor contains a rotor and a stator. The rotor is built from the MS ring (FliF) and FliG, and possibly also FliM and FliN (Figure 2). The stator is built from the proteins MotA and MotB (see [148, 185, 352, 460, 535] for elaborated reviews of the flagellar motor). MotA and MotB, observed in electron micrographs as 10 particles or studs arranged in a ring surrounding the rotor [185, 348], are integral membrane proteins involved in generating thrust. The drive shaft of the motor, termed the rod, is built from the proteins FlgB, FlgC, FlgF, and FlgG. The rod is surrounded and held by the L and P rings, probably serving as a bushing. The helical propeller is the filament, and the universal joint that connects it to the rod of the motor is the hook. The motor also has a gearshift, termed a switch. Its role is to shift the direction of rotation of the motor according to signals received from the receptors on the cell surface.

Function
The flagellar motor can rotate extremely fast, up to 270 or 600 revolutions per second for a H^+- or Na^+-driven motor, respectively [390, 440, 527, 528]. (Rotation as fast as 1700 r.p.s. [102 000 rpm!] was measured for the polar flagellum of *V. alginolyticus* at 35° C [465].) Several force-generating units, which can function independently and which, in the case of H^+-driven motors, are made of the MotA and MotB proteins, drive each of these motors. This was elegantly demonstrated by Block and Berg [113] and Blair and Berg [102]. Following an earlier observation of Silverman *et al.* [659] that paralyzed *mot* mutants can be resurrected by Mot protein synthesis, Berg's group gradually expressed MotA

or MotB in paralyzed *motA* or *motB* mutants, respectively, and found gradual restoration of flagellar rotation. The speed of rotation increased in steps, up to eight steps, suggesting that there are up to eight force-generating units per motor [102, 113]. Using a somewhat opposite approach for Na^+-driven motors, Muramoto *et al.* [526] inhibited the rotation of alkalophilic *Bacillus* cells with a photoactivated derivative of amiloride (a potent inhibitor of Na^+ channels) and found that the inhibition occurs in 5–9 steps. This suggested that there are 5–9 force-generating units per Na^+-driven motor.

In H^+-driven motors, MotA has four transmembrane α-helices with polar residues in the cytoplasm and only two short segments in the periplasm [103–105, 181, 529, 644, 819, 820]. MotB, with only a single transmembrane domain [690], is mostly located in the periplasm [162, 529]. It was, therefore, assumed that MotA might form a H^+-conducting channel, with MotB as an anchor to the cell wall [102, 162]. Subsequently, by comparing the H^+ permeabilities of vesicles containing wild-type or mutant MotA proteins and finding that the H^+ conductance of the latter was ~sixfold lower than that of the former, Blair and Berg [103] provided direct evidence that MotA conducts protons. These data taken with the finding that MotA and MotB interact with each other [251, 707, 781] indicated that these two proteins constitute a transmembrane H^+-channel complex, which is anchored to the cell wall by a peptidoglycan-binding domain of MotB.

More recent studies suggested that the structure and function of Na^+-driven motors may be similar to those of H^+-driven motors, though with some subtle differences [39]. Residues known to be critical for torque generation in H^+-driven motors are conserved in the Na^+-driven motor, suggesting a common rotation mechanism regardless of the different driving force [118]. In the case of *V. alginolyticus*, four proteins may be involved in Na^+-conductance: MotX [487], MotY [486], PomA, and PomB [39]. The latter two have been demonstrated to possess Na^+-conducting properties [625] and to have sequence similarities to MotA and MotB, respectively. Furthermore, MotA of *R. sphaeroides* can replace PomA of *Vibrio*, generating torque by Na^+ flux [40]. Similarly, a chimeric protein constructed of the N-terminal MotB (including the entire transmembrane region) and the C-terminal PomB, gave rise to a Na^+-driven motor with MotA [41]. MotX and MotY interact with each other [487]. They have no similarity to MotA and MotB, respectively, but expression of MotX results in Na^+ conductance [487]. All these observations suggest either that all the four proteins form a

Na$^+$-channel complex, with MotX and PomA constituting the Na$^+$-conducting part, or that there are two different Na$^+$ channels in *Vibrio*, one made of MotX and MotY, and the other of PomA and PomB [39, 487]. Recently, Gosink and Häse [262] replaced the *pomA*, *pomB*, *motX*, and *motY* genes of *V. cholerae* by the *motA* and *motB* genes of *E. coli*. The resulting hybrid motor was functional and driven by H$^+$, not by Na$^+$. (For a recent review of the Na$^+$-driven motor, see [805].)

It is not known how the flux of protons or sodium ions through the H$^+$- or Na$^+$-channel, respectively, actually rotates the motor. The few facts that we do know about torque generation and the function of the flagellar motor are about H$^+$-driven motors. They can be summarized as follows:

(a) In addition to the components of the H$^+$ channel (MotA and MotB), torque generation requires the protein FliG (but not the other two switch proteins FliM and FliN) [430].

(b) Specific charged residues on FliG (Arg281, Asp288, and Asp289) [431, 821], MotA (Arg90 and Glu98) [820], and MotB (Asp32) [822] are necessary for torque generation. In addition, proline residues 173 and 222 of MotA appear to have pivotal roles in coupling proton flow to motor rotation [130].

(c) FliG interacts with the MS ring [479], MotA [251, 252, 718, 821], FliM [478, 479, 718], and FliN [718] (Figure 2). The functionally important charged residues of FliG interact with those of MotA [821].

(d) The motor speed is not constant but rather fluctuates [211, 333, 390, 616]. This is probably a reflection of the fact that the force-generating units of the motor step independently (i.e., the flagellar motor is a stepping motor) [617].

(e) The motor can rotate in either direction. There is no barrier to backward rotation [94, 95].

(f) At low load, the speed of the motor is independent of the number of force-generating units; at a load close to zero, one unit turns the motor as fast as many [613].

(g) The motor operates in two dynamic regimes. At 23° C, the torque is approximately constant up to a speed of nearly 200 Hz. As the motor's speed further increases, the torque rapidly falls. At ~350 Hz, the torque is essentially zero. In the low-speed regime, torque is insensitive to changes in temperature. In the high-speed regime, it decreases markedly at low temperature [154].

In addition, a number of observations have been made whose significance, on the basis of current knowledge, is obscure. For example, it was unexpectedly found that H-NS (a transcription factor, histone-like nucleoid-structuring protein, which regulates the flagellar master operon *flhDC* [371, 680]) binds to FliG [479], and that the tightness of the binding determines the speed of flagellar rotation [195]. It is possible that FliG plays a regulatory role by allowing H-NS to activate downstream class 3 genes (Table 3) [479, 681], or that H-NS is somehow involved in torque generation through its interactions with FliG [195]. It has been subsequently demonstrated that H-NS indeed affects flagellar function (as well as biogenesis) and that the genes *ycgR* and *yhjH*, which contain the consensus sequence found among the class III promoters of the flagellar regulon, are involved in this function too [370]. Other unexplained findings include the inhibition of motor rotation by arsenate [346, 474, 775], by the sweet-tasting protein thaumatin [66], or by some chemorepellents in the absence of chemotaxis proteins [211].

Over the years, many models have been put forward for the coupling between proton flow and flagellar rotation (see [148] for a comprehensive review). In view of facts (a)–(c) above, the mechanism of function of the motor probably involves electrostatic interactions between the rotor and stator [821]. A number of models involving electrostatic interactions have been proposed (see [92, 215, 757, 758] for the more recent models). In one such model [215] torque is generated by electrostatic interaction between fixed negative charges on the channels and fixed positive charges arranged in tilted rows on the rotor, with the protons playing a modulating role in that they screen the negative charges when located appropriately. In an alternative model which reproduces observed behavior far more closely [758], torque is generated by direct electrostatic interaction between a proton in a channel and fixed positive and negative charges arranged in alternating tilted rows on the rotor. The recently revealed crystal structure of the carboxy-terminal domain of FliG [432] is well in line with this model. In both models switching from one direction of rotation to the other requires a change in the angle of tilt. A number of recent models, which do not rely on electrostatic interactions, have also been proposed. Thus, Oplatka [553, 554] suggested that the motor may act like a water turbine, i.e., that rotation is the outcome of water minijets impinging tangentially on the rotor. According to this suggestion, the minijets are formed when the protons (or sodium ions) lose part or all of their hydration water upon interacting with the proper site on the rotor. Atsumi [49] proposed an ultrasonic motor model, according to which the influx of ions through the ion channel causes conformational changes of the channel itself, resulting in traveling waves in the C ring of the motor. This wave stabilizes the cyclical movement of the channel that generates the rotating force. On the basis of the apparently different symmetries of the C and M rings (34 and 26, respectively), Thomas *et al.* [730] proposed yet another model. According to this model, 26 levers of the C ring bind to 26 equivalent sites on the M ring, and the excess 8 levers bind to proton-pore complexes and form 8 torque generators. Rotation results from the swapping of pore-bound levers with M-ring bound levers.

Functional states of the motor
The flagella of bacteria such as *E. coli* and *Salmonella* can rotate counterclockwise and clockwise (the direction of rotation defined for a flagellum viewed from its distal end towards the bacterial cell), and they can also pause [209, 405, 407, 591, 658]. A pause seems to result from a futile switching attempt from counterclockwise to clockwise [211]. Under non-stimulated conditions, the flagella rotate mostly counterclockwise with brief intermissions of clockwise rotation and pauses. The flagellar motors of other bacterial species may similarly have three functional states, or they may only have two states: rotation in one direction and pausing (e.g., in *R. sphaeroides*), or rotation in both directions without pausing (e.g., in *Pseudomonas*) (Table 1).

2.2.6. *The flagellar switch*

Definition
The flagellar motor switches from one direction of rotation to the other either spontaneously or in response to signals received from the receptors or other mediators. The element that receives the signal and determines the probability of the motor to rotate in a certain direction is the switch.

Location
Initially, the switch was thought to be associated with the cytoplasmic membrane. This is because of two main reasons: (a) The switch is a part of the motor that should be exposed to the cytoplasm for receiving signals (e.g., [570, 794]). (b) Phenotypes of mutations in *fliM* and *fliG* are retained in cytoplasm-free envelopes of *Salmonella* [208, 592]. Later, when the C ring of the motor was identified, the switch was demonstrated to be in this structure [234, 813] and to be linked to the motor via FliG [479] (Figure 2).

Structure

Very little is known about the structure of the switch. The switch is a complex made of three proteins—FliM, FliN, and FliG—that interact with each other ([358, 447, 478, 479, 552, 718, 741, 794, 814] and references cited therein[a]; Figure 2). The first identification of these three proteins as switch components was made by the observation that they, unlike other chemotaxis and motility proteins, can have four different mutant phenotypes: lack of flagella, paralyzed flagella, flagellar rotation biased to clockwise direction, and rotation biased to counterclockwise direction [180, 315, 345, 570, 672]. On the basis of the mutual interaction and the number of copies of each protein per flagellum (Table 3), a reasonable model of the switch-complex structure, proposed by Mathews *et al.* [481], is shown in Figure 5. The intraprotein organization of the switch is in the process of being revealed. It is known, for example, that the N terminus of FliM is the main binding domain of the signaling protein CheY (to be discussed in Section 7.5) [135, 481, 740, 741], that the C terminus is the binding domain of FliN [481, 741], and that the mid-protein segment between residues 140 and 200 apparently contains the FliG-binding site [741]. FliM might possess additional binding sites for CheY and FliG [481].

Function

An intriguing question is how the switch complex causes the motor to reverse direction even though the polarity of the proton flux, which drives the motor, is unchanged. It seems certain that the symmetry of the switch-motor complex must undergo a transformation to allow switching from one direction of rotation to another. This transformation probably includes conformational changes in at least part of the rotor (which constituted the switch proteins). Such a conformational change should be fast, because the rotational reversal is accomplished within less than 1 ms [390]. Unfortunately, no conformational studies of the switch-motor complex have been published. A number of models for the mechanism of the switch have been put forward (for example, in references [18, 142, 459, 628, 758]). However, the mechanism of function of the switch will probably not be completely understood until the mechanism of function of the motor is resolved.

---

[a] Note that papers published before 1988 use the old nomenclature of flagellar genes. The reader is referred to Iino *et al.* [312] for a nomenclature conversion table.

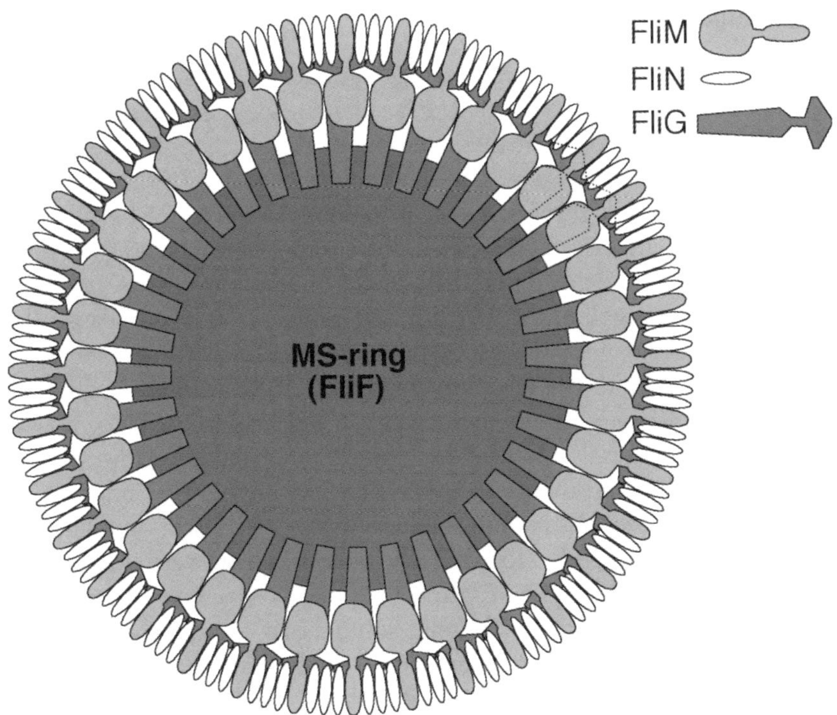

Figure 5. Model of the arrangement of the switch proteins within the switch-motor complex. The figure shows the way FliM, FliN, and FliG are mounted on the MS ring as viewed from the cytoplasm. The actual FliN:FliM stoichiometry might not be 3:1 as shown (Blair, personal communication). Taken with permission from Mathews *et al.* [481].

The switch can be directly modified by at least two intracellular constituents: the response regulator CheY and fumarate. (The CheY-switch interaction and its outcome will be discussed in Section 7.5. The effect of fumarate will be discussed in Section 8.2.8.) Other factors that affect switching are the proton-motive force and the temperature. As the proton-motive force is reduced, the motor becomes more and more counterclockwise biased, until at about 70% of the maximal motor speed the flagella rotate exclusively counterclockwise [346]. Conversely, in a process that is independent of the presence of CheY and fumarate, switching increases when the temperature decreases [748]. The reason is that the standard free energy difference ($\Delta G°$) between the clockwise and counterclockwise states of the switch becomes lower at low temperatures. The molecular mechanism underlying this change is not known.

2.2.7. *Assembly of flagella*

In vitro, under proper conditions, flagellin and hook monomers can be spontaneously polymerized into flagellar filaments and hooks, respectively [42, 460]. *In vivo*, however, the assembly of flagella is tightly regulated and coupled with gene expression. It is synchronized with the cell cycle and it depends on cell division and growth phase [12, 460, 588]. About 2% of the cell's biosynthetic energy expenditure is for flagellar synthesis [460].

Years of studies revealed that the flagellar filament is not assembled at the base of the flagellum (as pili do), but rather at its tip. Because the larger part of the flagellum is extracellular and yet many proteins that are the building stones of the flagellum are synthesized within the cell, they should cross the cytoplasmic membrane and be exported. If so, how is large loss of material into the bulk medium avoided? How do the exported proteins find their way to the tip? The answer lies in a central, hollow channel running the entire length of the rod, hook, and filament [462]. The proteins needed for assembly are pushed outward through the channel by a flagellum-specific "export apparatus," located at the center of the C ring [15, 460]. The flagellin subunits are exported in this way directly to the assembly site rather than to the bulk medium. This export is probably at the expense of ATP, and it probably involves FliI, a flagellum-specific ATPase [15, 197, 224, 460, 656]. (Most flagellar proteins do not have the cleavable signal sequence that is recognized by the primary protein export pathway, the Sec pathway [15, 460].) The sequence and timing of the export are tightly controlled. However, the mechanism underlying this control is poorly understood.

Interestingly, similarities have been found between some proteins involved in the flagellar export apparatus and type III secretion proteins involved in the export of virulence factors [243, 460, 462]. Also, the supramolecular structure of the type III secretion machinery somewhat resembles the structure of the flagellar basal body [389]. Recently it has been demonstrated experimentally that, in the pathogenic bacterium *Yersinia enterocolitica*, the flagellar export apparatus has the ability to function also as a secretion system for the transport of several extracellular proteins not related to motility [807].

The stepwise assembly process of the flagellum has been concluded from incomplete flagellar structures produced by flagellar mutants [325, 387, 713, 714]. This sequence of steps was recently confirmed by real-time monitoring of the transcriptional activation of the flagellar operons by means of a panel of 14 reporter plasmids in which green-fluorescent protein (GFP) was under the control of one of the flagellar promoters [330]. The sequence is as follows (Figure 6):

(i)   The MS ring is the first observable structure in the assembly of the flagellum [15, 460, 535]. This suggests that first FliF protein incorporates into the cytoplasmic membrane and self-assembles to form the MS ring [388, 536, 751]. The occurrence of this step depends,

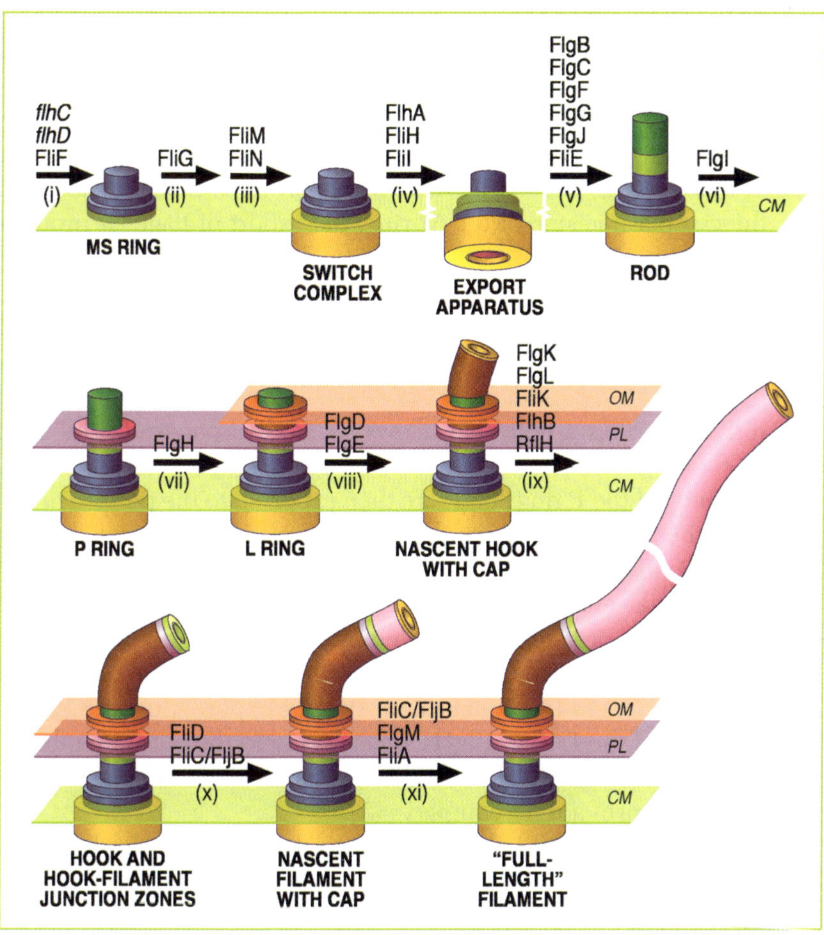

Figure 6. Schematic description of the stepwise assembly of *E. coli* and *Salmonella* flagella. The step numbers correspond to those described in the text. Abbreviations: OM, outer membrane; PL, peptidoglycan layer; CM, cytoplasmic membrane. Modified (with permission) from Aizawa [15].

of course, on the availability of FliF, the expression of which is regulated by the master genes *flhC* and *flhD*. A tetrameric complex of FlhC and FlhD (at a 2:2 ratio), which has been shown to function as a transcriptional activator of a few class 2 operons [426], is perhaps the transcriptional activator of the *fliFGHIJK* operon (Table 3). FlhC may be an allosteric effector that activates FlhD for the recognition of the DNA [146].

(ii) FliG binds to the cytoplasmic face of the MS ring, forming a partial switch structure. This binding does not require any other flagellar proteins [388].

(iii) FliM and FliN associate cooperatively with the FliG-FliF complex, thus completing the switch complex and assembling the C ring [388].

(iv) The proteins FlhA, FliH, and FliI form, at the center of the C ring, the export apparatus, which possibly interacts with the inner pore of the MS ring [15, 359]. Once the export apparatus is installed at the cytoplasmic side of the M ring, the export of flagellar proteins can begin [15].

(v) The rod of the basal body, composed of the proteins FlgB, FlgC, FlgF, and FlgG [303], is added. FliE is also required for the rod assembly [460]. FlgJ, which has a muramidase activity, probably hydrolyzes the peptidoglycan layer locally at the tip of the nascent rod and makes a hole that allows the rod to penetrate the layer [296, 536].

(vi) Subunits of the P ring (FlgI) are then exported to the periplasm and form the P ring in the peptidoglycan layer [15]. This ring formation may also involve the muramidase function of FlgJ [536]. In contrast to other flagellar proteins, FlgI (as well as FlgH that forms the L ring in the next step) is secreted by the primary (Sec) pathway in a signal peptide-dependent manner [15, 460]. FlgJ, on the other hand, is secreted by the flagellum-specific export apparatus [536].

(vii) Subunits of the L ring (FlgH) are exported to the outer membrane and form the L ring. The P and L rings join, by an unknown mechanism, to form a rigid complex [15].

(viii) The hook is assembled from FlgE molecules with the help of a scaffolding protein, FlgD [546]. This protein is associated with the tip of the growing hook. The initial growth rate of the hook is very fast (40 nm/min in the case of *Salmonella*). Then the rate exponentially slows until the length reaches a certain value (55 nm for *Salmonella*). Above this value the hook grows at a constant rate (8 nm/min for *Salmonella*) [379].

(ix)   When the hook reaches the desired length, FlgK (HAP1) is exported and replaces FlgD, forming the first hook-filament junction zone. Then a second hook-filament junction zone, made out of the FlgL (HAP3) protein, is assembled [460]. Initially it was believed that the hook's length is controlled by FliK and FlhB [295, 505, 530, 531, 780] as well as by RflH (a product of a gene located at ~52 min, where no flagellum-related genes have been found) [399]. However, when Makishima *et al.* [467] found that mutations in the switch genes (*fliM*, *fliN*, and *fliG*) may result in a shorter hook, they proposed that the C ring, which is built from the products of these genes, acts as a quantized measuring cup. Makishima *et al.* [467] suggested the following sequence of events: "The hook monomers (FlgE) accumulate to fill the C ring and are secreted en bloc to form the hook of a finite length. When the C ring is empty, FliK is secreted, which converts the mode of secretion into that for flagellin. Then, FlgD at the tip of a nascent hook is replaced by FlgK, which terminates the hook elongation."

(x)   A cap, made up of FliD (HAP2) subunits, is formed. The cap enables flagellin subunits (FliC in the case of *E. coli*; FliC or FljB in the case of *Salmonella*) to insert into the distal end of the nascent filament, similarly to the role fulfilled by the scaffolding protein FlgD in hook assembly. However, in contrast to the hook cap, the filament cap is retained indefinitely [460]. In a beautiful study, Yonekura *et al.* [804] recently demonstrated that the filament is assembled by stepwise rotation of the cap. As is the case with the hook, the growth rate of the filament appears to decrease exponentially as the length of the filament increases [12, 311].

(xi)   The filament grows over several generations until it slows to an almost negligible rate. In the case of *Salmonella*, this happens when the filament reaches a length of >20 μm [12]. Termination of filament assembly is believed to be partly regulated by FlgM. FlgM is an antisigma factor that binds to the sigma factor FliA and prevents its association with RNA polymerase core enzyme. A high intracellular concentration of FlgM represses the level-3 operons and thereby prevents or reduces the expression of the level-3 gene products. FlgM is apparently secreted from the cell through the central channel of the flagellum. When the assembly of the hook-basal body is completed, FlgM is exported outside the cell, its intracellular concentration is consequently maintained at a low level, and the expression of the level-3 gene products is relatively fast. It is believed that as the filament increases in length, the travel of the exported proteins, including FlgM, through the filament become increasingly more difficult and, as a result, the efflux of FlgM slows down. Consequently, the level of FlgM in the cell rises and the expression of level-3 gene products, including flagellin, is suppressed [308, 398]. (If, at any time, the cell is subjected to forces that shear off the flagella, intracellular FlgM would be immediately exported through the hook-basal body structures present, and transcription of the level-3 genes would be initiated as a result of the presence of the FliA sigma factor and the absence of the FlgM antisigma factor [308].)

Since flagellar assembly takes several generations, the number of flagella per cell goes down when the cell divides. Roughly a constant number of *de novo* filaments appears on each of the daughter cells, and it takes ~3 generations until these filaments can be detected by an electron microscope [12].

2.3. *Modes of swimming behavior*

2.3.1. *Swimming in the absence of chemotactic stimuli*

Bacteria such as *E. coli* and *Salmonella* have two main swimming patterns: smooth swimming in a rather straight line (a run) and a brief but abrupt turning motion (a tumble) [74, 451]. In the absence of stimuli the tumbles usually occur once every 1–10 s (depending on the bacterial strain and the growth phase [692]). Consequently, the bacterial cells execute a random walk, composed of runs and tumbles with essentially no net vectorial movement (Figure 7A).

The run is the consequence of counterclockwise rotation of the flagella [407, 658]. Because of the flagellar left-handed helicity, counterclockwise rotation exerts a pushing force on the cell. Since the flagella around the cell have different lengths and their distribution is not symmetric, the net force is not zero. Consequently, the cell moves in the direction of the net force and, due to the viscous drag of the medium, the flagella are swept to the rear of the cell, amplify the net force in the direction of movement, and form a left-handed bundle (aligned with the long axis of the cell) that pushes the cell forward [27, 453].

The tumble is the consequence of flagellar clockwise rotation [407] and pauses [211]. Unlike counterclockwise rotation, which stabilizes the left-handed form of the flagella, clockwise rotation destabilizes the

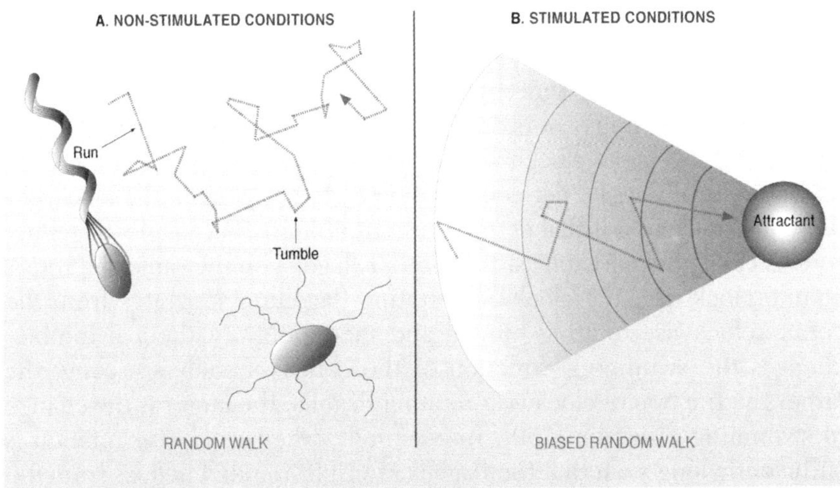

Figure 7. Swimming behavior of *E. coli* or *Salmonella* cells. (A) Nonstimulated conditions; (B) stimulated conditions. (Taken with permission from Eisenbach [214].)

left-handed helix. Consequently, the flagella undergo a transition from a left-handed helix to right-handed one, and the transition propagates from the flagellar junction with the cell body towards the distal end of the filament [454]. However, because the periods of clockwise rotation are relatively short and because of the occasional pauses, the transformation from left to right-handed helix is usually incomplete [211, 454]. The consequence is that some flagella have segments of opposite-handedness within the very same filament, resulting in a large angle between the segments (Figure 7A) [454]. This angle, which provides angular motion to the bacterial cell, prevents bundle formation, and forces each flagellum to act separately (each exerts force in a different direction), thus causing tumbling.

The conclusion from these observations is that the direction of flagellar rotation determines the bacterial swimming mode. (In other bacterial species, as well, the direction of flagellar rotation determines the bacterial swimming mode. It determines whether the cell moves forward, moves backward, or pauses (Table 1).) This raised the question of whether all the flagella on a given cell are synchronized, changing their direction of rotation at the same time. In cleverly designed experiments, Ishihara *et al.* [316] and Macnab and Han [456] demonstrated that different flagella on a given cell reverse and pause asynchronously, suggesting that the flagella are independent of each other. The resulting question—how a bacterial cell of *E. coli* or *Salmonella* behaves when some flagella rotate counterclockwise and others rotate clockwise—was unanswered for a long time. Recently Turner *et al.* [750] have demonstrated that tumbling occurs only when ~25% or more of the flagella on a given cell reverse to clockwise rotation. This means that, depending on the number of flagella per cell, rotation of 1–2 flagella in the clockwise direction may be sufficient to cause tumbling. If a single flagellum rotates clockwise and at least three other flagella on the same cell rotate counterclockwise, the clockwise-rotating flagellum separates from the counterclockwise-rotating bundle and moderately (without a tumble) changes the swimming direction of the cell. Generally speaking, the larger the fraction of clockwise-rotating flagella, the larger is the change in swimming direction [750]. Interestingly, when clockwise rotation is sufficiently long such that the flagella are right-handed helices from top to end, the cells swim in a rather straight line [750], a phenomenon previously observed only with clockwise-rotating mutants [345, 454].

2.3.2. *Swimming in a gradient of a chemotactic stimulus*

In species like *E. coli* and *Salmonella*, positive stimulation (an increasing chemoattractant gradient or a decreasing chemorepellent gradient) decreases the probability of clockwise rotation and, therefore, the probability of tumbles. In contrast, negative stimulation (a decreasing chemoattractant gradient or an increasing chemorepellent gradient) increases this probability [407]. This means that runs in the "right" direction are prolonged, and runs in the "wrong" direction are shorter. The outcome is a random walk of the bacterial cell, biased towards the chemoattractant (Figure 7B) or away from the chemorepellent [74, 407, 451]. Therefore, the question of how the chemotaxis process is carried out in bacteria can be reduced to the question of how the direction of flagellar rotation is regulated.

A change in the direction of flagellar rotation is not the only response to chemotactic stimuli. Depending on the varieties of flagellar motility (Table 1), some bacterial species react to changes in the concentration of chemical stimuli by changing the speed of swimming (chemokinesis) or by stopping. A few examples are shown in Table 4. Generally speaking, when a bacterial cell senses a positive chemotactic stimulus it continues to swim in the same direction. When it senses a negative chemotactic stimulus it ceases to move in the original direction and reorients itself.

Table 4. Examples of responses to chemotactic stimuli in bacteria with flagellar motility.[a]

Species	Response to positive stimuli	Response to negative stimuli	References
E. coli, Salmonella	Increased probability and rate of counterclockwise flagellar rotation. Consequently, runs are prolonged.	Increased probability of clockwise rotation and pausing. Consequently, the cell tumbles and reorients more frequently.	[74, 211, 407, 451]
B. subtilis	Increased probability of clockwise flagellar rotation. Consequently, runs are prolonged.	Increased probability of counterclockwise rotation. Consequently, the cell tumbles and reorients more frequently.	[261, 557]
S. meliloti	Increased speed of flagellar rotation. Consequently, runs are prolonged.	Decreased speed of flagellar rotation. Consequently, the bundle of rotating flagella separates to individual filaments rotating at different speeds, and the cell turns.	[675]
R. sphaeroides	Increased speed and decreased stopping probability of flagellar rotation. Consequently, runs are prolonged.	Increased stopping probability. Consequently, the cell reorients itself.	[33, 565, 566]
Azospirillum brasilense	Increased speed and decreased reversal probability of flagellar rotation. Consequently, runs are prolonged.	Presumably increased reversal probability of flagellar rotation. Consequently, the cell reverses and reorients itself.	[827]
Spirochetes	Flagella rotate without pausing, resulting in coordinated rotation of the two polar bundles. Consequently, the cell swims in a straight line.	Flagella pause frequently and extensively, disrupting the coordinated rotation of the two polar bundles. Consequently, the cell flexes and pauses.	[232]

[a] Modified with permission from Eisenbach [214].

2.4. *The gradient sensed by bacteria: temporal vs. spatial*

How does a bacterial cell sense a gradient of a chemotactic stimulus? Does it compare the stimulant concentrations between both ends of the cell or does it make the comparison between two consecutive time points? In other words, do bacteria monitor a spatial or a temporal gradient? In an elegant study, Macnab and Koshland [451] demonstrated that the latter possibility is the correct one. They rapidly mixed bacteria with a chemotactic stimulant and monitored the behavior of the bacteria within ~0.5 s, when there was no spatial gradient anymore. The cells responded to chemoattractants and chemorepellents even though the concentration change was uniformly distributed and there was no spatial gradient. The observation that the gradient sensed by bacteria is temporal means that bacteria possess a memory, which compares past information with present information to make a decision [381]. As shown in Figure 8, this memory is long enough so that the bacteria can make an accurate comparison between two points more distal than the bacterial body length (otherwise the memory would give the bacteria no advantage over the detection of a spatial gradient). On the other hand, the memory is short enough so that it signals the bacteria before they tumble and change the direction of their swimming. This arrangement

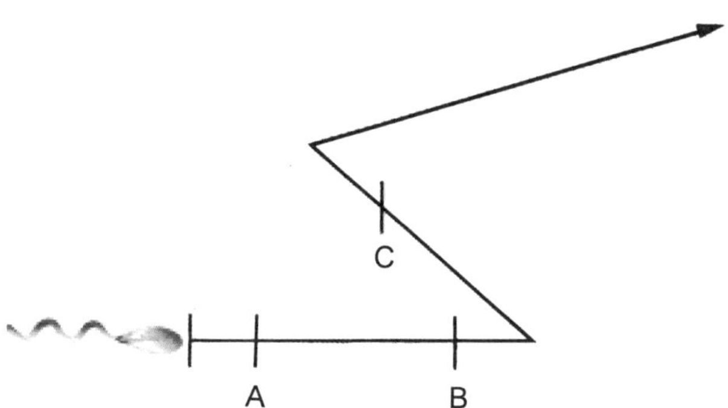

Figure 8. A scheme demonstrating an optimal memory for bacterial chemotaxis, in relation to the track made by the swimming cell. The line represents the track.

—A memory, which extends for a time period that is only sufficient for swimming from the initiation point of the track to point A, is too short. Such memory length, which is only sufficient for moving a distance comparable to the body length, gives no advantage over spatial detection, where the stimulus concentration is compared between both ends of the cell.

—A memory that extends for a time period, during which the cell reaches point C, is too long. In such a case, the cell remembers the stimulus concentration that existed before it tumbled and moved in a new direction. This information is no longer relevant.

—A memory that extends for a time period, during which the cell swims from the initiation point of the track to point B, is optimal. This memory is longer than the time required for swimming a body length and shorter than the time interval between two consecutive tumbles.

is optimal for bacteria of the size and shape of *E. coli* and *Salmonella*, taking into consideration that a change in receptor occupancy as small as 0.4% (e.g., a change from aspartate concentration of 60 nM to 71 nM) elicits a detectable chemotactic response [320, 638]. As for bacterial species with larger dimensions or different shapes, theoretical considerations suggest that they may well sense spatial gradients [199]. However, concrete examples are not yet known.

2.5. *Excitation and adaptation*

Following the finding that the gradient sensed by bacteria is temporal, time-dependent studies of the chemotactic response were carried out. These studies revealed that, like many other sensory systems, the chemotactic response involves two processes: excitation and adaptation (for a review, see [685]). When bacteria are stimulated, the direction of flagellar rotation and, hence, their swimming mode are changed instantaneously. This rapid, initial process, termed excitation, is completed within 0.07 s or less [320, 350, 354]. Later on, the bacteria resume their prestimulus behavior, even though the stimulus is still present. This

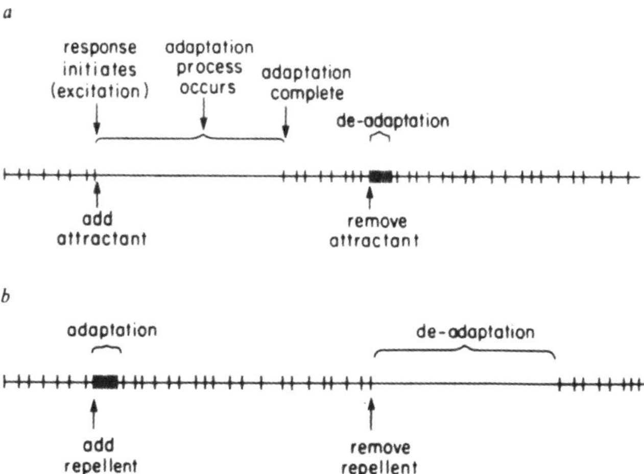

Figure 9. Schematic representation of excitation and adaptation. Each vertical line represents a tumble. The absence of vertical lines that follows addition of attractant or removal of repellent represents suppression of tumbling, while the increased frequency of lines that follows removal of attractant or addition of repellent represents continuous tumbling. (a) Effect of attractants; (b) effect of repellents. (Taken with permission from Springer *et al.* [685].)

process, termed adaptation, is relatively slow (in the range of seconds or minutes). Adaptation thus enables bacteria to adjust to changes in the stimulus intensity and respond to new stimuli. The excitation and adaptation processes in response to attractants and repellents are schematically shown in Figure 9.

3. Techniques to Measure Motility and Chemotaxis

A variety of techniques for measuring bacterial motility and chemotaxis have been developed. The main ones are listed in Table 5 and described in brief.

3.1. *Assays in which a gradient of the stimulant is established by diffusion*

3.1.1. *Capillary assay*

The first technique used to demonstrate bacterial chemotaxis, a technique that is still in use, is the capillary assay [5, 7, 499, 575, 576]. In this technique, a capillary tube containing a chemoattractant-free buffer (for motility measurement) or a chemoattractant (for measurement of positive chemotaxis) is placed in a suspension of bacteria. Bacteria then enter the capillary and accumulate in it. When the capillary contains a chemoattractant-free buffer, the extent of accumulation is a measure of the motility of the bacteria. When the capillary contains a chemoattractant, a spatial gradient of the chemoattractant is established by diffusion [5] (Figure 10A). The bacteria follow this gradient and accumulate in the capillary. When negative chemotaxis is measured, a chemorepellent-free capillary is immersed in a chemorepellent-containing suspension of bacteria. The bacteria escape from the chemorepellent into the capillary and accumulate there [747]. The bacterial accumulation can be watched under the microscope [429, 540], or the bacteria in the capillary can be counted by plating them on a culture medium [7] or by using radioactively labeled cells [777]. The extent of additional accumulation (beyond the accumulation observed in a chemoattractant-free buffer) is a measure of chemotaxis [5, 7]. The advantage of this technique is its high sensitivity. Its main disadvantage is that the gradient of the stimulant is time-dependent, because it is built and dissipated by diffusion. Also, in the case of metabolizable chemoattractants, the gradient might be distorted by metabolism [7]. It should be noted that when the capillary

Table 5. Techniques used to measure motility and chemotaxis.[a]

Technique	Able to measure			Enables isolation of mutants in		Reservations/comments	References
	Motility	**Positive chemotaxis**	**Negative chemotaxis**	**Motility**	**Chemotaxis**		
Capillary assay	+	+	+	−	−	Stimulant gradient changes with time; insensitive to high chemoattractant concentrations	[7]
Chemical-in-plug assay	−	+	+	+	+	Semiquantitative	[747]
Capillary arrays	+	+	+	+	+	Not in common use	[85]
Stable diffusion gradient assay	+	+	+	+	+	Not in common use	[217, 731]
Diffusion gradient over a membrane	+	+	+	−	−	Continuously changing gradient over a short distance; rarely used	[30]
Stopped-flow diffusion chamber assay	−	+	+	−	−	Not in common use	[229]
Preformed liquid gradient assay	+	+	+	+	+	Efficient but cumbersome; not in common use	[45, 175]
Ring forming assay on semisolid agar	+	+	−	+	+	Restricted to metabolizable chemoattractants; frequently used	[4]
Three-dimensional tracking	+	+	+	−	−	Complex; rarely used	[51, 73, 74, 82, 439]
Temporal assay	+	+	+	−	−	Simple; frequently used	[349, 350, 451, 687]
Tethering assay	+	+	+	−	−	Frequently used	[658]
Visualization of functional flagella	+	+	+	−	−	The only technique that enables visualization of rotating flagella on a swimming cell; not in common use	[116, 390, 452, 750]
χ phage assay [591, 726]	+	−	−	+	−	Macroscopic assay for incessant flagellar rotation; not in use	

[a] Bold letters indicate assays that are in common use.

contains too high concentration of a chemoattractant, the chemotaxis receptors might become saturated when the bacteria are still outside the capillary. In such a case the bacteria are unable to further sense the gradient and a sharp drop is observed in the number of bacteria accumulated in the capillary (Figure 10B).

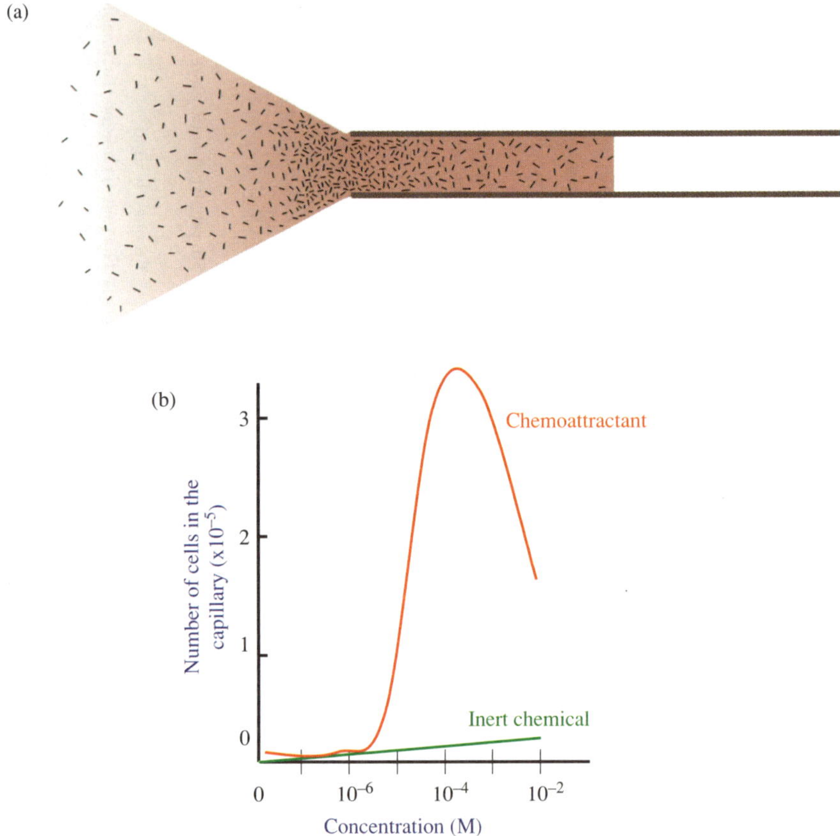

Figure 10. Capillary assay for positive chemotaxis. (a) A capillary containing a chemoattractant (reddish color) is placed within a bacterial suspension (the bacteria are represented by short lines). Initially the bacteria are randomly distributed in the suspension and their density is equal throughout. With time the cells concentrate near the capillary opening and then (usually after 30–60 min from the initiation of the assay in the case of *E. coli* and similar bacteria) they accumulate within the capillary. (b) Number of bacteria in the capillary as a function of the concentrations of a chemoattractant and of an inert chemical.

3.1.2. *Chemical-in-plug assay*

In this assay bacteria are suspended in semisolid agar (usually 0.2–0.3% agar) that is soft enough to allow motility. The bacteria are sufficiently concentrated to give visible turbidity. A plug of hard agar (usually 2–3% agar) [747] or an inert polymer [403] containing a stimulant is put within the semisolid agar. A concentric gradient of the stimulant is established by diffusion from the plug of hard agar or the polymer. If the stimulant is a chemoattractant, the bacteria congregate around the plug. (As in the case of the capillary assay, a high chemoattractant concentration may saturate the chemotaxis receptors when the bacteria are still distant from the plug. This situation may prevent the bacteria from getting close to the plug, thus forming a bacterial ring at a distance from the plug.) When the stimulant is a chemorepellent, the bacteria evacuate the zone around the plug. The higher the potency or the concentration of the chemorepellent, the larger the clear area around the plug [747]. Nonchemotactic or nonmotile mutants can be isolated from the clear zone. A number of modifications of the chemical-in-plug assay have been published:

(a) A chemical in agarose plug between a microscope slide and a coverslip [808].

(b) For chemorepellents, an inverted assay in which the chemorepellent-containing hard agar fills the plate and surrounds a hole filled with a bacterial suspension in soft agar. In this assay, bacteria escaping from the chemorepellent move towards the center of the well [285].

(c) The "drop assay," in which bacteria are suspended in a viscous medium and are attracted to a drop of a chemoattractant (or repelled from a chemorepellent) [220]. The advantage of this modification is the relatively short time required for an observable response ($\sim$15 min).

(d) In a more rapid technique, a bacterial suspension is put in a cuvette on top of an agarose gel layer containing a stimulant. The migration of the bacteria towards the chemoattractant or away from the chemorepellent is followed spectrophotometrically by the turbidity change of the suspension above the gel [826]. A similar spectrophotometric technique, suitable for anaerobic conditions, has been developed as well [538].

A general limitation of the chemical-in-plug assay is that it is semiquantitative.

3.1.3. *Capillary arrays*

In this technique, two stirred chambers are connected by a porous glass plate comprising a fused array of capillary tubes [85]. When one chamber contains a stimulant, a gradient is established in the capillaries by diffusion. This assay was used for positive chemotaxis, with bacteria put in one chamber and the chemoattractant in the other [85]. However, the assay can also be used, in principle, for negative chemotaxis with bacteria and a chemorepellent put in the same chamber. The number of bacteria that migrate from one chamber to the other is determined from the extent of laser light scattering caused by the bacteria in the other chamber. Nonchemotactic mutants can be selected because they move through the capillaries against a chemoattractant gradient [85, 86]. As in the case of the capillary assay, the assay can be used in the absence of a stimulant as a measure of motility.

3.1.4. *Stable diffusion gradient assay*

The principle of this assay is the establishment of a stable gradient of a stimulant between large reservoirs. The gradient can be linear or two-dimensional. A linear gradient is established on an agar bridge between two reservoirs [731]. For establishing a two-dimensional gradient, a polycarbonate box containing a large well (e.g., $5 \times 5 \times 2$ cm) is filled with semisolid growth medium. Continuously replenished solute reservoirs are positioned on each side of the well but separated from it by a porous membrane. These reservoirs enable the formation of multiple, intersecting, and well-defined gradients of solutes in two dimensions throughout the well [217, 779]. This technique, which is not in common use, provided, among other things, information about how the metabolism of chemoattractants affect the gradient established by diffusion [217].

3.1.5. *Diffusion gradient over a membrane*

In this technique, which is seldom used with bacteria, a stimulant gradient is established by diffusion between upper and lower wells, separated by a polycarbonate membrane [30]. Bacteria are placed in the lower well and a chemoattractant in the upper well. In the case of negative chemotaxis, the chemorepellent is included with the bacteria in the lower well. In either case, the bacteria follow the gradient, swim upwards through the polycarbonate membrane and accumulate in the upper well. The number of bacteria in the upper well is counted by a Coulter counter [30] or, when preradiolabeled cells are used, by their radioactivity [332]. One of the disadvantages of the technique is that the gradient is established over a very small distance (essentially the width of the polycarbonate membrane) and the gradient is continuously changing.

3.1.6. *Stopped-flow diffusion chamber assay*

Here two bacterial suspensions, differing only in stimulant concentrations, are contacted by impinging flow in a stopped-flow diffusion chamber. As long as there is a flow through the chamber, no mixing occurs between the two suspensions. Consequently, a step change in the stimulant concentration is imposed on the bacteria. When the flow is stopped, diffusion causes a transient chemical gradient to develop. The bacteria respond by forming a traveling band of high cell density, detected by light scattering [229]. This assay, which is not common, was used for quantitative characterization of the chemotactic response in terms of intrinsic cell properties [229].

3.2. *Population migration in a preformed liquid gradient*

In this technique the movement of bacteria in a defined gradient of a stimulant is measured [45, 175]. Bacteria are placed in a narrow band at the edge of a step gradient of a stimulant in a vertical column. A linear density gradient of glycerol is used to stabilize the stimulant gradient. The bacteria migrate according to the gradient and their distribution in the column is determined by laser light scattering. This technique, although cumbersome, is very efficient for the isolation of motility or chemotaxis mutants [45, 285], because the location and distribution of nonmotile, nonchemotactic, and wild-type cells are different [45].

3.3. *Ring forming assay on semisolid agar*

The ring forming assay on semisolid agar—a technique that was introduced by J. Adler in the nineteen sixties [4]—is perhaps the most commonly used technique even today. In this assay, bacteria are put at a certain spot in a plate containing semisolid agar and a homogeneous, low concentration of a metabolizable chemoattractant. The bacteria metabolize the chemoattractant and thereby produce a chemoattractant gradient. As soon as the bacteria use up the local supply of chemoattractant they follow the gradient that they themselves have produced. The consequence is a continuously expanding ring of dense bacteria. Due to the turbidity of the dense bacteria, the ring is easily visible with a naked eye. The ring marks the boundary between the region that has been depleted of chemoattractant and the region still containing the chemoattractant [4]. When the plate contains a number of chemoattractants, a number of expanding rings are formed. For example, bacteria placed in a semisolid agar plate containing tryptone broth, usually form three rings (Figure 11). The rings consist of bacteria "chasing" serine, aspartate and threonine [4]. (The second ring—the "aspartate" ring—is actually a doublet because the cells create an oxygen gradient in the process of

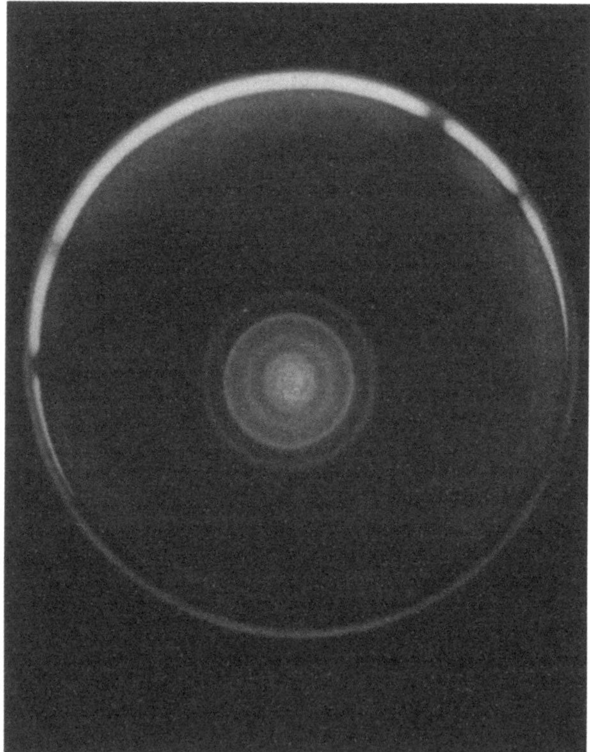

Figure 11. Expanding rings of bacteria in semisolid agar plate containing tryptone broth. See text for details. The photograph is from an experiment carried out in the author's laboratory with strain EW142 of *E. coli*. The plate contained 0.3% agar and 1% tryptone. The photograph was taken after 4 h inoculation at 35°C.

metabolizing aspartate. Aer⁻ mutants, which are chemotactically blind to oxygen, make a thinner, sharper aspartate ring [Parkinson, personal communication].) The big advantage of this technique is that it allows easy detection and isolation of nonmotile or nonchemotactic mutants. Nonmotile mutants, due to their inability to move, remain in their inoculation spot, and their colony only slightly expands by growth and proliferation. A colony of motile but nonchemotactic mutants expands by random motility. Due to the randomness, the extent of spreading is much lower than that of wild-type bacteria, and no rings are formed [35]. The main disadvantage of this technique is that it is restricted to metabolizable chemoattractants and that lack of ring formation may also result from defects in the chemoattractant metabolism. Another disadvantage is the dependence of the expansion speed (often taken as a measure of chemotaxis) on the motility and tumbling frequency of the cells [784].

3.4. *Tracking free-swimming bacteria (behavioral assays)*

Tracking free-swimming bacteria under a microscope provides direct information not only on the motility of the bacteria but also on their chemotactic behavior. In the case of aberrant chemotaxis, this technique can provide information about whether the excitation or the adaptation step of chemotaxis is defective. Two main approaches have been used:

3.4.1. *Three-dimensional tracking*

Sophisticated instruments have been designed to monitor the swimming behavior of bacteria in three dimensions. H.C. Berg built a microscope that automatically follows individual cells [73, 82]. A modified, manual version of a three-dimensional tracker was later developed by Lovely *et al.* [439]. Berg used his three-dimensional tracker for determining the swimming behavior of *E. coli* in the absence and presence of a stimulant gradient [74]. The chemical gradient was established by diffusion from a capillary. More recently, the instrument was used for determining the swimming behavior of *R. sphaeroides* in the absence of a gradient [34]. The disadvantage of the three-dimensional trackers is their complexity and the fact that only a single cell can be analyzed at one time. The latter limitation was resolved with the introduction of the three-dimensional video recording and analysis system, involving two charge-coupled device (CCD) cameras [51]. All these three-dimensional tracking techniques are very rarely used.

3.4.2. *Two-dimensional tracking—temporal assay*

The finding that the chemical gradient sensed by bacteria such as *E. coli* is temporal (Section 2.4 above) opened the way for a very convenient assay for measuring the bacterial response to stimuli—the temporal assay. In this assay, the stimulant concentration is changed over time rather than in space, and the bacterial swimming behavior is tracked in two dimensions. In the simple version of this assay, bacteria are quickly mixed (in a tube or on a microscope glass slide) with a stimulant, and the tumbling frequency of the bacteria is measured [451, 687]. In the more sophisticated versions of the assay, the bacterial response is analyzed by a computerized motion analysis system. This analysis, carried out either in real time (e.g., [334]) or during playback of video recordings (e.g., [23, 438, 614]), provides additional information such as curvelinear velocity, straight velocity, and track linearity (defined in Chapter 7, Section 2.3). A further improvement, which eliminates the dead time caused by mixing (usually between 0.5 s and a few seconds, depending on the mixing technique), was introduced by Khan *et al.* [349, 350]. The improvement involves the use of caged stimulants, i.e., photolabile compounds that release the stimulant upon photolysis. With stimulants that are generated anywhere in the solution within a few microseconds or a few tens of milliseconds, depending on the caged compound, the response of the cells can be monitored instantaneously [196, 350, 351]. Due to its simplicity, the temporal assay is frequently used for measuring both the bacterial motility and response to stimuli.

3.5. *Flagellar rotation*

Since changes in flagellar rotation are the basis of chemotaxis in bacteria, much information can be gained from studying the direction of flagellar rotation in response to stimuli. The problem is that bacterial flagella are too thin to be visible under regular light microscopy. This problem has been solved by the following techniques.

3.5.1. *Tethering assay*

A resourceful solution was provided in 1974 by M. Silverman and M. Simon [658]. In their technique, bacterial cells are tethered to a microscope glass slide by means of an antiflagellin antibody. The antibody binds to the flagella and happens to stick to glass. The antibody thus "glues" the flagella to the glass slide. (There are, however, bacterial species and mutants whose flagella stick to glass and, therefore, can be tethered without antibodies [469, 628].) As a consequence, the flagellar filament cannot rotate by the motor. Instead, the whole cell (which is easily visible under a light microscope) rotates in the opposite direction. This can occur when only a single flagellum of the cell is tethered. If more flagella are tethered, rotation is prevented. This condition can be achieved in bacteria such as *E. coli* and *Salmonella* that have flagella all around (peritrichous bacteria) by one of the following means:

(a) Mechanical shearing of flagella, e.g., by putting a bacterial suspension in a blender [204, 345] or by repeated passage of the suspension between two connected syringes [469].
(b) Bacterial growth under nutritive conditions that suppress flagellar synthesis (e.g., presence of glucose, a catabolic repressor of *E. coli*) [407].
(c) Use of mutants, which produce long hooks but no filaments, with antihook antibodies [76, 658].
(d) Use of mutants that produce small numbers of flagella (occasionally only one) [688].

In most of these approaches the resulting bacterial suspension is heterogeneous. Some cells have only one flagellum, as desired, but others have either no flagella or more than one flagellum. This is not a hindrance because the tethering assay is usually carried out in a flow chamber that allows continuous exchange of the medium [84]. Bacteria without flagella are washed away from the chamber, bacteria tethered by more than one flagellum do not rotate, and the only cells that rotate are those with a single tethered flagellum. The speed and direction of rotation as a function of time can be analyzed in real time or in playback of video recordings by a motion analysis system [350, 520] or by a linear-graded filter apparatus [87]. The tethering technique, coupled with an externally applied mechanical force (using optical tweezers—a single-beam gradient force optical trap) [114, 115] or electrical force [93, 95, 766], enables the measurement of the mechanical properties of the motor. Being very informative at the level of a single flagellar motor, the tethering assay is one of the most commonly used techniques in chemotaxis.

3.5.2. *Visualization of functional flagella*

Bacterial flagella are too thin to be visualized by regular light microscopy (bright field, dark field, or phase contrast microscopy). However, if the microscope's light source is very strong, the flagella scatter sufficient light to be observed under dark-field conditions [452]. The light source can be either a short-arc xenon lamp with extremely high brightness [452] or a He-Ne laser [390]. Alternatively, flagella can be visualized by video-enhanced differential interference-contrast (Nomarski DIC) microscopy coupled with computer-based image processing [116]. Although these techniques have the big advantage that they allow visualization of rotating flagella on a swimming cell, they are not simple and each of them has its own drawbacks. For example, in the high-intensity, dark-field microscopy the cell body scatters much more light than do the flagella. As a consequence, the cell is surrounded by a strong halo of light that masks a significant part of the flagellar filament [453]. When the laser beam is used, there is no halo (due to the small diameter of the beam) [390]. However, the filament is not seen in full but rather appears as a series of bright spots that move along the helical axis during rotation. In the Nomarski DIC microscopy the flagella cannot be observed, unless the video image is enhanced by the computer [116]. The laser dark-field microscopy appears to be advantageous over the other two observation techniques because it can follow and measure fast-rotating flagella [390, 466]. In both the high-intensity dark-field microscopy and the Nomarski DIC microscopy, flagellar rotation can be seen only when it is sufficiently slow [116, 454, 456]. Turner *et al.* [750] recently circumvented all these drawbacks by labeling *E. coli* and *Salmonella* cells with an amino-specific fluorescent dye (an Alexa Fluor dye). This dye stains the cell so that the filaments are extremely bright under a fluorescence microscope and the cell bodies are relatively dim.

3.5.3. *Chi phage assay*

Bacteriophage χ can only attach to bacterial cells with rotating flagella [75, 310, 501, 591, 618, 626]. Rotation of truncated flagella or elongated flagellar hooks with no filaments (in polyhook [*fliK*] mutants) appears to be sufficient for this [327, 618, 626, 793]. Therefore, the degree of irreversible χ adsorption can be used as a measure of flagellar rotation [726]. This technique is the only available macroscopic assay for flagellar rotation. It is effective even in the case of straight filaments, whose rotation does not result in swimming [310]. There are differences of opinion as to whether or not irreversible χ adsorption is dependent on the direction of flagellar rotation [591, 618]. It does seem to be dependent on incessant flagellar rotation: the higher the fraction of cells whose flagella rotate without pauses, the higher is the extent of irreversible χ adsorption [591].

(Reduced adsorption during clockwise rotation [618] might be due to pauses [591], known to accompany mainly rotation in the clockwise direction [211, 405].) This assay stopped being used when the more informative assays, discussed above, were introduced.

4. Chemotactic Stimuli for Bacteria

4.1. *Types of stimuli*

The chemical stimuli for bacteria are diverse. For bacteria such as *E. coli* and *Salmonella*, common stimuli are some sugars, amino acids, alcohols, ions, oxygen, and more. These stimuli are divided into chemoattractants and chemorepellents. Tables 6 and 7 list the known chemoattractants and chemorepellents for *E. coli*. These tables demonstrate that there are families of chemicals that only serve as chemoattractants, others that only serve as chemorepellents, and still others that include both types of stimuli. Sugars, dipeptides and weak bases can only serve as chemoattractants for *E. coli* (Table 6). Alcohols, weak organic acids, inorganic ions and extreme extracellular pH values only serve as chemorepellents

Table 6. Common chemoattractants for *E. coli*.

Class	Chemoattractants	References
D-sugars	*N*-Acetylglucosamine, 6-deoxyglucose, fructose, fucose (6-deoxygalactose), galactitol, galactose,1-glycerol-β-galactoside, glucosamine, glucose, glucose-1-phosphate, lactose, maltose, mannitol, mannose, methyl-β-galactoside, methyl-β-glucoside, ribose, sorbitol, trehalose	[8]
L-amino acids	Aspartate, serine, glutamate, alanine, asparagine, glycine, cysteine	[286, 498]
Amino-acid analogs and uncommon L-amino acids	*N*-acetyl-aspartate, *N*-acetyl-serine, α-amino-*n*-butyrate, β-aminoisobutyrate, aspartate ethyl or methyl ester, citrulline, cysteate, fumarate, glutathione, homoserine, *erythro*- and *threo*-β-hydroxyaspartate, isoasparagine, isoserine, malate, α- and β-methylaspartate, methylserine, phenol, serine amide, serine methyl ester, succinate	[498, 747]
Dipeptides	Glycine-glycine, glycine-L-leucine, glycine-L-proline, glycine-L-valine, L-leucine-glycine, L-leucine-L-proline, L-proline-L-glycine, L-proline-L-leucine, L-proline-L-phenylalanine, L-valine-L-glycine	[471]
Energy-linked chemicals	Oxygen at ~0.7 μM, glycerol at <50 mM, L-proline, pyrrolquinoline quinone	[163, 178, 498, 652, 830]
Weak bases	Ammonium, ethanolamine, methylamine, trimethylamine, Tris	[498, 598]

(Table 7). Amino acids and energy-linked chemicals are divided between chemoattractants and chemorepellents.

There is no obvious rule that distinguishes attractive from repulsive amino acids. Some amino acids (e.g., L-proline and L-threonine) possibly act as weak chemoattractants due to their effect on the energy metabolism of the cell [163, 496, 498]. Hydrophobic amino acids are all chemorepellents. The rule appears more obvious for energy-linked chemicals such as oxygen and glycerol. *E. coli* and *Salmonella*, like many other bacterial species, seek an optimal oxygen concentration. Therefore, an oxygen concentration at around $0.7\,\mu$M is an attractant stimulus, whereas a high concentration (e.g., $1\,$mM) acts as a repellent stimulus [652]. Likewise, glycerol is a chemoattractant for de-energized *E. coli* and *Salmonella* because it energizes the cells [830]. However, at high concentrations, it is a chemorepellent for *E. coli*, as are other polyalcohols [549, 830].

Table 7. Common chemorepellents for *E. coli*.

Class	Chemorepellents	References
Alcohols and analogs	Methanol, ethanol, *n*-propanol, isopropanol, *n*-butanol, isobutanol, isoamyl alcohol, mercaptoethanol, 2-propanethiol, benzyl alcohol, ethylene glycol	[210, 549, 747]
L-amino acids	Glutamine, histidine, isoleucine, leucine (L and D), norleucine, norvaline, phenylalanine (L and D), tryptophan, valine	[747]
Amino-acid analogs	DL-α-Amino-n-octanoate, D-α-aminophenylacetate, *p*-amino-DL-phenylalanine, *N*-benzoyl-DL-leucine, L-α-hydroxyisocaproate, α-ketoisocaproate, *N*-methyl-DL-leucine	[747]
Inorganic ions	Co^{2+}, Ni^{2+}, OCl^-, S^{2-}	[72, 747]
Energy-linked chemicals	Oxygen at $\sim 1\,$mM, glycerol at $>50\,$mM	[549, 652, 830]
Weak organic acids	Acetate, *p*-aminobenzoate, *p*-aminosalicylate, anthranilate, benzoate, butyrate, caprate, caproate, caprylate, *m*-chlorobenzoate, crotonate, formate, gentisate, heptanoate, *p*-hedroxybenzoate, L-2-hydroxy-isocaproate, isobutyrate, isovalerate, 2-keto-isocaproate, 2-mercaptoacetate, *p*-nalidixate, *m*-nitrobenzoate, phenylacetate, phthalate, propionate, saccharin, salicylate, sorbate, thioacetate, thiosalicylate, valerate	[747]
pH	Acid, alkali	[747]
Others	Acetamide, aspirin, *N*-chlorotaurine, ethylbenzoate, hydrogen peroxide, 5-hydroxyindole, indole, indole-5-carboxylate, isovaleramide, methylbenzoate, 2/3/5/7-methylindole, methylsalicylate, tryptophol	[72, 747]

The bacterial response to acidic or alkaline environments is another example of a response that can be attraction or repulsion, depending on the conditions. *E. coli* and *Salmonella* prefer an environment with a pH value around neutrality. This is the reason for their swimming away from acidic and alkaline surroundings, as demonstrated in capillary assays [747]. Surprisingly, however, the results in temporal assays are different: while a decreased pH acts as a repellent stimulus and causes tumbling, an increased pH acts as an attractant stimulus and represses tumbling [357, 598, 668]. The reason for the discrepancy between the capillary and temporal assays is not known. Bacteria also respond to changes in their intracellular pH (pH_{in}). Weak organic acids and ionophores that lower pH_{in} are repellent stimuli, whereas weak organic bases and ionophores that elevate pH_{in} are attractant stimuli [357, 598, 746, 747].

4.2. *General characteristics of stimuli*

A general rule that appears to emerge from the lists is that nutritious and useful chemicals are chemoattractants, whereas noxious and unfavorable chemicals are chemorepellents. In addition, a number of chemorepellents are excretory products of *E. coli* (e.g., acetate, ethanol, indole and S^{2-}). They may thus be indicators of a crowded (and, therefore, undesirable) environment [747]. Other chemicals (hydrogen peroxide, hypochlorite, and *N*-chlorotaurine) are products of the respiratory burst of phagocytic cells, suggesting that they contribute to the survival of bacteria by enabling them to escape the phagocytic cells [72]. In soil bacteria, seed and root exudates are chemoattractants that assist the bacteria-colonizing plant roots [645, 828]. However, there are exceptions to the rule. Not every nutrient is a chemoattractant and not every unfavorable chemical is a chemorepellent. (This is so also in higher organisms. For example, we sense saccharine, which has no metabolic value, and we find many valuable nutrients tasteless. Many toxic gases and chemicals are odorless, whereas many bad-smell odorants are not toxic [381].) A general answer may be that, in their current world, bacteria such as *E. coli* and *Salmonella* that reside in the guts are usually exposed to decomposed proteins containing the 20 common amino acids. It is, therefore, unnecessary for the bacteria to be attracted to all the amino acids. If serine is present, it is probable that tryptophan will also be present. Hence, a selective few amino acids can serve as appropriate indicators of favorable conditions without the expense of making 20 receptors for all the 20 amino acids [381]. However, bacteria in different habitats may respond to all naturally occurring amino acids (e.g., *B. subtilis* and *S. meliloti* [263, 561]).

An intriguing question is why some amino acids are chemorepellents, even though they are desirable. It has been proposed that, in some cases, the answer might lie in the similarity between the side chain of the amino acid and a toxic small molecule [381]. For example, the side chains of tyrosine and tryptophan are phenol and indole, respectively:

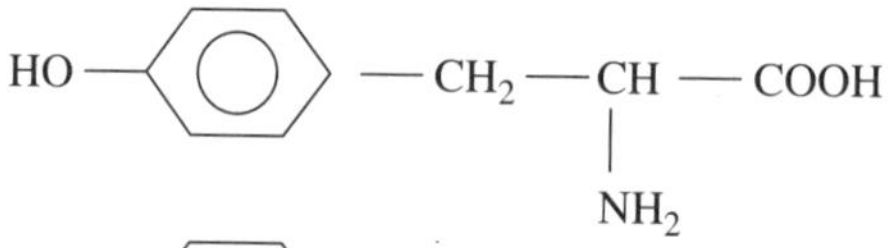

tyrosine

phenol

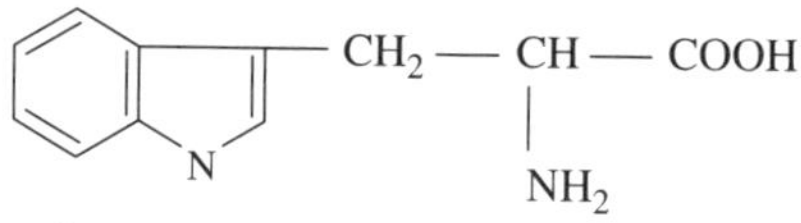

tryptophan

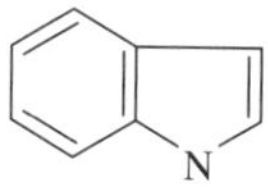

indole

Indole and phenol are chemorepellents for *Salmonella* [412, 746]. It is possible that, due to the overlap in structure, tryptophan and tyrosine activate the chemorepellent receptors of indole and phenol, respectively, resulting in repulsion. This will cause no harm to the bacteria, because the presence of serine and aspartate—extremely strong attractants—will clearly overwhelm any mild repulsive action of tryptophan and tyrosine [381].

4.3. *Diversity of stimuli in different species*

The chemical stimuli for bacteria depend on the habitat in which the bacteria live. Therefore, a certain chemoattractant for one bacterial species might act as a chemorepellent for another. Often such differences may even be found between species that live in the same habitat. A few examples are listed in Table 8. The molecular basis for at least some of the differences is understood. For example, *Salmonella* cells possess two receptors that are sensitive to phenol, the one mediating a repellent response and the other mediating an attractant response [313].

Table 8. Examples of stimulants that are chemoattractants for some species and chemorepellents for other species.

Stimulant	Chemoattractant for	Chemorepellent for	References
Acetate	*Chromatium vinosum*	*E. coli, Salmonella, R. sphaeroides*	[31, 746, 747]
Aspartate	*E. coli, Salmonella*	*Alcaligenes faecalis, Pseudomonas fluorescens*	[496, 498, 640]
Benzoate	*Pseudomonas putida*	*E. coli, Salmonella*	[281, 357, 747]
Leucine	*B. subtilis*	*E. coli, Salmonella*	[559, 746, 747]
Phenol	*E. coli*	*Salmonella*	[412, 746, 747]
Tryptophan	*B. subtilis, C. vinosum*	*E. coli, Salmonella, R. sphaeroides*	[31, 559, 746, 747]
Valine	*B. subtilis*	*E. coli, Salmonella*	[559, 746, 747]
H^+, OH^-	*R. sphaeroides*	*C. vinosum, E. coli*	[31, 747]

The former receptor dominates the latter receptor, for which reason phenol is a chemorepellent for *Salmonella*. *E. coli* lacks the repellent receptor, therefore, only attraction to phenol is observed [313]. Likewise, *Salmonella* but not *E. coli* is attracted to citrate, possibly because only the former can transport citrate [314, 356]. *E. coli* but not *Salmonella* is attracted to maltose [8, 381] because only the former has a chemotaxis receptor that can bind the complex of maltose with its binding protein (Section 6 below) [174, 515]. Similarly, Co^{2+} and Ni^{2+} are chemorepellents for *E. coli* but not for *Salmonella* [706]. Some examples of chemoattractants and chemorepellents for bacteria other than *E. coli* and *Salmonella* are listed in Table 9.

4.4. *Are the stimuli themselves detected or their metabolic products?*

Early studies of J. Adler have demonstrated that, in *E. coli*, the chemoattractants themselves are detected [6, 9] rather than their metabolic products. This conclusion was based on two types of observations. One was that mutant bacteria, which have lost the ability of metabolizing or transporting a chemical, can nevertheless be attracted to the chemical. The other type of observation was that nonmetabolizable analogs of metabolizable chemicals (e.g., D-fucose, α-methyl-DL-aspartate, or α-aminoisobutyrate—analogs of D-galactose, L-aspartate, or L-serine, respectively) attract bacteria. In contrast, some metabolizable chemicals

Table 9. Chemotactic stimulants in bacterial species other than *E. coli* and *Salmonella*.

Species	Chemoattractants	Chemorepellents	References
Agrobacterium tumefaciens	Arabinose, arginine, fructose, galactose, glucose, lactulose, maltose, raffinose, stachyose, sucrose, valine	Not found	[434]
Alcaligenes faecalis	Not found	Arabic acid (galactoaraban), L-aspartate, L-cysteine, La^{3+}, L-lysine, HCl, acetate	[193, 640]
Azospirillum brasilense	Citrate, malate, oxygen at low concentrations (3–5 μM), proline, succinate	Oxygen at high concentrations	[597, 827, 829]
Azotobacter vinelandii	N-acetyl-D-glucosamine (GlcNAc), N-acetyl-D-mannosamine, arabinose, arabitol, fructose, glucose, glycerol, maltose, mannitol, mannose, melibiose, ribitol, ribose, sorbitol, sucrose, xylitol	Not found	[275]
Bacillus licheniformis	L-Histidine, DL-isoleucine, L-leucine, D-mannose, L-methionine, L-proline, L-rhamnose, DL-serine, L-tryptophan	Arabic acid, $Al_2(SO_4)_3$, CuCl, $CuCl_2$, $Cu(NO_3)_2$, $CuSO_4$, $FeCl_3$, $Fe_2(SO_4)_3$, $La(NO_3)_3$, $NiCl_2$, L-arginine, L-lysine, HCl, acetate	[193, 640]
Bacillus megaterium	Alanine, asparagine, glutamine, malate, malonate, serine, threonine	Not found	[815]
Bacillus subtilis	All 20 naturally occurring amino acids, N-acetylmannosamine, 2-deoxy-D-glucose, D-fructose, gentiobiose, GlcNAc, D-glucose, maltose, D-mannitol, D-mannose, α-methyl-D-glucoside, β-methyl-D-glucoside, α-methyl-D-mannoside, D-sorbitol, L-sorbose, sucrose, trehalose, D-xylose	Indole, inhibitors of electron transport [e.g., 2-heptyl-4-hydroxyquinoline-N-oxide (HOQNO)], local anesthetics, organic acids (e.g., acetate, benzoate, butyrate), uncouplers of oxidative phosphorylation [e.g., trifluoromethoxycarbonylcyanide phenylhydrazone (FCCP)]	[557–561, 753]
Chromatium vinosum	Acetate, L-cysteine, ethanol, S^{2-}, L-tryptophan, urea	H^+, OH^-	[31]
Cytophaga johnsonae	Not found	N-chlorotaurine, H_2O_2, OCl^-, low pH	[428]

Erwinia carotovora	Glycerol, L-lyxose, D-xylose, D-fructose, D-galactose, D-glucose, D-mannitol, D-mannose, L-rhamnose, D-sorbitol, L-sorbose, D-lactose, α-methyl-D-glucoside, α-methyl-D-mannoside	Arabic acid, $Al_2(SO_4)_3$, CuCl, $CuCl_2$, $Cu(NO_3)_2$, $CuSO_4$, $FeCl_3$, $Fe_2(SO_4)_3$, $La(NO_3)_3$, L-lysine, HCl, acetate	[193, 640]
Halobacterium salinarium	Arginine, cysteine, isoleucine, leucine, methionine, valine, leucine-alanine, leucine-glycine, leucine-leucine, methionine-isoleucine, methionine-valine, methionine-arginine, leucine-glycine-leucine	Benzoate, indole, phenolate	[708]
Leptospirillum ferrooxidans	Cu^{2+}, Fe^{2+}, Ni^{2+}	Aspartate	[2]
Magnetotactic cocci	Oxygen	Not found	[236]
Myxococcus xanthus	Phosphatidylethanolamine	Butanol, dimethyl sulphoxide, ethanol, isoamyl alcohol, isobutanol, isopropanol, propanol	[200, 341, 647]
Proteus morganii	DL-alanine, glycerol, L-leucine, D-lyxose, L-lyxose, L-methionine, L-phenylalanine, L-rhamnose, DL-serine, L-threonine, L-tryptophan, DL-valine	Arabic acid, Fe^{3+}, La^{3+}, Ni^{2+}, SO_4^{2-}, L-lysine, HCl, acetate	[193, 640]
Pseudomonas aeruginosa	Adonitol, L-alanine, α-aminoisobutyrate, L-arginine, L-asparagine, L-cysteine, D-glucose, L-glutamine, glycine, L-histidine, L-isoleucine, L-leucine, L-lysine, L-methionine, L-phenylalanine, P_i (P_i-starved cells only), L-proline, sedoheptulose, L-serine, succinate, L-threonine, L-tryptophan, L-tyrosine, L-valine	Arabic acid, isothiocyanic esters (e.g., allyl isothiocyanate and methyl isothiocyanate), thiocyanic esters (e.g., ethyl thiocyanate and methyl thiocyanate), Cu^+, Fe^{3+}, La^{3+}, HCl, acetate	[170, 171, 193, 338, 396, 545, 547, 640]
Pseudomonas fluorescens	Arbutin, D-lyxose, L-methionine, α-methyl-D-mannoside, D-ribose, D-sedoheptulose, DL-serine, D-sorbitol, L-threonine, DL-valine	Arabic acid, L-aspartate, L-lysine, Al^{3+}, Cu^+, Fe^{3+}, La^{3+}, HCl, acetate	[193, 640]
Pseudomonas putida	Benzoate, benzoylformate, *p*-hydroxybenzoate, DL-mandelate, methylbenzoates, β-phenylpyruvate, salicylate, *m-/p-/o*-toluate	Not found	[281]
Rhizobium lupini	All common L-amino acids except for leucine, D-glucose, D-mannitol	Not found	[263]

(Continued)

Table 9. *Continued*

Species	Chemoattractants	Chemorepellents	References
Rhodobacter sphaeroides	L-Alanine, L-aspartate, dimethyl sulphoxide, fructose, D-galactose, D-glucose, D-mannitol, mannose, oxygen (when the photosynthetic electron transport does not function), succinate, H^+, OH^-	Acetate, L-cysteine, ethanol, L-leucine, malate, mercaptoacetate, oxygen (when the photosynthetic electron transport functions), L-tryptophan	[31, 254, 323, 566]
Sarcina ureae	Not found	Arabic acid, L-arginine, L-lysine, Fe^{3+}, HCl, acetate	[193, 640]
Serratia marcescens	Xylitol, 2-deoxy-D-glucose, D-lactose, α-methyl-D-glucoside	Arabic acid, L-lysine, Al^{3+}, Cu^{2+}, Fe^{3+}, La^{3+}, HCl, acetate	[193, 640]
Sinorhizobium meliloti	All common 20 amino acids (the strongest are arginine, glycine, leucine, lysine and proline), nodulation gene-inducing compounds (4′,7-dihydroxyflavone, 4′,7-dihydroxyflavanone, 4,4′-dihydroxy-2′-methoxychalcone)	Not found	[186, 263]
Spirillum serpens	Adonitol, L-alanine, D/L-arabinose, D/L-arabitol, D/L-lyxose, xylitol, D-fucose, D-mannose, D-mannitol, L-rhamnose, D-sorbitol, L-sorbose	Arabic acid, L-arginine, glycine, L-lysine, Al^{3+}, Fe^{3+}, La^{3+}, Ni^{2+}, HCl, acetate	[193, 640]
Spirochaeta aurantia	Cellobiose, 2-deoxy-D-glucose, D-fructose, D-fucose, D-galactose, D-glucosamine, D-glucose, α-methyl-D-glucoside, maltose, D-mannose, oxygen at low concentrations, D-xylose	Oxygen at high concentrations	[269]
Vibrio alginolyticus	Serine	Phenol	[304]
Vibrio furnissii	Chitin oligosaccharides $(GlcNAc)_n$, GlcNAc, glutamine	Not found	[69]

(e.g., gluconate, succinate, and fumarate) fail to attract bacteria. These observations led to the conclusion that bacteria like *E. coli* have receptors that sense the outside concentrations of the chemoattractants (chemorepellents will be discussed in Section 6.3). Over the years, this conclusion has been proven correct for the large majority of the stimulants of *E. coli* and *Salmonella*. A few chemoattractants (e.g., proline and glycerol), however, do act by being metabolized, thereby elevating the energy level of the bacteria [163, 830]. In species like *R. sphaeroides* and *A. brasilense*, the rule is different—metabolism of the chemoattractant (e.g., sugars [323] and alanine [580] in the case of *R. sphaeroides* or most of the chemoattractants in the case of *A. brasilense* [17]) is required for chemotaxis.

5. Chemotaxis-Related Genes

The ease by which nonchemotactic mutants can be isolated (Table 5) resulted in relatively rapid identification of genes involved in chemotaxis. Three groups of mutants were isolated: mutants that are nonchemotactic for specific stimuli only, mutants that are nonchemotactic for multiple stimuli, and mutants that are nonchemotactic for all the stimuli (Figure 12). The first group led to the identification of genes coding for chemotaxis receptors for specific chemicals (e.g., references [10, 282, 283, 555, 556]; for a review, see [11]), usually dual-function receptors that reside either in the periplasm or in the cytoplasmic membrane (Figure 12). (These receptors will be discussed in Section 6.2.) Mutants belonging to the second group were defective in chemotaxis to multiple stimuli (but not to all stimuli). This group of mutants identified genes coding for chemotaxis-specific, methylatable, trans-membrane receptors termed MCPs (for methyl-accepting chemotaxis proteins) or transducers (e.g., [595]). (These receptors will be discussed in Section 6.1.) The third group identified general chemotaxis genes (*che* genes) [35, 570], namely genes whose products are involved in transmitting information from the receptors to the flagella in a pathway common to all stimuli. (These gene products will be discussed in Section 7.) This three-group division indicated that chemotaxis receptors send signals to the flagella in a convergent pathway (Figure 12).

The chemotaxis genes of *E. coli* are listed in Table 10. They include the genes identified by the groups of mutants that are nonchemotactic for multiple stimuli and for all the stimuli, as well as flagellar genes that are involved in regulating the direction of flagellar rotation. The

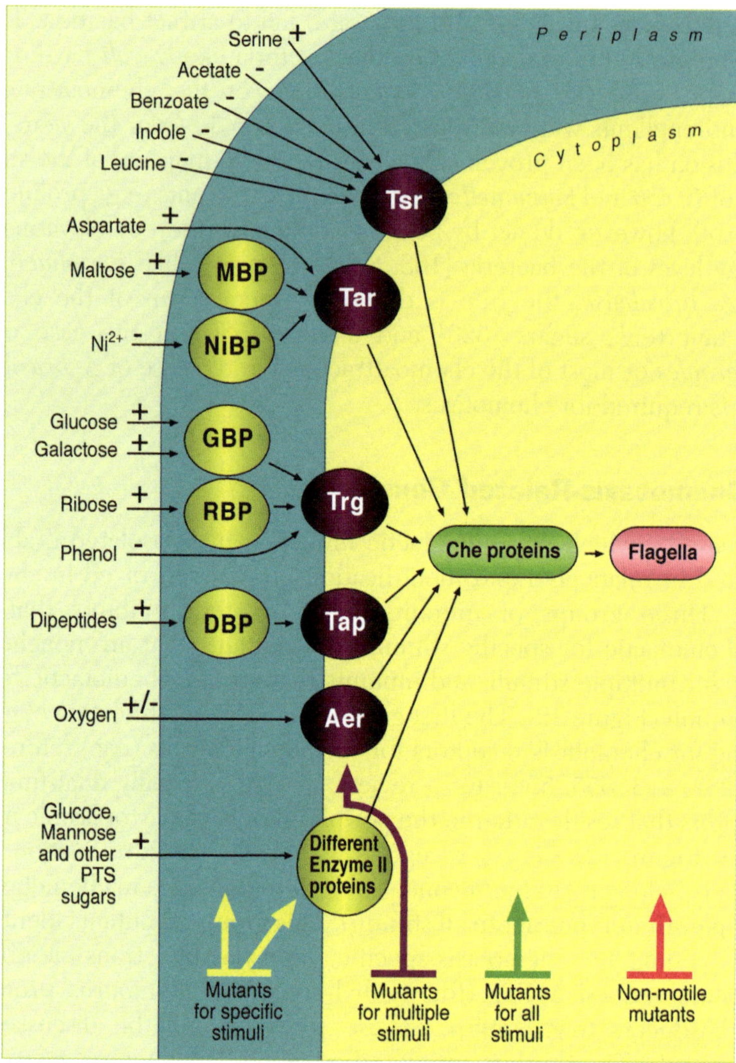

Figure 12. Schematic representation of the convergent information flow in chemotaxis of *E. coli*. Only representative stimuli are shown. The names of the MCPs are abbreviations of taxis to serine and away from certain repellents (Tsr), taxis to aspartate and away from certain repellents (Tar), taxis to ribose and galactose (Trg), taxis-associated protein (Tap), and aerotaxis and energy responses (Aer). MBP, NiBP, GBP, RBP, and DBP stand for maltose-binding protein, Ni^{2+}-binding protein, galactose-binding protein, ribose-binding protein, and dipeptide-binding protein, respectively. Plus and minus signs above arrows stand for chemoattractants and chemorepellents, respectively.

Table 10. Chemotaxis genes of *E. coli.*[a]

Gene	Operon	Mutant phenotype (flagellar rotation)	Gene product		Location	Molecules per cell[b]	Function
			Polymeric form	Monomeric size (kDa)			
aer	Unknown	WT[e]	Dimer	55	Cytoplasmic membrane	Presumably 150[c]	Receptor that mediates the chemotactic response to oxygen (aerotaxis) and to changes in the cell's energy level; a flavoprotein
cheA	*motAB-cheAW*	CCW	Dimer	73 & 67[f]	Cytoplasm, receptor-bound (via CheW)	6600 ± 1400[d]	Histidine kinase
cheB	*tar-tap-cheRBYZ*	CW-biased	Monomer	36	Cytoplasm	240 ± 50[d]	Methyl esterase
cheR	*tar-tap-cheRBYZ*	CCW	Monomer	32	Cytoplasm	140 ± 30[d]	Methyl transferase
cheW	*motAB-cheAW*	CCW	Monomer	18	Cytoplasm, receptor-bound	6800 ± 900[d]	A scaffolding protein that couples CheA to the MCPs?
cheY	*tar-tap-cheRBYZ*	CCW	Monomer	14	Cytoplasm	8200 ± 300[d]	Response regulator
cheZ	*tar-tap-cheRBYZ*	CW-biased	Dimer	24	Cytoplasm	3200 ± 90[d]	Phosphatase
fliG	*fliFGHIJK*	Any	In a complex	37	Cytoplasm, motor-bound	35 per flagellum[g]	Switch protein
fliM	*fliLMNOPQ*	Any	In a complex	38	Cytoplasm, motor-bound (via FliG)	35 per flagellum[g]	CheY-binding switch protein
fliN	*fliLMNOPQ*	Any	In a complex	15	Cytoplasm, motor-bound (via FliG)	100 per flagellum[g]	Switch protein
tap	*tar-tap-cheRBYZ*	WT	Dimer	~60	Cytoplasmic membrane	150[c]	Receptor for dipeptides; also mediates the response to temperature changes

(Continued)

Table 10. *Continued*

Gene	Operon	Mutant phenotype (flagellar rotation)	Gene product					Function
			Polymeric form	Monomeric size (kDa)	Location	Molecules per cell[b]		
tar	*tar-tap-cheRBYZ*	WT	Dimer	~60	Cytoplasmic membrane	900[c]		Receptor for some amino acids (e.g., aspartate, glutamate) and for maltose-bound, periplasmic binding protein; also mediates the response to some chemorepellents (Ni^{2+}, Co^{2+}) and to temperature changes
tsr	*tsr*	WT	Dimer	~60	Cytoplasmic membrane	1600[c]		Receptor for some amino acids (e.g., serine, alanine, glycine, cysteine); also mediates the response to some chemorepellents (indole, leucine, benzoate), to changes in the cell's energy level and to temperature changes
trg	*trg*	WT	Dimer	~60	Cytoplasmic membrane	150[c]		Receptor for some sugar-bound, periplasmic binding proteins (e.g., galactose, ribose); also mediates the response to some chemorepellents (phenol) and to temperature changes

[a] According to [18, 96, 256, 284, 353, 460, 482, 596, 627, 706, 814].

[b] Unless mentioned otherwise, the values are approximate. Note that the number of molecules of each of the proteins listed varies to a large extent among different strains and it depends on the conditions and phase of growth (M. Li and G.L. Hazelbauer, personal communication). Values calculated from the protein concentration in the cell are based on a cell volume of 10^{-15} l (the cell volume depends on the strain and growth conditions).

[c] The actual value is probably higher because the total number of all MCP molecules per cell is $15\,000 \pm 1700$ in strain RP437 at mid-log phase (M. Li and G.L. Hazelbauer, personal communication), i.e., higher than the sum of the values listed for all MCPs.

[d] This value is for strain RP437 at mid-log phase (M. Li and G.L. Hazelbauer, personal communication).

[e] Abbreviations: CCW, counterclockwise; CW, clockwise; WT, wild-type like;

[f] The gene *cheA* encodes for two polypeptides, long and short ($CheA_L$ and $CheA_S$, respectively), as a consequence of translational initiation at two distinct in-frame initiation sites.

[g] The data are for *Salmonella*.

properties of their gene products will be discussed in the following sections (Sections 6 and 7).

All the *che* genes are clustered within region II of the flagellar genes on the chromosome (42.3–42.5 min in *E. coli*), together with the *mot* genes and some flagellar genes. The genes *tap* and *tar* are also in this region. In contrast, *tsr*, *trg*, and *aer* are each located in different regions (in *E. coli* at 98.9, 32.1 and 69.3 min, respectively) [90]. All the chemotaxis genes belong to the third level of regulatory hierarchy [460] (Section 2.2.2 above). This means that the flagellar genes of the first and second levels must be expressed before the chemotaxis genes can be expressed, and that the master operon *flhDC* is indirectly involved in controlling chemotaxis in the sense that it regulates the synthesis of the chemotaxis machinery. As was discussed in Section 2.2.2, this master operon is itself positively regulated by the intracellular levels of cAMP and its receptor. It thus links the metabolic state of the cell to the expression of the motility and chemotaxis components. When the level of cAMP goes up (e.g., when glucose availability goes down), the *flhDC* operon is rapidly transcribed, the motility and chemotaxis machinery is synthesized, and the bacteria can navigate themselves to better locations [105, 460].

6. Chemotaxis receptors

In bacteria like *E. coli* and *Salmonella*, some chemotactic stimuli bind directly to chemotaxis-specific receptors, whereas others bind first to a primary receptor, which then interacts with a chemotaxis-specific receptor (Figure 12). The chemotaxis-specific receptors are the MCPs, mentioned in Section 5. The primary receptors are dual function in the sense that they are involved in both chemotaxis and transport of the stimulants.

Historically, the receptors for bacterial chemotaxis in *E. coli* as well as the chemoattractants detected by each were identified by isolation of mutants defective in individual receptor activities [11]. Ascription of stimuli to the identified receptors was, in addition, carried out by competition experiments and by examining (in the case of inducible receptors) which chemoattractants are detected only when the receptor is induced [11]. With time, most of the receptors have been cloned, isolated, and studied biochemically and structurally.

6.1. *Chemotaxis-specific receptors*

E. coli has five chemotaxis-specific receptors, MCPs, which sense a variety of stimuli. Of these receptors, Aer appears to be the most specific in the sense that it only perceives oxygen and the energy level of the cell [96, 596]. The other four receptors recognize multiple stimuli (Figure 12) [212, 386, 706, 795]. For example, Tar senses the chemoattractants aspartate and glutamate by direct binding, the chemoattractant maltose via the maltose-binding protein, the chemorepellents Co^{2+} and Ni^{2+} (probably via the Ni^{2+}-binding protein), the ambient pH, and the ambient temperature. The abundance of the MCPs varies, with Tsr and Tar being highly abundant, whereas Tap, Trg, and Aer being lowly abundant (Table 10). *Salmonella* also has five MCPs, but there the Tap and Aer receptors have not been found. On the other hand, this species has two other MCPs that are absent in *E. coli*. These are Tcp (for *t*axis to *c*itrate and away from *p*henol) [796] and Tip (for *t*axis-*i*nvolved *p*rotein), whose function is not known [612].

6.1.1. *Structure of the MCPs*

The MCPs are closely related in sequence and structure. Excluding Aer, each of them is built from the following parts: a short N-terminal extension in the cytoplasm, a hydrophobic transmembrane segment TM1, a ligand-binding domain in the periplasm, another hydrophobic transmembrane segment TM2, a linker region in the cytoplasm, a cytoplasmic domain CD1 which is a methylated helix, a signaling domain, another methylated helix CD2, and a variable C-terminal domain [706] (Figure 13). Aer is different in the sense that it lacks the periplasmic ligand-binding domain between TM1 and TM2 (the two transmembrane domains in Aer possibly form a tight hairpin) and that its N-terminal extension in the cytoplasm (possibly the oxygen-sensing domain) is long and contains an FAD-binding site [96, 596]. As might be expected, the signaling domain as well as the CD1 and CD2 domains are the most conserved among the MCPs, whereas the ligand-binding domain is the most divergent.

The helical content of the MCPs is very high [68, 233, 524]. The ligand-binding domain forms an antiparallel four-helix bundle [158, 636, 801]. The transmembrane domains TM1 and TM2 appear to be a continuation of the corresponding α-helices of the ligand-binding domain [419]. The cytoplasmic portions of the MCPs are, likewise, predominantly α-helical [67, 144, 153, 177, 362, 410, 523, 712, 790].

In a very elegant study involving site-directed cross-linking of the receptor Tar, Milligan and Koshland [503] have demonstrated that the MCPs are homodimers (Figure 13). The dimers appear to be the

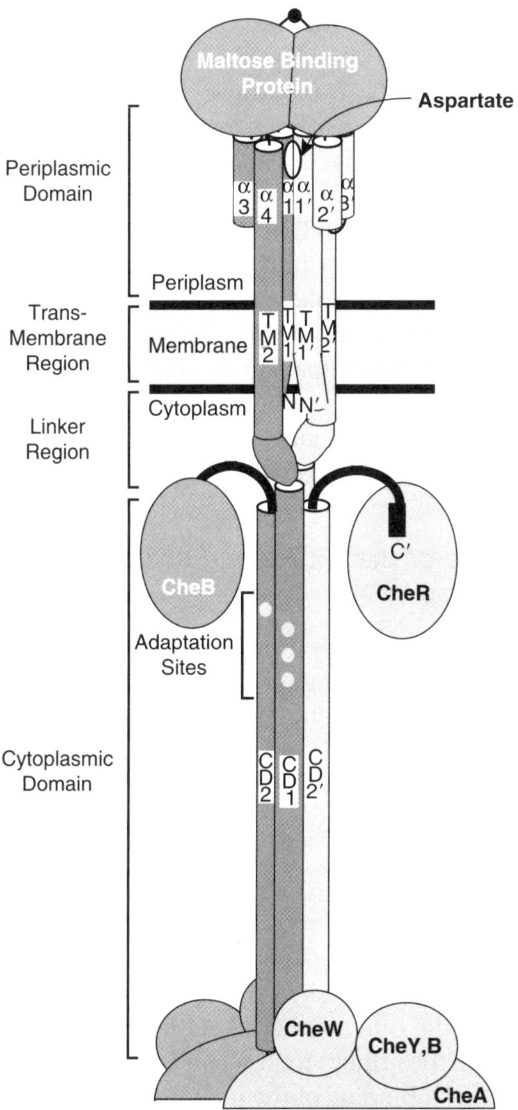

Figure 13. Schematic presentation of the structure of the best-characterized chemotaxis-specific receptor, Tar. The regions of interaction with the ligands and the proteins in the receptor complex are shown. For simplicity, helix supercoiling is omitted, and the α-helices are represented by cylinders. Components that dock to the receptor in the assembled complex are shown schematically as ellipsoids and spheres. Cytoplasmic sites of methylation and demethylation (adaptation sites, residues 295, 302, 309, and 491) are shown as small ovals. (Taken with permission from Falke and Hazelbauer [223], with slight modifications made by J.J. Falke.)

smallest functional unit of the MCP structure because the dimeric structure does not change upon ligand binding or methylation. Based on the crystal structure of the cytoplasmic domain of the Tsr receptor, it appears that the tails of three dimer receptors come together and form a trimeric structure [362]. The dimers constitute one part of a larger signaling complex that contains, in addition, the chemotaxis proteins CheW and CheA [235, 256, 635]. Most, if not all of the soluble cytoplasmic chemotaxis proteins (CheB, CheR, CheY, and CheZ) appear to interact with the ternary complex of MCP-CheA-CheW [147, 677] and to be in rapid equilibrium with it (dissociation constants around 2–3 μM [222]). This organization of signaling components in large complexes is not unique to chemotaxis; it is rather widely spread among signal transduction systems, probably because of the advantage provided by interacting components being in close proximity [132].

6.1.2. *Functions of the MCPs and structure-function relationship*

The MCPs fulfill at least three functions in bacterial chemotaxis: they bind the ligands (or sense changes in the ambient temperature and pH), they transduce the chemotactic signal across the cytoplasmic membrane, and, as part of their involvement in adaptation, they undergo methylation or demethylation. The focus here will be on Tar, which is the best-characterized MCP.

Ligand binding
Aspartate binding to Tar ($K_D \approx 1\,\mu$M [43, 98, 164, 176, 233, 504, 523]) is negatively cooperative [98, 376]: binding of the ligand to one site causes asymmetric conformational changes in the dimer that, in the case of *E. coli*, occlude the other site [382, 419, 502, 801] or, in the case of *Salmonella*, reduce the affinity for a second aspartate but do not prevent the binding [98, 504]. Binding of the maltose-binding protein to Tar similarly appears to be asymmetrical [246]. The binding sites on Tar for aspartate and for the maltose-bound maltose-binding protein are distinct (Figure 13). These two ligands can simultaneously bind to Tar, provided that each of them binds to a different subunit of the dimer [247, 811].

Negative cooperativity appears to be a general rule for the MCPs, as serine binding to Tsr was also found to be negatively cooperative [423]. Even though the amino acid sequences of the ligand-binding domains of the MCPs are not homologous, the serine and aspartate binding sites on Tsr and Tar, respectively, are structurally similar [761]. The affinity of the MCPs for their respective ligands appears to depend on their methylation level [420].

Transmembrane signal transduction
As discussed above, the conformational changes in the ligand-binding domain, induced by binding the ligand, are very effective [309, 563]. Nevertheless, they are surprisingly small, and one wonders how such small changes can efficiently affect the components of the signaling complex located at the cytoplasmic side of the membrane [419]. Many receptors (e.g., the insulin receptor [287] and the epidermal growth factor receptor [800]) signal by undergoing changes in their dimeric state (dissociation or association). This does not appear to be the case with the MCPs because covalent dimerization by engineered intersubunit disulfides does not activate the downstream signaling components [155, 221, 413] and, conversely, removal of the cytoplasmic domain of one of the dimer's subunits (leaving only one functional subunit) does not block transmembrane signaling [245, 719]. Based on crystallographic data of the Tar-binding domain in its free and aspartate-bound forms, on ^{19}F-NMR studies of this domain, and on EPR studies of labeled MCP, two models have been suggested for transmembrane signaling. According to one model, signaling occurs within the monomer and is accomplished by a piston-like, 1.6 Å displacement of the α_4 and TM2 helices of one of the two MCP subunits towards the cytoplasm as well as a 5° helix tilt [156, 222, 419, 450, 564] (for a recent review, see reference [223]). This notion of signaling within a subunit is supported by some additional studies (e.g., [309, 532]). According to the other model, the signal is between the subunits and is transduced by a 5–8°, scissors-like rotation of one subunit relative to the other subunit [158, 502, 801]. The dilemma as to how the subtle conformational changes result in large signals within the cytoplasm may be solved by assuming that the receptor-coupled enzymes can detect small changes in the receptor conformation [564] or that lateral signaling is involved [419].

Methylation
Years ago it was found that L-methionine-depleted cells of *E. coli* and *Salmonella* are unable to tumble and are nonchemotactic, indicating that methionine is essential for chemotaxis [5, 44, 684]. Later it was found that not methionine *per se* is required but rather its metabolite, S-adenosylmethionine (AdoMet) [36, 37, 46]. The fact that AdoMet is a common

methyl donor initiated a search for methylation reactions involved in chemotaxis. Soon after, chemoattractant-stimulated methylation was indeed found and the methylated membrane proteins, MCPs, were identified (at that time they were not yet recognized as receptors) [380].

The methylation reaction is carried out by the enzyme methyltransferase, CheR [686]. Demethylation of methylated MCPs is carried out by the enzyme methylesterase, CheB [704]. The methylation is on the carboxyl group of 4–6 glutamate residues [368, 754]. Two of these methylation sites are glutamine residues made available for methylation by CheB-mediated deamidation, which converts them to glutamate residues [343, 543, 544, 600, 725]. In Tar, the methylation sites are glutamate residues 302 and 491, and glutamine residues 295 and 309. Modulation of the MCPs' methylation level by CheR and CheB plays a major role in adaptation to chemotactic stimuli (Section 8.2).

Differences between high- and low-abundance receptors
The MCPs are not only different from each other in the sequences of their ligand-binding domains; they are also different in their abundance (Table 10). Interestingly, mutants lacking the high-abundance receptors (Tsr and Tar) tumble less frequently than their wild-type parents. They do so even when the other receptors are overexpressed from a multi-copy plasmid [227, 770]. They also need longer periods of time to adapt to stimuli of the other receptors and, therefore, their chemotactic response is diminished [64, 227, 228, 770]. The reason for these observations, which were unexpected at the time, is that the low-abundance receptors (Tap, Trg, and Aer) lack, at their extreme carboxy termini, 18–28 residues [96, 119, 385, 596] that contain the CheR- and CheB-binding site—an NWETF pentapeptide [65, 788]. It was nicely demonstrated that when an NWETF-containing peptide of the final 19 residues of Tsr is added to Trg (thus forming a Tar-Trg hybrid), Trg acquires the properties of a high-abundance receptor [228]. (A similar addition of the final 18 residues of Tar to Tap appeared to improve methylation but it neither improved the extreme counterclockwise bias of cells containing Tap as the sole MCP nor restored their chemotactic responsiveness [770]. The reasons for the different results in these experiments are not yet clear [228].) It was demonstrated for the Tsr [422] and Tar [411] receptors that inter-dimer methylation can occur. Apparently, in wild-type cells, methylation of a low-abundance receptor is catalyzed by CheR bound to a high-abundance receptor. Consequently, the low-abundance MCPs depend on the presence of Tsr and

Tar to be methylated and thereby to function in adaptation. This need for an interaction between the low- and high-abundance receptors may be one of the reasons for the organization of the receptors in clusters, discussed in the next section.

6.1.3. *Receptor clustering*

The presumption that prevailed until about a decade ago was that the chemotaxis receptors are randomly distributed around the cell. This presumption was based on both theoretical considerations for optimal chemotaxis [81] and experimental data indicating that the receptors are not localized at the vicinity of the flagellar motors [219]. It was, therefore, surprising when Maddock and Shapiro [463] found, by immuno-electron microscopy, that the MCPs are clustered at the bacterial poles—one (more frequently) or both of them. This led to the suggestion that *E. coli* might have a "nose," i.e., that the pole at which the receptors are clustered is the leading end of the moving cell [574]. This suggestion was not supported experimentally in the absence of a stimulus gradient, where there was no preference for any pole as a leading end [88]. However, there are no experimental data to exclude this possibility within a gradient of a stimulus, where a "nose" might be useful. Whether or not *E. coli* has a "nose," receptor clustering appears to play an important role in signal transduction and adaptation.

Subsequent to the finding that the MCPs are clustered at the poles, it was found that most proteins, with which they interact (excluding, perhaps, CheB and CheR, for which no experimental evidence has been published), are clustered at the poles as well [448, 449, 463, 667, 677]. This raised the possibility that the ternary complexes as a whole are clustered and that they form supramolecular complexes. Formation of active supramolecular complexes, consisting of about seven receptors, two or four CheW molecules, and one CheA dimer (Figure 14), was indeed demonstrated *in vitro* with purified proteins, giving rise to a seven-dimer model of the receptor supramolecular complex [427]. Another model, based on the recently resolved crystal structure of the cytoplasmic domain of the Tsr receptor, which revealed that the tails of three dimer receptors form a trimeric structure [362], was proposed by Shimizu *et al.* [649]. They elegantly examined plastic models, generated by 3-D technology, of the proteins involved in the receptor supramolecular complexes. The models predicted that these supramolecular complexes form a two-dimensional hexagonal lattice, built from trigonal

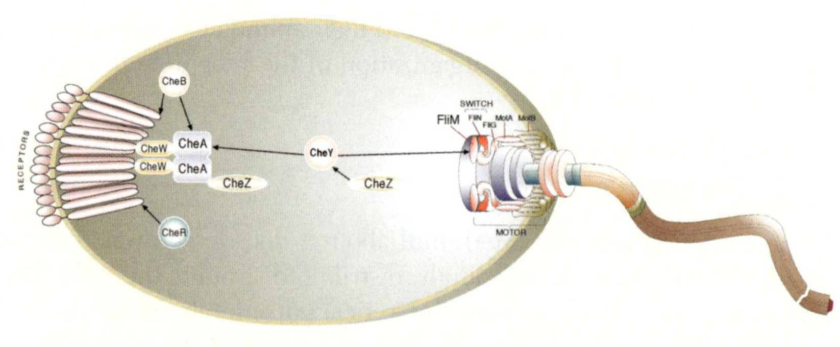

Figure 14. Simplified scheme of protein–protein interactions that transduce the sensory signal from the receptor supramolecular complex to the flagellar-motor supramolecular complex. Black arrows stand for regulated interactions. The scheme is not drawn to scale. (Taken with permission from Bren and Eisenbach [137].)

units (Figure 15). Each unit is composed of three MCP dimers, three molecules of CheW, and three monomers of CheA, joined to CheA monomers of another unit at their dimerization domain. The reason for the different MCP-CheA-CheW stoichiometries in the seven-dimer model [427] and the lattice model [649] is not known. According to another model, proposed by Kim *et al.* [363], each receptor dimer is in contact with four other receptor dimers forming a two-dimensional slab with the interconnected receptors acting as trusses, and with each ligand-binding domain dimer being in contact with two other ligand-binding domain dimers at the periplasmic side.

While it is believed that the high-order structure of the receptor supramolecular complex has a role in chemotactic signaling, the composition and stoichiometry of the five different MCPs within these receptor supramolecular complexes are not known. If the seven-dimer model [427] is correct, the supramolecular complexes must differ from each other with respect to their MCP compositions. This is because the low-abundance receptors must interact with the high-abundance ones for their normal function [411, 422] and for being a part of the cluster [449]. Had the MCP compositions in all the receptor supramolecular complexes been the same, each complex should have been composed of at least 19–20 receptor dimers according to the known stoichiometry between the different types of MCPs [10–11 Tsr : 6 Tar : 1 Aer : 1 Tap : 1 Trg (Table 10)]. If, on the other hand, the lattice model [649] is correct, any MCP combination may be possible. The flexibility that this model

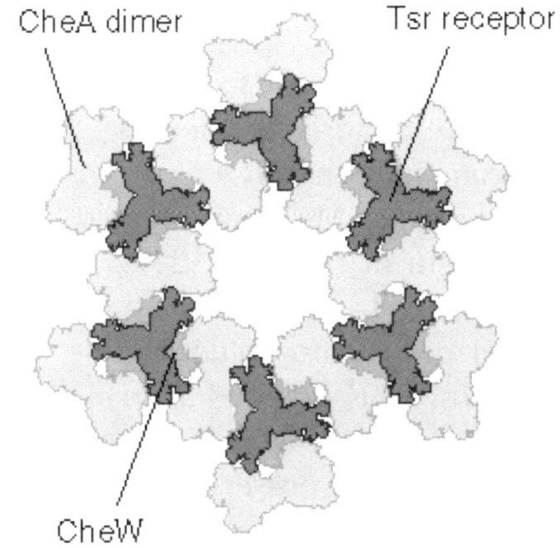

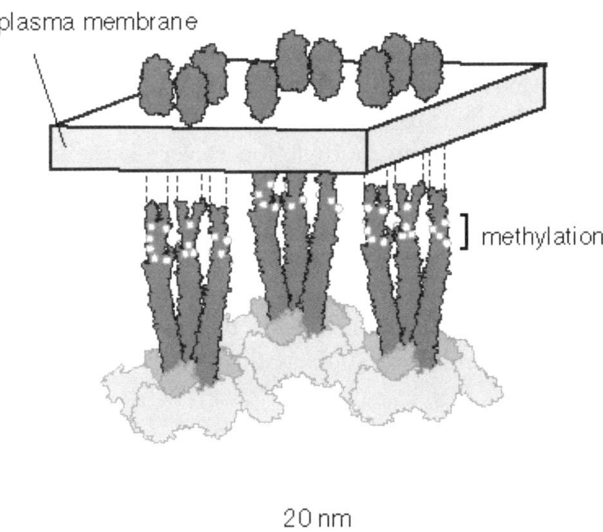

Figure 15. Hexagonal lattice model of the receptor supramolecular complex. Top: "Plan view," as seen from the cytoplasmic membrane looking into the cell. Bottom: Schematic side view of the network. (Kindly provided by D. Bray, Cambridge University.)

provides makes it a favorable model in my eyes. Recently, Ames *et al.* [22] isolated mutants that cannot form receptor clusters due to mutations at residues involved in the formation of trimers of dimer MCPs and, using crosslinking, demonstrated that the cluster contains mutually interacting Tsr and Tar.

6.1.4. *MCPs in species other than* E. coli *and* Salmonella

MCPs or MCP-like receptors appear to be involved in most, if not all, chemotactically responsive bacterial species (e.g., [410]). In all bacterial and archaeal species examined, the MCPs are clustered at the poles. In elongated bacterial species, the MCPs are located both at the poles and at regions along the length of the cells [257]. The number of MCPs varies a lot between species, from 5 in *E. coli* to 43 in *Vibrio cholerae* [288], which is the current record-holder. While the MCPs of many species are similar to those of *E. coli* and *Salmonella*, others are significantly different. A few examples follow.

R. sphaeroides and *S. meliloti* possess two MCP-like proteins, TlpA and TlpB (Tlp for *transducer-like protein*) [33]. Both proteins (~70 and ~40 kDa, respectively, in the case of *R. sphaeroides*, and 57 and 11 kDa in the case of *S. meliloti*) are primarily cytoplasmic proteins as revealed by antibody studies and as indicated by the absence of membrane-spanning domains in their primary structure. Recently, a third MCP-like protein, TlpC, was discovered in *R. sphaeroides* and found to localize to a discrete region in the cytoplasm [756]. Like MCPs, all the Tlp proteins can be methylated, but the methylation is very slow and occurs only after a long period of starvation. These soluble receptors may sense internal stimuli (yet unidentified) that reflect the metabolic or energy state of the cell [33]. Recently *R. sphaeroides'* MCPs were found to be localized at the poles and in clusters within the cytoplasm [279, 756]. Interestingly, the distribution of the MCPs was found to be dependent on environmental conditions such as the presence of oxygen and the light intensity.

The Archaeon *H. salinarium* possesses at least three subfamilies of MCPs (termed Htr's for *halobacterial transducers*) [611, 809]: *E. coli*-like MCPs that contain periplasmic and cytoplasmic domains connected by two transmembrane domains (e.g., HtrII [306]), MCPs that have two or more transmembrane domains but lacking periplasmic domains (e.g., HtrI [799] and HtrVIII [140]), and soluble MCPs (e.g., HtrXI [139] and Car [for *cytoplasmic arginine* transducer] [708]). All the *H. salinarium* MCPs possess the highly conserved, cytoplasmic region and the methylation sites. Some of the MCPs of this species interact with chemotaxis-specific binding proteins [375], a few others mediate both the chemotactic and phototactic responses. For example, HtrII is both a serine receptor and an MCP for the photoreceptor "sensory rhodopsin II" (SRII) involved in color vision of *H. salinarium* [306]. Aerotaxis of *H. salinarium* is mediated by HtrVIII [140].

M. xanthus possesses an MCP-like receptor, FrzCD, which is a methylatable soluble protein [765]. ("Frz" comes from mutants that form tangled *"frizzy"* filaments under fruiting conditions [831].) FrzCD senses an extracellular C-factor that induces FrzCD methylation and causes changes in the social motility of the gliding cells, culminating in cell aggregation [322, 673]. In addition, *M. xanthus* possesses a transmembrane MCP homologue, DifA (Dif for *defective in fruiting*), which is structurally similar to *H. salinarium*'s HtrI and, like FrzCD, appears to be essential for social motility [798].

B. subtilis possesses ten methylatable MCPs or MCP-like proteins: McpA, McpB, McpC, TlpA, TlpB, TlpC, YfmS, YoaH, YvaQ, and YhfV (renamed HemAT) [277, 278, 525] (see the genome in http://genolist.pasteur.fr/SubtiList/). Except for TlpC that has a molecular mass of 62 kDa, the other two Tlp proteins and the three Mcp proteins are 72 kDa in size. The other four MCPs are smaller, from 30 kDa (YfmS) to 63 kDa (YvaQ). The C-terminal domains of the MCPs and MCP-like proteins exhibit high homology to *E. coli*'s MCPs, *M. xanthus'* FrzCD, *H. salinarium*'s HtrI, and *Caulobacter crescentus'* McpA. While the stimuli sensed by McpA, McpB and McpC have been identified [250, 277], those sensed by the other MCPs are still obscure [277, 278].

The findings that some MCPs are soluble proteins that reside in the cytoplasm suggest that, in such cases, some chemoattractants are first internalized and only then detected by a soluble MCP [139, 708]. This appears to be different from the case of *E. coli* and *Salmonella*, where chemoattractants are mostly detected on the outside.

6.2. *Dual-function receptors*

Most of the dual-function receptors bind specific sugars, but a few of them bind other chemicals (e.g., dipeptides and Ni^{2+}; Figure 12). Some of the sugar receptors—sugar-binding proteins (e.g., the galactose-, maltose-, and ribose-binding proteins)—reside in the periplasm. Those, which sense carbohydrates that are transported by the phospho*enol*pyruvate-dependent carbohydrate phosphotransferase system (PTS)[b] (e.g., glucose, mannose, and mannitol), are in the cytoplasmic

---

[b] Abbreviations: Acs, acetyl coenzyme A synthetase; CheA~P, phosphorylated CheA; CheY~P, phosphorylated CheY; FRET, fluorescence resonance energy transfer; GFP, green-fluorescent protein; PTS, phospho*enol*pyruvate-dependent carbohydrate phosphotransferase system; YFP, yellow-fluorescent protein.

membrane (Figure 12). Like the MCPs, at least some of the dual-function receptors (e.g., the periplasmic maltose-binding protein [120, 187, 463]) are clustered at the bacterial poles, probably in order to allow direct interaction with the MCPs.

6.2.1. *Periplasmic binding proteins*

All the known periplasmic binding proteins are parts of the ATP-binding-cassette (ABC) transport system [121]. (The opposite is not correct; only a few of the ABC binding proteins are involved in chemotaxis.) They are soluble proteins that form a complex with the stimulant in the periplasm, and then the complex interacts with the proper chemotaxis-specific or transport-specific membrane receptors (MCP and permease, respectively). The binding sites for the ligand, the MCP, and the permease are different (e.g., [555] for the galactose-binding protein). The periplasmic receptors are inducible and, upon induction, their concentration in the periplasm may be as high as $\sim$1 mM [121, 470].

These receptors, ranging in size from $\sim$20 to 60 kDa, share relatively little sequence homology. Yet, they all fold to a similar global three-dimensional structure and they share some major common structural features [589]: (a) They consist of two distinct globular domains separated by a deep cleft or groove. (b) The ligand is bound within the cleft or groove and engulfed by both domains. (c) A hinge-bending motion between the two domains modulates access to and from the binding site. And (d) a small segment, 60 residues in length (corresponding to the N-terminal residues of the maltose-binding protein), is significantly conserved in sequence and structure among these receptors. However, this segment is involved in transport rather than in chemotaxis [524]. Within this resemblance, the receptors are divided into three structural classes [524]: the ribose- and the galactose-binding proteins, the maltose-binding protein, and the dipeptide- and the Ni^{2+}-binding proteins. The second and third classes are similar in structure, except for an extra domain at the N-terminus of the third class.

The structures of most of the binding proteins involved in chemotaxis have been solved both in the free and ligand-bound states (see [524] for a review). From these studies it became clear that, upon binding the ligand, the receptor undergoes a massive conformational change, as a result of which the probability of the protein to be in its open form is reduced and it converts to a closed form. The closed form of the receptor, unlike its open form, possesses a spatial structure that enables it to bind specifically to its cognate MCP (e.g., [361, 383, 643, 689, 810]).

6.2.2. *Phosphotransferase receptors*

The chemotaxis receptors for PTS carbohydrates are enzymes II [10]—membrane proteins that transport certain hexoses, hexosamines, polyhydric alcohols, and disaccharides [582, 602]. The mechanism of transport involves activation of enzyme II by phosphorylation, carried out by two cytoplasmic protein kinases: enzyme I and histidine-containing protein (HPr) [582]. Unlike the case of the periplasmic binding proteins, the PTS' primary receptor—enzyme II—does not interact directly with an MCP. As will be discussed in Section 8.2.7, it is enzyme I which links the occupancy of enzyme II with the chemotaxis system.

6.3. *Chemorepellent receptors*

The first hypothesis about how chemorepellents might be sensed by bacteria was proposed by R.N. Doetsch and colleagues in the early nineteen seventies. According to this hypothesis, chemorepellents modulate the membrane potential (directly, without the involvement of the chemotaxis machinery), and the negatively charged flagella orients themselves in relation to it [193, 194]. When, soon after, chemorepellents for *E. coli* and *Salmonella* were discovered, it was assumed (on the basis of competition and additivity experiments) that they are sensed by receptors [746, 747]. However, since binding of a chemorepellent to a receptor has never been demonstrated, doubts have been raised as to the mere existence of chemorepellent receptors [458]. The findings that there is no stereospecificity for chemorepellents (for example, the potencies of D- and L-leucine or D- and L-phenylalanine as chemorepellents are comparable [747]) and that chemorepellents are effective only at relatively high concentrations (in the millimolar range) [549, 747], further substantiated the doubts. An alternative that was proposed is that chemorepellents change one or more membrane properties in a way that is sensed by the chemotaxis system and elicits a repellent response [458, 558]. (This proposal was different from that of Doetsch [193, 194] in that the chemotaxis machinery is required for a response.) General membrane properties that may, in principle, affect bacterial behavior are the membrane potential and the membrane fluidity, but both of them have been demonstrated not to be involved. Membrane potential was ruled out by demonstrating lack of consistent changes in the membrane potential in response to chemorepellents [670] (for a review on chemotaxis-related measurements of membrane potential, see [207]). Membrane fluidity was eliminated by demonstrating (a) that most chemorepellents cause a repellent response at concentrations much below those at which any change in the membrane fluidity can be detected, and (b) lack of a consistent value of membrane fluidity at which a repellent response is elicited with chemorepellents that fluidize the membrane (e.g., alcohols) [210]. Thus, by way of elimination, it was suggested that chemorepellents are, after all, detected by receptors.

Since the response to most of the chemorepellents is mediated by specific MCPs (Figure 12; alcohols, ethylene glycol, and high concentrations of glycerol are exceptions in the sense that they can be perceived as chemorepellents by anyone of the MCPs Tsr, Tar, Tap, and Trg [210, 549, 550]), it is reasonable that the receptors are the MCPs themselves [210]. The finding that *E. coli*'s Trg possesses a specific recognition site for the chemorepellent phenol [795] is in line with this notion. Ni^{2+} may be an exception to this rule as it was found that a mutation in *nikA*, which encodes the periplasmic Ni^{2+}-binding protein, results in loss of chemotaxis away from Ni^{2+} [179]. It is, therefore, possible that Ni^{2+} first binds to NikA and only then the Ni^{2+}:NikA complex binds to its cognate MCP—Tar (Figure 12). It is also possible that Ni^{2+} binds to NikA for being transported into the cell and only then does it bind to Tar from the inside.

How can the lack of stereospecificity and the relatively high concentrations of chemorepellents, required for eliciting a behavioral response, be reconciled with binding of chemorepellents to receptors? Apparently, the affinity of the MCPs to chemorepellents is low, for which reason detection of chemorepellent binding to them is difficult and their specificity is low. Analogous cases are those of the olfactory and taste systems in eukaryotes, where the affinities of odorants and taste compounds (e.g., sugar) are rather low [318, 402, 724].

Some chemorepellents are sensed indirectly by causing some perturbation in the cell, and the perturbation, not the chemorepellent, is sensed by the MCPs. Prominent examples are weak organic acids such as acetate, benzoate, and propionate, which act by lowering pH_{in}. This change in pH, not the organic compound *per se*, is sensed by the MCPs [357, 598, 668]. Another example might be the chemorepellents indole and benzoate that, in addition to their MCP-dependent responses, inhibit the enzyme fumarase within the cell. The consequence is elevation of the fumarate level in the cell, resulting in increased probability of clockwise rotation [519]. (The way by which fumarate affects flagellar rotation will be discussed in Section 8.2.8.) However, it is not yet known whether or not this MCP-independent mechanism plays a role under physiological conditions.

7. Other Chemotaxis Proteins

7.1. *CheA*

CheA is a histidine kinase that belongs to the superfamily of two-component regulatory systems, each constituting a sensor (a histidine kinase) and a response regulator [573]. As mentioned above, CheA together with CheW and the MCPs forms a ternary complex at the bacterial poles [256, 463, 635].

7.1.1. *Function of CheA*

The activity of CheA is to transfer a γ-phosphoryl group from Mg^{2+}-ATP to the imidazole side chain of a specific histidine residue of the protein (residue 48 in *E. coli*'s CheA), i.e., to undergo autophosphorylation. It then phosphorylates the response regulators CheY and CheB [290, 292, 293, 792]. Purified CheA exhibits equilibrium between an inactive monomer and an active dimer [711]. Within the dimer, each subunit catalyzes the phosphorylation of the histidine residue on the other subunit [715, 785]. CheA is linked to the MCPs via CheW. In the MCP:CheA:CheW complex, the rate of CheA autophosphorylation is higher than the autophosphorylation rate of soluble CheA [123, 493, 541]. In the complex, the autophosphorylation activity of CheA is regulated (over a range of at least two orders of magnitude) by the occupancy and methylation state of the receptors [122–124, 541]. CheA binds CheW with a K_D of ~15 μM and a 1:1 stoichiometry [255], CheY with a K_D of 1–2 μM [421, 635], and ATP with a K_D of 0.3 mM [720].

7.1.2. *Structure of CheA*

The CheA monomer consists of five functionally and topologically independent domains, some of which are linked through flexible linkers (Figure 16) [99, 127, 192, 253, 418, 522, 706, 716]:

(a) The phosphotransfer domain, 134 residues in length and termed P1, is at the N terminus. It contains the autophosphorylation site, His-48 [292]. It is composed of an antiparallel five-helix bundle with the His-48 residue accessible to solvent on the outer surface of the second helix [99, 816, 818].

(b) A specific recognition domain for CheY, 68 residues in length and termed P2 [490, 491, 521, 717], which is connected to P1 by a flexible linker, 22 residues long [521, 817]. In solution it forms an

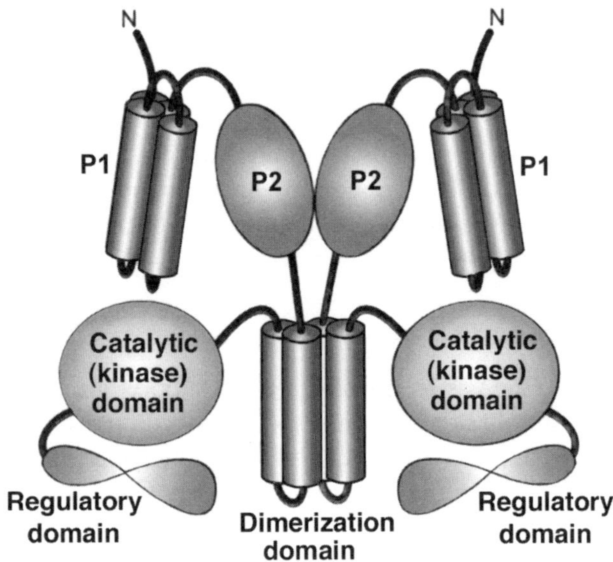

Figure 16. Schematic presentation of the structure of dimeric CheA. P1 and P2 are the phosphotransfer and CheY-binding (or CheB-binding) domains, respectively. The phosphorylation site is in the P1 domain. The ATP-binding site is in the kinase domain. CheW binds to the regulatory domain. Cylinders represent α-helices. (Taken with slight modifications and with permission from Stock [702].)

open-faced β-sandwich domain, with two antiparallel helices flanking one face of an antiparallel, four-stranded β-sheet [192, 490]. The crystal structure of the complex CheY-P2 revealed that the interface of P2 with CheY is formed by these two helices [492, 776].

(c) A dimerization domain, about 65 residues long, connected to P2 by another flexible linker, 25 residues long [99, 521, 522]. The dimerization determinants appear to be entirely localized to this dimerization domain [418, 702]. This domain is built from two antiparallel α-helices that form the interface for dimerization with the corresponding domain of another CheA molecule [99, 702]. It accounts for over 97% of the monomer's surface area buried in the dimer interface [99].

(d) A catalytic (kinase) domain, about 180 residues long, for autophosphorylation of the P1 domain [99, 521, 522]. This catalytic domain is the most highly conserved part of CheA structure [573, 702]. The domain contains an ATP-binding site [695] and "G boxes"—the characteristic histidine kinase sequence fingerprints [573, 702].

(e) The regulatory coupling domain, about 147 residues in length, is at the C terminus [99, 418]. This domain binds CheW—the scaffolding protein that couples CheA to the MCPs. The structure of this domain resembles "Src homology 3" (SH3) domains in tandem and provides different protein recognition surfaces at each end of the molecule [99].

7.1.3. *Two forms of CheA*

The *cheA* gene encodes two protein products, $CheA_L$ and $CheA_S$ (long and short, respectively; Table 10). Both proteins are translated in the same reading frame from different start points, the additional one being at Met-98 [372]. This means that $CheA_S$ lacks 97 residues at the N terminus, including the phosphorylation site (His-48). Both proteins are present in wild-type cells, but their relative amounts vary with the growth phase. The variations are from a $CheA_L : CheA_S$ molar ratio of 1:1 during optimal motility conditions (mid-log phase) to 4:1 [760]. It appears that, within the cell, $CheA_S$ exists in complexes of $CheA_L$-$CheA_S$-CheW (1:1:1 molar ratio) and $CheA_S$-CheZ ($\sim$1:5) [493, 759, 760]. In line with this notion, most bacterial species that lack CheZ also lack $CheA_S$ [494]. Although $CheA_S$ lacks the ability to autophosphorylate, it can mediate phosphorylation of kinase-deficient variants of $CheA_L$, consistent with the notion of trans-phosphorylation within the CheA dimer [715, 785]. The function of $CheA_S$ in chemotaxis is still obscure. It is known, however, that $CheA_S$ links between CheZ and the receptor supramolecular complex [147].

7.1.4. *CheA in species other than* E. coli *and* Salmonella

The CheA protein of *B. subtilis* (originally termed CheN) and that of the archaeon *H. salinarium* share (each) 33% amino-acid identity with *E. coli*'s CheA and, like the latter, they autophosphorylate and transfer the phosphoryl group to CheY [240, 249, 609, 610]. *B. subtilis* and *H. salinarium* do not possess $CheA_S$ [248, 609]. *R. sphaeroides*, as well, does not possess $CheA_S$, but it contains four $CheA_L$ proteins encoded by different operons and participating in distinct chemotactic pathways [274, 581]. The gliding bacterium, *M. xanthus*, possesses two CheA analogs, FrzE [765] and DifE, which share 37% amino-acid identity [798]. These proteins are probably involved in different signal transduction pathways. The N-terminal portion of FrzE shares homology with the respective portion of CheA; its C-terminal portion is homologous to that of CheY. The CheA-like portion has an autophosphorylation activity. It also has a phosphotransfer activity that results in phosphorylation of the CheY-like portion [1].

7.2. *CheB*

CheB is a response regulator that belongs to the superfamily of two-component regulatory systems [573]. It is a specific methylesterase that, together with the methyltransferase CheR, controls the methylation level of the MCPs [671, 704, 737]. In addition, CheB has an amidase activity that converts, by irreversible deamidation, two nonmethylatable glutamine residues on each MCP to methylatable glutamate residues [342, 343, 543, 600, 725]. It is composed of two distinct domains: an N-terminal regulatory domain, 120 residues in length, homologous to the entire length of CheY (and likewise includes a phosphorylation site on an aspartate residue—Asp-56), and a C-terminal catalytic domain with amidase and methylesterase activities [664, 696]. A linker of 10–20 residues, susceptible to proteolysis, connects the two domains (Figure 17). When separated, the catalytic domain has, on the one hand, methylesterase activity $\sim$10 fold higher than that of the intact protein [664]. On the other hand, when the N-terminal domain of the intact protein is phosphorylated, the methylesterase activity of the intact protein is significantly higher than that of the isolated N-terminal domain [24, 444, 699]. This suggests that the N-terminal domain fulfills dual regulatory roles: when nonphosphorylated, it inhibits the methylesterase activity of the protein; when phosphorylated, it stimulates this activity [24, 192].

The crystal structure of the catalytic domain of CheB revealed that it consists of a central seven-stranded parallel β-sheet flanked by six α-helices [192, 778] (Figure 17). The active site consists of a catalytic triad involving Ser-164, His-190 and Asp-286. The crystal structure of intact, nonphosphorylated CheB revealed that the N-terminal, CheY-like, regulatory domain is positioned above the C-terminal catalytic domain with its β-sheets almost perpendicular to those of the other domain (Figure 17) [190, 192]. It appears that the C-terminal catalytic domain is rigid whereas the N-terminal domain is relatively flexible. On the basis of the structural data, a model for CheB activation has been proposed [24]. According to the model, phosphorylation of the regulatory domain results in a reorganization of the domain interface, exposing

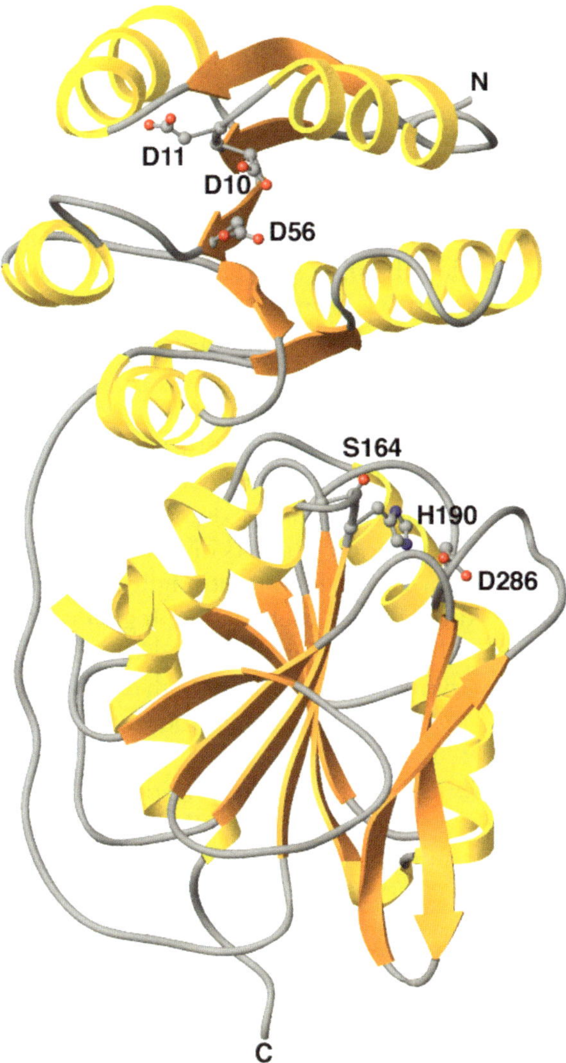

Figure 17. Ribbon diagram of the 3-D structure of CheB. The diagram shows the active site of the N-terminal regulatory domain, including the site of phosphorylation, Asp-56, and the Ser-His-Asp catalytic triad of the methylesterase C-terminal domain. Gray and red spheres represent carbon and oxygen atoms, respectively. (Taken with permission from Djordjevic and Stock [192].)

the active site to the MCP, and simultaneously stimulating the methylesterase activity of CheB. This model is supported by the more recent finding that, upon phosphorylation, CheB undergoes conformational changes [25].

7.3. *CheR*

CheR, a 32 kDa soluble monomeric protein [665], is a specific methyltransferase that, with AdoMet as a precursor, methylates the MCPs [165, 686]. In the cell, it is bound to the specific CheR/CheB-binding sequence (NWETF) at the cytoplasmic C-terminus of each of the two high-abundance receptors, Tsr and Tar [788]. From there it methylates neighboring receptor molecules, including low-abundance receptors that lack the CheR/CheB-binding site (Section 6.1.2 above). CheR is a relatively slow enzyme [665]. Its methyltransferase activity can apparently be regulated by chemotactic stimuli [369], but the mechanism is not known.

The crystal structures of CheR in complex with its product, *S*-adenosylhomocysteine, and in complex with both its product and the C-terminal pentapeptide of Tar (the CheR/CheB-binding site), revealed a two-domain protein. The N-terminal domain, which appears to be involved in substrate recognition, consists of four α-helices (Figure 18). It contains positively charged residues that might complement the negatively charged residues in the methylation region of the receptors [192]. The actual binding, however, is carried out by the C-terminal domain (Figure 18). This domain consists of both α-helices and β-sheets, and it contains the common features of AdoMet-dependent methyltransferases [189, 191, 192].

CheR of *B. subtilis* shares 29% amino acid identity with *E. coli*'s CheR [367]. Like in *E. coli*, it has a methyltransferase activity that methylates glutamate residues on the MCPs and activates CheA. However, unlike in *E. coli*, the consequence of these activities is smooth swimming [248, 367]. As will be discussed in Section 7.5.4, this is probably due to the fact that, in *B. subtilis*, CheA-mediated phosphorylation of CheY is a signal for smooth swimming.

7.4. *CheW*

Relatively little information is available on CheW. It is apparently a scaffolding protein that couples CheA to the MCPs [255, 256, 424, 425, 493, 635]. (Note, however, that CheA can apparently be also coupled to the MCPs directly [424].) CheW may also be involved in controlling the rate of CheA autophosphorylation and phosphotransfer to CheY [122, 123, 541]. Based on a consensus sequence, CheW may have a nucleotide-binding site [697], but none has thus far been demonstrated.

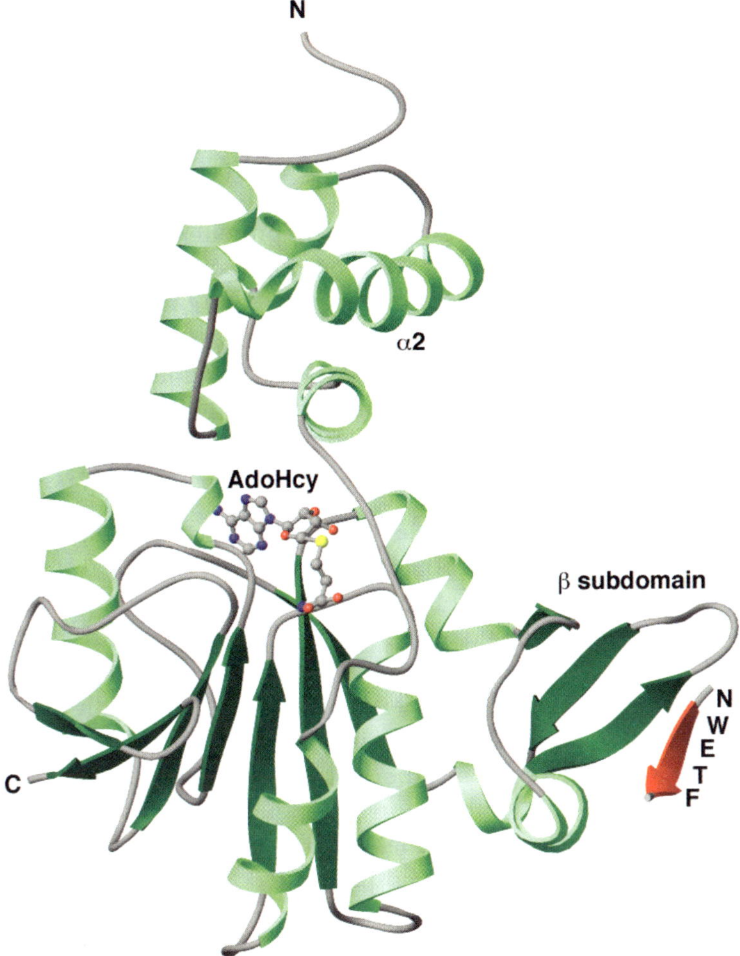

Figure 18. Ribbon diagram of the 3-D structure of CheR bound to *S*-adenosylhomo-cysteine (AdoHcy) and a synthetic peptide (NWETF, shown in red) corresponding to the C-terminal five residues of Tar. The diagram shows the β-subdomain, which constitutes a distinct structural motif for binding to the MCPs, and helix α2, which interacts with the methylation region of the MCPs. Gray, red, and yellow spheres represent carbon, oxygen, and sulfur atoms, respectively. (Taken with permission from Djordjevic and Stock [192].)

CheW of *B. subtilis* shares 29% amino acid identity with *E. coli's* CheW [276]. Together with CheV (a protein, unique to *B. subtilis*, that contains N-terminal and C-terminal domains analogous to CheW and CheY, respectively [604]), it might link CheA to the MCPs [248].

R. sphaeroides possesses four CheW proteins, but their function is not yet clear [274, 581].

7.5. *CheY*

7.5.1. *Function of CheY*

CheY, perhaps the most investigated Che protein, is a multifunctional, key response regulator whose role is to bridge between two membrane-associated, supramolecular complexes: the receptor complex and the switch-motor complex (Figure 14). As will be discussed in Section 8.2, its main role is to shift, in response to signals received from the receptors, the direction of flagellar rotation from the default direction, counterclockwise, to clockwise. This role is fulfilled by the following specific functions.

Phosphorylation
CheY undergoes relatively rapid phosphorylation on a specific aspartate residue (Asp-57[c]) by the histidine kinase CheA [126, 293, 621, 792] or, much slower, by small phosphodonors such as acetyl phosphate, phosphoramidate, phosphoimidazole, or carbamoyl phosphate [173, 443, 484, 661]. At least one of the small phosphodonors, acetyl phosphate, is present within the cell, and its level varies with the growth phase and conditions [488, 586]. This CheA-independent phosphorylation indicates that CheY, like other response regulators, can catalyze its own phosphorylation. CheY phosphorylation has hitherto been demonstrated only *in vitro*. However, its occurrence *in vivo* has been established beyond any doubt. The phosphorylated form of CheY (CheY~P) undergoes spontaneous autodephosphorylation, as a result of which its lifetime is relatively short ($t_{1/2} \approx 15$–$20\,\mathrm{s}$) [134, 291, 293]. (The lifespan of denatured CheY~P in sodium dodecyl sulfate [SDS] is, however, much longer, $t_{1/2} \approx 2\,\mathrm{h}$ [694].) In the presence of CheZ—a specific phosphatase that will be discussed in Section 7.6—the lifespan of CheY~P decreases by an order of magnitude [110, 134, 307, 442, 622].

The following additional facts are known about the phosphorylation and dephosphorylation reactions of CheY:

(a) Both processes are Mg^{2+}-dependent [441, 443].
(b) The phosphotransfer reaction from CheA~P to CheY is reversible [694].

---

[c] In mutants that cannot be phosphorylated at site 57 due to substitution (e.g., CheY57DN), the protein is phosphorylated at an alternative site—Ser-56 [28].

(c) The phosphotransfer is very fast, at least 10-fold faster than CheA autophosphorylation. CheA autophosphorylation is, therefore, the rate-limiting step in CheY phosphorylation [694].

(d) Chemoattractants bound to a chemotaxis-specific receptor inhibit CheA autophosphorylation, thereby lowering the phosphorylation level of CheY. (No reports regarding the *in-vitro* effect of repellents on modulation of the kinase activity are available.) It is, therefore, generally assumed that, *in vivo*, attractants and repellents inhibit and activate the kinase autophosphorylation, respectively, and accordingly modulate the phosphorylation level of CheY [122, 123, 541]. Likewise, methylation of the receptors enhances CheA autophosphorylation, thereby raising the phosphorylation level of CheY [124, 541].

(e) The phosphotransfer reaction itself is not modulated by the occupancy and methylation state of the receptors [123, 541].

(f) In *E. coli*, under nonstimulated conditions, the steady-state level of CheY~P is estimated to be ~30% of the intracellular pool of CheY [18]. In the presence of excess of externally added acetyl phosphate, the phosphorylation level rises to ~86% [110].

The phosphorylation level of CheY regulates all the functions of this protein: binding to the switch, clockwise generation, binding to CheA, and binding to CheZ.

Binding to the switch with a resultant clockwise rotation

The first indication that CheY interacts with the switch came from genetic second-site suppression analyses made by Parkinson *et al.* [572], followed by other groups [315, 464, 603, 672, 738, 794]. In these studies pseudorevertants were isolated from nonchemotactic *cheY* mutants and were found to have a second mutation that phenotypically compensated for the original mutation in *cheY* (for a recent review of the second-site suppression approach as a tool to study protein–protein interactions, see [472]). The second mutation mapped mostly at the switch gene *fliM* and, to a lesser extent, at *fliG* [315, 672]. Similarly, mutations in *fliM* and *fliG* were found to be compensated by mutations in *cheY* [603]. These observations suggested that CheY interacts mainly with FliM and possibly also with FliG. However, Macnab and coworkers found that the compensating mutations are not allele-specific (i.e., a given mutation in *cheY* could be suppressed by a number of different mutations in a switch protein). These findings suggested that the restoration of chemotaxis by a second mutation might be achieved by nonspecific adjustment of the switch bias rather than by specific structural compensation [315, 672]. These observations seemed "to pull the rug from under the feet" of the second-site suppression approach with respect to CheY and the switch proteins, and they could not be taken as evidence for physical interaction between CheY and the switch proteins [459].

Another line of evidence for CheY-switch interaction came from studies with strains lacking the cytoplasmic chemotaxis proteins and some of the membrane receptors (denoted as "gutted strains"). When CheY was overproduced in these strains (which contain no chemotaxis proteins that can potentially mediate between this protein and the switch), the probability of clockwise rotation increased [168, 392, 669, 782]. A somewhat more direct evidence for CheY-switch interaction came from the incorporation of purified CheY into cytoplasm-free bacterial envelopes possessing functional flagella [594]. The presence of CheY in the envelopes caused some of them to rotate clockwise [208, 594]. The absence of cytoplasm (and hence the absence of the cytoplasmic chemotaxis proteins) was verified in each envelope studied, indicating that the interaction of CheY with the switch is direct, without mediators. Later biochemical studies with purified CheY and switch proteins provided direct evidence that CheY indeed binds to the switch [773], that its docking site on the switch is the N terminus of FliM [135, 481, 773], and that the extent of binding is positively dependent on the phosphorylation level of CheY [489, 773, 774]. Recently, binding of CheY to the switch was also demonstrated *in vivo*: phosphorylation-dependent binding of CheY to overproduced switch complexes was demonstrated by fluorescence microscopy of functional CheY fused with GFP [355], and stimulus-dependent changes in CheY-FliM binding were demonstrated by fluorescence resonance energy transfer (FRET) employing FliM coupled to cyan fluorescent protein (CFP) and CheY fused with yellow fluorescent protein (YFP) [679].

The ability of CheY~P to generate clockwise rotation was demonstrated *in vivo* by expressing CheY from a low-copy-number plasmid (under the control of the *lac* promoter) in an *E. coli* strain deleted for the genes *cheB*, *cheY*, and *cheZ* [18, 169]. The absence of CheB resulted in fully methylated MCPs and, consequently, in highly active CheA and in rapid phosphorylation rate of CheY. The absence of CheZ greatly reduced the rate of CheY dephosphorylation. As a result of the enhanced phosphorylation and reduced dephosphorylation rates of CheY, all of the CheY molecules expressed in this strain were essentially phosphorylated. Clockwise rotation (and, hence, the tumbling frequency) increased in this strain with the concentration of intracellularly produced CheY~P, reaching almost 100% clockwise rotation. When, as a negative control, CheY was expressed, instead, in a gutted strain and, consequently, the CheY produced was essentially non-phosphorylated, clockwise rotation was not observed [18]. (Qualitatively similar results were obtained in earlier studies using cells with undefined kinase activity [392, 393, 782], or cells expressing an active CheY mutant protein [628].) Similar studies with individual cells demonstrated that the increase in clockwise rotation is very steep, suggesting high cooperativity of CheY~P binding or switching [169]. In this way, these results confirmed the semi-*in-vitro* studies [54] and endorsed the conclusion that phosphorylation activates the clockwise-generating activity of CheY.

Binding to CheA and CheZ
The binding of CheY to the enzymes that modify its activity—the kinase CheA and the phosphatase CheZ—is phosphorylation-dependent as well. However, while the extent of CheY binding to CheZ increases with the phosphorylation level [106, 489], the extent of CheY binding to CheA decreases upon phosphorylation [421, 716]. Unlike the case of CheY-FliM binding [774], the binding of CheY or CheY~P to CheZ depends on the presence of Mg^{2+} [106]. It appears that CheY competes with CheB for CheA [421].

Acetylation
An intriguing phenomenon, discovered in 1988 by Wolfe *et al.* [783], was that acetate causes a strong and prolonged clockwise bias in a gutted strain containing CheY. This phenomenon (denoted as the "acetate effect") was attributed to the effect of an intermediate of acetate metabolism on CheY. Acetate can be activated to acetyl-coenzyme A by two different pathways (Figure 19). One utilizes the acetate-inducible enzyme Acs [143] and proceeds through an acetyladenylate (AcAMP) intermediate [89]. The other pathway uses two enzymes, acetate kinase and phosphotransacetylase, and proceeds through an acetyl phosphate intermediate [607]. Recently it became clear that both intermediates can activate CheY *in vivo* [58]. As mentioned above, acetyl phosphate can phosphorylate CheY and thereby generate clockwise rotation [443].

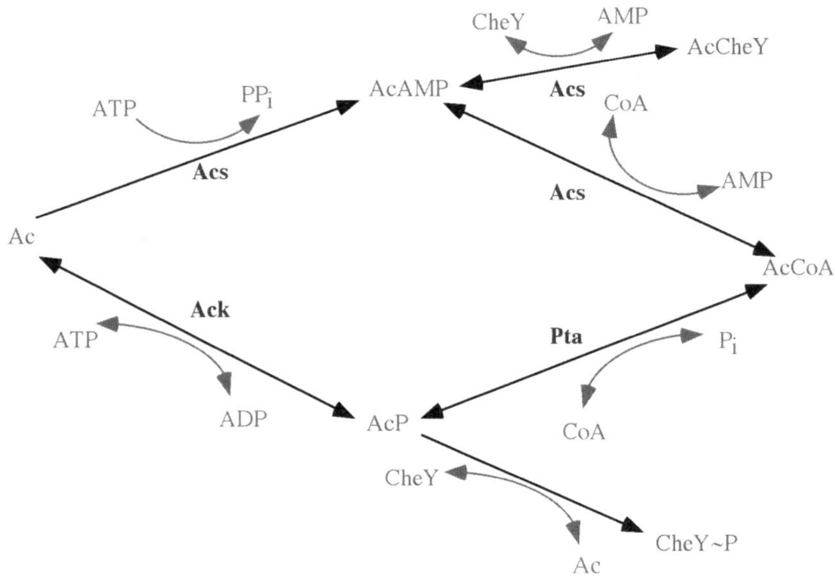

Figure 19. Two pathways of acetate metabolism to acetyl coenzyme A and their links with CheY. Abbreviations: Ac, acetate; AcAMP, acetyladenylate; AcCheY, acetylated CheY; AcCoA, acetyl coenzyme A; Ack, acetate kinase; AcP, acetyl phosphate; Acs, AcCoA synthetase; CheY~P, phosphorylated CheY; CoA, coenzyme A; PPi, pyrophosphate; Pta, phosphotransacetylase. (Taken with permission from Barak *et al.* [58].)

Acetyladenylate in the presence of Acs was found to acetylate CheY (at up to five sites—lysine residues 4, 26, 45, 92, and 109 [61, 590]) (Figure 19) with a consequent large increase in the clockwise-causing activity of CheY [55]. Mutants lacking the acetylating enzyme Acs, or *cheY* mutants that cannot be fully acetylated due to defects in the acetylation sites, are all defective in their chemotactic response, indicating that CheY acetylation is essential for chemotaxis [60]. Both the role that CheY acetylation fulfills in chemotaxis and the mechanism, by which acetylated CheY modulates flagellar rotation, are still a mystery. In view of the relatively slow kinetics of CheY acetylation and in view of the dependence of the intracellular concentrations of acetate, AcCoA, and Acs on the metabolic state of the cell [586, 650], it was proposed that the acetylation level of CheY links the metabolic state of the cell to the chemotaxis system [60]. Since the acetylation of CheY does not seem to affect its binding to FliM, it has been proposed that the acetylation is involved in a post-FliM binding step [590]. If this is so, it might be speculated that, while phosphorylation regulates the extent of binding of

CheY to the switch [773], acetylation modulates the "productivity" or the outcome of the binding [60]. This is not without precedent. A number of key transcription factors in mammals (e.g., p53 [159, 401]) and histones [157, 166, 433] should be both phosphorylated and acetylated for being active.

7.5.2. *Structure of CheY*

In spite of being multifunctional, CheY is a relatively small protein (only 128 amino acids), consisting of a single regulatory domain. Its three-dimensional structure has been resolved by both NMR and X-ray crystallography (for reviews, see [192, 706]). It has a doubly wound α/β topology, consisting of a five-stranded parallel β-sheet surrounded by five α-helices (Figure 20). The phosphorylation site (Asp-57) and other highly conserved residues (Asp-12, Asp-13, and Lys-109), which form

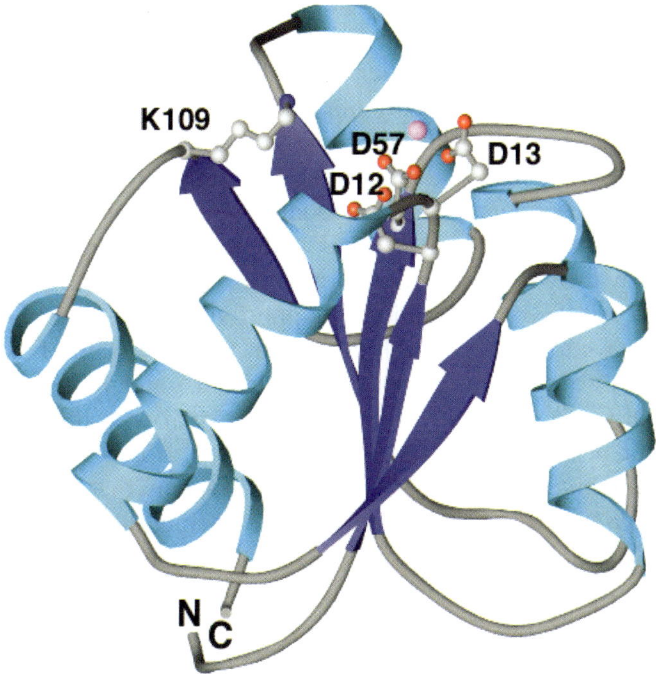

Figure 20. Ribbon diagram of the 3-D structure of CheY. The diagram shows conserved residues that form the active site for phosphoryl transfer, including the site for phosphorylation, Asp-57, and a catalytically essential Mg^{2+} (magenta sphere). Gray and red spheres represent carbon and oxygen atoms, respectively. (Taken with permission from Djordjevic and Stock [192].)

the active site for the catalysis of the Mg^{2+}-dependent phosphorylation and dephosphorylation, are clustered in a cleft atop the C-terminal edge of the β-sheet [192]. From the many structures of CheY deposited in the Brookhaven Protein Data Bank [91], including structures of different CheY mutant proteins, CheY bound to different metal ions, and CheY in complexes with other proteins, CheY appears to be a relatively malleable protein, capable of adopting numerous, subtly different, conformations [192]. Generally speaking, when CheY binds to the CheY-binding domain of CheA, its phospho-accepting site is exposed and ready to accept a phosphoryl group from the phospho-histidine side chain in CheA's phospho-accepting domain helix [702, 776]. Phosphorylation induces conformational changes in CheY, the extent of which is not yet clear.

A prominent difference between the various studied forms of CheY, as revealed from their X-ray structures, is the orientation of the side chain of Tyr-106, located on the face of the molecule. This side chain appears in wild-type CheY as a mixture of inward and outward conformations, whereas in all the other studied mutant proteins and analogs, the side chain is only in one orientation (for a review, see [137]). It was proposed that phosphorylation of Asp-57 initiates a conversion of Tyr-106 from a solvent-exposed orientation to a more internal position, possibly as a consequence of repositioning of residues Thr-87 (which appears to form a hydrogen bond with Asp-57 [161]) and Lys-109 (which may form a hydrogen bond with the oxygen atoms of the phosphoryl group) [273]. This notion is supported by a number of observations, reviewed in [137]. The finding, that the side chain of Tyr-106 is oriented outwardly in the complex between CheY and the CheY-binding domain of CheA [492, 776], further suggests that repositioning of Tyr-106 might be involved in the release of CheY from CheA and in its subsequent binding to FliM. It, therefore, seems that the rotameric state of Tyr-106 may be important for determining the activity of CheY [161, 273, 824]. However, the Tyr-106 orientations in the available structures of CheY are not always consistent with the above notions, for which reason the situation still seems to be somewhat ambiguous (for a review, see [137]). Apparently, more experiments are required for deducing the mechanism of CheY activation and for resolving the involvement of Tyr-106 in this activation. Recently, on the basis of random mutagenesis of *E. coli* followed by selection of *cheY* mutants that are constitutively active in the absence of phosphorylation, Da Re *et al.* [172]

proposed that the activation is not due to subtle conformational changes in CheY. Rather, the activation may be caused by an intermolecular mechanism that involves a dynamic interplay between CheY and the switch. Based on studies of the effects of binding of peptides of CheA, CheZ and FliM to CheY on CheY phosphorylation, Schuster *et al.* [634] suggested that CheY is activated in multiple steps.

The binding interfaces on CheY for FliM and CheZ, which are localized to the C-terminal regions of CheY, overlap [489, 622, 655, 825]. Therefore, CheY cannot be bound simultaneously to both FliM and CheZ, for which reason CheZ can exert its phosphatase activity on CheY~P only when the latter is not bound to the switch [134]. Furthermore, the C-terminal regions of CheY that interact with CheZ and FliM also overlap with the binding interface for CheA [654, 717, 825]. This suggests that, when bound to CheA, CheY cannot bind to CheZ or FliM.

CheY apparently belongs to a superfamily of structurally similar proteins in prokaryotes [38, 755]. In addition, it has a significant sequence homology and three-dimensional structural similarity to the eukaryotic GTP-binding protein Ras p21 [152], as well as a three-dimensional structural resemblance to the superfamily of hydrolases, which consists of P-type ATPases, phosphatases, and more [601].

7.5.3. *Active CheY mutants and analogs*

Structural information on the phosphorylated, active state of CheY is not available because of the short life span of this state. This situation triggered a number of attempts both to design constitutively active mutant CheY proteins and to synthesize active CheY analogs. As shown in Table 11, which lists selected mutant and analog proteins, none of the currently available mutant and analog proteins can fulfill all the functions of CheY~P. (The beryllofluoride analog of CheY, for which data on its ability to generate clockwise rotation are still not available, might be an exception.) Nevertheless, these proteins provide important insight into the structure-function relationship in CheY. For example, the finding that CheY13DK106YW, which does not contain a phosphoryl group, "looks to CheZ like nonphosphorylated CheY, but to FliM it looks like CheY~P" [627] suggests that it is the conformation of CheY~P, not the phosphoryl group *per se*, that counts for FliM binding [136]. However, proper conformation is apparently not the only requirement, because the binding of this double mutant protein to FliM$_{1-16}$ (a peptide built

Table 11. Constitutively active mutant CheY proteins and active CheY analogs.

Mutant/analog	Activities	Comments
Mutant CheY proteins		
CheY13DK	Generates clockwise rotation [18, 126, 128, 632] although it binds poorly to purified FliM [774].	Cannot be phosphorylated by CheA [126]. Its conformation is more similar to that of nonphosphorylated CheY and does not represent the activated state of CheY [324]. Might behave differently *in vivo* and *in vitro* [18].
CheY95IV	Generates more phosphorylation-dependent, clockwise rotation than wild-type CheY, and has an increased binding affinity to FliM [632].	Clockwise activity can be observed only in the presence of CheA [632].
CheY106YW	Generates more phosphorylation-dependent, clockwise rotation than wild-type CheY but the phosphorylation and FliM-binding activities are normal [823, 824].	In this mutant, the side chain of Trp-106 stays exclusively in the inside position [824].
CheY13DK106YW	Generates clockwise rotation without phosphorylation [628]. Binds like CheY~P to FliM [136, 627].	Has no binding activity to CheZ and does not induce CheZ oligomerization [136, 627].
CheY analogs		
Phosphono-CheY	Binds to FliM and CheZ, as does CheY~P [272].	Unlike CheY~P, does not generate clockwise rotation in semi-envelopes.[a] Stable for months [272].
Beryllofluoride (BeF$_3^-$) CheY	Binds to FliM [414, 797] and CheZ [161], as does CheY~P.	Forms an acyl-phosphate analog [797]. Was not tested for clockwise generation.

[a] Bren, A., Halkides, C.J., Dahlquist, F.W., and Eisenbach, M., unpublished data.

from the first 16 residues of FliM, including the CheY-binding domain [135]) is lower than the binding of wild-type CheY~P to this peptide [273, 633]. It, therefore, seems that the phosphoryl group contributes, after all, to the binding and, consequently, to CheY~P activation. Even though the CheZ- and FliM-binding surfaces of CheY overlap, and there is similarity between the CheY-binding domains of FliM and CheZ [135, 489], it appears that, for binding CheZ, the phosphoryl group on CheY is essential. Apparently, the requirements and constraints for binding CheY by FliM and CheZ are different.

7.5.4. *CheY in species other than* E. coli *and* Salmonella

CheY of *B. subtilis* (termed CheB) shares 36% amino acid identity with *E. coli*'s CheY [100] and, like the latter, it is phosphorylated by CheA [101, 249]. However, unlike null *cheY* or *cheA* mutants of *E. coli*, such null mutants of *B. subtilis* tumble incessantly, suggesting that phosphorylation of CheY in *B. subtilis* is a signal for smooth swimming, not for tumbling as in *E. coli* [100, 240]. Furthermore, in this species, CheY~P is apparently involved in a feedback mechanism of remethylation of the MCPs following stimulus-induced demethylation [364].

Some bacterial species possess a number of different CheY proteins. The record for the highest number of CheY proteins seems to belong to *R. sphaeroides*, which possess no less than seven CheY proteins [274, 581] (Armitage, personal communication). *S. meliloti* [675] and *A. tumefaciens* [787] possess two CheY proteins each. In *S. meliloti*, one of the CheY proteins (CheY$_1$) apparently acts as a phosphate sink for the other CheY (CheY$_2$). It may thereby enhance the dephosphorylation of CheY$_2$, fulfilling in this manner a role similar to that of CheZ in *E. coli* [676]. *M. xanthus* appears to possess three different CheY homologs: FrzE, FrzZ and DifD [765, 798]. As was indicated above, FrzE is a kind of a CheA-CheY hybrid, in which the CheA domain at the N-terminal portion autophosphorylates, and then phosphorylates the CheY domain at the C-terminal portion [1]. FrzZ consists of two domains, each of which is homologous to CheY [745]. Its function is not yet known. The function of DifD, which is over 60% identical in its amino-acid sequence to CheY of *B. subtilis*, is also not yet known [798].

7.6. *CheZ*

CheZ is a unique protein that, of the known bacterial chemotaxis systems and two-component regulatory systems, exists only in chemotactic enteric bacteria such as *E. coli* and *Salmonella* [494] as well as in *P. aeruginosa* [339, 709]. Its role is to enhance CheY~P dephosphorylation [293]. (The term phosphatase is used in this chapter to describe, in the broader sense, the activity of CheZ. It does not connote the mechanism of CheZ action.) CheZ does so with a high specificity, as evident from its failure to dephosphorylate CheB [293], even though the N-terminal portion of the latter is homologous to CheY as a whole [696]. Initially it was assumed, on the basis of second-site suppression analysis in which mutations in *cheZ* were phenotypically suppressed by mutations in any one of the three switch genes [569, 572], that CheZ, like CheY, interacts with the switch. Later, however, it was found that the suppression of the mutations in *cheZ* was not allele-specific (i.e., a given mutation in *cheZ* could be suppressed by a number of different mutations in a switch protein) [315, 672]. They could not, therefore, be taken as evidence for physical interaction between CheZ and the switch proteins [459]. Today we know that CheZ does not interact with any of the switch proteins [134].

CheZ binds to two proteins: CheY and CheA$_S$. A direct interaction of CheZ with CheY was biochemically demonstrated by immobilizing CheY onto a solid support either in a column [482, 493] or in a batch [106]. The binding of CheZ to CheY is phosphorylation-dependent, with binding to CheY$\sim$P being two orders of magnitude higher than the binding to nonphosphorylated CheY [106]. Binding of CheZ to CheA$_S$ was demonstrated by coprecipitation of CheA$_S$ and CheZ with anti-CheZ antibody. Coprecipitation was obtained both with a bacterial lysate and with purified proteins [759]. Recently, the binding of CheZ to CheA$_S$ was also demonstrated *in vivo*, employing CheZ-GFP fusion protein [147]. These interactions have a role in regulating the phosphatase activity of CheZ.

Until a few years ago, it was believed that the phosphorylation level of CheY is only regulated by modulation of the kinase activity of CheA in response to changes in receptor occupancy (Section 7.1.1 above). The activity of CheZ was thought to be unaffected, at least to a first approximation, by any chemotaxis component [706]. Today it is known that the phosphatase activity of CheZ is regulated by both CheY$\sim$P [106–108, 110] and CheA$_S$ [759]. Thus, by measuring the kinetics of CheZ-mediated dephosphorylation of CheY$\sim$P in the millisecond time scale, Blat *et al.* demonstrated that the phosphatase activity of CheZ is tightly dependent on the phosphorylation level of CheY [110]. They further found that CheZ is a dimer which, upon binding to CheY$\sim$P, undergoes oligomerization, possibly making a tetramer [107]. Two lines of evidence suggested that, as a result of this oligomerization, the phosphatase is activated. The first one was a good correlation between the ability of mutant CheZ proteins to undergo oligomerization and their ability to stimulate CheY$\sim$P dephosphorylation [108, 117]. The other line came from kinetic data, which demonstrated that the oligomerization precedes the dephosphorylation process [110]. Blat *et al.* further found that the specific activity of the phosphatase is cooperative with respect to CheY$\sim$P concentration [110]. This means that under nonstimulated conditions, when the cellular CheY~P level is low, the phosphatase activity of CheZ is almost negligible because of the cooperative nature of the dephosphorylation reaction. (Lack of CheZ activity at very low concentrations of CheY~P was demonstrated experimentally [106].) This cooperativity may function as a mechanism which ensures that a basal level of CheY~P—sufficient to mediate occasional tumbling events under nonstimulated conditions—is maintained in spite of the presence of CheZ.

Like CheY~P, CheA$_S$ (only under reducing conditions) binds to CheZ and enhances its dephosphorylating activity [759]. Although CheA$_S$-CheZ binding was also demonstrated *in vivo* [147], it is not yet clear whether this CheA$_S$-mediated regulation of CheZ occurs *in vivo* under physiological conditions and, if so, what its role is.

Since the function of CheZ is to dephosphorylate CheY~P, thereby terminating its interaction with the switch, one could expect that CheZ will act primarily on switch-bound CheY~P and will be randomly distributed in the cell or localized near the switch. However, none of these expectations was found to be correct. First, Bren *et al.* found that CheZ cannot act on switch-bound CheY~P; it can only act on free CheY~P [134]. This is the consequence of the overlap, discussed in Section 7.5.2, between the interfaces of CheY that bind FliM and CheZ. Second, the distribution of CheZ in the cell is not random: by analyzing cells expressing a functional, full-length CheZ fused with GFP or YFP, it was recently found that at least some of the CheZ molecules are localized in clusters at the cell's poles [147, 677], CheZ being bound to CheA$_S$ [147]. This observation suggests that CheZ, like all the other cytoplasmic chemotaxis proteins, can be attached to the receptor supramolecular complex via CheA$_S$. This situation suggests that CheZ may have two different functions in chemotaxis: one—still unknown—fulfilled by CheZ localized at the receptor supramolecular complex, and one—fulfilled by non-localized CheZ—to terminate the interaction of CheY~P with the switch.

The amino acid sequence of CheZ in *E. coli*, *Salmonella*, and *P. aeruginosa* [480, 533, 698] suggests that it has three conserved domains [109]. The C-terminal domain (residues 202–214) is involved in CheY binding [109]. The central domain, which includes residues 138–148, is apparently involved in CheZ oligomerization and activation [108, 623]. The function of the third conserved domain (residues 50–62) is not known.

The three-dimensional structure of the *E. coli* CheZ dimer in complex with the active analog of CheY, CheY-BeF$_3^-$, has recently been revealed [812]. CheZ is a long four-helix bundle composed of two

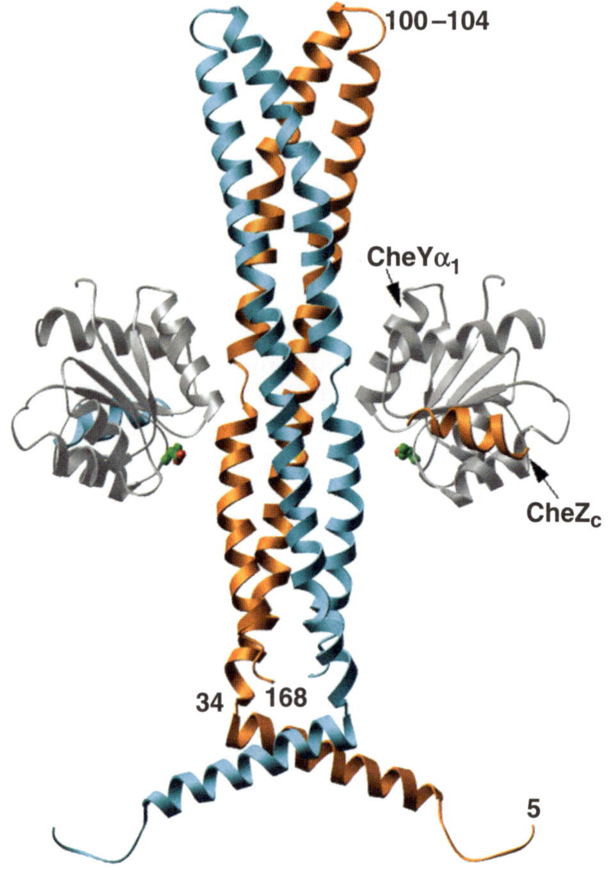

Figure 21. Ribbon diagram of the 3-D structure of the complex between dimeric CheZ and CheY-BeF$_3^-$ [(CheY-BeF$_3^-$-Mg^{2+})$_2$CheZ$_2$]. The CheZ$_2$ chains are cyan and orange and the CheY molecules are gray. BeF$_3^-$ (green) and Mg^{2+} (red) are in space-filling representation. (Taken with permission from Zhao *et al.* [812].)

helices from each monomer (Figure 21). Each CheY binds to CheZ through two distinct, large interaction surfaces that cause the hinged CheZ molecule to "clamp down" on the globular CheY molecule [812]. This is probably the reason for the very tight binding between the two molecules [106, 663].

A long-standing question in bacterial chemotaxis was whether CheZ is a specific phosphatase of CheY~P or whether it is an allosteric regulator of the intrinsic dephosphorylation activity of CheY~P. The structure of the CheZ:CheY-BeF$_3^-$ complex suggests that none of these mechanisms functions exclusively. Rather, the mechanism appears to involve features of both [812], thus being analogous to the mechanism by which the GTPase-activating proteins insert an arginine residue into the active site of the oncogene product, Ras [629].

8. Signal Transduction During Chemotaxis

8.1. *History*

The mechanism of signal transduction in bacterial chemotaxis was a mystery until the mid-1980s. Initially it was hypothesized that the signal, in analogy to nerve excitation, is electrical in nature [167]. It was proposed that conformational changes of the receptors lead to changes of the membrane potential, resulting in changes in the direction of swimming ([194] and references cited therein). However, although relatively slow membrane-potential changes in response to chemotactic stimuli were detected under specific conditions ([205, 206] and references cited therein), they seemed to be the consequence of an efflux of a charged organic molecule (probably a polyamine) not specific to chemotaxis (M. Eisenbach, unpublished data). More recent studies excluded electrical signaling altogether [207, 473, 637]. Exceptions are large bacterial species, where membrane-potential changes might be involved in signaling (Section 8.5.6).

Other potential modes of signaling, as well, were examined and eliminated [207]. Direct interaction between the receptors and the flagella was excluded by the findings that the area surrounding the motor is not enriched with receptors [219] and that cytoplasm-free envelopes do not respond to chemotactic stimuli [592]. Likewise, signaling by specific ions or small molecules was shown experimentally to be improbable [207, 637]. By way of elimination [207] and by demonstrating that, in filamentous cells of *E. coli*, the signal is inactivated as it moves away from its point of origin [637], it was concluded that the signal involves chemical modifications of a diffusible protein(s) within the cytoplasm. At about the same time, CheY was identified as the signal molecule that generates clockwise rotation [168, 207, 594]. The studies that led to this identification and the mechanism of signaling by CheY are described in this section.

8.2. *Mechanism of excitation*

In Section 6.1.2 above I discussed the possibilities for the way of inward cross membrane signaling the information about the extent of receptor occupancy. How does the signal propagate from there, reach the switch-motor complex, and dictate a change in the direction of flagellar rotation? Two major approaches—a whole cell approach and a reductionist approach—were taken to address this question. The whole-cell approach involved preparation and isolation of chemotaxis mutants and studying their phenotypes (for a review, see [570]). The reductionist approach involved insertion of purified chemotaxis proteins into cytoplasm-free envelopes of bacteria and studying the effect of the proteins on the direction of flagellar rotation (for a review, see [208]).

8.2.1. *The flagellar motor has a default direction of rotation, counterclockwise*

Both approaches have indicated that the flagellar motor has a preferred direction of rotation, a default direction, which is counterclockwise. Thus, the flagella of a gutted strain rotate exclusively counterclockwise [571, 782]. Similarly, "empty" wild-type envelopes rotate their flagella exclusively counterclockwise, even though the flagella have a mechanical ability to rotate clockwise [204, 592, 594]. In other words, the motor always rotates counterclockwise, unless it receives a signal to do otherwise. (This statement is restricted to wild-type motors and it holds for the temperature range 20–37°C. As was discussed in Section 2.2.5 above, at lower temperatures, $\Delta G°$ between the clockwise and counterclockwise states of the switch becomes sufficiently low to allow clockwise rotation. Also, some switch mutants can rotate their flagella clockwise even in the absence of cytoplasm [592].)

8.2.2. *CheY is the clockwise signal*

The chemotaxis protein CheY was subsequently identified by both the whole-cell and reductionist approaches as the clockwise signal, i.e., as the molecule that interacts with the switch, thereby promoting clockwise rotation. In the whole-cell approach, second-site suppression analysis implied that CheY interacts with the switch (for reviews, see [56, 459]), and overproduction of CheY in a gutted strain generated clockwise rotation ([168]; for a review, see reference [212]). In the reductionist approach, purified CheY inserted into cytoplasm-free envelopes caused some of the envelopes to rotate clockwise [594]. The absence of cytoplasmic chemotaxis proteins in the intact cells and the absence of cytoplasm in the envelopes indicated that the interaction of CheY with the switch is direct. Later biochemical studies demonstrated direct binding of CheY to FliM [773], with the docking site being at the N terminus of FliM [135].

8.2.3. *Regulation of CheY activity*

The identification of CheY as the clockwise signal raised the question of whether the mechanism of changing the direction of flagellar rotation involves changes in the level of CheY or whether it involves some modification of the CheY molecules in the cell. Generally speaking, regulation of a key function by the concentration level of a regulating protein is not sufficiently sensitive. This is especially so in the case of CheY, which appears to be the most abundant chemotaxis protein in the cell (Table 10). It was known for years that, in the absence of ATP, bacteria fail to tumble and the rotation of their flagella is restricted to counterclockwise (this ATP dependence of clockwise rotation is beyond the ATP requirement for synthesis of the precursor of receptor methylation—AdoMet) [29, 46, 244, 346, 378, 653, 684, 721]. Based on this knowledge, a number of research groups looked for CheY phosphorylation *in vivo* but were unable to detect any. The turning point came when M. Simon's group looked for phosphorylation of purified cytoplasmic chemotaxis proteins involved in signal transduction (CheA, CheW, CheY, and CheZ) and, as was discussed in Section 7.1.1, found that only CheA can be autophosphorylated by ATP [290]. This group as well as J. Stock's group further found that CheA transfers the phosphoryl group to CheY, and that CheY~P undergoes spontaneous dephosphorylation ($t_{0.5} = 20\,\text{s}$) [293, 792], which is greatly enhanced by CheZ [293]. These findings, supported by studies with phosphorylation-defective mutants [292, 551], suggested that phosphorylation might regulate CheY activity. However, due to the inability to detect CheY phosphorylation *in vivo*, there was no way to examine this presumption directly. Such a direct proof was provided by Barak and Eisenbach [54]. They found that CheY phosphorylation had no effect on the direction of flagellar rotation in cytoplasm-free envelopes, unless the envelopes contained, in addition to CheY and the phosphorylating agents, cytoplasmic constituents other than the known chemotaxis proteins (such a preparation is termed "semienvelopes"). In this preparation CheY phosphorylation was effective and the probability of clockwise rotation increased. This was demonstrated with either CheA and ATP [54] or acetyl phosphate (Y. Blat, R. Barak, and M. Eisenbach, unpublished observations) as phosphorylating agents. These observations indicated (a) that phosphorylation of CheY activates the protein in the sense that it can now bind to a much larger extent to the switch protein FliM and generate clockwise rotation, and (b) that binding of CheY~P to FliM may not be sufficient for switching the motor from its default state to its clockwise state.

As was discussed in Section 7.5.1, the ligand-occupancy level of the receptors and their methylation level regulate the phosphorylation level of CheY, and this phosphorylation level regulates all the binding activities of CheY. CheY~P binds weaker than does CheY to its kinase CheA [421, 716], as a result of which CheY is released from the receptor supramolecular complex [635] (Figure 14). However, CheY~P binds stronger than does CheY to the switch protein FliM [135, 355, 489, 773], with a consequent increased clockwise rotation of the flagellar motor [54]. It also binds stronger to the phosphatase CheZ [106, 109, 489], with a consequent delayed oligomerization of CheZ and increased phosphatase activity [107, 108, 110].

This major mechanism of regulation of CheY activity is not the only chemical modification that CheY undergoes and not the only one that activates the protein. As was discussed in Section 7.5.1, CheY also undergoes lysine acetylation [55, 590], which increases, to a large extent, the clockwise-causing activity of CheY both *in vitro* (in cytoplasm-free envelopes) [55] and *in vivo* [58, 590]. Although it was recently demonstrated that this chemical modification is involved in chemotaxis [60], its role is still an open question.

8.2.4. *Mechanism of switching*

It has been proposed that each molecule of CheY~P that binds to the switch reduces the free energy difference between the clockwise and counterclockwise states of the switch-motor complex by a fixed amount [749]. Whether or not this is the case and, if so, how this is done and what happens at the switch subsequent to CheY~P binding—are still obscure. Another question is how many CheY~P molecules should bind to the switch for shifting the rotation to clockwise. This question has recently been addressed by Bren and Eisenbach [138], who used molecules of mutant FliM, which are almost locked in the clockwise state (FliM$_{cw}$), as representatives of CheY~P-bound, wild-type FliM molecules [138]. They found that, for clockwise rotation, at least ~80% of the FliM molecules at the switch should be in their clockwise state. Furthermore, even though CheY~P binding to FliM was not involved in this experiment, the gain of clockwise rotation beyond the 80% point was very steep. Therefore, if $FliM_{CW}$ correctly reflects wild-type FliM occupied by CheY~P, these findings suggest that the steep gain of clockwise bias upon CheY binding, observed in earlier studies [18, 169, 393, 628], results from a postbinding step [138]. Subsequent studies, carried out *in vivo* [679] and *in vitro* [615], confirmed this conclusion by demonstrating that the mere binding of CheY~P to FliM within the switch complex is hardly cooperative.

Two fundamentally different sorts of switching mechanisms have been proposed: "deterministic" and "stochastic" [628]. According to the former, the direction of rotation is completely determined by the degree of CheY~P binding to the switch, i.e., CheY~P throws the switch. According to the latter, both directions of rotation are in thermal equilibrium, and CheY~P binding to the switch increases the *probability* of clockwise rotation. In other words, switching can occur even in the absence of CheY, and CheY~P just changes the relative stabilities of the rotational states (for reviews, see references [213, 662]). A resulting major difference between these sorts of models is that the deterministic model involves a stepwise separation between CheY~P binding and switching, whereas such a separation is not made in the stochastic model [662]. The observation that a prefixed mixture of $FliM_{CW}$ and $FliM_{CCW}$ molecules in a single switch can promote an intermediate rotational probability [138] is in favor of stochastic switching.

8.2.5. *Sequence of events*

From all the information, summarized above, it seems quite probable that signal transduction during excitation involves three major steps (Figure 14): (a) stimulus-induced modulation of the receptor supramolecular complex; (b) information transfer to the flagellar-motor supramolecular complex by the messenger protein CheY; and (c) modulation of this complex and, consequently, the direction of flagellar rotation. Accordingly, the signal-transduction pathway involves the following sequential steps:

Signal transduction in response to chemoattractants
Binding of a chemoattractant, say aspartate, to the receptor (Tar in the case of aspartate), results in subtle conformational changes of Tar with a consequent rearrangement of most, if not all, of the constituents of the receptor supramolecular complex. This rearrangement, which possibly involves stronger packing of the receptors [437], is sensed by the

receptor-binding domain of CheA [127]. Consequently, the autophosphorylation activity of CheA is inhibited [123] and the phosphorylation levels of CheA and CheY drop. Pending the resolution of some intriguing questions, CheZ molecules within the receptor supramolecular complex might be involved in enhancing dephosphorylation of CheY~P molecules that reside within the receptor supramolecular complex [137]. The drop in the CheY~P level results in its dissociation from the motor supramolecular complex. In a mechanism within the switch complex that is still obscure, the outcome of the dissociation is an increased probability of counterclockwise rotation and, hence, a decreased frequency of tumbling. At the same time, dephosphorylated CheY binds to CheA within the receptor supramolecular complex.

Signal transduction in response to chemorepellents
Chemorepellents apparently bind weakly to their cognate MCPs (Section 6.3) and stimulate the autophosphorylation of CheA. (I say "apparently" because neither the binding nor the stimulation of autophosphorylation has been demonstrated experimentally.) Consequently, the phosphorylation level of CheY rises. CheY molecules that have become phosphorylated acquire a reduced affinity to CheA and a high affinity to FliM, possibly as a result of the rotation of Tyr106 of CheY from its outward to its inward position. Binding of CheY~P to FliM increases, in an unknown mechanism, the clockwise probability of the motor with a resultant increased tumbling frequency.

8.2.6. *Coordinating signals of different intensity*

Some years ago, Ames and Parkinson isolated *tsr* mutants locked in either clockwise or counterclockwise rotation [20]. The finding of such mutants raised the possibility that each receptor exists in one of two active signaling states, clockwise or counterclockwise, and that the receptor output is controlled by modulating the ratio between these states [20, 21]. However, without some additional mechanism, such a situation cannot explain how signals from a small fraction of receptors (e.g., signals generated by low abundance receptors or by lowly occupied high abundance receptors) can be "heard" on top of much stronger, conflicting signals from nonstimulated receptors.

Such an additional mechanism may be provided by modulation of the interreceptor interactions within the supramolecular complexes.

Indirect supportive evidence for changes in the aggregation state at different signaling states of the receptors was provided by Long and Weis [436, 437], who studied locked mutants of the Tar receptor. They found a correlation between the swimming phenotype of *tar* mutants and the oligomerization properties of the receptor cytoplasmic fragment. Most of the studied cytoplasmic fragments, derived from mutants locked in the counterclockwise state, formed oligomers at neutral pH. In contrast, cytoplasmic fragments derived from clockwise-locked mutants or from the wild-type strain, did not exhibit any significant oligomer formation. Accordingly, Long and Weis suggested that subunit interactions within the cytoplasmic region are stronger in the attractant-bound form of the receptor than in the attractant-free form. Cooperativity of the kinase activity of CheA with respect to the stimulus concentration [125, 420] is also in line with the possibility of a transmembrane signaling mechanism that involves receptor aggregates and stimulus-induced changes in the aggregation states of the receptors.

How can such modulation of the interreceptor interactions make the signals from a small fraction of receptors be "heard" on top of conflicting signals from the nonstimulated receptors? Y. Blat proposed in 1995 a hypothesis (unpublished), according to which stimulation of a receptor might change the packing of the receptors within the cluster. This change might act as a "shut off" mechanism for other receptors in the cluster, resulting in modulation of the kinase activity according to the signal from the stimulated receptor. Levit *et al.* [419] suggested that ligand binding changes the packing of the receptors and CheA within the cluster and, consequently, CheA acquires an active or an inactive conformation, depending on the signal. Bray *et al.* [133] proposed that changes in interreceptor interactions might be involved in amplification of the signals from the stimulated receptors. They calculated that a mechanism, by which ligand binding changes the activity of a receptor and then this change propagates to neighboring receptors in the cluster, can account quantitatively for the high sensitivity and response range of *E. coli*. Interestingly, it was found *in vitro* that subsaturating attractant concentrations accelerate formation of active supramolecular complexes, implying that not only the interactions within the receptor-kinase complexes, but also the assembly/disassembly of the complexes may contribute to signal regulation [420]. Recently, in a very elegant study that employed synthetic multivalent ligands, Gestwicki and Kiessling [258] demonstrated that receptor clustering indeed causes signal amplification by at least two orders of magnitude and that the entire MCP array, including both high- and low-abundance receptors, is involved in this amplification.

8.2.7. *Signal amplification*

One of the puzzles in bacterial chemotaxis is how bacteria can sense stimuli over a wide concentration range and, in spite of the wide range, do so with very high sensitivity. For example, a change of a single molecule in the occupancy of Tar of *E. coli* by aspartate results in a measurable change in the direction of flagellar rotation [638]. Furthermore, *E. coli* can respond to a <1% change in Tar occupancy over a concentration range of three orders of magnitude [321]! Much effort, both experimental and computational, has been made in order to identify the amplification step(s) in the chemotactic signal transduction (e.g., [18, 131, 321, 628, 682, 703] and references cited therein). As mentioned above, part of the amplification may be provided by propagation of the excitation signal from the stimulated receptor molecule to neighboring receptor molecules [133, 198]. Other possibilities that have been raised are activation of a number CheA molecules within the receptor supramolecular complex by a single receptor molecule [124], release of receptor-sequestered CheZ molecules [245] or their activation in a cooperative manner [682], and cooperativity of CheY binding to the switch [169, 393, 459, 682]. Recent experimental data, which demonstrated very steep dependence of clockwise rotation on the CheY~P level in individual cells (a Hill coefficient of ~10) [169], suggested that the amplification is, to a large part, at the level of the switch-motor complex. Subsequent studies confined the steepness (and, therefore, the amplification) to a post-binding step at the switch-motor complex [138, 615, 679]. By using mutants lacking CheZ or lacking CheB and/or CheR, Kim *et al.* [360] have elegantly demonstrated that changes in the rate of CheY~P dephosphorylation are not involved in amplification of the excitation signal, but that CheB and/or CheR are involved. They suggested that differential binding of CheB and/or CheR to distinct MCP signaling conformations controls the sensitivity of the response. Similar conclusion with respect to the involvement of CheB in the sensitivity was reached experimentally by Sourjik and Berg [678], employing FRET, and, on the basis of modeling, by Barkai *et al.* [63]. Based on these recently accumulated data, it seems reasonable that there are two amplification steps, one at the level of the receptor supramolecular complex and the other at the level of the flagellar-motor supramolecular complex. The mediator, CheY, is apparently not involved in amplification.

8.2.8. *Variations on excitatory signal transduction in* E. coli

A number of stimuli are not sensed directly by the MCPs and, therefore, have somewhat modified signal transduction pathways. The most common examples follow.

PTS carbohydrates

As has been mentioned in Section 6.2, some carbohydrates (e.g., glucose, mannitol, lactose, and mannose) are chemotactically sensed by the same system that transports them inwardly—the PTS [417]. There are at least 15 PTSs in *E. coli*, each recognizing one or more different carbohydrates [583]. The PTSs consist of three main proteins: one membrane-bound protein and two soluble proteins. The membrane-bound protein—Enzyme II—is substrate specific, and each PTS has its own Enzyme II. The soluble proteins—Enzyme I, which is a phospho-*enol*pyruvate-dependent histidine kinase that phosphorylates Enzyme II, and HPr, which is a phosphohistidine carrier protein—are common to all the Enzyme II proteins. During transport, there is a cascade of phosphorylation starting with phosphorylation of Enzyme I by phospho-*enol*pyruvate (Figure 22, bottom). The phosphorylated enzyme phosphorylates HPr, HPr phosphorylates Enzyme II, and Enzyme II phosphorylates the carbohydrate substrate while being transported. All these three proteins are essential for chemotaxis to PTS substrates [416, 446]. *In vitro* studies demonstrated that CheA autophosphorylation is inhibited by the nonphosphorylated form of Enzyme I, but not by its phosphorylated form [445]. *In vivo* studies demonstrated a retarded chemotactic response to PTS substrates in *cheZ* mutants and a reduced response in mutants with defective MCPs [446]. Taken together, the results suggested that the substrate-dependent dephosphorylation of Enzyme I inhibits the autophosphorylation of CheA within the receptor supramolecular complex (Figure 22). The reduced level of CheA~P results in a reduced level of CheY~P, and the regular cascade of events (Section 8.2.5) follows.

Oxygen

Aerotaxis is a type of energy taxis, in which bacteria respond to changes in the respiratory electron transport that result from changes in the

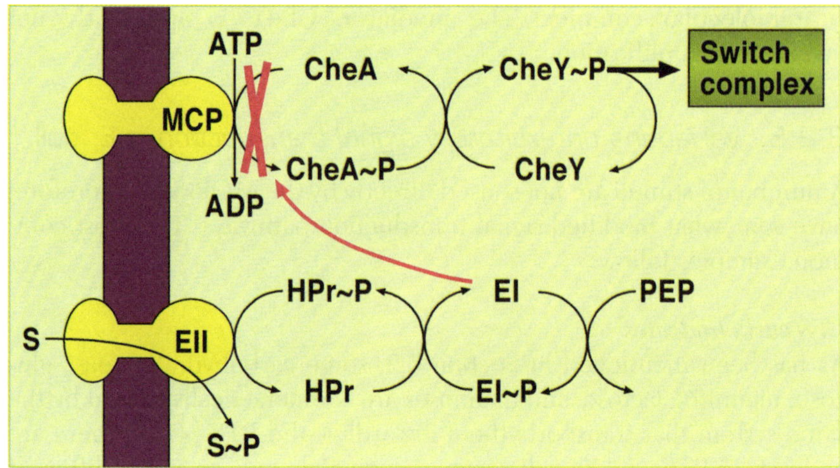

Figure 22. The PTS chemotactic signal-transduction pathway. Bold black arrow, interaction between proteins; red arrow, inhibitory interaction. Abbreviations: EI, Enzyme I; EII, Enzyme II; EI~P, phosphorylated Enzyme I; HPr~P, phosphorylated HPr; PEP, phospho*enol*pyruvate; S, PTS substrate; S~P, phosphorylated PTS substrate. See text for details.

oxygen concentration [723]. When oxygen—the preferred electron acceptor—is absent, other electron acceptors (e.g., fumarate or nitrate) are used by the cell. How is the respiratory electron transport system linked to the chemotaxis signal transduction system? The linkage is apparently made by Aer (Figure 12) [96, 596]. The N-terminal domain of Aer is essentially a PAS domain, to which the cofactor FAD is bound. ("PAS is an acronym formed from the names of the proteins in which imperfect repeat sequences were first recognized: the *Drosophila* period clock protein (PER), vertebrate aryl hydrocarbon receptor nuclear translocator (ARNT), and *Drosophila* single-minded protein (SIM). PAS domains are important signaling modules that monitor changes in light, redox potential, oxygen, small ligands, and overall energy level of a cell. Unlike most other sensor modules, PAS domains are located in the cytosol." [722]) Substitution of specific residues within the PAS domain results in loss of aerotaxis or in aberrant aerotaxis [599]. It appears that the redox state of the FAD cofactor, which reflects the redox state of the cell, modulates the conformation of the Aer protein as a whole, then the regular sequence of signaling steps (Section 8.2.5) occurs [97, 722].

Some chemorepellents

As was discussed in Section 6.3, some chemorepellents are apparently not sensed by receptors directly but rather by causing a specific perturbation in the cell. For example, weak organic acids lower pH_{in} and this change in pH is recognized by the cell as a repellent response [357, 598, 668]. The change in pH_{in} is sensed by specific MCPs [386, 668], and the MCPs transmit the signal to the flagella by the conventional pathway (Section 8.2.5).

It, therefore, seems that all the chemotactic responses, including those in which the stimuli are not sensed directly by the MCPs, are mediated by MCPs and they all share a common signal-transduction pathway.

8.2.9. *Some open questions*

The excitatory signal transduction, as reviewed in the preceding paragraphs, appears to be relatively well worked out. However, there are a number of components and processes that have been shown to be involved in this transduction but whose function and site of action are not yet known. In addition, there are several intriguing questions with respect to major signal transduction processes. A few examples follow.

(a) A number of lines of evidence suggest that Ca^{2+} is involved in chemotaxis of *E. coli*: (i) An artificially imposed elevation of the intracellular concentration of free Ca^{2+} ($[Ca^{2+}_{in}]$) causes tumbling [548, 732]. This phenomenon requires CheA, CheW, and CheY, but not the Tsr, Tar, Trg, and Tap receptors [732]. (ii) Mutants with elevated $[Ca^{2+}_{in}]$ due to a defective Ca^{2+}-transport system are tumbly and non-chemotactic [735]. (iii) Lowering $[Ca^{2+}_{in}]$ in these mutants and in wild-type cells by electroporation of Ca^{2+} chelators represses tumbling [735]. (iv) Ca^{2+}-channel blockers prevent tumbling and inhibit chemotaxis [733, 736], probably by inhibiting Ca^{2+} uptake [735]. (v) Chemoattractants cause a transient decrease in $[Ca^{2+}_{in}]$, whereas chemorepellents cause a transient increase [734, 768]. The process(es) or protein(s) for which Ca^{2+} is essential have not yet been identified.

(b) Fumarate, a metabolite of the citric-acid cycle, was found first in *H. salinarium* [476, 477, 518] and then in *E. coli* and *Salmonella* [53, 57, 520, 584] to possess the potential to enhance switching from counterclockwise to clockwise rotation. Of special interest was the observation that CheY-containing, cytoplasm-free envelopes of *E. coli* and *Salmonella*, which never switch in the absence of fumarate, switched back and forth between the two directions of rotation in the presence of fumarate [53, 57]. Fumarate analogs (succinate, malate, and maleate) were found to be effective as well, though to a lesser extent [57]. More recent studies demonstrated that fumarate similarly functions in intact *E. coli* cells [520, 584], that it is effective even in the absence of CheY (indicating that the target of fumarate is the switch, not CheY) [584], and that it acts by lowering the standard free energy of the clockwise state relative to that of the counterclockwise state [584]. Furthermore, mutants with aberrant fumarate concentrations appear to be defective in their chemotactic response to chemorepellents [585]. How fumarate is linked to the signal-transduction pathway, its role in chemotaxis, and its molecular mechanism of function are still a mystery.

(c) CheY can be activated (in the sense of acquiring the potential to generate clockwise rotation) by either phosphorylation or acetylation (Section 7.5.1). It has been shown that, in the absence of acetylation, chemotaxis is defective [60]. However, it is not known how CheY acetylation is involved in chemotaxis and why two different chemical modifications of CheY are necessary.

(d) In recent years evidence has been accumulated for activated forms of CheY (e.g., acetylated CheY or the mutant protein CheY13DK) that generate clockwise rotation without increasing the affinity between CheY and purified FliM. Although it is not known how this is accomplished, a few possibilities have been raised [662]: (i) Purified FliM might not be identical in function or structure to FliM within the switch complex. Perhaps the modified CheY forms do have higher affinity to FliM *in vivo*. (ii) While purified FliM might always be in its default counterclockwise state, perhaps the modified CheY forms can only bind to the clockwise state of FliM. (iii) The modified CheY forms might activate a postbinding switching step. It is not known which of these possibilities is correct.

(e) As discussed above, CheA$_S$ forms a complex with CheZ at the cell's poles, probably being a part of the receptor supramolecular complex. CheZ-CheA$_S$ interaction may thus play a role in the rapid CheY~P dephosphorylation that presumably occurs during an attractant response. According to this speculation, CheZ at the receptor supramolecular complex may be involved in counterclockwise generation in response to attractants, whereas free CheZ may be involved, after a delay, in clockwise termination following a repellent response. This possibility, however, raises a number of intriguing questions. For example: (i) It was found that mutants, which do not express $CheA_S$, are unimpaired in their ability to respond to attractants under standard assay conditions [620]. If indeed $CheZ$–$CheA_S$ interaction is involved in counterclockwise generation, why does the absence of $CheA_S$ have no effect on the response to attractants? (ii) If $CheZ$ and $CheA_S$ are in constant interaction with each other within the receptor supramolecular complex, how is the activity of $CheZ$ modulated by attractants rather than being constitutively active? (As was discussed in Section 7.6, $CheZ$ is activated upon interaction with $CheA_S$, at least *in vitro*.) If, on the other hand, $CheZ$ and $CheA_S$ interact only when a ligand is bound to the receptor supramolecular complex, how is the activation of $CheZ$ sufficiently fast, in view of the fact that the oligomerization-dependent activation of $CheZ$ is relatively slow (Section 7.6)? These questions are still open and the roles of $CheZ$ and $CheA_S$ within the receptor supramolecular complex are still obscure.

(f) Other questions that I have mentioned in different sections are: How are the sensory signals transduced across the membrane by the chemotaxis receptors? How do differently abundant receptors generate signals of similar strength? How is the signal propagated within the switch subsequent to $CheY{\sim}P$ binding?

8.3. *Mechanism of adaptation*

Adaptation—the process of recovery from a stimulated behavior when the stimulus is still present—is essential for every behavioral system. It allows detection of small changes in the stimulus level on top of existing, constant level of stimulation. In the case of bacterial chemotaxis, adaptation enables bacteria to respond to new stimuli in the presence of constant levels of chemoattractants and/or chemorepellents.

8.3.1. *Involvement of MCP methylation in adaptation*

One of the first steps in revealing the mechanism of adaptation was the finding that CheR-mediated MCP methylation (Section 7.3) and CheB-mediated demethylation (Section 7.2) are involved in adaptation (for a review, see reference [685]). The basis for the linkage between the methylation level and adaptation was the observations that (a) wild-type cells starved for methionine (the precursor of methylation) as well as *cheR* and *cheB* null mutants fail to adapt [267, 568, 685, 803], and (b) the chemoattractant-triggered rise and the chemorepellent-triggered fall in the methylation level are each synchronous with the timing of adaptation (Figure 23). Initially it was believed that each MCP functions independently of the other MCPs and mediates adaptation to stimuli sensed by it. Later, when it became apparent that the low-abundance receptors lack the CheR- and CheB-binding sites, it became clear that CheR or CheB, docked on a high-abundance receptor, catalyzes the methylation or demethylation, respectively, of a low-abundance receptor (Section 6.1.2).

It has been shown in a cell-free system that the CheR-mediated methylation reaction is enhanced by chemoattractants and inhibited by chemorepellents, but the mechanism underlying these effects is not known [369]. The outcome of this methylation is an enhancement of CheA autophosphorylation and, thereby, transmission of a clockwise signal [124, 541]. Conversely, the demethylation reaction is enhanced by chemorepellents and inhibited by chemoattractants [369]. The outcome of this demethylation is inhibition of CheA autophosphorylation and, thereby, transmission of a counterclockwise signal [124, 541]. In this

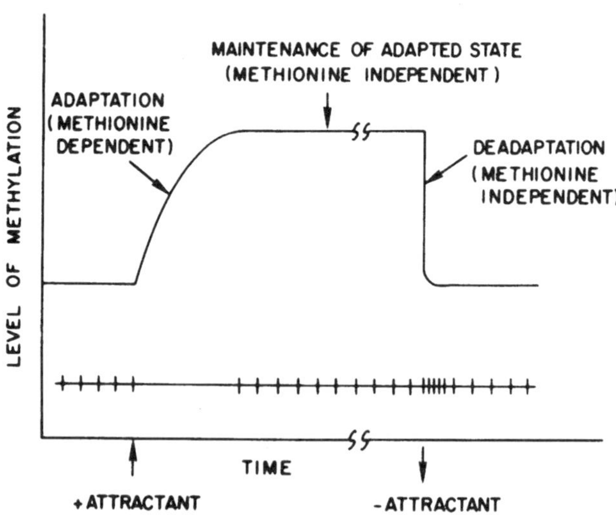

Figure 23. Correlation between methylation and adaptation. Intersections with the line at the bottom each indicate a tumble. (Taken with permission from Adler [11].)

manner the relative rates of the methylation and demethylation reactions determine the steady-state level of receptor methylation, and this level regulates the kinase activity of CheA. Recently it was demonstrated that a high methylation level decreases the affinity of the receptor supramolecular complex to chemoattractants [125, 420] (up to 10 000-fold in the case of serine [420]), suggesting that the methylation level regulates ligand binding to the receptor supramolecular complexes. This modulation of ligand-binding affinity extends the range of the chemotactic response and suggests that the cells adapt not only by methylation-dependent modulation of the kinase activity but also by decreasing the extent of stimulant binding to the receptor complex.

8.3.2. *The signal for adaptation*

What is the signal for adaptation? Does MCP stimulation trigger simultaneously both the excitation and adaptation processes, or does adaptation commence in a feedback mechanism? Early studies, demonstrating that it is possible to adjust the magnitude of positive and negative stimuli so that no behavioral response is observed even though the adaptation time of a positive stimulus is longer than that of an equivalent negative stimulus, suggested that excitation and adaptation are sequential processes [80, 344, 693]. Later studies suggested that, indeed, adaptation occurs in a feedback mechanism and that the activity of CheB (and possibly also of CheR, though the mechanism is yet unknown) is regulated only after the initiation of the chemotactic response. The feedback of CheB starts with CheA autophosphorylation. It was shown that when CheB is nonphosphorylated, its esterase activity is inhibited, and when phosphorylated, its activity is stimulated [24, 26, 190, 444, 664, 699]. On the basis of structural and biochemical data, it was proposed that phosphorylation of the regulatory domain of CheB results in reorganization of its interface, exposing the active site to the receptor, and simultaneously stimulating the methylesterase activity of the protein [24, 190].

Accordingly, the sequence of events appears to be as follows: Negative stimulation results in enhanced autophosphorylation of CheA, which, in turn, increases the steady-state phosphorylation level of CheY and, more slowly, of CheB. The increased phosphorylation level of CheY results in increased probability of clockwise rotation. As CheB becomes phosphorylated it is activated, the methylation level of the proper MCP is reduced, and the probability of clockwise rotation decreases to the prestimulus level. Conversely, positive stimulation inhibits CheA autophosphorylation. The resulting reduced phosphorylation level of CheY decreases the probability of clockwise rotation. CheR, more slowly, methylates the MCP. The elevated methylation level of the proper MCP enhances CheA autophosphorylation and the end result is increased probability of clockwise rotation and restoration of the prestimulus level.

8.3.3. *Exactness and robustness of adaptation*

Years ago Spudich and Koshland [688] demonstrated that there is non-genetic individuality in a bacterial population, and that the chemotactic response varies between individuals in a given population. This non-genetic variation is the consequence of different levels of chemotaxis proteins in different cells. Very interestingly, Barkai and Leibler [62] proposed, and then Alon *et al.* [19] demonstrated experimentally that unlike the steady-state tumbling frequency and the adaptation time, which vary with the intracellular concentrations of the chemotaxis proteins, adaptation in bacterial chemotaxis is precise (i.e., the post-adaptation swimming behavior is exactly like the prestimulus behavior) and independent of the exact level of the proteins involved in adaptation. They termed this independence "robustness." Why is the adaptation precision a robust property whereas the steady-state tumbling frequency and the adaptation time are not robust and vary with protein concentrations? According to Alon *et al.* [19], properties critical for function should be robust so as to withstand natural variations. These authors suggested that exact adaptation is critical for chemotaxis, whereas the steady-state tumbling frequency and the adaptation time may be inexact without a severe effect on the chemotactic responsiveness of the cells. Intriguingly, studies with cells deleted for chromosomal *cheB* but expressing plasmid-borne *cheB* of activated mutant CheB, which can demethylate the MCPs but lacks the domain phosphorylated by CheA, suggest that although CheB phosphorylation is essential for normal chemotaxis, neither the exact adaptation nor the adaptation time are dependent on this phosphorylation [19].

8.3.4. *Methylation-independent adaptation*

Several observations, made in the nineteen eighties, that strains deleted for both *cheR* and *cheB* can partially adapt ([700, 701, 705, 772] and references therein), suggested that methylation and demethylation might not be the only processes involved in adaptation and that an additional, methylation-independent process(es) might be involved (for a review, see [212]). It was suggested that CheZ, which is able to set the CheY~P concentration back to the prestimulus level, might be involved in this adaptive response [110]. The basis for this suggestion was the finding that both the activation and the deactivation of the phosphatase activity of CheZ are delayed. The apparent consequence of this delay is that the modulation of the phosphatase activity occurs only after the excitatory signal is complete. Therefore, the delayed activation and deactivation might constitute an adaptation mechanism, which ensures that the phosphorylation level is partially set close to the prestimulus level. Blat *et al.* [110] raised the possibility that CheZ mediates the first step of adaptation while the second, slower step, which includes the precise tuning of the direction of flagellar rotation, is mediated by the methylation system. However, it is not known whether the delayed activation and deactivation of CheZ are indeed involved in the mechanism of the methylation-independent adaptation or whether they are involved in a mechanism that shuts off or reduces the excitatory signal after a while. In line with the possibility of involvement in adaptation, it was demonstrated that *cheZ* mutants adapt slower than wild type [112, 638, 672]. This finding, however, cannot be taken as a proof of involvement in adaptation.

A common denominator for the methylation-dependent adaptation and the CheZ-dependent termination or adaptation is that both processes are triggered by changes in the phosphorylation level of CheY. It appears that the complete mechanism of adaptation, methylation-dependent and independent, is still not fully resolved. Methylation-independent adaptation was also demonstrated in *B. subtilis*, where it was found to be very effective [366].

8.4. *A nonconventional signal transduction pathway in* E. coli

It has been known for some time that some two-component regulatory systems of *E. coli* (e.g., PhoR/PhoB [415, 762], NtrB/NtrC [226], and EnvZ/OmpR [230, 231, 488]) can function even in the absence of their histidine kinase. Each of these systems may possibly undergo cross-regulation with another signal transduction system or it may possess an additional (yet-unidentified) signal transduction pathway [763]. Barak and Eisenbach [59] were surprised to realize that such a phenomenon had not been observed in bacterial chemotaxis. Assuming that the extremely biased flagellar rotation of nonchemotactic mutants might have perturbed the possibility to detect chemotaxis in cells lacking the kinase, they studied gutted strains in which CheY was expressed from a plasmid. They found that these strains exhibited a chemotactic-like response to chemoattractants and chemorepellents, though the response was not as good as that of wild-type cells. This apparent chemotaxis in the absence of most of the chemotactic signal transduction components suggested that, at least when the conventional signal transduction components are missing, a nonconventional chemotactic signal transduction pathway might be functional in *E. coli*. The components involved in this pathway have not been identified. It was proposed that the substrate-recognition components of the appropriate transport systems might serve as chemoreceptors for this pathway [59] but no evidence is yet available. More experiments are required for addressing this question and for revealing the physiological role, if any, of this pathway. The nonconventional pathway might be, in wild-type cells, a backup system that detects low concentrations of stimuli under certain conditions, e.g., in stationary phase or under anaerobic conditions. It might also be a primitive mechanism that, during evolution, became masked by a better, more efficient system [59]. It might be mentioned, in a parenthetical clause, that the inhibition of the enzyme fumarase by the chemorepellents indole and benzoate, discussed in Section 6.3, might also be related to the nonconventional chemotactic signal transduction pathway. It seems unlikely that this fumarase-mediated response is a part of the regular chemotactic response, for two main reasons:

(a) The response, measured in gutted strains, is delayed for a number of seconds [585], whereas a chemotactic response is (and must be) immediate.

(b) This response does not involve the MCPs, whereas wild-type cells must have functional Tsr for responding to indole or benzoate [747].

8.5. *Variations in signal transduction pathways in other bacterial species*

8.5.1. *More than one signal transduction pathway*

As the genome sequences of more and more bacterial species become available, it turns out that, unlike *E. coli*, which has only one set of *che* genes [111], a significant proportion of the bacterial species have two or more sets of *che* genes [33]. This suggests that these species possess at least two signal transduction pathways. For example, *R. sphaeroides* has 12 MCPs (some of which are in the cytoplasm), 4 CheA, 2 CheB, 3 CheR, 4 CheW, and 7 CheY; in contrast, it has no CheZ ([581, 641, 642] and references cited therein; J. Armitage, personal communication). By studying proper mutants of *R. sphaeroides* it was demonstrated that this species indeed has two (or more) pathways [274]. It was found that five of the CheY proteins can be phosphorylated by $CheA_2$, but only three of them by $CheA_1$. Also, only $CheY_4{\sim}P$ and $CheY_5{\sim}P$ can bind to the motor under normal circumstances [642]. It has been proposed that most other CheY proteins act in *R. sphaeroides* as phosphate sinks that facilitate the dephosphorylation of $CheY_4{\sim}P$ and $CheY_5{\sim}P$ [642]. It is not yet known whether the different pathways work in parallel or whether each is functional (or expressed) under different conditions. The major chemotaxis genes of *R. sphaeroides* are arranged in 3 loci, one close to the flagellar genes. Locus 1 has 2 MCPs, 1 Tlp, 1 CheD, 3 CheY, 1 CheA, 1 CheW, and 1 CheR. Locus 2 has 1 CheY, 1 CheA, 2 CheW, 1 CheR, 1 CheB, and 1 Tlp. Locus 3 has 2 CheA, 1 CheY, 1 CheW, 1 CheR, 1 CheB, and 1 Tlp. The other genes are scattered in other sites. *R. sphaeroides* possesses two chromosomes. All the loci are on the major chromosome, but the MCP genes are on both. All but one of the CheW proteins and all the CheA proteins encoded on both the second and third loci are essential for chemotaxis and are not redundant [581]. Other species in which an additional signal-transduction pathway has been implied are *B. subtilis* [238], *P. aeruginosa* [709], *V. cholerae* [288] and *H. salinarium* [149]. It has been proposed that the sensory pathway of the latter involves an oscillatory signal [630]. A comparison between the components of the signal transduction systems of selected bacterial species is shown in Table 12.

8.5.2. *Lack of CheZ*

As shown in Table 12, many bacterial species do not contain CheZ. It has been suggested that, with the exception of *P. aeruginosa*, CheZ only exists in enteric bacteria that contain $CheA_S$ [494], suggesting that the $CheZ$–$CheA_S$ interaction plays a role in chemotaxis of CheZ-containing species (Section 8.2.8). Species that do not contain CheZ usually possess a few CheY proteins. One or more of the CheY proteins apparently acts as a phosphate sink and thereby fulfills the "phosphatase" role of CheZ [642, 676].

Table 12. Composition of signal transduction components in selected bacterial species.

Species	MCPs	CheA	CheB	CheR	CheW	CheY	CheZ	Other proteins	References
A. tumefaciens	1	1	1	1	1	2	0		[787]
B. subtilis	10	1 CheA$_L$	1	1	1	1	0	CheC, CheD, CheV	[100, 240, 276–278, 367, 391, 525, 604, 605, 727–729]
E. coli	5	1	1	1	1	1	1		Table 10
H. salinarium	18	1 CheA$_L$	1	1	2	1	0	2 CheJ (CheC of *B. subtilis*), CheD	[374, 539, 609–611, 708, 809]
M. xanthus	2	1	1	1	2	1	0	FrzA, FrzB, FrzE (CheAY), FrzZ (CheYY)	[765, 798]
P. aeruginosa	26	2	4	4	5	5	1	2 CheAY, CheD, CheV, unidentified ORFs[a]	[339, 480, 709]
R. sphaeroides	12	4 CheA$_L$	2	3	4	7	0	CheD, CheABR	[274, 641, 642]
S. meliloti	2	1 CheA$_L$	1	1	1	2	0	CheD	[268, 675, 764]

[a] ORFs, open reading frames.

8.5.3. *Variations in Che proteins*

The difference between *E. coli* and other species is not restricted to the presence or absence of Che proteins or to their number. Some species have Che proteins that are absent in *E. coli* (Table 12). For example, *B. subtilis* contains the protein CheV, which is homologous to both CheW and CheY and can, in part, functionally replace CheW [604]. It also contains CheC and CheD, which share no significant homology with other Che proteins [605]. These proteins are likely involved in methylation-dependent adaptation of *B. subtilis*, which appears to be totally different from that of *E. coli* [606]. Two forms of CheC and a single form of CheD are also present in *H. salinarium* [539]. CheD appears in *S. meliloti* as well [268]. *M. xanthus* contains FrzE, which is a kind of a CheA-CheY hybrid [1], and FrzZ, which consists of two domains homologous to CheY [745].

8.5.4. *Different inputs*

A few examples of inputs different from those of *E. coli* are as follows:

(a) In contrast to *E. coli*, where chemoattractants are sensed on the bacterial surface, there are bacterial species (e.g., *R. sphaeroides* [279], *H. salinarium* [708], and *M. xanthus* [765]) in which some of the MCPs reside in the cytoplasm. The chemoattractants (or their metabolites) for these MCPs are detected intracellularly [323, 580].

(b) In *H. salinarium*, the MCPs serve not only as chemo-transducers but also as photo-transducers. For example, changes in light intensity sensed by sensory rhodopsin II (SRII) are processed by HtrII—an MCP that is also involved in the chemotactic response of this species to serine [306].

(c) Another input, observed in *H. salinarium*, is the membrane potential. In this species, upon changes in the light intensity, the proton pumping activity of bacteriorhodopsin changes the membrane potential and these changes are sensed by the bacterial flagella [270].

(d) The aerotactic transducers of *B. subtilis* and *H. salinarium*, unlike those of *E. coli*, are myoglobin-like, haem-containing proteins [305].

(e) In *A. brasilense*, unlike in *E. coli*, the dominant behavior is energy taxis: chemotaxis to most chemoattractants is metabolism-dependent and requires a functional electron-transport system [17].

8.5.5. *Different outputs*

In some species the outcome of CheY~P interaction with the switch is different from the outcome of this interaction in *E. coli*. In *H. salinarium*, for example, CheY~P appears to increase the switching probability rather than the clockwise probability of the motor [610]. In *S. meliloti*, an interaction of CheY~P with the flagellar motor appears to slow down the rotation [675] instead of changing its direction (the flagella of *S. meliloti* rotate only in one direction—Table 1). In *B. subtilis*, phosphorylation of CheY apparently decreases (rather than increases) the clockwise probability [248].

8.5.6. *Signal transduction in large bacterial species*

The signal transduction pathways discussed above are essentially networks of interacting enzymes, resulting in a relatively short signaling range. Therefore, they are not suitable for large (longer than 20 μm) bacterial species. Indirect evidence suggests that in such species (e.g., *S. volutans*, *Rhodospirillum rubrum*, *Thiospirillum jenense*, and cyanobacteria) the signal is electrical in nature (for a review, see [212]). Perhaps the most convincing evidence was obtained in Spirochetes, where neurotoxins, which affect the action potential in excitable eukaryotic cells, were found to inhibit chemotaxis [266], and where clamping the membrane potential at ~0 mV had a similar inhibiting effect [265].

References

1. Acuña, G., Shi, W.Y., Trudeau, K. and Zusman, D.R. (1995). The "CheA" and "CheY" domains of *Myxococcus xanthus* FrzE function independently *in vitro* as an autokinase and a phosphate acceptor, respectively. *FEBS Lett.* **358**, 31–33.
2. Acuña, J., Rojas, J., Amaro, A.M., Toledo, H. and Jerez, C.A. (1992). Chemotaxis of *Leptospirillum ferrooxidans* and other acidophilic chemolithotrophs: comparison with the *Escherichia coli* chemosensory system. *FEMS Microbiol. Lett.* **96**, 37–42.
3. Adams, D.G., Ashworth, D. and Nelmes, B. (1999). Fibrillar array in the cell wall of a gliding filamentous cyanobacterium. *J. Bacteriol.* **181**, 884–892.
4. Adler, J. (1966). Chemotaxis in bacteria. *Science* **153**, 708–716.
5. Adler, J. and Dahl, M.M. (1967). A method for measuring the motility of bacteria and for comparing random and non-random motility. *J. Gen. Microbiol.* **46**, 161–173.
6. Adler, J. (1969). Chemoreceptors in bacteria. *Science* **166**, 1588–1597.
7. Adler, J. (1973). A method for measuring chemotaxis and use of the method to determine optimum conditions for chemotaxis by *Escherichia coli*. *J. Gen. Microbiol.* **74**, 77–91.

8. Adler, J., Hazelbauer, G.L. and Dahl, M.M. (1973). Chemotaxis towards sugars in *Escherichia coli. J. Bacteriol.* **115**, 824–847.

9. Adler, J. (1974). Chemotaxis in bacteria, in *Biochemistry of Sensory Functions* (L. Jaenicke, ed.), pp. 107–131. Springer-Verlag, Berlin, Heidelberg.

10. Adler, J. and Epstein, W. (1974). Phosphotransferase-system enzymes as chemoreceptors for certain sugars in *Escherichia coli* chemotaxis. *Proc. Natl. Acad. Sci. U.S.A.* **71**, 2895–2899.

11. Adler, J. (1978). Chemotaxis in bacteria. *Harvey Lect.* **72**, 195–230.

12. Aizawa, S. and Kubori, T. (1998). Bacterial flagellation and cell division. *Genes Cells* **3**, 625–634.

13. Aizawa, S.-I., Dean, G.E., Jones, C.J., Macnab, R.M. and Yamaguchi, S. (1985). Purification and characterization of the flagellar hook-basal body complex of *Salmonella typhimurium. J. Bacteriol.* **161**, 836–849.

14. Aizawa, S.-I. (2001). Flagella, in *Molecular Medical Microbiology* (M.R. Sussman, ed.), pp. 155–175. Academic Press.

15. Aizawa, S.I. (1996). Flagellar assembly in *Salmonella typhimurium. Mol. Microbiol.* **19**, 1–5.

16. Alam, M. and Oesterhelt, D. (1984). Morphology, function and isolation of halobacterial flagella. *J. Mol. Biol.* **176**, 459–475.

17. Alexandre, G., Greer, S.E. and Zhulin, I.B. (2000). Energy taxis is the dominant behavior in *Azospirillum brasilense. J. Bacteriol.* **182**, 6042–6048.

18. Alon, U., Camarena, L., Surette, M.G., Arcas, B.A.Y., Liu, Y., Leibler, S. and Stock, J.B. (1998). Response regulator output in bacterial chemotaxis. *EMBO J.* **17**, 4238–4248.

19. Alon, U., Surette, M.G., Barkai, N. and Leibler, S. (1999). Robustness in bacterial chemotaxis. *Nature* **397**, 168–171.

20. Ames, P. and Parkinson, J.S. (1988). Transmembrane signaling by bacterial chemoreceptors: *E. coli* transducers with locked signal output. *Cell* **55**, 817–826.

21. Ames, P. and Parkinson, J.S. (1994). Constitutively signaling fragments of Tsr, the *Escherichia coli* serine chemoreceptor. *J. Bacteriol.* **176**, 6340–6348.

22. Ames, P., Studdert, C.A., Reiser, R.H. and Parkinson, J.S. (2002). Collaborative signaling by mixed chemoreceptor teams in *Escherichia coli. Proc. Natl. Acad. Sci. U.S.A.* **99**, 7060–7065.

23. Amsler, C.D. (1996). Use of computer-assisted motion analysis for quantitative measurements of swimming behavior in peritrichously flagellated bacteria. *Analyt. Biochem.* **235**, 20–25.

24. Anand, G.S., Goudreau, P.N. and Stock, A.M. (1998). Activation of methylesterase CheB: Evidence of a dual role for the regulatory domain. *Biochemistry* **37**, 14038–14047.

25. Anand, G.S., Goudreau, P.N., Lewis, J.K. and Stock, A.M. (2000). Evidence for phosphorylation-dependent conformational changes in methylesterase CheB. *Protein Sci.* **9**, 898–906.

26. Anand, G.S. and Stock, A.M. (2002). Kinetic basis for the stimulatory effect of phosphorylation on the methylesterase activity of CheB. *Biochemistry* **41**, 6752–6760.

27. Anderson, R.A. (1975). Formation of the bacterial flagellar bundle, in *Swimming and Flying in Nature* (T.Y.T. Wu, C.J. Brokaw and C. Brennen, eds.), pp. 45–56. Plenum, New York.

28. Appleby, J.L. and Bourret, R.B. (1999). Activation of CheY mutant D57N by phosphorylation at an alternative site, Ser-56. *Mol. Microbiol.* **34**, 915–925.

29. Arai, T. (1981). Effect of arsenate on chemotactic behavior of *Escherichia coli.* *J. Bacteriol.* **145**, 803–807.

30. Armitage, J.P., Josey, D.P. and Smith, D.G. (1977). A simple, quantitative method for measuring chemotaxis and motility in bacteria. *J. Gen. Microbiol.* **102**, 199–202.

31. Armitage, J.P., Keighley, P. and Evans, M.C.W. (1979). Chemotaxis in photosynthetic bacteria. *FEMS Microbiol. Lett.* **6**, 99–102.

32. Armitage, J.P. and Macnab, R.M. (1987). Unidirectional, intermittent rotation of the flagellum of *Rhodobacter sphaeroides.* *J. Bacteriol.* **169**, 514–518.

33. Armitage, J.P. and Schmitt, R. (1997). Bacterial chemotaxis: *Rhodobacter sphaeroides* and *Sinorhizobium meliloti*—variations on a theme? *Microbiology* **143**, 3671–3682.

34. Armitage, J.P., Pitta, T.P., Vigeant, M.A.-S., Packer, H. and Ford, R.M. (1999). Transformations in flagellar structure of *Rhodobacter sphaeroides* and possible relationship to changes in swimming speed. *J. Bacteriol.* **181**, 4825–4833.

35. Armstrong, J.B., Adler, J. and Dahl, M.M. (1967). Nonchemotactic mutants of *Escherichia coli.* *J. Bacteriol.* **93**, 390–398.

36. Armstrong, J.B. (1972). Chemotaxis and methionine metabolism in *Escherichia coli.* *Can. J. Microbiol.* **18**, 591–596.

37. Armstrong, J.B. (1972). An S-adenosylmethionine requirement for chemotaxis in *Escherichia coli.* *Can. J. Microbiol.* **18**, 1695–1701.

38. Artymiuk, P.J., Rice, D.W., Mitchell, E.M. and Willett, P. (1990). Structural resemblance between the families of bacterial signal-transduction proteins and of G proteins revealed by graph theoretical techniques. *Prot. Eng.* **4**, 39–43.

39. Asai, Y., Kojima, S., Kato, H., Nishioka, N., Kawagishi, I. and Homma, M. (1997). Putative channel components for the fast-rotating sodium-driven flagellar motor of a marine bacterium. *J. Bacteriol.* **179**, 5104–5110.

40. Asai, Y., Kawagishi, I., Sockett, R.E. and Homma, M. (1999). Hybrid motor with H^+- and Na^+-driven components can rotate *Vibrio* polar flagella by using sodium ions. *J. Bacteriol.* **181**, 6332–6338.

41. Asai, Y., Kawagishi, I., Sockett, R.E. and Homma, M. (2000). Coupling ion specificity of chimeras between H^+- and Na^+-driven motor proteins, MotB and PomB, in *Vibrio* polar flagella. *EMBO J.* **19**, 3639–3648.

42. Asakura, S. (1970). Polymerization of flagellin and polymorphism of flagella. *Adv. Biophys.* **1**, 99–155.

43. Askamit, R.R., Howlett, B.J. and Koshland, D.E., Jr. (1975). Soluble and membrane-bound aspartate-binding activities in *Salmonella typhimurium.* *J. Bacteriol.* **123**, 1000–1005.

44. Aswad, D. and Koshland, D.E., Jr. (1974). Role of methionine in bacterial chemotaxis. *J. Bacteriol.* **118**, 640–645.

45. Aswad, D. and Koshland, D.E., Jr. (1975). Isolation, characterization and complementation of *Salmonella typhimurium* chemotaxis mutants. *J. Mol. Biol.* **97**, 225–235.

46. Aswad, D.W. and Koshland, D.E., Jr. (1975). Evidence for an S-adenosylmethionine requirement in the chemotactic behavior of *Salmonella typhimurium.* *J. Mol. Biol.* **97**, 207–223.

47. Atsumi, T., McCarter, L. and Imae, Y. (1992). Polar and lateral flagellar motors of marine *Vibrio* are driven by different ion-motive forces. *Nature* **355**, 182–184.

48. Atsumi, T., Maekawa, Y., Yamada, T., Kawagishi, I., Imae, Y. and Homma, M. (1996). Effect of viscosity on swimming by the lateral and polar flagella of *Vibrio alginolyticus*. *J. Bacteriol.* **178**, 5024–5026.

49. Atsumi, T. (2001). An ultrasonic motor model for bacterial flagellar motors. *J. Theor. Biol.* **213**, 31–51.

50. Auvray, F., Thomas, J., Fraser, G.M. and Hughes, C. (2001). Flagellin polymerisation control by a cytosolic export chaperone. *J. Mol. Biol.* **308**, 221–229.

51. Baba, S.A., Inomata, S., Ooya, M., Mogami, Y. and Izumi-Kurotani, A. (1991). Three-dimensional recording and measurement of swimming paths of micro-organisms with two synchronized monochrome cameras. *Rev. Sci. Instrum.* **62**, 540–541.

52. Baracchini, O. and Sherris, J.C. (1959). The chemotactic effect of oxygen on bacteria. *J. Path. Bact.* **77**, 565–574.

53. Barak, R. and Eisenbach, M. (1992). Fumarate or a fumarate metabolite restores switching ability to rotating flagella of bacterial envelopes. *J. Bacteriol.* **174**, 643–645.

54. Barak, R. and Eisenbach, M. (1992). Correlation between phosphorylation of the chemotaxis protein CheY and its activity at the flagellar motor. *Biochemistry* **31**, 1821–1826.

55. Barak, R., Welch, M., Yanovsky, A., Oosawa, K. and Eisenbach, M. (1992). Acetyladenylate or its derivative acetylates the chemotaxis protein CheY *in vitro* and increases its activity at the flagellar switch. *Biochemistry* **31**, 10099–10107.

56. Barak, R. and Eisenbach, M. (1996). Regulation of interaction between signaling protein CheY and flagellar motor during bacterial chemotaxis. *Curr. Top. Cell. Reg.* **34**, 137–158.

57. Barak, R., Giebel, I. and Eisenbach, M. (1996). The specificity of fumarate as a switching factor of the bacterial flagellar motor. *Mol. Microbiol.* **19**, 139–144.

58. Barak, R., Abouhamad, W.N. and Eisenbach, M. (1998). Both acetate kinase and acetyl Coenzyme A synthetase are involved in acetate-stimulated change in the direction of flagellar rotation in *Escherichia coli*. *J. Bacteriol.* **180**, 985–988.

59. Barak, R. and Eisenbach, M. (1999). Chemotactic-like response of *Escherichia coli* cells lacking the known chemotaxis machinery but containing overexpressed CheY. *Mol. Microbiol.* **31**, 1125–1137.

60. Barak, R. and Eisenbach, M. (2001). Acetylation of the response regulator, CheY, is involved in bacterial chemotaxis. *Mol. Microbiol.* **40**, 731–743.

61. Barak, R., Prasad, K., Shainskaya, A., Wolfe, A.J. and Eisenbach, M. (2003). Acetylation of the chemotaxis response regulator CheY by acetyl-CoA synthetase from *Escherichia coli*. Submitted.

62. Barkai, N. and Leibler, S. (1997). Robustness in simple biochemical networks. *Nature* **387**, 913–917.

63. Barkai, N., Alon, R. and Leibler, S. (2001). Robust amplification in adaptive signal transduction networks. *C. R. Acad. Sci. Paris Série IV* **2**, 871–877.

64. Barnakov, A.N., Barnakova, L.A. and Hazelbauer, G.L. (1998). Comparison in vitro of a high- and a low-abundance chemoreceptor of *Escherichia coli*: Similar kinase activation but different methyl-accepting activities. *J. Bacteriol.* **180**, 6713–6718.

65. Barnakov, A.N., Barnakova, L.A. and Hazelbauer, G.L. (1999). Efficient adaptational demethylation of chemoreceptors requires the same enzyme-docking site as efficient methylation. *Proc. Natl. Acad. Sci. U.S.A.* **96**, 10667–10672.

66. Barreau, C. and van der Wel, H. (1983). The effect of thaumatins on the chemotactic behavior of *Escherichia coli. Chem. Senses* **8**, 71–79.

67. Bass, R.B. and Falke, J.J. (1998). Detection of a conserved α-helix in the kinase-docking region of the aspartate receptor by cysteine and disulfide scanning. *J. Biol. Chem.* **273**, 25006–25014.

68. Bass, R.B. and Falke, J.J. (1999). The aspartate receptor cytoplasmic domain: *in situ* chemical analysis of structure, mechanism and dynamics. *Structure* **7**, 829–840.

69. Bassler, B.L., Gibbons, P.J., Yu, C. and Roseman, S. (1991). Chitin utilization by marine bacteria. Chemotaxis to chitin oligosaccharides by *Vibrio furnissii. J. Biol. Chem.* **266**, 24268–24275.

70. Bayley, D.P. and Jarrell, K.F. (1998). Further evidence to suggest that archaeal flagella are related tab bacterial type IV pili. *J. Mol. Evol.* **46**, 370–373.

71. Bennett, J.C.Q., Thomas, J., Fraser, G.M. and Hughes, C. (2001). Substrate complexes and domain organization of the *Salmonella* flagellar export chaperones FlgN and FliT. *Mol. Microbiol.* **39**, 781–791.

72. Benov, L. and Fridovich, I. (1996). *Escherichia coli* exhibits negative chemotaxis in gradients of hydrogen peroxide, hypochlorite, and *N*-chlorotaurine: Products of the respiratory burst of phagocytic cells. *Proc. Natl. Acad. Sci. U.S.A.* **93**, 4999–5002.

73. Berg, H.C. (1971). How to track bacteria. *Rev. Sci. Instrum.* **42**, 868–871.

74. Berg, H.C. and Brown, D.A. (1972). Chemotaxis in *Escherichia coli* analysed by three-dimensional tracking. *Nature* **239**, 500–504.

75. Berg, H.C. and Anderson, R.A. (1973). Bacteria swim by rotating their flagellar filaments. *Nature* **245**, 380–382.

76. Berg, H.C. (1974). Dynamic properties of bacterial flagellar motors. *Nature* **249**, 77–79.

77. Berg, H.C. (1975). How bacteria swim. *Sci. Am.* **233** (8), 36–44.

78. Berg, H.C. (1975). Bacterial behaviour. *Nature* **254**, 389–392.

79. Berg, H.C. (1975). Bacterial movement, in *Swimming and Flying in Nature* (T.Y.-T. Wu, C.J. Brokaw and C. Brennen, eds.), pp. 1–11. Plenum Press, New York and London.

80. Berg, H.C. and Tedesco, P.M. (1975). Transient response to chemotactic stimuli in *Escherichia coli. Proc. Natl. Acad. Sci. U.S.A.* **72**, 3235–3239.

81. Berg, H.C. and Purcell, E.M. (1977). Physics of chemoreception. *Biophys. J.* **20**, 193–219.

82. Berg, H.C. (1978). The tracking microscope. *Adv. Opt. Electron Microsc.* **7**, 1–15.

83. Berg, H.C., Manson, M.D. and Conley, M.P. (1982). Dynamics and energetics of flagellar rotation in bacteria. *Soc. Exp. Biol. Symp.* **35**, 1–31.

84. Berg, H.C. and Block, S.M. (1984). A miniature flow cell designed for rapid exchange of media under high-power microscope objectives. *J. Gen. Microbiol.* **130**, 2915–2920.

85. Berg, H.C. and Turner, L. (1990). Chemotaxis of bacteria in glass capillary arrays. *Biophys. J.* **58**, 919–930.

86. Berg, H.C. and Turner, L. (1991). Selection of motile nonchemotactic mutants of *Escherichia coli* by field-flow fractionation. *Proc. Natl. Acad. Sci. U.S.A.* **88**, 8145–8148.

87. Berg, H.C. and Turner, L. (1993). Torque generated by the flagellar motor of *Escherichia coli*. *Biophys. J.* **65**, 2201–2216.

88. Berg, H.C. and Turner, L. (1995). Cells of *Escherichia coli* swim either end forward. *Proc. Natl. Acad. Sci. U.S.A.* **92**, 477–479.

89. Berg, P. (1956). Acyl adenylates: an enzymatic mechanism of acetate activation. *J. Biol. Chem.* **222**, 991–1013.

90. Berlyn, M.K.B. (1998). Linkage map of *Escherichia coli* K-12, edition 10: the traditional map. *Microbiol. Mol. Biol. Rev.* **62**, 814–984.

91. Bernstein, F.C., Koetzle, T.F., Williams, G.J., Meyer, E.E., Jr., Brice, M.D., Rodgers, J.R., Kennard, O., Shimanouchi, T. and Tasumi, M. (1977). The Protein Data Bank: a computer-based archival file for macromolecular structures. *J. Mol. Biol.* **112**, 535–542.

92. Berry, R.M. (1993). Torque and switching in the bacterial flagellar motor. An electrostatic model. *Biophys. J.* **64**, 961–973.

93. Berry, R.M., Turner, L. and Berg, H.C. (1995). Mechanical limits of bacterial flagellar motors probed by electrorotation. *Biophys. J.* **69**, 280–286.

94. Berry, R.M. and Berg, H.C. (1997). Absence of a barrier to backwards rotation of the bacterial flagellar motor demonstrated with optical tweezers. *Proc. Natl. Acad. Sci. U.S.A.* **94**, 14433–14437.

95. Berry, R.M. and Berg, H.C. (1999). Torque generated by the flagellar motor of *Escherichia coli* while driven backward. *Biophys. J.* **76**, 580–587.

96. Bibikov, S.I., Biran, R., Rudd, K.E. and Parkinson, J.S. (1997). A signal transducer for aerotaxis in *Escherichia coli*. *J. Bacteriol.* **179**, 4075–4079.

97. Bibikov, S.I., Barnes, L.A., Gitin, Y. and Parkinson, J.S. (2000). Domain organization and flavin adenine dinucleotide-binding determinants in the aerotaxis signal transducer Aer of *Escherichia coli*. *Proc. Natl. Acad. Sci. U.S.A.* **97**, 5830–5835.

98. Biemann, H.P. and Koshland, D.E., Jr. (1994). Aspartate receptors of *Escherichia coli* and *Salmonella typhimurium* bind ligand with negative and half-of-the-sites cooperativity. *Biochemistry* **33**, 629–634.

99. Bilwes, A.M., Alex, L.A., Crane, B.R. and Simon, M.I. (1999). Structure of CheA, a signal-transducing histidine kinase. *Cell* **96**, 131–141.

100. Bischoff, D.S. and Ordal, G.W. (1991). Sequence and characterization of *Bacillus subtilis* CheB, a homolog of *Escherichia coli* CheY, and its role in a different mechanism of chemotaxis. *J. Biol. Chem.* **266**, 12301–12305.

101. Bischoff, D.S., Bourret, R.B., Kirsch, M.L. and Ordal, G.W. (1993). Purification and characterization of *Bacillus subtilis* CheY. *Biochemistry* **32**, 9256–9261.

102. Blair, D.F. and Berg, H.C. (1988). Restoration of torque in defective flagellar motors. *Science* **242**, 1678–1681.

103. Blair, D.F. and Berg, H.C. (1990). The MotA protein of *Escherichia coli* is a proton-conducting component of the flagellar motor. *Cell* **60**, 439–449.

104. Blair, D.F. and Berg, H.C. (1991). Mutations in the MotA protein of Escherichia coli reveal domains critical for proton conduction. *J. Mol. Biol.* **221**, 1433–1442.

105. Blair, D.F. (1995). How bacteria sense and swim. *Annu. Rev. Microbiol.* **49**, 489–522.

106. Blat, Y. and Eisenbach, M. (1994). Phosphorylation-dependent binding of the chemotaxis signal molecule CheY to its phosphatase, CheZ. *Biochemistry* **33**, 902–906.

107. Blat, Y. and Eisenbach, M. (1996). Oligomerization of the phosphatase CheZ upon interaction with the phosphorylated form of CheY, the signal protein of bacterial chemotaxis. *J. Biol. Chem.* **271**, 1226–1231.

108. Blat, Y. and Eisenbach, M. (1996). Mutants with defective phosphatase activity show no phosphorylation-dependent oligomerization of CheZ, the phosphatase of bacterial chemotaxis. *J. Biol. Chem.* **271**, 1232–1236.

109. Blat, Y. and Eisenbach, M. (1996). Conserved C-terminus of the phosphatase CheZ is a binding domain for the chemotactic response regulator CheY. *Biochemistry* **35**, 5679–5683.

110. Blat, Y., Gillespie, B., Bren, A., Dahlquist, F.W. and Eisenbach, M. (1998). Regulation of phosphatase activity in bacterial chemotaxis. *J. Mol. Biol.* **284**, 1191–1199.

111. Blattner, F.R., Plunkett, G., Bloch, C.A., Perna, N.T., Burland, V., Riley, M., Collado-Vides, J., Glasner, J.D., Rode, C.K., Mayhew, G.F., Gregor, J., Davis, N.W., Kirkpatrick, H.A., Goeden, M.A., Rose, D.J., Mau, B. and Shao, Y. (1997). The complete genome sequence of *Escherichia coli* K-12. *Science* **277**, 1453–1462.

112. Block, S.M., Segall, J.E. and Berg, H.C. (1982). Impulse responses in bacterial chemotaxis. *Cell* **31**, 215–226.

113. Block, S.M. and Berg, H.C. (1984). Successive incorporation of force-generating units in the bacterial rotary motor. *Nature (London)* **309**, 470–472.

114. Block, S.M. (1990). Optical tweezers: a new tool for biophysics, in *Noninvasive Techniques in Cell Biology*, pp. 375–402. Wiley-Liss.

115. Block, S.M., Blair, D.F. and Berg, H.C. (1991). Compliance of bacterial polyhooks measured with optical tweezers. *Cytometry* **12**, 492–496.

116. Block, S.M., Fahrner, K.A. and Berg, H.C. (1991). Visualization of bacterial flagella by video-enhanced light microscopy. *J. Bacteriol.* **173**, 933–936.

117. Boesch, K.C., Silversmith, R.E. and Bourret, R.B. (2000). Isolation and characterization of nonchemotactic CheZ mutants of *Escherichia coli*. *J. Bacteriol.* **182**, 3544–3552.

118. Boles, B.R. and McCarter, L.L. (2000). Insertional inactivation of genes encoding components of the sodium-type flagellar motor and switch of *Vibrio parahaemolyticus*. *J. Bacteriol.* **182**, 1035–1045.

119. Bollinger, J., Park, C., Harayama, S. and Hazelbauer, G.L. (1984). Structure of the Trg protein: Homologies with and differences from other sensory transducers of *Escherichia coli*. *Proc. Natl. Acad. Sci. U.S.A.* **81**, 3287–3291.

120. Boos, W. and Staehelin, A.L. (1981). Ultrastructure localization of the maltose-binding protein within the cell envelope of *Escherichia coli*. *Arch. Microbiol.* **129**, 240–246.

121. Boos, W. and Lucht, J.M. (1996). Periplasmic binding protein-dependent ABC transporters, in *Escherichia coli and Salmonella: Cellular and Molecular Biology* (F.C. Neidhardt, R. Curtis III, J.L. Ingraham, E.C.C. Lin, K.B. Low, B. Magasanik, W.S. Reznikoff, M. Riley, M. Schaechter and H.E. Umbarger, eds.), pp. 1175–1209. American Society for Microbiology, Washington, DC.

122. Borkovich, K.A., Kaplan, N., Hess, J.F. and Simon, M.I. (1989). Transmembrane signal transduction in bacterial chemotaxis involves ligand-dependent activation of phosphate group transfer. *Proc. Natl. Acad. Sci. U.S.A.* **86**, 1208–1212.

123. Borkovich, K.A. and Simon, M.I. (1990). The dynamics of protein phosphorylation in bacterial chemotaxis. *Cell* **63**, 1339–1348.

124. Borkovich, K.A., Alex, L.A. and Simon, M.I. (1992). Attenuation of sensory receptor signaling by covalent modification. *Proc. Natl. Acad. Sci. U.S.A.* **89**, 6756–6760.

125. Bornhorst, J.A. and Falke, J.J. (2000). Attractant regulation of the aspartate receptor-kinase complex: Limited cooperative interactions between receptors and effects of the receptor modification state. *Biochemistry* **39**, 9486–9493.

126. Bourret, R.B., Hess, J.F. and Simon, M.I. (1990). Conserved aspartate residues and phosphorylation in signal transduction by the chemotaxis protein CheY. *Proc. Natl. Acad. Sci. U.S.A.* **87**, 41–45.

127. Bourret, R.B., Davagnino, J. and Simon, M.I. (1993). The carboxy-terminal portion of the CheA kinase mediates regulation of autophosphorylation by transducer and CheW. *J. Bacteriol.* **175**, 2097–2101.

128. Bourret, R.B., Drake, S.K., Chervitz, S.A., Simon, M.I. and Falke, J.J. (1993). Activation of the phosphosignaling protein CheY II. Analysis of activated mutants by ^{19}F NMR and protein engineering. *J. Biol. Chem.* **268**, 13 089–13 096.

129. Brahamsha, B. and Greenberg, E.P. (1988). A biochemical and cytological analysis of the complex periplasmic flagella from *Spirochaeta aurantia*. *J. Bacteriol.* **170**, 4023–4032.

130. Braun, T.F., Poulson, S., Gully, J.B., Empey, J.C., Van Way, S., Putnam, A. and Blair, D.F. (1999). Function of proline residues of MotA in torque generation by the flagellar motor of *Escherichia coli*. *J. Bacteriol.* **181**, 3542–3551.

131. Bray, D., Bourret, R.B. and Simon, M.I. (1993). Computer simulation of the phosphorylation cascade controlling bacterial chemotaxis. *Mol. Biol. Cell* **4**, 469–482.

132. Bray, D. (1998). Signaling complexes: biophysical constraints on intracellular communication. *Annu. Rev. Biophys. Biomol. Str.* **27**, 59–75.

133. Bray, D., Levin, M.D. and Morton-Firth, C.J. (1998). Receptor clustering as a cellular mechanism to control sensitivity. *Nature* **393**, 85–88.

134. Bren, A., Welch, M., Blat, Y. and Eisenbach, M. (1996). Signal termination in bacterial chemotaxis: CheZ mediates dephosphorylation of free rather than switch-bound CheY. *Proc. Natl. Acad. Sci. U.S.A.* **93**, 10 090–10 093.

135. Bren, A. and Eisenbach, M. (1998). The N terminus of the flagellar switch protein, FliM, is the binding domain for the chemotactic response regulator, CheY. *J. Mol. Biol.* **278**, 507–514.

136. Bren, A. and Eisenbach, M. (1999). The phosphate group on CheY appears to be essential for binding to CheZ but not to FliM. *Bacterial Locomotion and Signal Transduction.* Abst.

137. Bren, A. and Eisenbach, M. (2000). How signals are heard during bacterial chemotaxis: Protein–protein interactions in sensory signal propagation. *J. Bacteriol.* **182**, 6865–6873.

138. Bren, A. and Eisenbach, M. (2001). Changing the direction of flagellar rotation in bacteria by modulating the ratio between the rotational states of the switch protein FliM. *J. Mol. Biol.* **312**, 699–709.

139. Brooun, A., Zhang, W.S. and Alam, M. (1997). Primary structure and functional analysis of the soluble transducer protein HtrXI in the archaeon *Halobacterium salinarium*. *J. Bacteriol.* **179**, 2963–2968.

140. Brooun, A., Bell, J., Freitas, T., Larsen, R.W. and Alam, M. (1998). An archaeal aerotaxis transducer combines subunit I core structures of eukaryotic cytochrome *c* oxidase and eubacterial methyl-accepting chemotaxis proteins. *J. Bacteriol.* **180**, 1642–1646.

141. Brown, II and Hase, C.C. (2001). Flagellum-independent surface migration of *Vibrio cholerae* and *Escherichia coli*. *J. Bacteriol.* **183**, 3784–3790.

142. Brown, P.N., Hill, C.P. and Blair, D.F. (2002). Crystal structure of the middle and C-terminal domains of the flagellar rotor protein FliG. *EMBO J.* **21**, 3225–3234.

143. Brown, T.D.K., Jones-Mortimer, M.C. and Kornberg, H.L. (1977). The enzymatic interconversion of acetate and acetyl-coenzyme A in *Escherichia coli*. *J. Gen. Microbiol.* **102**, 327–336.

144. Butler, S.L. and Falke, J.J. (1998). Cysteine and disulfide scanning reveals two amphiphilic helices in the linker region of the aspartate chemoreceptor. *Biochemistry* **37**, 10746–10756.

145. Cameron, L.A., Giardini, P.A., Soo, F.S. and Theriot, J.A. (2000). Secrets of actin-based motility revealed by a bacterial pathogen. *Nature Rev. Mol. Cell Biol.* **1**, 110–119.

146. Campos, A. and Matsumura, P. (2001). Extensive alanine scanning reveals protein-protein and protein-DNA interaction surfaces in the global regulator FlhD from *Escherichia coli*. *Mol. Microbiol.* **39**, 581–594.

147. Cantwell, B.J., Draheim, R.R., Weart, R.B., Nguyen, C., Stewart, R.C. and Manson, M.D. (2003). CheZ phosphatase localizes to chemoreceptor patches via CheA-short. *J. Bacteriol.* **185**, 2354–2361.

148. Caplan, S.R. and Kara-Ivanov, M. (1993). The bacterial flagellar motor. *Internat'l. Rev. Cytol.* **147**, 97–164.

149. Cercignani, G., Lucia, S. and Petracchi, D. (1998). Photoresponses of *Halobacterium salinarum* to repetitive pulse stimuli. *Biophys. J.* **75**, 1466–1472.

150. Chang, B.Y. and Dworkin, M. (1994). Isolated fibrils rescue cohesion and development in the Dsp mutant of *Myxococcus xanthus*. *J. Bacteriol.* **176**, 7190–7196.

151. Charon, N.W. and Goldstein, S.F. (2002). Genetics of motility and chemotaxis of a fascinating group of bacteria: the spirochetes. *Annu. Rev. Genet.* **36**, 47–73.

152. Chen, J.M., Lee, G., Murphy, R.B., Brandt-Rauf, P. and Pincus, M.R. (1990). Comparisons between the three-dimensional structures of the chemotactic protein CheY and the normal Gly 12-p21 protein. *Int. J. Peptide Protein Res.* **36**, 1–6.

153. Chen, X. and Koshland, D.E., Jr. (1997). Probing the structure of the cytoplasmic domain of the aspartate receptor by targeted disulfide cross-linking. *Biochemistry* **36**, 11858–11864.

154. Chen, X. and Berg, H.C. (2000). Torque-speed relationship of the flagellar rotary motor of *Escherichia coli*. *Biophys. J.* **78**, 1036–1041.

155. Chervitz, S.A., Lin, C.M. and Falke, J.J. (1995). Transmembrane signaling by the aspartate receptor: engineered disulfides reveal static regions of the subunit interface. *Biochemistry* **34**, 9722–9733.

156. Chervitz, S.A. and Falke, J.J. (1996). Molecular mechanism of transmembrane signaling by the aspartate receptor: A model. *Proc. Natl. Acad. Sci. U.S.A.* **93**, 2545–2550.

157. Cheung, P., Tanner, K.G., Cheung, W.L., Sassone-Corsi, P., Denu, J.M. and Allis, C.D. (2000). Synergistic coupling of histone H3 phosphorylation and acetylation in response to epidermal growth factor stimulation. *Mol. Cell.* **5**, 905–915.

158. Chi, Y.I., Yokota, H. and Kim, S.H. (1997). Apo structure of the ligand-binding domain of aspartate receptor from *Escherichia coli* and its comparison with ligand-bound or pseudoligand-bound structures. *FEBS Lett.* **414**, 327–332.

159. Chiarugi, V., Cinelli, M. and Magnelli, L. (1998). Acetylation and phosphorylation of the carboxy-terminal domain of p53: regulative significance. *Oncol. Res.* **10**, 55–57.

160. Chilcott, G.S. and Hughes, K.T. (2000). Coupling of flagellar gene expression to flagellar assembly in *Salmonella enterica* serovar typhimurium and *Escherichia coli*. *Microbiol. Mol. Biol. Rev.* **64**, 694–708.

161. Cho, H.S., Lee, S.Y., Yan, D., Pan, X., Parkinson, J.S., Kustu, S., Wemmer, D.E. and Pelton, J.G. (2000). NMR Structure of Activated CheY. *J .Mol. Biol.* **297**, 543–551.

162. Chun, S.Y. and Parkinson, J.S. (1988). Bacterial motility: membrane topology of the *Escherichia coli* MotB protein. *Science* **239**, 276–278.

163. Clancy, M., Madill, K.A. and Wood, J.M. (1981). Genetic and biochemical requirements for chemotaxis to L-proline in *Escherichia coli*. *J. Bacteriol.* **146**, 902–906.

164. Clarke, S. and Koshland, D.E., Jr. (1979). Membrane receptors for aspartate and serine in bacterial chemotaxis. *J. Biol. Chem.* **254**, 9695–9702.

165. Clarke, S., Sparrow, K., Panasenko, S. and Koshland, D.E., Jr. (1980). *In vitro* methylation of bacterial chemotaxis proteins: Characterization of protein methyltransferase activity in crude extracts of *Salmonella typhimurium*. *J. Supramol. Struct.* **13**, 315–328.

166. Clayton, A.L., Rose, S., Barratt, M.J. and Mahadevan, L.C. (2000). Phospho-acetylation of histone H3 on c-fos- and c-jun-associated nucleosomes upon gene activation. *Embo. J.* **19**, 3714–3726.

167. Clayton, R.K. (1958). On the interplay of environmental factors affecting taxis and motility in *Rhodospirillum rubrum*. *Arch. Mikrobiol.* **29**, 189–212.

168. Clegg, D.O. and Koshland, D.E., Jr. (1984). The role of a signaling protein in bacterial sensing: Behavioral effects of increased gene expression. *Proc. Natl. Acad. Sci. U.S.A.* **81**, 5056–5060.

169. Cluzel, P., Surette, M. and Leibler, S. (2000). An ultrasensitive bacterial motor revealed by monitoring signaling proteins in single cells. *Science* **287**, 1652–1655.

170. Craven, R. and Montie, T.C. (1985). Regulation of *Pseudomonas aeruginosa* chemotaxis by the nitrogen source. *J. Bacteriol.* **164**, 544–549.

171. Craven, R.C. and Montie, T.C. (1981). Motility and chemotaxis of three strains of *Pseudomonas aeruginosa* used for virulence studies. *Can. J. Microbiol.* **27**, 458–460.

172. Da Re, S., Tolstykh, T., Wolanin, P.M. and Stock, J.B. (2002). Genetic analysis of response regulator activation in bacterial chemotaxis suggests an intermolecular mechanism. *Protein Sci.* **11**, 2644–2654.

173. Da Re, S.S., Deville-Bonne, D., Tolstykh, T., Véron, M. and Stock, J.B. (1999). Kinetics of CheY phosphorylation by small molecule phosphodonors. *FEBS Lett.* **457**, 323–326.

174. Dahl, M.K. and Manson, M.D. (1985). Interspecific reconstitution of maltose transport and chemotaxis in *Escherichia coli* with maltose-binding protein from various enteric bacteria. *J. Bacteriol.* **164**, 1057–1063.

175. Dahlquist, F.W., Lovely, P. and Koshland, D.E., Jr. (1972). Quantitative analysis of bacterial migration in chemotaxis. *Nature New Biol.* **236**, 120–123.

176. Danielson, M.A., Biemann, H.-P., Koshland, D.E., Jr. and Falke, J.J. (1994). Attractant- and disulfide-induced conformational changes in the ligand binding domain of the chemotaxis aspartate receptor: a ^{19}F NMR study. *Biochemistry* **33**, 6100–6109.

177. Danielson, M.A., Bass, R.B. and Falke, J.J. (1997). Cysteine and disulfide scanning reveals a regulatory α-helix in the cytoplasmic domain of the aspartate receptor. *J. Biol. Chem.* **272**, 32878–32888.

178. de Jonge, R., de Mattos, M.J.T., Stock, J.B. and Neijssel, O.M. (1996). Pyrroloquinoline quinone, a chemotactic attractant for *Escherichia coli.* *J. Bacteriol.* **178**, 1224–1226.

179. De Pina, K., Navarro, C., Mcwalter, L., Boxer, D.H., Price, N.C., Kelly, S.M., Mandrand-Berthelot, M.-A. and Wu, L.-F. (1995). Purification and characterization of the periplasmic nickel-binding protein NikA of *Escherichia coli* k12. *Eur. J. Biochem.* **227**, 857–865.

180. Dean, G.E., Aizawa, S.-I. and Macnab, R.M. (1983). *flaAII* (*motC, cheV*) of *Salmonella typhimurium* is a structural gene involved in energization and switching of the flagellar motor. *J. Bacteriol.* **154**, 84–91.

181. Dean, G.E., Macnab, R.M., Stader, J., Matsumura, P. and Burks, C. (1984). Gene sequence and predicted amino acid sequence of the MotA protein, a membrane-associated protein required for flagellar rotation in *Escherichia coli.* *J. Bacteriol.* **159**, 991–999.

182. DePamphilis, M.L. and Adler, J. (1971). Attachment of flagellar basal bodies to the cell envelope: Specific attachment to the outer, lipopolysaccharide membrane and the cytoplasmic membrane. *J. Bacteriol.* **105**, 396–407.

183. DePamphilis, M.L. and Adler, J. (1971). Fine structure and isolation of the hook-basal body complex of flagella from *Escherichia coli* and *Bacillus subtilis.* *J. Bacteriol.* **105**, 384–395.

184. DeRosier, D.J. (1995). Spinning tails. *Curr. Opin. Struct. Biol.* **5**, 187–193.

185. DeRosier, D.J. (1998). The turn of the screw: The bacterial flagellar motor. *Cell* **93**, 17–20.

186. Dharmatilake, A.J. and Bauer, W.D. (1992). Chemotaxis of *Rhizobium meliloti* towards nodulation gene-inducing compounds from alfalfa roots. *Appl. Environ. Microbiol.* **58**, 1153–1158.

187. Dietzel, I., Kolb, V. and Boos, W. (1978). Pole cap formation in *Escherichia coli* following induction of the maltose-binding protein. *Arch. Microbiol.* **118**, 207–218.

188. Dimmit, K. and Simon, M. (1972). Purification and thermal stability of intact *Bacillus subtilis* flagella. *J. Bacteriol.* **105**, 369–375.

189. Djordjevic, S. and Stock, A.M. (1997). Crystal structure of the chemotaxis receptor methyltransferase CheR suggests a conserved structural motif for binding S-adenosylmethionine. *Structure* **5**, 545–558.

190. Djordjevic, S., Goudreau, P.N., Xu, Q.P., Stock, A.M. and West, A.H. (1998). Structural basis for methylesterase CheB regulation by a phosphorylation-activated domain. *Proc. Natl. Acad. Sci. U.S.A.* **95**, 1381–1386.

191. Djordjevic, S. and Stock, A.M. (1998). Chemotaxis receptor recognition by protein methyltransferase CheR. *Nature Struct. Biol.* **5**, 446–450.

192. Djordjevic, S. and Stock, A.M. (1998). Structural analysis of bacterial chemotaxis proteins: Components of a dynamic signaling system. *J. Struct. Biol.* **124**, 189–200.

193. Doetsch, R.N. and Seymour, W.F. (1970). Negative chemotaxis in bacteria. *Life Sci.* **9**, 1029–1037.

194. Doetsch, R.N. (1972). A unified theory of bacterial motile behavior. *J. Theor. Biol.* **35**, 55–66.

195. Donato, G.M. and Kawula, T.H. (1998). Enhanced binding of altered H-NS protein to flagellar rotor protein FliG causes increased flagellar rotational speed and hypermotility in *Escherichia coli*. *J. Biol. Chem.* **273**, 24030–24036.

196. Dowd, J.P. and Matsumura, P. (1997). The use of flash photolysis for a high-resolution temporal and spatial analysis of bacterial chemotactic behaviour: CheZ is not always necessary for chemotaxis. *Mol. Microbiol.* **25**, 295–302.

197. Dreyfus, G., Williams, A.W., Kawagishi, I. and Macnab, R.M. (1993). Genetic and biochemical analysis of *Salmonella typhimurium* FliI, a flagellar protein related to the catalytic subunit of the F_0F_1 ATPase and to virulence proteins of mammalian and plant pathogens. *J. Bacteriol.* **175**, 3131–3138.

198. Duke, T.A.J. and Bray, D. (1999). Heightened sensitivity of a lattice of membrane receptors. *Proc. Natl. Acad. Sci. U.S.A.* **96**, 10104–10108.

199. Dusenbery, D.B. (1998). Spatial sensing of stimulus gradients can be superior to temporal sensing for free-swimming bacteria. *Biophys. J.* **74**, 2272–2277.

200. Dworkin, M. (1996). Recent advances in the social and developmental biology of the myxobacteria. *Microbiol. Rev.* **60**, 70–102.

201. Dworkin, M. (1999). Fibrils as extracellular appendages of bacteria: their role in contact-mediated cell-cell interactions in *Myxococcus xanthus*. *Bioessays* **21**, 590–595.

202. Eberl, L., Molin, S. and Givskov, M. (1999). Surface motility of *Serratia liquefaciens* MG1. *J. Bacteriol.* **181**, 1703–1712.

203. Egile, C., Loisel, T.P., Laurent, V., Li, R., Pantaloni, D., Sansonetti, P.J. and Carlier, M.F. (1999). Activation of the CDC42 effector N-WASP by the *Shigella flexneri* IcsA protein promotes actin nucleation by Arp2/3 complex and bacterial actin-based motility. *J. Cell Biol.* **146**, 1319–1332.

204. Eisenbach, M. and Adler, J. (1981). Bacterial cell envelopes with functional flagella. *J. Biol. Chem.* **256**, 8807–8814.

205. Eisenbach, M. (1982). Changes in membrane potential of *Escherichia coli* in response to temporal gradients of chemicals. *Biochemistry* **21**, 6818–6825.

206. Eisenbach, M., Margolin, Y., Ciobotariu, A. and Rottenberg, H. (1984). Distinction between changes in membrane potential and surface charge upon chemotactic stimulation of *Escherichia coli*. *Biophys. J.* **45**, 463–467.

207. Eisenbach, M., Margolin, Y. and Ravid, S. (1985). Excitatory signaling in bacterial chemotaxis, in *Sensing and Response in Microorganisms* (M. Eisenbach and M. Balaban, eds.), pp. 43–61. Elsevier, Amsterdam.

208. Eisenbach, M. and Matsumura, P. (1988). *In vitro* approach to bacterial chemotaxis. *Botan. Acta* **101**, 105–110.

209. Eisenbach, M. (1990). Functions of the flagellar modes of rotation in bacterial motility and chemotaxis. *Mol. Microbiol.* **4**, 161–167.

210. Eisenbach, M., Constantinou, C., Aloni, H. and Shinitzky, M. (1990). Repellents for *Escherichia coli* operate neither by changing membrane fluidity nor by

being sensed by periplasmic receptors during chemotaxis. *J. Bacteriol.* **172**, 5218–5224.

211. Eisenbach, M., Wolf, A., Welch, M., Caplan, S.R., Lapidus, I.R., Macnab, R.M., Aloni, H. and Asher, O. (1990). Pausing, switching and speed fluctuation of bacterial flagellar motor and their relation to motility and chemotaxis. *J. Mol. Biol.* **211**, 551–563.

212. Eisenbach, M. (1991). Signal transduction in bacterial chemotaxis. *Modern Cell Biol.* **10**, 137–208.

213. Eisenbach, M. and Caplan, S.R. (1998). Bacterial chemotaxis: Unsolved mystery of the flagellar switch. *Curr. Biol.* **8**, R444–R446.

214. Eisenbach, M. (2002). Bacterial chemotaxis, in *Nature Encyclopedia of Life Sciences* (www.els.net) (S. Robertson, ed.), pp. 506–519. Nature Publishing Group, London.

215. Elston, T.C. and Oster, G. (1997). Protein turbines I: The bacterial flagellar motor. *Biophys. J.* **73**, 703–721.

216. Ely, B., Ely, T.W., Crymes, W.B. and Minnich, S.A. (2000). A family of six flagellin genes contributes to the *Caulobacter crescentus* flagellar filament. *J. Bacteriol.* **182**, 5001–5004.

217. Emerson, D., Worden, R.M. and Breznak, J.A. (1994). A diffusion gradient chamber for studying microbial behavior and separating microorganisms. *Appl. Environ. Microbiol.* **60**, 1269–1278.

218. Engelmann, W. (1882). *Bot. Ztg.* **40**, 419–426.

219. Engström, P. and Hazelbauer, G.L. (1982). Methyl-accepting chemotaxis proteins are distributed in the membrane independently from basal ends of bacterial flagella. *Biochim. Biophys. Acta* **686**, 19–26.

220. Fahrner, K.A., Block, S.M., Krishnaswamy, S., Parkinson, J.S. and Berg, H.C. (1994). A mutant hook-associated protein (HAP3) facilitates torsionally induced transformations of the flagellar filament of *Escherichia coli. J. Mol. Biol.* **238**, 173–186.

221. Falke, J.J. and Koshland, D.E., Jr. (1987). Global flexibility in a sensory receptor: a site-directed cross-linking approach. *Science* **237**, 1596–1600.

222. Falke, J.J., Bass, R.B., Butler, S.L., Chervitz, S.A. and Danielson, M.A. (1997). The two-component signaling pathway of bacterial chemotaxis: A molecular view of signal transduction by receptors, kinases, and adaptation enzymes. *Annu. Rev. Cell Dev. Biol.* **13**, 457–512.

223. Falke, J.J. and Hazelbauer, G.L. (2001). Transmembrane signaling in bacterial chemoreceptors. *Trends Biochem. Sci.* **26**, 257–265.

224. Fan, F. and Macnab, R.M. (1996). Enzymatic characterization of FliI—an ATPase involved in flagellar assembly in *Salmonella typhimurium. J. Biol. Chem.* **271**, 31 981–31 988.

225. Fan, F., Ohnishi, K., Francis, N.R. and Macnab, R.M. (1997). The FliP and FliR proteins of *Salmonella typhimurium*, putative components of the type III flagellar export apparatus, are located in the flagellar basal body. *Mol. Microbiol.* **26**, 1035–1046.

226. Feng, J., Atkinson, M.R., McCleary, W., Stock, J.B., Wanner, B.L. and Ninfa, A.J. (1992). Role of phosphorylated metabolic intermediates in the regulation of glutamine synthetase synthesis in *Escherichia coli. J. Bacteriol.* **174**, 6061–6070.

227. Feng, X.H., Baumgartner, J.W. and Hazelbauer, G.L. (1997). High- and low-abundance chemoreceptors in *Escherichia coli*: Differential activities associated with closely related cytoplasmic domains. *J. Bacteriol.* **179**, 6714–6720.

228. Feng, X.H., Lilly, A.A. and Hazelbauer, G.L. (1999). Enhanced function conferred on low-abundance chemoreceptor Trg by a methyltransferase-docking site. *J. Bacteriol.* **181**, 3164–3171.

229. Ford, R.M., Phillips, B.R., Quinn, J.A. and Lauffenburger, D.A. (1991). Measurement of bacterial random motility and chemotaxis coefficients. 1. Stopped-flow diffusion chamber assay. *Biotechnol. Bioengineer.* **37**, 647–660.

230. Forst, S., Delgado, J., Ramakrishnan, G. and Inouye, M. (1988). Regulation of *ompC* and *ompF* expression in *Escherichia coli* in the absence of *envZ*. *J. Bacteriol.* **170**, 5080–5085.

231. Forst, S., Delgado, J. and Inouye, M. (1989). DNA-binding properties of the transcription activator (OmpR) for the upstream sequences of *ompF* in *Escherichia coli* are altered by *envZ* mutations and medium osmolarity. *J. Bacteriol.* **171**, 2949–2955.

232. Fosnaugh, K. and Greenberg, E.P. (1988). Motility and chemotaxis of *Spriochaeta aurantia*: Computer-assisted motion analysis. *J. Bacteriol.* **170**, 1768–1774.

233. Foster, D.L., Mowbray, S.L., Jap, B.K. and Koshland, D.E., Jr. (1985). Purification and characterization of the aspartate chemoreceptor. *J. Biol. Chem.* **260**, 11706–11710.

234. Francis, N.R., Sosinsky, G.E., Thomas, D. and DeRosier, D.J. (1994). Isolation, characterization and structure of bacterial flagellar motors containing the switch complex. *J. Mol. Biol.* **235**, 1261–1270.

235. Francis, N.R., Levit, M.N., Shaikh, T.R., Melanson, L.A., Stock, J.B. and DeRosier, D.J. (2002). Subunit organization in a soluble complex of Tar, CheW, and CheA by electron microscopy. *J. Biol. Chem.* **277**, 36755–36759.

236. Frankel, R.B., Bazylinski, D.A., Johnson, M.S. and Taylor, B.L. (1997). Magneto-aerotaxis in marine coccoid bacteria. *Biophys. J.* **73**, 994–1000.

237. Fraser, G.M., Bennett, J.C.Q. and Hughes, C. (1999). Substrate-specific binding of hook-associated proteins by FlgN and FliT, putative chaperones for flagellum assembly. *Mol. Microbiol.* **32**, 569–580.

238. Fredrick, K.L. and Helmann, J.D. (1994). Dual chemotaxis signaling pathways in *Bacillus subtilis*: a σ^D-dependent gene encodes a novel protein with both CheW and CheY homologous domains. *J. Bacteriol.* **176**, 2727–2735.

239. Freese, A., Reichenbach, H. and Lünsdorf, H. (1997). Further characterization and in situ localization of chain-like aggregates of the gliding bacteria *Myxococcus fulvus* and *Myxococcus xanthus*. *J. Bacteriol.* **179**, 1246–1252.

240. Fuhrer, D.K. and Ordal, G.W. (1991). *Bacillus subtilis* CheN, a homolog of CheA, the central regulator of chemotaxis in *Escherichia coli*. *J. Bacteriol.* **173**, 7443–7448.

241. Fung, D.C. and Berg, H.C. (1995). Powering the flagellar motor of *Escherichia coli* with an external voltage source. *Nature* **375**, 809–812.

242. Fuqua, C. and Greenberg, E.P. (1998). Self perception in bacteria: quorum sensing with acylated homoserine lactones. *Curr. Opin. Microbiol.* **1**, 183–189.

243. Galán, J.E. and Collmer, A. (1999). Type III secretion machines: bacterial devices for protein delivery into host cells. *Science* **284**, 1322–1328.

244. Galloway, R.J. and Taylor, B.L. (1980). Histidine starvation and adenosine 5-triphosphate depletion in chemotaxis of *Salmonella typhimurium*. *J. Bacteriol.* **144**, 1068–1075.

245. Gardina, P.J. and Manson, M.D. (1996). Attractant signaling by an aspartate chemoreceptor dimer with a single cytoplasmic domain. *Science* **274**, 425–426.

246. Gardina, P.J., Bormans, A.F., Hawkins, M.A., Meeker, J.W. and Manson, M.D. (1997). Maltose-binding protein interacts simultaneously and asymmetrically with both subunits of the Tar chemoreceptor. *Mol. Microbiol.* **23**, 1181–1191.

247. Gardina, P.J., Bormans, A.F. and Manson, M.D. (1998). A mechanism for simultaneous sensing of aspartate and maltose by the Tar chemoreceptor of *Escherichia coli*. *Mol. Microbiol.* **29**, 1147–1154.

248. Garrity, L.F. and Ordal, G.W. (1995). Chemotaxis in *Bacillus subtilis*: how bacteria monitor environmental signals. *Pharmacol. Ther.* **68**, 87–104.

249. Garrity, L.F. and Ordal, G.W. (1997). Activation of the CheA kinase by asparagine in *Bacillus subtilis* chemotaxis. *Microbiology* **143**, 2945–2951.

250. Garrity, L.F., Schiel, S.L., Merrill, R., Reizer, J., Saier, M.H. and Ordal, G.W. (1998). Unique regulation of carbohydrate chemotaxis in *Bacillus subtilis* by the phosphoenolpyruvate-dependent phosphotransferase system and the methyl-accepting chemotaxis protein McpC. *J. Bacteriol.* **180**, 4475–4480.

251. Garza, A.G., Harris-Haller, L.W., Stoebner, R.A. and Manson, M.D. (1995). Motility protein interactions in the bacterial flagellar motor. *Proc. Natl. Acad. Sci. U.S.A.* **92**, 1970–1974.

252. Garza, A.G., Biran, R., Wohlschlegel, J.A. and Manson, M.D. (1996). Mutations in *motB* suppressible by changes in stator or rotor components of the bacterial flagellar motor. *J. Mol. Biol.* **258**, 270–285.

253. Garzón, A. and Parkinson, J.S. (1996). Chemotactic signaling by the P1 phosphorylation domain liberated from the CheA histidine kinase of *Escherichia coli*. *J. Bacteriol.* **178**, 6752–6758.

254. Gauden, D.E. and Armitage, J.P. (1995). Electron transport-dependent taxis in *Rhodobacter sphaeroides*. *J. Bacteriol.* **177**, 5853–5859.

255. Gegner, J.A. and Dahlquist, F.W. (1991). Signal transduction in bacteria: CheW forms a reversible complex with the protein kinase CheA. *Proc. Natl. Acad. Sci. U.S.A.* **88**, 750–754.

256. Gegner, J.A., Graham, D.R., Roth, A.F. and Dahlquist, F.W. (1992). Assembly of an MCP receptor, CheW, and kinase CheA complex in the bacterial chemotaxis signal transduction pathway. *Cell* **70**, 975–982.

257. Gestwicki, J.E., Lamanna, A.C., Harshey, R.M., McCarter, L.L., Kiessling, L.L. and Adler, J. (2000). Evolutionary conservation of methyl-accepting chemotaxis protein location in *Bacteria* and *Archaea*. *J. Bacteriol.* **182**, 6499–6502.

258. Gestwicki, J.E. and Kiessling, L.L. (2002). Inter-receptor communication through arrays of bacterial chemoreceptors. *Nature* **415**, 81–84.

259. Gilad, R., Porat, A. and Trachtenberg, S. (2003). Motility modes of *Spiroplasma melliferum* BC3: a helical, wall-less bacterium driven by a linear motor. *Mol. Microbiol.* **47**, 657–669.

260. Glagolev, A.N. and Skulachev, V.P. (1978). The proton pump is a molecular engine of motile bacteria. *Nature* **272**, 280–282.

261. Goldman, D.J. and Ordal, G.W. (1981). Sensory adaptation and deadaptation by *Bacillus subtilis*. *J. Bacteriol*. **147**, 267–270.
262. Gosink, K.K. and Häse, C.C. (2000). Requirements for conversion of the Na$^+$-driven flagellar motor of *Vibrio cholerae* to the H$^+$-driven motor of *Escherichia coli*. *J. Bacteriol*. **182**, 4234–4240.
263. Götz, R., Limmer, N., Ober, K. and Schmitt, R. (1982). Motility and chemotaxis in two strains of *Rhizobium* with complex flagella. *J. Gen. Microbiol*. **128**, 789–798.
264. Gouin, E., Gantelet, H., Egile, C., Lasa, I., Ohayon, H., Villiers, V., Gounon, P., Sansonetti, P.J. and Cossart, P. (1999). A comparative study of the actin-based motilities of the pathogenic bacteria *Listeria monocytogenes*, *Shigella flexneri* and *Rickettsia conorii*. *J. Cell Sci*. **112**, 1697–1708.
265. Goulbourne, E.A. and Greenberg, E.P. (1983). A voltage clamp inhibits chemotaxis of *Spirochaeta aurantia*. *J. Bacteriol*. **153**, 916–920.
266. Goulbourne, E.A. and Greenberg, E.P. (1983). Inhibition of *Spirochaeta aurantia* chemotaxis by neurotoxins. *J. Bacteriol*. **155**, 1443–1445.
267. Goy, M.F., Springer, M.S. and Adler, J. (1978). Failure of sensory adaptation in bacterial mutants that are defective in a protein methylation reaction. *Cell* **15**, 1231–1240.
268. Greck, M., Platzer, J., Sourjik, V. and Schmitt, R. (1995). Analysis of a chemotaxis operon in *Rhizobium meliloti*. *Mol. Microbiol*. **15**, 989–1000.
269. Greenberg, E.P. and Canale-Parola, E. (1977). Chemotaxis in *Spirochaeta aurentia*. *J. Bacteriol*. **130**, 485–494.
270. Grishanin, R.N., Bibikov, S.I., Altschuler, I.M., Kaulen, A.D., Kazimirchuk, S.B., Armitage, J.P. and Skulachev, V.P. (1996). $\Delta\psi$-mediated signaling in the bacteriorhodopsin-dependent photoresponse. *J. Bacteriol*. **178**, 3008–3014.
271. Halfen, L.N. (1979). Gliding movements, in *Encyclopedia of Plant Physiology: Physiology of Movements* (W. Haupt and M.E. Feinleib, eds.), pp. 250–267. Springer-Verlag, Berlin, Heidelberg, New York.
272. Halkides, C.J., Zhu, X.Y., Phillion, D.P., Matsumura, P. and Dahlquist, F.W. (1998). Synthesis and biochemical characterization of an analogue of CheY-phosphate, a signal transduction protein in bacterial chemotaxis. *Biochemistry* **37**, 13 674–13 680.
273. Halkides, C.J., McEvoy, M.M., Casper, E., Matsumura, P., Volz, K. and Dahlquist, F.W. (2000). The 1.9 Å resolution crystal structure of phosphono-CheY, an analogue of the active form of the response regulator, CheY. *Biochemistry* **39**, 5280–5286.
274. Hamblin, P.A., Maguire, B.A., Grishanin, R.N. and Armitage, J.P. (1997). Evidence for two chemosensory pathways in *Rhodobacter sphaeroides*. *Mol. Microbiol*. **26**, 1083–1096.
275. Haneline, S., Connelly, C.J. and Melton, T. (1991). Chemotactic behavior of *Azotobacter vinelandii*. *Appl. Environmen. Microbiol*. **57**, 825–829.
276. Hanlon, D.W., Marquez-Magana, L.M., Carpenter, P.B., Chamberlin, M.J. and Ordal, G.W. (1992). Sequence and characterization of *Bacillus subtilis* CheW. *J. Biol. Chem*. **267**, 12 055–12 060.
277. Hanlon, D.W. and Ordal, G.W. (1994). Cloning and characterization of genes encoding methyl-accepting chemotaxis proteins in *Bacillus subtilis*. *J. Biol. Chem*. **269**, 14 038–14 046.

278. Hanlon, D.W., Rosario, M.M.L., Ordal, G.W., Venema, G. and Vansinderen, D. (1994). Identification of TlpC, a novel 62 kDa MCP-like protein from *Bacillus subtilis*. *Microbiology* **140**, 1847–1854.

279. Harrison, D.M., Skidmore, J., Armitage, J.P. and Maddock, J.R. (1999). Localization and environmental regulation of MCP-like proteins in *Rhodobacter sphaeroides*. *Mol. Microbiol.* **31**, 885–892.

280. Harshey, R.M. (1994). Bees aren't the only ones: swarming in Gram-negative bacteria. *Mol. Microbiol.* **13**, 389–394.

281. Harwood, C.S., Rivelli, M. and Ornston, L.N. (1984). Aromatic acids are chemoattractants for *Pseudomonas putida*. *J. Bacteriol.* **160**, 622–628.

282. Hazelbauer, G.L., Mesibov, R.E. and Adler, J. (1969). *Escherichia coli* mutants defective in chemotaxis toward specific chemicals. *Proc. Natl. Acad. Sci. U.S.A.* **64**, 1300–1307.

283. Hazelbauer, G.L. and Adler, J. (1971). Role of the galactose binding protein in chemotaxis of *Escherichia coli* toward galactose. *Nature New Biol.* **230**, 101–104.

284. Hazelbauer, G.L. and Harayama, S. (1983). Sensory transduction in bacterial chemotaxis. *Internat. Rev. Cytol.* **81**, 33–70.

285. Hedblom, M.L. and Adler, J. (1980). Genetic and biochemical properties of *Escherichia coli* mutants with defects in serine chemotaxis. *J. Bacteriol.* **144**, 1048–1060.

286. Hedblom, M.L. and Adler, J. (1983). Chemotactic response of *Escherichia coli* to chemically synthesized amino acids. *J. Bacteriol.* **155**, 1463–1466.

287. Heffetz, D. and Zick, Y. (1986). Receptor aggregation is necessary for activation of the soluble insulin receptor kinase. *J. Biol. Chem.* **261**, 889–894.

288. Heidelberg, J.F., Eisen, J.A., Nelson, W.C., Clayton, R.A., Gwinn, M.L., Dodson, R.J., Haft, D.H., Hickey, E.K., Peterson, J.D., Umayam, L., Gill, S.R., Nelson, K.E., Read, T.D., Tettelin, H., Richardson, D., Ermolaeva, M.D., Vamathevan, J., Bass, S., Qin, H., Dragoi, I., Sellers, P., McDonald, L., Utterback, T., Fleishmann, R.D., Nierman, W.C. and White, O. (2000). DNA sequence of both chromosomes of the cholera pathogen *Vibrio cholerae*. *Nature* **406**, 477–483.

289. Henrichsen, J. (1983). Twitching motility. *Ann. Rev. Microbiol.* **37**, 81–93.

290. Hess, J.F., Oosawa, K., Matsumura, P. and Simon, M.I. (1987). Protein phosphorylation is involved in bacterial chemotaxis. *Proc. Natl. Acad. Sci. U.S.A.* **84**, 7609–7613.

291. Hess, J.F., Bourret, R.B., Oosawa, K., Matsumura, P. and Simon, M.I. (1988). Protein phosphorylation and bacterial chemotaxis. *Cold Spring Harbor Symp. Quant. Biol.* **53**, 41–48.

292. Hess, J.F., Bourret, R.B. and Simon, M.I. (1988). Histidine phosphorylation and phosphoryl group transfer in bacterial chemotaxis. *Nature* **336**, 139–143.

293. Hess, J.F., Oosawa, K., Kaplan, N. and Simon, M.I. (1988). Phosphorylation of three proteins in the signaling pathway of bacterial chemotaxis. *Cell* **53**, 79–87.

294. Hildebrandt, K., Eastman, D. and Dworkin, M. (1997). ADP-ribosylation by the extracellular fibrils of *Myxococcus xanthus*. *Mol. Microbiol.* **23**, 231–235.

295. Hirano, T., Yamaguchi, S., Oosawa, K. and Aizawa, S.-I. (1994). Roles of FliK and FlhB in determination of flagellar hook length in *Salmonella typhimurium*. *J. Bacteriol.* **176**, 5439–5449.

296. Hirano, T., Minamino, T. and Macnab, R.M. (2001). The role in flagellar rod assembly of the N-terminal domain of *Salmonella* FlgJ, a flagellum-specific muramidase. *J. Mol. Biol.* **312**, 359–369.

297. Hirota, N., Kitada, M. and Imae, Y. (1981). Flagellar motors of alkalophilic *Bacillus* are powered by an electrochemical potential gradient of Na$^+$. *FEBS Lett.* **132**, 278–280.

298. Hirota, N. and Imae, Y. (1983). Na$^+$-driven flagellar motors of an alkalophilic *Bacillus* strain YN-1. *J. Biol. Chem.* **258**, 10577–10581.

299. Hodgkin, J. and Kaiser, D. (1979). Genetics of gliding motility in *Myxococcus xanthus* (Myxobacterales): genes controlling movement of single cells. *Mol. Gen. Genet.* **171**, 167–176.

300. Hodgkin, J. and Kaiser, D. (1979). Genetics of gliding motility in *Myxococcus xanthus* (Myxobacterales): two gene systems control movement. *Mol. Gen. Genet.* **171**, 177–191.

301. Hoiczyk, E. and Baumeister, W. (1998). The junctional pore complex, a prokaryotic secretion organelle, is the molecular motor underlying gliding motility in cyanobacteria. *Curr. Biol.* **8**, 1161–1168.

302. Hoiczyk, E. (2000). Gliding motility in cyanobacteria: observations and possible explanations. *Arch. Microbiol.* **174**, 11–17.

303. Homma, M., Kutsukake, K., Hasebe, M., Iino, T. and Macnab, R.M. (1990). FlgB, FlgC, FlgF, and FlgG. A family of structurally related proteins in the flagellar basal body of *Salmonella typhimurium*. *J. Mol. Biol.* **211**, 465–477.

304. Homma, M., Oota, H., Kojima, S., Kawagishi, I. and Imae, Y. (1996). Chemotactic responses to an attractant and a repellent by the polar and lateral flagellar systems of *Vibrio alginolyticus*. *Microbiology (UK)* **142**, 2777–2783.

305. Hou, S., Larsen, R.W., Boudko, D., Riley, C.W., Karatan, E., Zimmer, M., Ordal, G.W. and Alam, M. (2000). Myoglobin-like aerotaxis transducers in Archaea and Bacteria. *Nature* **403**, 540–544.

306. Hou, S.B., Brooun, A., Yu, H.S., Freitas, T. and Alam, M. (1998). Sensory rhodopsin II transducer HtrII is also responsible for serine chemotaxis in the archaeon *Halobacterium salinarum*. *J. Bacteriol.* **180**, 1600–1602.

307. Huang, C.X. and Stewart, R.C. (1993). CheZ mutants with enhanced ability to dephosphorylate CheY, the response regulator in bacterial chemotaxis. *Biochim. Biophys. Acta* **1202**, 297–304.

308. Hughes, K.T., Gillen, K.L., Semon, M.J. and Karlinsey, J.E. (1993). Sensing structural intermediates in bacterial flagellar assembly by export of a negative regulator. *Science* **262**, 1277–1280.

309. Hughson, A.G. and Hazelbauer, G.L. (1996). Detecting the conformational change of transmembrane signaling in a bacterial chemoreceptor by measuring effects on disulfide cross-linking *in vivo*. *Proc. Natl. Acad. Sci. U.S.A.* **93**, 11546–11551.

310. Iino, T. and Mitani, M. (1967). Infection of *Serratia marcescens* by bacteriophage χ. *J. Virol.* **1**, 445–447.

311. Iino, T. (1974). Assembly of *Salmonella* flagellin *in vitro* and *in vivo*. *J. Supramol. Struct.* **2**, 372–384.

312. Iino, T., Komeda, Y., Kutsukake, K., Macnab, R.M., Matsumura, P., Parkinson, J.S., Simon, M.I. and Yamaguchi, S. (1988). New unified nomenclature for the flagellar

genes of *Escherichia coli* and *Salmonella typhimurium*. *Microbiol. Rev.* **52**, 533–535.

313. Imae, Y., Oosawa, K., Mizuno, T., Kihara, M. and Macnab, R.M. (1987). Phenol: a complex chemoeffecter in bacterial chemotaxis. *J. Bacteriol.* **169**, 371–379.

314. Ingolia, T.D. and Koshland, D.E., Jr. (1979). Response to a metal ion-citrate complex in bacterial sensing. *J. Bacteriol.* **140**, 798–804.

315. Irikura, V.M., Kihara, M., Yamaguchi, S., Sockett, H. and Macnab, R.M. (1993). *Salmonella typhimurium fliG* and *fliN* mutations causing defects in assembly, rotation, and switching of the flagellar motor. *J. Bacteriol.* **175**, 802–810.

316. Ishihara, A., Segall, J.E., Block, S.M. and Berg, H.C. (1983). Coordination of flagella on filamentous cells of *Escherichia coli*. *J. Bacteriol.* **155**, 228–237.

317. Itoh, T., Aiba, H., Baba, T., Fujita, K., Hayashi, K., Inada, T., Isono, K., Kasai, H., Kimura, S., Kitakawa, M., Kitagawa, M., Makino, K., Miki, T., Mizobuchi, K., Mori, H., Mori, T., Motomura, K., Nakade, S., Nakamura, Y., Nashimoto, H., Nishio, Y., Oshima, T., Saito, N., Sampei, G., Seki, Y., Sivasundaram, S., Tagami, H., Takeda, J., Takemoto, K., Wada, C., Yamamoto, Y. and Horiuchi, T. (1996). A 460-kb DNA sequence of the *Escherichia coli* K-12 genome corresponding to the 40.1-50.0 min region on the linkage map. *DNA Res.* **3**, 379–392.

318. Jakinovich, W. (1981). Comparative study of sweet taste specificity, in *Biochemistry of Taste and Olfaction* (R.H. Cagan and M.R. Kare, eds.), pp. 117–138. Academic Press, New York.

319. Jarrell, K.F., Bayley, D.P. and Kostyukova, A.S. (1996). The archaeal flagellum: a unique motility structure. *J. Bacteriol.* **178**, 5057–5064.

320. Jasuja, R., Keyoung, J., Reid, G.P., Trentham, D.R. and Khan, S. (1999). Chemotactic responses of *Escherichia coli* to small jumps of photoreleased L-aspartate. *Biophys. J.* **76**, 1706–1719.

321. Jasuja, R., Yu-Lin, Trentham, D.R. and Khan, S. (1999). Response tuning in bacterial chemotaxis. *Proc. Natl. Acad. Sci. U.S.A.* **96**, 11 346–11 351.

322. Jelsbak, L. and Sogaard-Andersen, L. (1999). The cell surface-associated intercellular C-signal induces behavioral changes in individual *Myxococcus xanthus* cells during fruiting body morphogenesis. *Proc. Natl. Acad. Sci. U.S.A.* **96**, 5031–5036.

323. Jeziore-Sassoon, Y., Hamblin, P.A., Bootle-Wilbraham, C.A., Poole, P.S. and Armitage, J.P. (1998). Metabolism is required for chemotaxis to sugars in *Rhodobacter sphaeroides*. *Microbiology* **144**, 229–239.

324. Jiang, M.Y., Bourret, R.B., Simon, M.I. and Volz, K. (1997). Uncoupled phosphorylation and activation in bacterial chemotaxis: The 2.3 Å structure of an aspartate to lysine mutant at position 13 of CheY. *J. Biol. Chem.* **272**, 11 850–11 855.

325. Jones, C.J. and Macnab, R.M. (1990). Flagellar assembly in *Salmonella typhimurium* analysis with temperature-sensitive mutants. *J. Bacteriol.* **172**, 1327–1339.

326. Jones, C.J., Macnab, R.M., Okino, H. and Aizawa, S.-I. (1990). Stoichiometric analysis of the flagellar hook-(basal-body) complex of *Salmonella typhimurium*. *J. Mol. Biol.* **212**, 377–387.

327. Kagawa, H., Ono, N., Enomoto, M. and Komeda, Y. (1984). Bacteriophage chi sensitivity and motility of *Escherichia coli* K-12 and *Salmonella typhimurium Fla*: mutants possessing the hook structure. *J. Bacteriol.* **157**, 649–654.

328. Kaiser, D. (1979). Social gliding is correlated with the presence of pili in *Myxococcus xanthus*. *Proc. Natl. Acad. Sci. U.S.A.* **76**, 5952–5956.

329. Kaiser, D. (2000). Bacterial motility: How do pili pull? *Curr. Biol.* **10**, R777–R780.

330. Kalir, S., McClure, J., Pabbaraju, K., Southward, C., Ronen, M., Leibler, S., Surette, M.G. and Alon, U. (2001). Ordering genes in a flagella pathway by analysis of expression kinetics from living bacteria. *Science* **292**, 2080–2083.

331. Kami-ike, N., Kudo, S. and Hotani, H. (1991). Rapid changes in flagellar rotation induced by external electric pulses. *Biophys. J.* **60**, 1350–1355.

332. Kangatharalingam, N., Wang, L.Z. and Priscu, J.C. (1991). Evidence for bacterial chemotaxis to cyanobacteria from a radioassay technique. *App. Environ. Microbiol.* **57**, 2395–2398.

333. Kara-Ivanov, M., Eisenbach, M. and Caplan, S.R. (1995). Fluctuations in rotation rate of the flagellar motor of *Escherichia coli. Biophys. J.* **69**, 250–263.

334. Karim, Q.N., Logan, R.P.H., Puels, J., Karnholz, A. and Worku, M.L. (1998). Measurement of motility of *Helicobacter pylori*, *Campylobacter jejuni*, and *Escherichia coli* by real time computer tracking using the Hobson BacTracker. *J. Clin. Pathol.* **51**, 623–628.

335. Karlinsey, J.E., Pease, A.J., Winkler, M.E., Bailey, J.L. and Hughes, K.T. (1997). The *flk* gene of *Salmonella typhimurium* couples flagellar P- and L-ring assembly to flagellar morphogenesis. *J. Bacteriol.* **179**, 2389–2400.

336. Karlinsey, J.E., Tsui, H.C.T., Winkler, M.E. and Hughes, K.T. (1998). *Flk* couples *flgM* translation to flagellar ring assembly in *Salmonella typhimurium. J. Bacteriol.* **180**, 5384–5397.

337. Katayama, E., Shiraishi, T., Oosawa, K., Baba, N. and Aizawa, S.-I. (1996). Geometry of the flagellar motor in the cytoplasmic membrane of *Salmonella typhimurium* as determined by stereo-photogrammetry of quick-freeze deep-etch replica images. *J. Mol. Biol.* **255**, 458–475.

338. Kato, J., Ito, A., Nikata, T. and Ohtake, H. (1992). Phosphate taxis in *Pseudomonas aeruginosa. J. Bacteriol.* **174**, 5149–5151.

339. Kato, J., Nakamura, T., Kuroda, A. and Ohtake, H. (1999). Cloning and characterization of chemotaxis genes in *Pseudomonas aeruginosa. Biosci. Biotechnol. Biochem.* **63**, 155–161.

340. Kawagishi, I., Maekawa, Y., Atsumi, T., Homma, M. and Imae, Y. (1995). Isolation of the polar and lateral flagellum-defective mutants in *Vibrio alginolyticus* and identification of their flagellar driving energy sources. *J. Bacteriol.* **177**, 5158–5160.

341. Kearns, D.B. and Shimkets, L.J. (1998). Chemotaxis in a gliding bacterium. *Proc. Natl. Acad. Sci. U.S.A.* **95**, 11 957–11 962.

342. Kehry, M.R. and Dahlquist, F.W. (1982). Adaptation in bacterial chemotaxis: *cheB*-dependent modification permits additional methylations of sensory transducer proteins. *Cell* **29**, 761–772.

343. Kehry, M.R., Bond, M.W., Hunkapiller, M.W. and Dahlquist, F.W. (1983). Enzymatic deamidation of methyl-accepting chemotaxis proteins in *Escherichia coli* catalyzed by the *cheB* gene product. *Proc. Natl. Acad. Sci. U.S.A.* **80**, 3599–3603.

344. Kehry, M.R., Doak, T.G. and Dahlquist, F.W. (1985). Sensory adaptation in bacterial chemotaxis: Regulation of demethylation. *J. Bacteriol.* **163**, 983–990.

345. Khan, S., Macnab, R.M., DeFranco, A.L. and Koshland, D.E., Jr. (1978). Inversion of a behavioral response in bacterial chemotaxis: Explanation at the molecular level. *Proc. Natl. Acad. Sci. U.S.A.* **75**, 4150–4154.

346. Khan, S. and Macnab, R.M. (1980). The steady-state counterclockwise/clockwise ratio of bacterial flagellar motors is regulated by proton-motive force. *J. Mol. Biol.* **138**, 563–597.

347. Khan, S. and Macnab, R.M. (1980). Proton chemical potential, proton electrical potential, and bacterial motility. *J. Mol. Biol.* **138**, 599–614.

348. Khan, S., Dapice, M. and Reese, T. (1988). Effects of *mot* gene expression on the structure of the flagellar motor. *J. Mol. Biol.* **202**, 575–584.

349. Khan, S., Amoyaw, K., Spudich, J.L., Reid, G.P. and Trentham, D.R. (1992). Bacterial chemoreceptor signaling probed by flash photorelease of a caged serine. *Biophys. J.* **62**, 67–68.

350. Khan, S., Castellano, F., Spudich, J.L., McCray, J.A., Goody, R.S., Reid, G.P. and Trentham, D.R. (1993). Excitatory signaling in bacteria probed by caged chemo-effectors. *Biophys. J.* **65**, 2368–2382.

351. Khan, S., Spudich, J.L., Mccray, J.A. and Trentham, D.R. (1995). Chemotactic signal integration in bacteria. *Proc. Natl. Acad. Sci. U.S.A.* **92**, 9757–9761.

352. Khan, S. (1997). Rotary chemiosmotic machines. *Biochim. Biophys. Acta* **1322**, 86–105.

353. Khan, S., Zhao, R.B. and Reese, T.S. (1998). Architectural features of the Salmonella typhimurium flagellar motor switch revealed by disrupted C-rings. *J. Struct. Biol.* **122**, 311–319.

354. Khan, S. (2000). On bacterial tactic response times and latencies. *Biophys. J.* **78**, 2186–2187.

355. Khan, S., Pierce, D. and Vale, R.D. (2000). Interactions of the chemotaxis signal protein CheY with bacterial flagellar motors visualized by evanescent wave microscopy. *Curr. Biol.* **10**, 927–930.

356. Kihara, M. and Macnab, R.M. (1979). Chemotaxis of *Salmonella typhimurium* toward citrate. *J. Bacteriol.* **140**, 297–300.

357. Kihara, M. and Macnab, R.M. (1981). Cytoplasmic pH mediates pH taxis and weak-acid repellent taxis of bacteria. *J. Bacteriol.* **145**, 1209–1221.

358. Kihara, M., Francis, N.R., DeRosier, D.J. and Macnab, R.M. (1996). Analysis of a FliM-FliN flagellar switch fusion mutant of *Salmonella typhimurium*. *J. Bacteriol.* **178**, 4582–4589.

359. Kihara, M., Minamino, T., Yamaguchi, S. and Macnab, R.M. (2001). Intergenic suppression between the flagellar MS ring protein FliF of *Salmonella* and FlhA, a membrane component of its export apparatus. *J. Bacteriol.* **183**, 1655–1662.

360. Kim, C., Jackson, M., Lux, R. and Khan, S. (2001). Determinants of chemotactic signal amplification in *Escherichia coli*. *J. Mol. Biol.* **307**, 119–135.

361. Kim, C.H., Jung, K.H., Eym, Y.B. and Park, C.Y. (1996). Tactic interaction of ribose-binding protein with the membrane receptor Trg. *Mol. Cells* **6**, 133–138.

362. Kim, K.K., Yokota, H. and Kim, S.-H. (1999). Four-helical-bundle structure of the cytoplasmic domain of a serine chemotaxis receptor. *Nature* **400**, 787–792.

363. Kim, S.-H., Wang, W.R. and Kim, K.K. (2002). Dynamic and clustering model of bacterial chemotaxis receptors: Structural basis for signaling and high sensitivity. *Proc. Natl. Acad. Sci. U.S.A.* **99**, 11611–11615.

364. Kirby, J.R., Saulmon, M.M., Kristich, C.J. and Ordal, G.W. (1999). CheY-dependent methylation of the asparagine receptor, McpB, during chemotaxis in *Bacillus subtilis*. *J. Biol. Chem.* **274**, 11092–11100.

365. Kirchhoff, H. (1992). Motility, in *Mycoplasmas: Molecular Biology and Pathogenesis* (J. Maniloff, R.N. McElhaney, L.R. Finch and J.B. Baseman, eds.), pp. 289–306. American Society for Microbiology, Washington DC.

366. Kirsch, M.L., Peters, P.D., Hanlon, D.W., Kirby, J.R. and Ordal, G.W. (1993). Chemotactic methylesterase promotes adaptation to high concentrations of attractant in *Bacillus subtilis*. *J. Biol. Chem.* **268**, 18610–18616.

367. Kirsch, M.L., Zuberi, A.R., Henner, D., Peters, P.D., Yazdi, M.A. and Ordal, G.W. (1993). Chemotactic methyltransferase promotes adaptation to repellents in *Bacillus subtilis*. *J. Biol. Chem.* **268**, 25350–25356.

368. Kleene, S.J., Toews, M.L. and Adler, J. (1977). Isolation of glutamic acid methyl ester from an *Escherichia coli* membrane protein involved in chemotaxis. *J. Biol. Chem.* **252**, 3214–3218.

369. Kleene, S.J., Hobson, A.C. and Adler, J. (1979). Attractants and repellents influence methylation and demethylation of methyl-accepting chemotaxis proteins in an extract of *Escherichia coli*. *Proc. Natl. Acad. Sci. U.S.A.* **76**, 6309–6313.

370. Ko, M. and Park, C. (2000). Two novel flagellar components and H-NS are involved in the motor function of *Escherichia coli*. *J. Mol. Biol.* **303**, 371–382.

371. Ko, M. and Park, C. (2000). H-NS-Dependent regulation of flagellar synthesis is mediated by a LysR family protein. *J. Bacteriol.* **182**, 4670–4672.

372. Kofoid, E.C. and Parkinson, J.S. (1991). Tandem translation starts in the *cheA* locus of *Escherichia coli*. *J. Bacteriol.* **173**, 2116–2119.

373. Kojima, S., Yamamoto, K., Kawagishi, I. and Homma, M. (1999). The polar flagellar motor of *Vibrio cholerae* is driven by an Na$^+$ motive force. *J. Bacteriol.* **181**, 1927–1930.

374. Kokoeva, M.V. and Oesterhelt, D. (2000). BasT, a membrane-bound transducer protein for amino acid detection in *Halobacterium salinarum*. *Mol. Microbiol.* **35**, 647–656.

375. Kokoeva, M.V., Storch, K.-F., Klein, C. and Oesterhelt, D. (2002). A novel mode of sensory transduction in archaea: binding protein-mediated chemotaxis towards osmoprotectants and amino acids. *EMBO J.* **21**, 2312–2322.

376. Kolodziej, A.F., Tan, T. and Koshland, D.E., Jr. (1996). Producing positive, negative, and no cooperativity by mutations at a single residue located at the subunit interface in the aspartate receptor of *Salmonella typhimurium*. *Biochemistry* **35**, 14782–14792.

377. Komeda, Y. (1986). Transcriptional control of flagellar genes in *Escherichia coli* K-12. *J. Bacteriol.* **168**, 1315–1318.

378. Kondoh, H. (1980). Tumbling chemotaxis mutants of *Escherichia coli*: Possible gene-dependent effect of methionine starvation. *J. Bacteriol.* **142**, 527–534.

379. Koroyasu, S., Yamazato, M., Hirano, T. and Aizawa, S.I. (1998). Kinetic analysis of the growth rate of the flagellar hook in *Salmonella typhimurium* by the population balance method. *Biophys. J.* **74**, 436–443.

380. Kort, E.N., Goy, M.F., Larsen, S.H. and Adler, J. (1975). Methylation of a membrane protein involved in bacterial chemotaxis. *Proc. Natl. Acad. Sci. U.S.A.* **72**, 3939–3943.

381. Koshland, D.E., Jr. (1980). *Bacterial Chemotaxis as a Model Behavioral System.* Raven Press, New York.

382. Koshland, D.E., Jr. (1996). The structural basis of negative cooperativity: receptors and enzymes. *Curr. Opin. Struct. Biol.* **6**, 757–761.

383. Kossmann, M., Wolff, C. and Manson, M.D. (1988). Maltose chemoreceptor of *Escherichia coli*: Interaction of maltose-binding protein and the Tar signal transducer. *J. Bacteriol.* **170**, 4516–4521.

384. Krieg, N.R., Tomelty, J.P. and Wells, J.S. (1967). Inhibition of flagellar coordination in Spirillum volutans. *J. Bacteriol.* **94**, 1431–1436.

385. Krikos, A., Mutoh, N., Boyd, A. and Simon, M.I. (1983). Sensory transducers of *E. coli* are composed of discrete structural and functional domains. *Cell* **33**, 615–622.

386. Krikos, A., Conley, M.P., Boyd, A., Berg, H.C. and Simon, M.I. (1985). Chimeric chemosensory transducers of *Escherichia coli. Proc. Natl. Acad. Sci. U.S.A.* **82**, 1326–1330.

387. Kubori, T., Shimamoto, N., Yamaguchi, S., Namba, K. and Aizawa, S.-I. (1992). Morphological pathway of flagellar assembly in *Salmonella typhimurium. J. Mol. Biol.* **226**, 433–446.

388. Kubori, T., Yamaguchi, S. and Aizawa, S.-I. (1997). Assembly of the switch complex onto the MS ring complex of *Salmonella typhimurium* does not require any other flagellar proteins. *J. Bacteriol.* **179**, 813–817.

389. Kubori, T., Matsushima, Y., Nakamura, D., Uralil, J., Lara-Tejero, M., Sukhan, A., Galán, J.E. and Aizawa, S.-I. (1998). Supramolecular structure of the *Salmonella typhimurium* type III protein secretion system. *Science* **280**, 602–605.

390. Kudo, S., Magariyama, Y. and Aizawa, S.-I. (1990). Abrupt changes in flagellar rotation observed by laser dark-field microscopy. *Nature* **346**, 677–680.

391. Kunst, F., Ogasawara, N., Moszer, I., Albertini, A.M., Alloni, G., Azevedo, V., Bertero, M.G., Bessieres, P., Bolotin, A., Borchert, S., Borriss, R., Boursier, L., Brans, A., Braun, M., Brignell, S.C., Bron, S., Brouillet, S., Bruschi, C.V., Caldwell, B., Capuano, V., Carter, N.M., Choi, S.K., Codani, J.J., Connerton, I.F., Danchin, A., *et al.* (1997). The complete genome sequence of the gram-positive bacterium *Bacillus subtilis. Nature* **390**, 249–256.

392. Kuo, S.C. and Koshland, D.E., Jr. (1987). Roles of *cheY* and *cheZ* gene products in controlling flagellar rotation in bacterial chemotaxis of *Escherichia coli. J. Bacteriol.* **169**, 1307–1314.

393. Kuo, S.C. and Koshland, D.E., Jr. (1989). Multiple kinetic states for the flagellar motor switch. *J. Bacteriol.* **171**, 6279–6287.

394. Kuo, S.C. and McGrath, J.L. (2000). Steps and fluctuations of *Listeria monocytogenes* during actin-based motility. *Nature* **407**, 1026–1029.

395. Kupper, J., Marwan, W., Typke, D., Grünberg, H., Uwer, U., Gluch, M. and Oesterhelt, D. (1994). The flagellar bundle of *Halobacterium salinarium* is inserted into a distinct polar cap structure. *J. Bacteriol.* **176**, 5184–5187.

396. Kuroda, A., Kumano, T., Taguchi, K., Nikata, T., Kato, J. and Ohtake, H. (1995). Molecular cloning and characterization of a chemotactic transducer gene in *Pseudomonas aeruginosa. J. Bacteriol.* **177**, 7019–7025.

397. Kutsukake, K., Ohya, Y. and Iino, T. (1990). Transcriptional analysis of the flagellar regulon of *Salmonella typhimurium. J. Bacteriol.* **172**, 741–747.

398. Kutsukake, K. (1994). Excretion of the anti-sigma factor through a flagellar substructure couples flagellar gene expression with flagellar assembly in *Salmonella typhimurium. Mol. Gen. Genet.* **243**, 605–612.

399. Kutsukake, K. (1997). Hook-length control of the export-switching machinery involves a double-locked gate in *Salmonella typhimurium* flagellar morphogenesis. *J. Bacteriol.* **179**, 1268–1273.

400. Kutsukake, K., Ikebe, T. and Yamamoto, S. (1999). Two novel regulatory genes, *fliT* and *fliZ*, in the flagellar regulon of *Salmonella*. *Genes Genet. Syst.* **74**, 287–292.

401. Lambert, P.F., Kashanchi, F., Radonovich, M.F., Shiekhattar, R. and Brady, J.N. (1998). Phosphorylation of p53 serine increases interaction with CBP. *J. Biol. Chem.* **273**, 33 048–33 053.

402. Lancet, D. and Pace, U. (1987). The molecular basis of odor recognition. *Trends Biochem. Sci.* **12**, 63–66.

403. Langer, R., Fefferman, M., Gryska, P. and Bergman, K. (1980). A simple method for studying chemotaxis using sustained release of attractants from inert polymers. *Can. J. Microbiol.* **26**, 274–278.

404. Lapidus, I.R. and Berg, H.C. (1982). Gliding motility of Cytophaga sp. strain U67. *J. Bacteriol.* **151**, 384–398.

405. Lapidus, I.R., Welch, M. and Eisenbach, M. (1988). Pausing of flagellar rotation is a component of bacterial motility and chemotaxis. *J. Bacteriol.* **170**, 3627–3632.

406. Larsen, S.H., Adler, J., Gargus, J.J. and Hogg, R.W. (1974). Chemomechanical coupling without ATP: The source of energy for motility and chemotaxis in bacteria. *Proc. Natl. Acad. Sci. U.S.A.* **71**, 1239–1243.

407. Larsen, S.H., Reader, R.W., Kort, E.N., Tso, W.-W. and Adler, J. (1974). Change in direction of flagellar rotation is the basis of the chemotactic response in *Escherichia coli. Nature* **249**, 74–77.

408. Lasa, I., Dehoux, P. and Cossart, P. (1998). Actin polymerization and bacterial movement. *Biochim. Biophys. Acta* **1402**, 217–228.

409. Laurent, V., Loisel, T.P., Harbeck, B., Wehman, A., Grobe, L., Jockusch, B.M., Wehland, J., Gertler, F.B. and Carlier, M.F. (1999). Role of proteins of the Ena/VASP family in actin-based motility of Listeria monocytogenes. *J. Cell Biol.* **144**, 1245–1258.

410. Le Moual, H. and Koshland, D.E., Jr. (1996). Molecular evolution of the C-terminal cytoplasmic domain of a superfamily of bacterial receptors involved in taxis. *J. Mol. Biol.* **261**, 568–585.

411. Le Moual, H., Quang, T. and Koshland, D.E., Jr. (1997). Methylation of the *Escherichia coli* chemotaxis receptors: Intra- and interdimer mechanisms. *Biochemistry* **36**, 13 441–13 448.

412. Lederberg, J. (1956). Linear inheritance in transductional clones. *Genetics* **41**, 845–871.

413. Lee, G.F., Lebert, M.R., Lilly, A.A. and Hazelbauer, G.L. (1995). Transmembrane signaling characterized in bacterial chemoreceptors by using sulfhydryl cross-linking in *vivo. Proc. Natl. Acad. Sci. U.S.A.* **92**, 3391–3395.

414. Lee, S.-Y., Cho, H.S., Pelton, J.G., Yan, D.L., Henderson, R.K., King, D.S., Huang, L.-S., Kustu, S., Berry, E.A. and Wemmer, D.E. (2001). Crystal structure of an activated response regulator bound to its target. *Nature Struct. Biology* **8**, 52–56.

415. Lee, T.-Y., Makino, K., Shinagawa, H. and Nakata, A. (1990). Overproduction of acetate kinase activates the phosphate regulation in the absence of the *phoR phoM* functions in *Escherichia coli. J. Bacteriol.* **172**, 2245–2249.

416. Lengeler, J., Auburger, A.-M., Mayer, R. and Pecher, A. (1981). The phosphoenolpyruvate-dependent carbohydrate: Phosphotransferase system enzymes II as chemoreceptors in chemotaxis of *Escherichia coli* K12. *Mol. Gen. Genet.* **183**, 163–170.

417. Lengeler, J.W. and Jahreis, K. (1996). Phosphotransferase systems or PTS as carbohydrate transport and as signal transduction systems, in *Handbook of Biological Physics* (W.N. Konings, H.R. Kaback and J.S. Lolkema, eds.), pp. 573–598. Elsevier Science, Amsterdam.

418. Levit, M., Liu, Y., Surette, M. and Stock, J. (1996). Active site interference and asymmetric activation in the chemotaxis protein histidine kinase CheA. *J. Biol. Chem.* **271**, 32 057–32 063.

419. Levit, M.N., Liu, Y. and Stock, J.B. (1998). Stimulus response coupling in bacterial chemotaxis: receptor dimers in signaling arrays. *Mol. Microbiol.* **30**, 459–466.

420. Li, G. and Weis, R.M. (2000). Covalent modification regulates ligand binding to receptor complexes in the chemosensory system of *Escherichia coli*. *Cell* **100**, 357–365.

421. Li, J.Y., Swanson, R.V., Simon, M.I. and Weis, R.M. (1995). The response regulators CheB and CheY exhibit competitive binding to the kinase CheA. *Biochemistry* **34**, 14 626–14 636.

422. Li, J.Y., Li, G.Y. and Weis, R.M. (1997). The serine chemoreceptor from *Escherichia coli* is methylated through an inter-dimer process. *Biochemistry* **36**, 11 851–11 857.

423. Lin, L.N., Li, J.Y., Brandts, J.F. and Weiss, R.M. (1994). The serine receptor of bacterial chemotaxis exhibits half-site saturation for serine binding. *Biochemistry* **33**, 6564–6570.

424. Liu, J. and Parkinson, J.S. (1989). Role of CheW protein in coupling membrane receptors to the intracellular signaling system of bacterial chemotaxis. *Proc. Natl. Acad. Sci. U.S.A.* **86**, 8703–8707.

425. Liu, J.D. and Parkinson, J.S. (1991). Genetic evidence for interaction between the CheW and Tsr proteins during chemoreceptor signaling by *Escherichia coli*. *J. Bacteriol.* **173**, 4941–4951.

426. Liu, X.Y. and Matsumura, P. (1994). The FlhD/FlhC complex, a transcriptional activator of the *Escherichia coli* flagellar class II operons. *J. Bacteriol.* **176**, 7345–7351.

427. Liu, Y., Levit, M., Lurz, R., Surette, M.G. and Stock, J.B. (1997). Receptor-mediated protein kinase activation and the mechanism of transmembrane signaling in bacterial chemotaxis. *EMBO J.* **16**, 7231–7240.

428. Liu, Z.-X. and Fridovich, I. (1996). Negative chemotaxis in *Cytophaga johnsonae*. *Can. J. Microbiol.* **42**, 515–518.

429. Liu, Z.W. and Papadopoulos, K.D. (1996). A method for measuring bacterial chemotaxis parameters in a microcapillary. *Biotechnol. Bioengineer.* **51**, 120–125.

430. Lloyd, S.A., Tang, H., Wang, X., Billings, S. and Blair, D.F. (1996). Torque generation in the flagellar motor of *Escherichia coli*: evidence of a direct role for FliG but not for FliM or FliN. *J. Bacteriol.* **178**, 223–231.

431. Lloyd, S.A. and Blair, D.F. (1997). Charged residues of the rotor protein FliG essential for torque generation in the flagellar motor of *Escherichia coli*. *J. Mol. Biol.* **266**, 733–744.

432. Lloyd, S.A., Whitby, F.G., Blair, D.F. and Hill, C.P. (1999). Structure of the C-terminal domain of FliG, a component of the rotor in the bacterial flagellar motor. *Nature* **400**, 472–475.

433. Lo, W.S., Trievel, R.C., Rojas, J.R., Duggan, L., Hsu, J.Y., Allis, C.D., Marmorstein, R. and Berger, S.L. (2000). Phosphorylation of serine 10 in histone

H3 is functionally linked in vitro and in vivo to Gcn5-mediated acetylation at lysine 14. *Mol. Cell.* **5**, 917–926.

434. Loake, G.J., Ashby, A.M. and Shaw, C.H. (1988). Attraction of Agrobacterium tumefaciens C58C: towards sugars involves a highly sensitive chemotaxis system. *J. Gen. Microbiol.* **134**, 1427–1432.

435. Loisel, T.P., Boujemaa, R., Pantaloni, D. and Carlier, M.F. (1999). Reconstitution of actin-based motility of *Listeria* and *Shigella* using pure proteins. *Nature* **401**, 613–616.

436. Long, D.G. and Weis, R.M. (1992). *Escherichia coli* aspartate receptor. Oligomerization of the cytoplasmic fragment. *Biophys. J.* **62**, 69–71.

437. Long, D.G. and Weis, R.M. (1992). Oligomerization of the cytoplasmic fragment from the aspartate receptor of *Escherichia coli*. *Biochemistry* **31**, 9904–9911.

438. Lopez-de-Victoria, G., Zimmer-Faust, R.K. and Lovell, C.R. (1995). Computer-assisted video motion analysis: A powerful technique for investigating motility and chemotaxis. *J. Microbiol. Meth.* **23**, 329–341.

439. Lovely, P., Dahlquist, F.W., Macnab, R. and Koshland, D.E., Jr. (1974). An instrument for recording the motions of microorganisms in chemical gradients. *Rev. Sci. Instrum.* **45**, 683–686.

440. Lowe, G., Meister, M. and Berg, H.C. (1987). Rapid rotation of flagellar bundles in swimming bacteria. *Nature* **325**, 637–640.

441. Lukat, G.S., Stock, A.M. and Stock, J.B. (1990). Divalent metal ion binding to the CheY protein and its significance to phosphotransfer in bacterial chemotaxis. *Biochemistry* **29**, 5436–5442.

442. Lukat, G.S., Lee, B.H., Mottonen, J.M., Stock, A. and Stock, J.B. (1991). Roles of the highly conserved aspartate and lysine residues in the response regulator of bacterial chemotaxis. *J. Biol. Chem.* **266**, 8348–8354.

443. Lukat, G.S., McCleary, W.R., Stock, A.M. and Stock, J.B. (1992). Phosphorylation of bacterial response regulator proteins by low molecular weight phospho-donors. *Proc. Natl. Acad. Sci. U.S.A.* **89**, 718–722.

444. Lupas, A. and Stock, J. (1989). Phosphorylation of an N-terminal regulatory domain activates the CheB methylesterase in bacterial chemotaxis. *J. Biol. Chem.* **264**, 17337–17342.

445. Lux, R., Jahreis, K., Bettenbrock, K., Parkinson, J.S. and Lengeler, J.W. (1995). Coupling the phosphotransferase system and the methyl-accepting chemotaxis protein-dependent chemotaxis signaling pathways of *Escherichia coli*. *Proc. Natl. Acad. Sci. U.S.A.* **92**, 11583–11587.

446. Lux, R., Munasinghe, V.R.N., Castellano, F., Lengeler, J.W., Corrie, J.E.T. and Khan, S. (1999). Elucidation of a PTS-carbohydrate chemotactic signal pathway in *Escherichia coli* using a time-resolved behavioral assay. *Mol. Biol. Cell* **10**, 1133–1146.

447. Lux, R., Kar, N. and Khan, S. (2000). Overproduced *Salmonella typhimurium* flagellar motor switch complexes. *J. Mol. Biol.* **298**, 577–583.

448. Lybarger, S.R. and Maddock, J.R. (1999). Clustering of the chemoreceptor complex in *Escherichia coli* is independent of the methyltransferase CheR and the methylesterase CheB. *J. Bacteriol.* **181**, 5527–5529.

449. Lybarger, S.R. and Maddock, J.R. (2000). Differences in the polar clustering of the high- and low-abundance chemoreceptors of *Escherichia coli*. *Proc. Natl. Acad. Sci. U.S.A.* **97**, 8057–8062.

450. Lynch, B.A. and Koshland, D.E., Jr. (1992). Structural similarities between the aspartate receptor of bacterial chemotaxis and the trp repressor of *E. coli*. Implications for transmembrane signaling. *FEBS Lett.* **307**, 3–9.

451. Macnab, R.M. and Koshland, D.E., Jr. (1972). The gradient-sensing mechanism in bacterial chemotaxis. *Proc. Natl. Acad. Sci. U.S.A.* **69**, 2509–2512.

452. Macnab, R.M. (1976). Examination of bacterial flagellation by dark-field microscopy. *J. Clin. Microbiol.* **4**, 258–265.

453. Macnab, R.M. (1977). Bacterial flagella rotating in bundles: A study in helical geometry. *Proc. Natl. Acad. Sci. U.S.A.* **74**, 221–225.

454. Macnab, R.M. and Ornston, M.K. (1977). Normal-to-curly flagellar transitions and their role in bacterial tumbling: Stabilization of an alternative quaternary structure by mechanical force. *J. Mol. Biol.* **112**, 1–30.

455. Macnab, R.M. (1979). Bacterial flagella, in *Encyclopedia of Plant Physiology: Physiology of Movements* (W. Haupt and M.E. Feinleib, eds.), pp. 207–223. Springer-Verlag, Berlin, Heidelberg, New York.

456. Macnab, R.M. and Han, D.P. (1983). Asynchronous switching of flagellar motors on a single bacterial cell. *Cell* **32**, 109–117.

457. Macnab, R.M. and Aizawa, S.-I. (1984). Bacterial motility and the bacterial flagellar motor. *Ann. Rev. Biophys. Bioeng.* **13**, 51–83.

458. Macnab, R.M. (1985). Biochemistry of sensory transduction in bacteria, in *Sensory Perception and Transduction in Aneural Organisms* (G. Colombetti, F. Lenci and P.-S. Song, eds.), pp. 31–46. Plenum Press, New York.

459. Macnab, R.M. (1995). Flagellar switch, in *Two-Component Signal Transduction* (J.A. Hoch and T.J. Silhavy, eds.), pp. 181–199. American Society for Microbiology, Washington, DC.

460. Macnab, R.M. (1996). Flagella and Motility, in Escherichia coli *and* Salmonella: *Cellular and Molecular Biology* (F.C. Neidhardt, R. Curtis III, J.L. Ingraham, E.C.C. Lin, K.B. Low, B. Magasanik, W.S. Reznikoff, M. Riley, M. Schaechter and H.E. Umbarger, eds.), pp. 123–145. American Society for Microbiology, Washington, DC.

461. Macnab, R.M. (1999). The bacterial flagellum: Reversible rotary propeller and type III export apparatus. *J. Bacteriol.* **181**, 7149–7153.

462. Macnab, R.M. (2000). Type III protein pathway exports *Salmonella* flagella. *ASM News* **66**, 738–745.

463. Maddock, J.R. and Shapiro, L. (1993). Polar location of the chemoreceptor complex in the *Escherichia coli* cell. *Science* **259**, 1717–1723.

464. Magariyama, Y., Yamaguchi, S. and Aizawa, S.-I. (1990). Genetic and behavioral analysis of flagellar switch mutants of *Salmonella typhimurium*. *J. Bacteriol.* **172**, 4359–4369.

465. Magariyama, Y., Sugiyama, S., Muramoto, K., Maekawa, Y., Kawagishi, I., Imae, Y. and Kudo, S. (1994). Very fast flagellar rotation. *Nature* **371**, 752.

466. Magariyama, Y., Sugiyama, S., Muramoto, K., Kawagishi, I., Imae, Y. and Kudo, S. (1995). Simultaneous measurement of bacterial flagellar rotation rate and swimming speed. *Biophys. J.* **69**, 2154–2162.

467. Makishima, S., Komoriya, K., Yamaguchi, S. and Aizawa, S.-I. (2001). Length of the flagellar hook and the capacity of the type III export apparatus. *Science* **291**, 2411–2413.

468. Manson, M.D., Tedesco, P., Berg, H.C., Harold, F.M. and van-der-Drift, C. (1977). A protonmotive force drives bacterial flagella. *Proc. Natl. Acad. Sci. U.S.A.* **74**, 3060–3064.

469. Manson, M.D., Tedesco, P.M. and Berg, H.C. (1980). Energetics of flagellar rotation in bacteria. *J. Mol. Biol.* **138**, 541–561.

470. Manson, M.D., Boos, W., Bassford, P.J. and Rasmussen, B.A. (1985). Dependence of maltose transport and chemotaxis on the amount of maltose-binding protein. *J. Biol. Chem.* **260**, 9727–9733.

471. Manson, M.D., Blank, V., Brade, G. and Higgins, C.F. (1986). Peptide chemotaxis in *E. coli* involves the Tap signal transducer and the dipeptide permease. *Nature (London)* **321**, 253–256.

472. Manson, M.D. (2000). Allele-specific suppression as a tool to study protein-protein interactions in bacteria. *Methods* **20**, 18–34.

473. Margolin, Y. and Eisenbach, M. (1984). Voltage-clamp effects on bacterial chemotaxis. *J. Bacteriol.* **159**, 605–610 [Errata: 161 (1985) 823].

474. Margolin, Y., Barak, R. and Eisenbach, M. (1994). Arsenate arrests flagellar rotation in cytoplasm-free envelopes of bacteria. *J. Bacteriol.* **176**, 5547–5549.

475. Martínez, A., Torello, S. and Kolter, R. (1999). Sliding motility in mycobacteria. *J. Bacteriol.* **181**, 7331–7338.

476. Marwan, W., Schafer, W. and Oesterhelt, D. (1990). Signal transduction in *Halobacterium* depends on fumarate. *EMBO J.* **9**, 355–362.

477. Marwan, W. and Oesterhelt, D. (1991). Light-induced release of the switch factor during photophobic responses of *Halobacterium halobium*. *Naturwissenschaften* **78**, 127–129.

478. Marykwas, D.L. and Berg, H.C. (1996). A mutational analysis of the interaction between FliG and FliM, two components of the flagellar motor of *Escherichia coli*. *J. Bacteriol.* **178**, 1289–1294.

479. Marykwas, D.L., Schmidt, S.A. and Berg, H.C. (1996). Interacting components of the flagellar motor of *Escherichia coli* revealed by the two-hybrid system in yeast. *J. Mol. Biol.* **256**, 564–576.

480. Masduki, A., Nakamura, J., Ohga, T., Umezaki, R., Kato, J. and Ohtake, H. (1995). Isolation and characterization of chemotaxis mutants and genes of *Pseudomonas aeruginosa*. *J. Bacteriol.* **177**, 948–952.

481. Mathews, M.A.A., Tang, H.L. and Blair, D.F. (1998). Domain analysis of the FliM protein of *Escherichia coli*. *J. Bacteriol.* **180**, 5580–5590.

482. Matsumura, P., Roman, S., Volz, K. and McNally, D. (1990). Signaling complexes in bacterial chemotaxis. *Soc. Gen. Microbiol. Symp.* **46**, 135–154.

483. Matsuura, S., Shioi, J.-I. and Imae, Y. (1977). Motility in *Bacillus subtilis* driven by an artificial protonmotive force. *FEBS Lett.* **82**, 187–190.

484. Mayover, T.L., Halkides, C.J. and Stewart, R.C. (1999). Kinetic characterization of CheY phosphorylation reactions: comparison of P-CheA and small-molecule phosphodonors. *Biochemistry* **38**, 2259–2271.

485. McCarter, L. and Silverman, M. (1990). Surface-induced swarmer cell differentiation of *Vibrio parahaemolyticus*. *Molec. Microbiol.* **4**, 1057–1062.

486. McCarter, L.L. (1994). MotY, a component of the sodium-type flagellar motor. *J. Bacteriol.* **176**, 4219–4225.

487. McCarter, L.L. (1994). MotX, the channel component of the sodium-type flagellar motor. *J. Bacteriol.* **176**, 5988–5998.

488. McCleary, W.R. and Stock, J.B. (1994). Acetyl phosphate and the activation of two-component response regulators. *J. Biol. Chem.* **269**, 31567–31572.

489. McEvoy, M., Bren, A., Eisenbach, M. and Dahlquist, F.W. (1999). Identification of the binding interfaces on CheY for two of its targets, the phosphatase CheZ and the flagellar switch protein FliM. *J. Mol. Biol.* **289**, 1423–1433.

490. McEvoy, M.M., Zhou, H.J., Roth, A.F., Lowry, D.F., Morrison, T.B., Kay, L.E. and Dahlquist, F.W. (1995). Nuclear magnetic resonance assignments and global fold of a CheY-binding domain in CheA, the chemotaxis-specific kinase of *Escherichia coli*. *Biochemistry* **34**, 13871–13880.

491. McEvoy, M.M., Muhandiram, D.R., Kay, L.E. and Dahlquist, F.W. (1996). Structure and dynamics of a CheY-binding domain of the chemotaxis kinase CheA determined by nuclear magnetic resonance spectroscopy. *Biochemistry* **35**, 5633–5640.

492. McEvoy, M.M., Hausrath, A.C., Randolph, G.B., Remington, S.J. and Dahlquist, F.W. (1998). Two binding modes reveal flexibility in kinase/response regulator interactions in the bacterial chemotaxis pathway. *Proc. Natl. Acad. Sci. U.S.A.* **95**, 7333–7338.

493. McNally, D.F. and Matsumura, P. (1991). Bacterial chemotaxis signaling complexes: Formation of a CheA/CheW complex enhances autophosphorylation and affinity for CheY. *Proc. Natl. Acad. Sci. U.S.A.* **88**, 6269–6273.

494. McNamara, B.P. and Wolfe, A.J. (1997). Coexpression of the long and short forms of CheA, the chemotaxis histidine kinase, by members of the family *Enterobacteriaceae*. *J. Bacteriol.* **179**, 1813–1818.

495. Meister, M., Lowe, G. and Berg, H.C. (1987). The proton flux through the bacterial flagellar motor. *Cell* **49**, 643–650.

496. Melton, T., Hartman, P.E., Stratis, J.P., Lee, T.L. and Davis, A.T. (1978). Chemotaxis of *Salmonella typhimurium* to amino acids and some sugars. *J. Bacteriol.* **133**, 708–716.

497. Merz, A.J., So, M. and Sheetz, M.P. (2000). Pilus retraction powers bacterial twitching motility. *Nature* **407**, 98–102.

498. Mesibov, R. and Adler, J. (1972). Chemotaxis towards amino acids in *Escherichia coli*. *J. Bacteriol.* **112**, 315–326.

499. Mesibov, R., Ordal, G.W. and Adler, J. (1973). The range of attractant concentrations for bacterial chemotaxis and the threshold and size of response over this range: Weber law and related phenomena. *J. Gen. Physiol.* **62**, 203–223.

500. Metzner, P. (1919). Die Bewegung und Reizbeantwortung der bipolar gegeisselten Spirillen. *Jahrb. Wiss. Botan.* **59**, 325–412.

501. Meynell, E.W. (1961). A phage, $\phi\chi$, which attacks motile bacteria. *J. Gen. Microbiol.* **25**, 253–290.

502. Milburn, M.V., Prive, G.G., Milligan, D.L., Scott, W.G., Yeh, J., Jancarik, J., Koshland, D.E., Jr. and Kim, S.H. (1991). Three-dimensional structures of the ligand-binding domain of the bacterial aspartate receptor with and without a ligand. *Science* **254**, 1342–1347.

503. Milligan, D.L. and Koshland, D.E., Jr. (1988). Site-directed cross-linking: establishing the dimeric structure of the aspartate receptor of bacterial chemotaxis. *J. Biol. Chem.* **263**, 6268–6275.

504. Milligan, D.L. and Koshland, D.E., Jr. (1993). Purification and characterization of the periplasmic domain of the aspartate chemoreceptor. *J. Biol. Chem.* **268**, 19991–19997.

505. Minamino, T., González-Pedrajo, B., Yamaguchi, K., Aizawa, S.-I. and Macnab, R.M. (1999). FliK, the protein responsible for flagellar hook length control in *Salmonella*, is exported during hook assembly. *Mol. Microbiol.* **34**, 295–304.

506. Minamino, T. and Macnab, R.M. (1999). Components of the *Salmonella* flagellar export apparatus and classification of export substrates. *J. Bacteriol.* **181**, 1388–1394.

507. Minamino, T., Chu, R., Yamaguchi, S. and Macnab, R.M. (2000). Role of FliJ in flagellar protein export in *Salmonella*. *J. Bacteriol.* **182**, 4207–4215.

508. Minamino, T. and Macnab, R.M. (2000). FliH, a soluble component of the type III flagellar export apparatus of *Salmonella*, forms a complex with FliI and inhibits its ATPase activity. *Mol. Microbiol.* **37**, 1494–1503.

509. Minamino, T., Yamaguchi, S. and Macnab, R.M. (2000). Interaction between FliE and FlgB, a proximal rod component of the flagellar basal body of *Salmonella*. *J. Bacteriol.* **182**, 3029–3036.

510. Mitchell, J.G., Pearson, L. and Dillon, S. (1996). Clustering of marine bacteria in seawater enrichments. *Appl. Environ. Microbiol.* **62**, 3716–3721.

511. Miyata, M. and Seto, S. (1999). Cell reproduction cycle of mycoplasma. *Biochimie* **81**, 873–878.

512. Miyata, M., Yamamoto, H., Shimizu, T., Uenoyama, A., Citti, C. and Rosengarten, R. (2000). Gliding mutants of *Mycoplasma mobile*: relationships between motility and cell morphology, cell adhesion, and microcolony formation. *Microbiology* **146**, 1311–1320.

513. Miyata, M., Ryu, W.S. and Berg, H.C. (2002). Force and velocity of *Mycoplasma mobile* gliding. *J. Bacteriol.* **184**, 1827–1831.

514. Miyata, M. and Uenoyama, A. (2002). Movement on the cell surface of the gliding bacterium, *Mycoplasma mobile*, is limited to its head-like structure. *FEMS Microbiol. Lett.* **215**, 285–289.

515. Mizuno, T., Mutoh, N., Panasenko, S.M. and Imae, Y. (1986). Acquisition of maltose chemotaxis in *Salmonella typhimurium* by the introduction of the *Escherichia coli* chemosensory transducer gene. *J. Bacteriol.* **165**, 890–895.

516. Moens, S., Michiels, K. and Vanderleyden, J. (1995). Glycosylation of the flagellin of the polar flagellum of *Azospirillum brasilense*, a Gram-negative nitrogen-fixing bacterium. *Microbiology* **141**, 2651–2657.

517. Moens, S. and Vanderleyden, J. (1996). Functions of bacterial flagella. *Crit. Rev. Microbiol.* **22**, 67–100.

518. Montrone, M., Marwan, W., Grunberg, H., Musseleck, S., Starostzik, C. and Oesterhelt, D. (1993). Sensory rhodopsin-controlled release of the switch factor fumarate in *Halobacterium salinarium*. *Mol. Microbiol.* **10**, 1077–1085.

519. Montrone, M., Oesterhelt, D. and Marwan, W. (1996). Phosphorylation-independent bacterial chemoresponses correlate with changes in the cytoplasmic level of fumarate. *J. Bacteriol.* **178**, 6882–6887.

520. Montrone, M., Eisenbach, M., Oesterhelt, D. and Marwan, W. (1998). Regulation of switching frequency and bias of the bacterial flagellar motor by CheY and fumarate. *J. Bacteriol.* **180**, 3375–3380.

521. Morrison, T.B. and Parkinson, J.S. (1994). Liberation of an interaction domain from the phosphotransfer region of CheA, a signaling kinase from *Escherichia coli*. *Proc. Natl. Acad. Sci. U.S.A* **91**, 5485–5489.

522. Morrison, T.B. and Parkinson, J.S. (1997). A fragment liberated from the *Escherichia coli* CheA kinase that blocks stimulatory, but not inhibitory, chemoreceptor signaling. *J. Bacteriol.* **179**, 5543–5550.

523. Mowbray, S.L., Foster, D.L. and Koshland, D.E., Jr. (1985). Proteolytic fragments identified with domains of the aspartate chemoreceptor. *J. Biol. Chem.* **260**, 11 711–11 718.

524. Mowbray, S.L. and Sandgren, M.O.J. (1998). Chemotaxis receptors: A progress report on structure and function. *J. Struct. Biol.* **124**, 257–275.

525. Müller, J., Schiel, S., Ordal, G.W. and Saxild, H.H. (1997). Functional and genetic characterization of *mcpC*, which encodes a third methyl-accepting chemotaxis protein in *Bacillus subtilis*. *Microbiology* **143**, 3231–3240.

526. Muramoto, K., Sugiyama, S., Cragoe, E.J. and Imae, Y. (1994). Successive inactivation of the force-generating units of sodium-driven bacterial flagellar motors by a photoreactive amiloride analog. *J. Biol. Chem.* **269**, 3374–3380.

527. Muramoto, K., Kawagishi, I., Kudo, S., Magariyama, Y., Imae, Y. and Homma, M. (1995). High-speed rotation and speed stability of the sodium-driven flagellar motor in *Vibrio alginolyticus*. *J. Mol. Biol.* **251**, 50–58.

528. Muramoto, K., Magariyama, Y., Homma, M., Kawagishi, I., Sugiyama, S., Imae, Y. and Kudo, S. (1996). Rotational fluctuation of the sodium-driven flagellar motor of *Vibrio alginolyticus* induced by binding of inhibitors. *J. Mol. Biol.* **259**, 687–695.

529. Muramoto, K. and Macnab, R.M. (1998). Deletion analysis of MotA and MotB, components of the force-generating unit in the flagellar motor of *Salmonella*. *Mol. Microbiol.* **29**, 1191–1202.

530. Muramoto, K., Makishima, S., Aizawa, S.-I. and Macnab, R.M. (1998). Effect of cellular level of FliK on flagellar hook and filament assembly in *Salmonella typhimurium*. *J. Mol. Biol.* **277**, 871–882.

531. Muramoto, K., Makishima, S., Aizawa, S.-I. and Macnab, R.M. (1999). Effect of hook subunit concentration on assembly and control of length of the flagellar hook of *Salmonella*. *J. Bacteriol.* **181**, 5808–5813.

532. Murphy, O.J., Kovacs, F.A., Sicard, E.L. and Thompson, L.K. (2001). Site-directed solid-state NMR measurement of a ligand-induced conformational change in the serine bacterial chemoreceptor. *Biochemistry* **40**, 1358–1366.

533. Mutoh, N. and Simon, M.I. (1986). Nucleotide sequence corresponding to five chemotaxis genes in *Escherichia coli*. *J. Bacteriol.* **165**, 161–166.

534. Mytelka, D.S. and Chamberlin, M.J. (1996). *Escherichia coli fliAZY* operon. *J. Bacteriol.* **178**, 24–34.

535. Namba, K. and Vonderviszt, F. (1997). Molecular architecture of bacterial flagellum. *Quart. Rev. Biophys.* **30**, 1–65.

536. Nambu, T., Minamino, T., Macnab, R.M. and Kutsukake, K. (1999). Peptidoglycan-hydrolyzing activity of the FlgJ protein, essential for flagellar rod formation in *Salmonella typhimurium*. *J. Bacteriol.* **181**, 1555–1561.

537. Nambu, T. and Kutsukake, K. (2000). The *Salmonella* FlgA protein, a putative periplasmic chaperone essential for flagellar P ring formation. *Microbiology* **146**, 1171–1178.

538. Nealson, K., Saffarini, D., Moser, D. and Smith, M.J. (1994). A spectrophotometric method for monitoring tactic responses of bacteria under anaerobic conditions. *J. Microbiol. Meth.* **20**, 211–218.

539. Ng, W.V., Kennedy, S.P., Mahairas, G.G., Berquist, B., Pan, M., Shukla, H.D., Lasky, S.R., Baliga, N.S., Thorsson, V., Sbrogna, J., Swartzell, S., Weir, D., Hall, J., Dahl, T.A., Welti, R., Goo, Y.A., Leithauser, B., Keller, K., Cruz, R., Danson, M.J., Hough, D.W., Maddocks, D.G., Jablonski, P.E., Krebs, M.P., Angevine, C.M., Dale, H., Isenbarger, T.A., Peck, R.F., Pohlschroder, M., Spudich, J.L., Jung, K.W., Alam, M., Freitas, T., Hou, S., Daniels, C.J., Dennis, P.P., Omer, A.D., Ebhardt, H., Lowe, T.M., Liang, P., Riley, M., Hood, L. and DasSarma, S. (2000). Genome sequence of *Halobacterium* species NRC-1. *Proc. Natl. Acad. Sci. U.S.A.* **97**, 12176–12181.

540. Nikata, T., Sumida, K., Kato, J. and Ohtake, H. (1992). Rapid method for analyzing bacterial behavioral responses to chemical stimuli. *Appl. Environ. Microbiol.* **58**, 2250–2254.

541. Ninfa, E.G., Stock, A., Mowbray, S. and Stock, J. (1991). Reconstitution of the bacterial chemotaxis signal transduction system from purified components. *J. Biol. Chem.* **266**, 9764–9770.

542. Nishimura, A. and Hirota, Y. (1989). A cell division regulatory mechanism controls the flagellar regulon in *Escherichia coli*. *Mol. Gen. Genet.* **216**, 340–346.

543. Nowlin, D.M., Bollinger, J. and Hazelbauer, G.L. (1987). Sites of covalent modification in Trg, a sensory transducer of *Escherichia coli*. *J. Biol. Chem.* **262**, 6039–6045.

544. Nowlin, D.M., Bollinger, J. and Hazelbauer, G.L. (1988). Site-directed mutations altering methyl-accepting residues of a sensory transducer protein. *Proteins Struct. Funct. Genet.* **3**, 102–112.

545. Ohga, T., Masduki, A., Kato, J. and Ohtake, H. (1993). Chemotaxis away from thiocyanic and isothiocyanic esters in *Pseudomonas aeruginosa*. *FEMS Microbiol. Lett.* **113**, 63–66.

546. Ohnishi, K., Ohto, Y., Aizawa, S.-I., Macnab, R.M. and Iino, T. (1994). FlgD is a scaffolding protein needed for flagellar hook assembly in *Salmonella typhimurium*. *J. Bacteriol.* **176**, 2272–2281.

547. Ohtake, H., Kato, J., Kuroda, A., Wu, H. and Ikeda, T. (1998). Regulation of bacterial phosphate taxis and polyphosphate accumulation in response to phosphate starvation stress. *J. Biosci.* **23**, 491–499.

548. Omirbekova, N.G., Gabai, V.L., Sherman, M.Y., Vorobyeva, N.V. and Glagolev, A.N. (1985). Involvement of Ca^+ and cGMP in bacterial taxis. *FEMS Microbiol. Lett.* **28**, 259–263.

549. Oosawa, K. and Imae, Y. (1983). Glycerol and ethylene glycol: members of a new class of repellents of *Escherichia coli* chemotaxis. *J. Bacteriol.* **154**, 104–112.

550. Oosawa, K. and Imae, Y. (1984). Demethylation of methyl-accepting chemotaxis proteins in *Escherichia coli* induced by the repellents glycerol and ethylene glycol. *J. Bacteriol.* **157**, 576–581.

551. Oosawa, K., Hess, J.F. and Simon, M.I. (1988). Mutants defective in bacterial chemotaxis show modified protein phosphorylation. *Cell* **53**, 89–96.

552. Oosawa, K., Ueno, T. and Aizawa, S. (1994). Overproduction of the bacterial flagellar switch proteins and their interactions with the MS ring complex *in vitro*. *J. Bacteriol.* **176**, 3683–3691.

553. Oplatka, A. (1998). Do the bacterial flagellar motor and ATP synthase operate as water turbines? *Biochem. Biophys. Res. Commun.* **249**, 573–578.

554. Oplatka, A. (1998). Are rotors at the heart of all biological motors? *Biochem. Biophys. Res. Commun.* **246**, 301–306.

555. Ordal, G.W. and Adler, J. (1974). Properties of mutants in galactose taxis and transport. *J. Bacteriol.* **117**, 517–526.

556. Ordal, G.W. and Adler, J. (1974). Isolation and complementation of mutants in galactose taxis and transport. *J. Bacteriol.* **117**, 509–516.

557. Ordal, G.W. and Goldman, D.J. (1975). Chemotaxis away from uncouplers of oxidative phosphorylation in *Bacillus subtilis*. *Science* **189**, 802–805.

558. Ordal, G.W. and Goldman, D.J. (1976). Chemotactic repellents of *Bacillus subtilis*. *J. Mol. Biol.* **100**, 103–108.

559. Ordal, G.W. and Gibson, K.J. (1977). Chemotaxis toward amino acids by *Bacillus subtilis*. *J. Bacteriol.* **129**, 151–155.

560. Ordal, G.W., Villani, D.P. and Rosendahl, M.S. (1979). Chemotaxis towards sugars by *Bacillus subtilis*. *J. Gen. Microbiol.* **115**, 167–172.

561. Ordal, G.W. and Nettleton, D.O. (1985). Chemotaxis in *Bacillus subtilis*, in *The Molecular Biology of the Bacilli*, pp. 53–72. Academic Press, New York.

562. Ottemann, K.M. and Miller, J.F. (1997). Roles for motility in bacterial–host interactions. *Mol. Microbiol.* **24**, 1109–1117.

563. Ottemann, K.M., Thorgeirsson, T.E., Kolodziej, A.F., Shin, Y.-K. and Koshland, D.E., Jr. (1998). Direct measurement of small ligand-induced conformational changes in the aspartate chemoreceptor using EPR. *Biochemistry* **37**, 7062–7069.

564. Ottemann, K.M., Xiao, W., Shin, Y.K. and Koshland, D.E., Jr. (1999). A piston model for transmembrane signaling of the aspartate receptor. *Science* **285**, 1751–1754.

565. Packer, H.L. and Armitage, J.P. (1994). The chemokinetic and chemotactic behavior of *Rhodobacter sphaeroides*: Two independent responses. *J. Bacteriol.* **176**, 206–212.

566. Packer, H.L., Gauden, D.E. and Armitage, J.P. (1996). The behavioural response of anaerobic *Rhodobacter sphaeroides* to temporal stimuli. *Microbiology* **142**, 593–599.

567. Pantaloni, D., Le Clainche, C. and Carlier, M.-F. (2001). Mechanism of actin-based motility. *Science* **292**, 1502–1506.

568. Parkinson, J.S. and Revello, P.T. (1978). Sensory adaptation mutants of *Escherichia coli*. *Cell* **15**, 1221–1230.

569. Parkinson, J.S. and Parker, S.R. (1979). Interaction of the *cheC* and *cheZ* gene products is required for chemotactic behavior in *Escherichia coli*. *Proc. Natl. Acad. Sci. U.S.A.* **76**, 2390–2394.

570. Parkinson, J.S. (1981). Genetics of bacterial chemotaxis. *Soc. Gen. Microbiol. Symp.* **31**, 265–290.

571. Parkinson, J.S. and Houts, S.E. (1982). Isolation and behavior of *Escherichia coli* deletion mutants lacking chemotaxis functions. *J. Bacteriol.* **151**, 106–113.

572. Parkinson, J.S., Parker, S.R., Talbert, P.B. and Houts, S.E. (1983). Interactions between chemotaxis genes and flagellar genes in *Escherichia coli*. *J. Bacteriol.* **155**, 265–274.

573. Parkinson, J.S. and Kofoid, E.C. (1992). Communication modules in bacterial signaling proteins. *Annu. Rev. Genet.* **26**, 71–112.

574. Parkinson, J.S. and Blair, D.F. (1993). Does *E. coli* have a nose. *Science* **259**, 1701–1702.

575. Pfeffer, W. (1884). Lokomotorische richtungsbewegungen durch chemische reize. *Untersuch. aus d. Botan. Inst. Tübingen* **1**, 363–482.

576. Pfeffer, W. (1888). Ueber chemotaktische bewegungen von bakterien, flagellaten und volvocineen. *Unters. Botan. Inst. Tubingen* **2**, 582–661.

577. Pitta, T.P. and Berg, H.C. (1995). Self-electrophoresis is not the mechanism for motility in swimming cyanobacteria. *J. Bacteriol.* **177**, 5701–5703.

578. Pitta, T.P., Sherwood, E.E., Kobel, A.M. and Berg, H.C. (1997). Calcium is required for swimming by the nonflagellated cyanobacterium *Synechococcus* strain WH8113. *J. Bacteriol.* **179**, 2524–2528.

579. Pollard, T.D. (1995). Actin cytoskeleton: Missing link for intracellular bacterial motility? *Curr. Biol.* **5**, 837–840.

580. Poole, P.S., Smith, M.J. and Armitage, J.P. (1993). Chemotactic signaling in *Rhodobacter sphaeroides* requires metabolism of attractants. *J. Bacteriol.* **175**, 291–294.

581. Porter, S.L., Warren, A.V., Martin, A.C. and Armitage, J.P. (2002). The third chemotaxis locus of *Rhodobacter sphaeroides* is essential for chemotaxis. *Mol. Microbiol.* **46**, 1081–1094.

582. Postma, P.W. and Lengeler, J.W. (1985). Phosphoenolpyruvate: carbohydrate phosphotransferase system of bacteria. *Microbiol. Rev.* **49**, 232–269.

583. Postma, P.W., Lengeler, J.W. and Jacobson, G.R. (1996). Phosphoenolpyruvate: carbohydrate phosphotransferase systems, in Escherichia coli *and* Salmonella: *Cellular and Molecular Biology* (F.C. Neidhardt, R. Curtis III, J.L. Ingraham, E.C.C. Lin, K.B. Low, B. Magasanik, W.S. Reznikoff, M. Riley, M. Schaechter and H.E. Umbarger, eds.), pp. 1149–1174. American Society for Microbiology, Washington, DC.

584. Prasad, K., Caplan, S.R. and Eisenbach, M. (1998). Fumarate modulates bacterial flagellar rotation by lowering the free energy difference between the clockwise and counterclockwise states of the motor. *J. Mol. Biol.* **280**, 821–828.

585. Prasad, K. (2001). Function of fumarate in bacterial chemotaxis. Ph.D. thesis, Weizmann Institute of Science, Rehovot, Israel.

586. Prüss, B.M. and Wolfe, A.J. (1994). Regulation of acetyl phosphate synthesis and degradation, and the control of flagellar expression in *Escherichia coli*. *Mol. Microbiol.* **12**, 973–984.

587. Prüss, B.M. and Matsumura, P. (1996). A regulator of the flagellar regulon of *Escherichia coli*, *flhD*, also affects cell division. *J. Bacteriol.* **178**, 668–674.

588. Prüss, B.M., Markovic, D. and Matsumura, P. (1997). The *Escherichia coli* flagellar transcriptional activator *flhD* regulates cell division through induction of the acid response gene *cadA*. *J. Bacteriol.* **179**, 3818–3821.

589. Quiocho, F.A. and Ledvina, P.S. (1996). Atomic structure and specificity of bacterial periplasmic receptors for active transport and chemotaxis: variation of common themes. *Mol. Microbiol.* **20**, 17–25.

590. Ramakrishnan, R., Schuster, M. and Bourret, R.B. (1998). Acetylation at Lys-92 enhances signaling by the chemotaxis response regulator protein CheY. *Proc. Natl. Acad. Sci. U.S.A.* **95**, 4918–4923.

591. Ravid, S. and Eisenbach, M. (1983). Correlation between bacteriophage *chi* adsorption and mode of flagellar rotation of *Escherichia coli* chemotaxis mutants. *J. Bacteriol.* **154**, 604–611.

592. Ravid, S. and Eisenbach, M. (1984). Direction of flagellar rotation in bacterial cell envelopes. *J. Bacteriol.* **158**, 222–230 [Erratum: 159 (1984) 433].

593. Ravid, S. and Eisenbach, M. (1984). Minimal requirements for rotation of bacterial flagella. *J. Bacteriol.* **158**, 1208–1210.

594. Ravid, S., Matsumura, P. and Eisenbach, M. (1986). Restoration of flagellar clockwise rotation in bacterial envelopes by insertion of the chemotaxis protein CheY. *Proc. Natl. Acad. Sci. U.S.A.* **83**, 7157–7161.

595. Reader, R.W., Tso, W.-W., Springer, M.S., Goy, M.F. and Adler, J. (1979). Pleiotropic aspartate taxis and serine taxis mutants of *Escherichia coli. J. Gen. Microbiol.* **111**, 363–374.

596. Rebbapragada, A., Johnson, M.S., Harding, G.P., Zuccarelli, A.J., Fletcher, H.M., Zhulin, I.B. and Taylor, B.L. (1997). The Aer protein and the serine chemoreceptor Tsr independently sense intracellular energy levels and transduce oxygen, redox, and energy signals for *Escherichia coli* behavior. *Proc. Natl. Acad. Sci. U.S.A.* **94**, 10541–10546.

597. Reiner, O. and Okon, Y. (1986). Oxygen recognition in aerotactic behaviour of *Azospirillum brasilense* Cd. *Can. J. Microbiol.* **32**, 829–834.

598. Repaske, D.R. and Adler, J. (1981). Change in intracellular pH of *Escherichia coli* mediates the chemotactic response to certain attractants and repellents. *J. Bacteriol.* **145**, 1196–1208.

599. Repik, A., Rebbapragada, A., Johnson, M.S., Haznedar, J.Ö., Zhulin, I.B. and Taylor, B.L. (2000). PAS domain residues involved in signal transduction by the Aer redox sensor of *Escherichia coli. Mol. Microbiol.* **36**, 806–816.

600. Rice, M.S. and Dahlquist, F.W. (1991). Sites of deamidation and methylation in Tsr, a bacterial chemotaxis sensory transducer. *J. Biol. Chem.* **266**, 9746–9753.

601. Ridder, I.S. and Dijkstra, B.W. (1999). Identification of the Mg^{2+}-binding site in the P-type ATPase and phosphatase members of the HAD (haloacid dehalogenase) superfamily by structural similarity to the response regulator protein CheY. *Biochem. J.* **339**, 223–226.

602. Robillard, G.T. and Broos, J. (1999). Structure/function studies on the bacterial carbohydrate transporters, enzymes II, of the phosphoenolpyruvate-dependent phosphotransferase system. *Biochim. Biophys. Acta* 1422, 73–104.

603. Roman, S.J., Meyers, M., Volz, K. and Matsumura, P. (1992). A chemotactic signaling surface on CheY defined by suppressors of flagellar switch mutations. *J. Bacteriol.* **174**, 6247–6255.

604. Rosario, M.M.L., Fredrick, K.L., Ordal, G.W. and Helmann, J.D. (1994). Chemotaxis in *Bacillus subtilis* requires either of two functionally redundant CheW homologs. *J. Bacteriol.* **176**, 2736–2739.

605. Rosario, M.M.L., Kirby, J.R., Bochar, D.A. and Ordal, G.W. (1995). Chemotactic methylation and behavior in *Bacillus subtilis*: role of two unique proteins, CheC and CheD. *Biochemistry* **34**, 3823–3831.

606. Rosario, M.M.L. and Ordal, G.W. (1996). CheC and CheD interact to regulate methylation of *Bacillus subtilis* methyl-accepting chemotaxis proteins. *Mol. Microbiol.* **21**, 511–518.

607. Rose, I.A., Grunberg-Manago, M., Korey, S.R. and Ochoa, S. (1954). Enzymatic phosphorylation of acetate. *J. Biol. Chem.* **211**, 737–756.

608. Rosenbluh, A. and Eisenbach, M. (1992). Effect of mechanical removal of pili on gliding motility of *Myxococcus xanthus*. *J. Bacteriol.* **174**, 5406–5413.

609. Rudolph, J. and Oesterhelt, D. (1995). Chemotaxis and phototaxis require a CheA histidine kinase in the archaeon *Halobacterium salinarium. EMBO J.* **14**, 667–673.

610. Rudolph, J., Tolliday, N., Schmitt, C., Schuster, S.C. and Oesterhelt, D. (1995). Phosphorylation in halobacterial signal transduction. *EMBO J.* **14**, 4249–4257.

611. Rudolph, J., Nordmann, B., Storch, K.-F., Gruenberg, H., Rodewald, K. and Oesterhelt, D. (1996). A family of halobacterial transducer proteins. *FEMS Microbiol. Lett.* **139**, 161–168.

612. Russo, A.F. and Koshland, D.E., Jr. (1986). Identification of the *tip*-encoded receptor in bacterial sensing. *J. Bacteriol.* **165**, 276–282.

613. Ryu, W.S., Berry, R.M. and Berg, H.C. (2000). Torque-generating units of the flagellar motor of *Escherichia coli* have a high duty ratio. *Nature* **403**, 444–447.

614. Sager, B.M., Sekelsky, J.J., Matsumura, P. and Adler, J. (1988). Use of a computer to assay motility in bacteria. *Anal. Biochem.* **173**, 271–277.

615. Sagi-Yosef, Y., Khan, S. and Eisenbach, M. (2003). Binding of the chemotaxis response regulator CheY to isolated, intact switch complex of the bacterial flagellar motor: lack of cooperativity. Submitted.

616. Samuel, A.D.T. and Berg, H.C. (1995). Fluctuation analysis of rotational speeds of the bacterial flagellar motor. *Proc. Natl. Acad. Sci. U.S.A.* **92**, 3502–3506.

617. Samuel, A.D.T. and Berg, H.C. (1996). Torque-generating units of the bacterial flagellar motor step independently. *Biophys. J.* **71**, 918–923.

618. Samuel, A.D.T., Pitta, T.P., Ryu, W.S., Danese, P.N., Leung, E.C.W. and Berg, H.C. (1999). Flagellar determinants of bacterial sensitivity to χ-phage. *Proc. Natl. Acad. Sci. U.S.A.* **96**, 9863–9866.

619. Samuel, A.D.T., Petersen, J.D. and Reese, T.S. (2001). Envelope structure of *Synechococcus* sp. WH8113, a nonflagellated swimming cyanobacterium. *BMC Microbiol.* **1**, 4 (http://www.biomedcentral.com/1471–2180/1/4).

620. Sanatinia, H., Kofoid, E.C., Morrison, T.B. and Parkinson, J.S. (1995). The smaller of two overlapping *cheA* gene products is not essential for chemotaxis in *Escherichia coli. J. Bacteriol.* **177**, 2713–2720.

621. Sanders, D.A., Gillece-Castro, B.L., Stock, A.M., Burlingame, A.L. and Koshland, D.E., Jr. (1989). Identification of the site of phosphorylation of the chemotaxis response regulator protein, CheY. *J. Biol. Chem.* **264**, 21 770–21 778.

622. Sanna, M.G., Swanson, R.V., Bourret, R.B. and Simon, M.I. (1995). Mutations in the chemotactic response regulator, CheY, that confer resistance to the phosphatase activity of CheZ. *Mol. Microbiol.* **15**, 1069–1079.

623. Sanna, M.G. and Simon, M.I. (1996). In vivo and in vitro characterization of *Escherichia coli* protein CheZ gain- and loss-of-function mutants. *J. Bacteriol.* **178**, 6275–6280.

624. Sar, N., McCarter, L., Simon, M. and Silverman, M. (1990). Chemotactic control of the two flagellar systems of *Vibrio parahaemolyticus. J. Bacteriol.* **172**, 334–341.

625. Sato, K. and Homma, M. (2000). Functional reconstitution of the Na^+-driven polar flagellar motor component of *Vibrio alginolyticus. J. Biol. Chem.* **275**, 5718–5722.

626. Schade, S.Z., Adler, J. and Ris, H. (1967). How bacteriophage χ attacks motile bacteria. *J. Virol.* **1**, 599–609.

627. Scharf, B.E., Fahrner, K.A. and Berg, H.C. (1998). CheZ has no effect on flagellar motors activated by CheY[13DK106YW]. *J. Bacteriol.* **180**, 5123–5128.

628. Scharf, B.E., Fahrner, K.A., Turner, L. and Berg, H.C. (1998). Control of direction of flagellar rotation in bacterial chemotaxis. *Proc. Natl. Acad. Sci. U.S.A.* **95**, 201–206.

629. Scheffzek, K., Ahmadian, M.R., Kabsch, W., Wiesmuller, L., Lautwein, A., Schmitz, F. and Wittinghofer, A. (1997). The Ras-RasGAP complex: structural basis for GTPase activation and its loss in oncogenic Ras mutants. *Science* **277**, 333–338.

630. Schimz, A. and Hildebrand, E. (1998). Oscillating signals in the sensory pathway of halobacteria induced by periodic light stimuli. *Eur. Biophys. J.* **27**, 646–650.

631. Schoenhals, G.J. and Macnab, R.M. (1999). FliL is a membrane-associated component of the flagellar basal body of *Salmonella*. *Microbiology* **145**, 1769–1775.

632. Schuster, M., Abouhamad, W.N., Silversmith, R.E. and Bourret, R.B. (1998). Chemotactic response regulator mutant CheY95IV exhibits enhanced binding to the flagellar switch and phosphorylation-dependent constitutive signaling. *Mol. Microbiol.* **27**, 1065–1075 [Erratum: 28 (1998) 863].

633. Schuster, M., Zhao, R., Bourret, R.B. and Collins, E.J. (2000). Correlated switch binding and signaling in bacterial chemotaxis. *J. Biol. Chem.* **275**, 19 752–19 758.

634. Schuster, M., Silversmith, R.E. and Bourret, R.B. (2001). Conformational coupling in the chemotaxis response regulator CheY. *Proc. Natl. Acad. Sci. U.S.A.* **98**, 6003–6008.

635. Schuster, S.C., Swanson, R.V., Alex, L.A., Bourret, R.B. and Simon, M.I. (1993). Assembly and function of a quaternary signal transduction complex monitored by surface plasmon resonance. *Nature* **365**, 343–347.

636. Scott, W.G., Milligan, D.L., Milburn, M.V., Prive, G.G., Yeh, J., Koshland, D.E., Jr. and Kim, S.H. (1993). Refined structures of the ligand-binding domain of the aspartate receptor from *Salmonella typhimurium*. *J. Mol. Biol.* **232**, 555–573.

637. Segall, J.E., Ishihara, A. and Berg, H.C. (1985). Chemotactic signaling in filamentous cells of *Escherichia coli*. *J. Bacteriol.* **161**, 51–59.

638. Segall, J.E., Block, S.M. and Berg, H.C. (1986). Temporal comparisons in bacterial chemotaxis. *Proc. Natl. Acad. Sci. U.S.A.* **83**, 8987–8991.

639. Semmler, A.B.T., Whitchurch, C.B. and Mattick, J.S. (1999). A re-examination of twitching motility in *Pseudomonas aeruginosa*. *Microbiology* **145**, 2863–2873.

640. Seymour, F.W.K. and Doetsch, R.N. (1973). Chemotactic responses by motile bacteria. *J. Gen. Microbiol.* **78**, 287–296.

641. Shah, D.S.H., Porter, S.L., Harris, D.C., Wadhams, G.H., Hamblin, P.A. and Armitage, J.P. (2000). Identification of a fourth cheY gene in *Rhodobacter sphaeroides* and interspecies interaction within the bacterial chemotaxis signal transduction pathway. *Mol. Microbiol.* **35**, 101–112.

642. Shah, D.S.H., Porter, S.L., Martin, A.C., Hamblin, P.A. and Armitage, J.P. (2000). Fine-tuning bacterial chemotaxis: analysis of *Rhodobacter sphaeroides* behaviour under aerobic and anaerobic conditions by mutation of the major chemotaxis operons and cheY genes. *EMBO J.* **19**, 4601–4613.

643. Sharff, A.J., Rodseth, L.E., Spurlino, J.C. and Quiocho, F.A. (1992). Crystallographic evidence of a large ligand-induced hinge-twist motion between the two domains of the maltodextrin-binding protein involved in active transport and chemotaxis. *Biochemistry* **31**, 10 657–10 663.

644. Sharp, L.L., Zhou, J.D. and Blair, D.F. (1995). Features of MotA proton channel structure revealed by tryptophan-scanning mutagenesis. *Proc. Natl. Acad. Sci. U.S.A.* **92**, 7946–7950.

645. Shaw, C.H. (1991). Swimming against the tide—chemotaxis in *Agrobacterium*. *Bioessays* **13**, 25–29.

646. Shea, C., Nunley, J.W. and Smith-Somerville, H.E. (1991). Variable expression of gliding and swimming motility in *Deleya marina*. *Can. J. Microbiol.* **37**, 808–814.

647. Shi, W.Y., Kohler, T. and Zusman, D.R. (1993). Chemotaxis plays a role in the social behaviour of *Myxococcus xanthus*. *Mol. Microbiol.* **9**, 601–611.

648. Shi, W.Y. and Zusman, D.R. (1993). The two motility systems of *Myxococcus xanthus* show different selective advantages on various surfaces. *Proc. Natl. Acad. Sci. U.S.A.* **90**, 3378–3382.

649. Shimizu, T.S., Le Novère, N., Levin, M.D., Beavil, A.J., Sutton, B.J. and Bray, D. (2000). Molecular model of a lattice of signaling proteins involved in bacterial chemotaxis. *Nature Cell Biol.* **2**, 792–796.

650. Shin, S., Song, S.G., Lee, D.S., Pan, J.G. and Park, C. (1997). Involvement of *iclR* and *rpoS* in the induction of *acs*, the gene for acetyl coenzyme A synthetase of *Escherichia coli* K-12. *FEMS Microbiol. Lett.* **146**, 103–108.

651. Shioi, J., Imae, Y. and Oosawa, F. (1978). Protonmotive force and motility of *Bacillus subtilis*. *J. Bacteriol.* **133**, 1083–1088.

652. Shioi, J., Dang, C.V. and Taylor, B.L. (1987). Oxygen as attractant and repellent in bacterial chemotaxis. *J. Bacteriol.* **169**, 3118–3123.

653. Shioi, J.-I., Galloway, R.J., Niwano, M., Chinnock, R.E. and Taylor, B.L. (1982). Requirement of ATP in bacterial chemotaxis. *J. Biol. Chem.* **257**, 7969–7975.

654. Shukla, D. and Matsumura, P. (1995). Mutations leading to altered CheA binding cluster on a face of CheY. *J. Biol. Chem.* **270**, 24 414–24 419.

655. Shukla, D., Zhu, X.Y. and Matsumura, P. (1998). Flagellar motor-switch binding face of CheY and the biochemical basis of suppression by CheY mutants that compensate for motor-switch defects in *Escherichia coli*. *J. Biol. Chem.* **273**, 23 993–23 999.

656. Silva-Herzog, E. and Dreyfus, G. (1999). Interaction of FliI, a component of the flagellar export apparatus, with flagellin and hook protein. *Biochim. Biophys. Acta* **1431**, 374–383.

657. Silverman, M. and Simon, M. (1973). Genetic analysis of flagellar mutants in *Escherichia coli*. *J. Bacteriol.* **113**, 105–113.

658. Silverman, M. and Simon, M. (1974). Flagellar rotation and the mechanism of bacterial motility. *Nature* **249**, 73–74.

659. Silverman, M., Matsumura, P. and Simon, M. (1976). The identification of the mot gene product with *Escherichia coli*–lambda hybrids. *Proc. Natl. Acad. Sci. U.S.A.* **73**, 3126–3130.

660. Silverman, M. and Simon, M. (1980). Phase variation: genetic analysis of switching mutants. *Cell* **19**, 845–854.

661. Silversmith, R.E., Appleby, J.L. and Bourret, R.B. (1997). Catalytic mechanism of phosphorylation and dephosphorylation of CheY: kinetic characterization of imidazole phosphates as phosphodonors and the role of acid catalysis. *Biochemistry* **36**, 14 965–14 974.

662. Silversmith, R.E. and Bourret, R.B. (1999). Throwing the switch in bacterial chemotaxis. *Trends Microbiol.* **7**, 16–22.

663. Silversmith, R.E., Smith, J.G., Guanga, G.P., Les, J.T. and Bourret, R.B. (2001). Alteration of a nonconserved active site residue in the chemotaxis response regulator CheY affects phosphorylation and interaction with CheZ. *J. Biol. Chem.* **276**, 18 478–18 484.

664. Simms, S.A., Keane, M.G. and Stock, J. (1985). Multiple forms of the CheB methylesterase in bacterial chemosensing. *J. Biol. Chem.* **260**, 10 161–10 168.

665. Simms, S.A., Stock, A.M. and Stock, J.B. (1987). Purification and characterization of the S-adenosylmethionine: glutamyl methyltransferase that modifies membrane chemoreceptor proteins in bacteria. *J. Biol. Chem.* **262**, 8537–8543.

666. Skerker, J.M. and Berg, H.C. (2001). Direct observation of extension and retraction of type IV pili. *Proc. Natl. Acad. Sci. U.S.A.* **98**, 6901–6904.

667. Skidmore, J.M., Ellefson, D.D., McNamara, B.P., Couto, M.M.P., Wolfe, A.J. and Maddock, J.R. (2000). Polar clustering of the chemoreceptor complex in *Escherichia coli* occurs in the absence of complete CheA function. *J. Bacteriol.* **182**, 967–973.

668. Slonczewski, J.L., Macnab, R.M., Alger, J.R. and Castle, A.M. (1982). Effects of pH and repellent tactic stimuli on protein methylation levels in *Escherichia coli*. *J. Bacteriol.* **152**, 384–399.

669. Smith, J.M., Rowsell, E.H., Shioi, J. and Taylor, B.L. (1988). Identification of a site of ATP requirement for signal processing in bacterial chemotaxis. *J. Bacteriol.* **170**, 2698–2704.

670. Snyder, M.A., Stock, J.B. and Koshland, D.E., Jr. (1981). Role of membrane potential and calcium in chemotactic sensing by bacteria. *J. Mol. Biol.* **149**, 241–257.

671. Snyder, M.A., Stock, J.B. and Koshland, D.E., Jr. (1984). Carboxymethyl esterase of bacterial chemotaxis. *Methods Enzymol.* **106**, 321–330.

672. Sockett, H., Yamaguchi, S., Kihara, M., Irikura, V.M. and Macnab, R.M. (1992). Molecular analysis of the flagellar switch protein FliM of *Salmonella typhimurium*. *J. Bacteriol.* **174**, 793–806.

673. Sogaard-Andersen, L. and Kaiser, D. (1996). C factor, a cell-surface-associated intercellular signaling protein, stimulates the cytoplasmic Frz signal transduction system in *Myxococcus xanthus*. *Proc. Natl. Acad. Sci. U.S.A.* **93**, 2675–2679.

674. Sosinsky, G.E., Francis, N.R., DeRosier, D.J., Wall, J.S., Simon, M.N. and Hainfeld, J. (1992). Mass determination and estimation of subunit stoichiometry of the bacterial hook-basal body flagellar complex of *Salmonella typhimurium* by scanning transmission electron microscopy. *Proc. Natl. Acad. Sci. U.S.A.* **89**, 4801–4805.

675. Sourjik, V. and Schmitt, R. (1996). Different roles of CheY1 and CheY2 in the chemotaxis of *Rhizobium meliloti*. *Mol. Microbiol.* **22**, 427–436.

676. Sourjik, V. and Schmitt, R. (1998). Phosphotransfer between CheA, CheY1, and CheY2 in the chemotaxis signal transduction chain of *Rhizobium meliloti*. *Biochemistry* **37**, 2327–2335.

677. Sourjik, V. and Berg, H.C. (2000). Localization of components of the chemotaxis machinery of *Escherichia coli* using fluorescent protein fusions. *Mol. Microbiol.* **37**, 740–751.

678. Sourjik, V. and Berg, H.C. (2002). Receptor sensitivity in bacterial chemotaxis. *Proc. Natl. Acad. Sci. U.S.A.* **99**, 123–127.

679. Sourjik, V. and Berg, H.C. (2002). Binding of the *Escherichia coli* response regulator CheY to its target measured *in vivo* by fluorescence resonance energy transfer. *Proc. Natl. Acad. Sci. U.S.A.* **99**, 12669–12674.

680. Soutourina, O., Kolb, A., Krin, E., Laurent-Winter, C., Rimsky, S., Danchin, A. and Bertin, P. (1999). Multiple control of flagellum biosynthesis in *Escherichia coli*: Role of H-NS protein and the cyclic AMP-catabolite activator protein complex in transcription of the *flhDC* master operon. *J. Bacteriol.* **181**, 7500–7508.

681. Soutourina, O.A., Krin, E., Laurent-Winter, C., Hommais, F., Danchin, A. and Bertin, P.N. (2002). Regulation of bacterial motility in response to low pH in *Escherichia coli*: the role of H-NS protein. *Microbiology* **148**, 1543–1551.

682. Spiro, P.A., Parkinson, J.S. and Othmer, H.G. (1997). A model of excitation and adaptation in bacterial chemotaxis. *Proc. Natl. Acad. Sci. U.S.A.* **94**, 7263–7268.

683. Spormann, A.M. and Kaiser, A.D. (1995). Gliding movements in *Myxococcus xanthus*. *J. Bacteriol.* **177**, 5846–5852.

684. Springer, M.S., Kort, E.N., Larsen, S.H., Ordal, G.W., Reader, R.W. and Adler, J. (1975). Role of methionine in bacterial chemotaxis: Requirement for tumbling and involvement in information processing. *Proc. Natl. Acad. Sci. U.S.A.* **72**, 4640–4644.

685. Springer, M.S., Goy, M.F. and Adler, J. (1979). Protein methylation in behavioral control mechanisms and in signal transduction. *Nature* **280**, 279–284.

686. Springer, W.R. and Koshland, D.E., Jr. (1977). Identification of a protein methyltransferase as the *cheR* gene product in the bacterial sensing system. *Proc. Natl. Acad. Sci. U.S.A.* **74**, 533–537.

687. Spudich, J.L. and Koshland, D.E., Jr. (1975). Quantitation of the sensory response in bacterial chemotaxis. *Proc. Natl. Acad. Sci. U.S.A.* **72**, 710–713.

688. Spudich, J.L. and Koshland, D.E., Jr. (1976). Non-genetic individuality: chance in the single cell. *Nature* **262**, 467–471.

689. Spurlino, J.C., Lu, G.Y. and Quiocho, F.A. (1991). The 2.3-Å resolution structure of the maltose-binding or maltodextrin-binding protein, a primary receptor of bacterial active transport and chemotaxis. *J. Biol. Chem.* **266**, 5202–5219.

690. Stader, J., Matsumura, P., Vacante, D., Dean, G.E. and Macnab, R.M. (1986). Nucleotide sequence of the *Escherichia coli motB* gene and site-limited incorporation of its product into the cytoplasmic membrane. *J. Bacteriol.* **166**, 244–252.

691. Stallmeyer, M.J.B., Aizawa, S.-I., Macnab, R.M. and DeRosier, D.J. (1989). Image reconstruction of the flagellar basal body of *Salmonella typhimurium*. *J. Mol. Biol.* **205**, 519–528.

692. Staropoli, J.F. and Alon, U. (2000). Computerized analysis of chemotaxis at different stages of bacterial growth. *Biophys. J.* **78**, 513–519.

693. Stewart, R.C. and Dahlquist, F.W. (1987). Molecular components of bacterial chemotaxis. *Chem. Rev.* **87**, 997–1025.

694. Stewart, R.C. (1997). Kinetic characterization of phosphotransfer between CheA and CheY in the bacterial chemotaxis signal transduction pathway. *Biochemistry* **36**, 2030–2040.

695. Stewart, R.C., VanBruggen, R., Ellefson, D.D. and Wolfe, A.J. (1998). TNP-ATP and TNP-ADP as probes of the nucleotide binding site of CheA, the histidine protein kinase in the chemotaxis signal transduction pathway of *Escherichia coli*. *Biochemistry* **37**, 12269–12279.

696. Stock, A., Koshland, D.E., Jr. and Stock, J. (1985). Homologies between *Salmonella typhimurium* CheY protein and proteins involved in the regulation of chemotaxis, membrane protein synthesis, and sporulation. *Proc. Natl. Acad. Sci. U.S.A.* **82**, 7989–7993.

697. Stock, A., Mottonen, J., Chen, T. and Stock, J. (1987). Identification of a possible nucleotide binding site in CheW, a protein required for sensory transduction in bacterial chemotaxis. *J. Biol. Chem.* **262**, 535–537.

698. Stock, A. and Stock, J.B. (1987). Purification and characterization of the CheZ protein of bacterial chemotaxis. *J. Bacteriol.* **169**, 3301–3311.

699. Stock, A.M., Wylie, D.C., Mottonen, J.M., Lupas, A.N., Ninfa, E.G., Ninfa, A.J., Schutt, C.E. and Stock, J.B. (1988). Phosphoproteins involved in bacterial signal transduction. *Cold Spring Harbor Symp. Quant. Biol.* **53**, 49–57.

700. Stock, J., Borczuk, A., Chiou, F. and Burchenal, J.E.B. (1985). Compensatory mutations in receptor function: A reevaluation of the role of methylation in bacterial chemotaxis. *Proc. Natl. Acad. Sci. U.S.A.* **82**, 8364–8368.

701. Stock, J., Kersulis, G. and Koshland, D.E., Jr. (1985). Neither methylating nor demethylating enzymes are required for bacterial chemotaxis. *Cell* **42**, 683–690.

702. Stock, J. (1999). Signal transduction: gyrating protein kinases. *Curr. Biol.* **9**, R364–R367.

703. Stock, J. (1999). Sensitivity, cooperativity and gain in chemotaxis signal transduction. *Trends Microbiol.* **7**, 1–4.

704. Stock, J.B. and Koshland, D.E., Jr. (1978). A protein methylesterase involved in bacterial sensing. *Proc. Natl. Acad. Sci. U.S.A.* **75**, 3659–3663.

705. Stock, J.B., Maderis, A.M. and Koshland, D.E., Jr. (1981). Bacterial chemotaxis in the absence of receptor carboxymethylation. *Cell* **27**, 37–44.

706. Stock, J.B. and Surette, M.G. (1996). Chemotaxis, in Escherichia coli *and* Salmonella: *Cellular and Molecular Biology* (F.C. Neidhardt, R. Curtis III, J.L. Ingraham, E.C.C. Lin, K.B. Low, B. Magasanik, W.S. Reznikoff, M. Riley, M. Schaechter and H.E. Umbarger, eds.), pp. 1103–1129. American Society for Microbiology, Washington, DC.

707. Stolz, B. and Berg, H.C. (1991). Evidence for interactions between MotA and MotB, torque-generating elements of the flagellar motor of *Escherichia coli*. *J. Bacteriol.* **173**, 7033–7037.

708. Storch, K.-F., Rudolph, J. and Oesterhelt, D. (1999). Car: a cytoplasmic sensor responsible for arginine chemotaxis in the archaeon *Halobacterium salinarum*. *EMBO J.* **18**, 1146–1158.

709. Stover, C.K., Pham, X.Q., Erwin, A.L., Mizoguchi, S.D., Warrener, P., Hickey, M.J., Brinkman, F.S., Hufnagle, W.O., Kowalik, D.J., Lagrou, M., Garber, R.L., Goltry, L., Tolentino, E., Westbrock-Wadman, S., Yuan, Y., Brody, L.L., Coulter, S.N., Folger, K.R., Kas, A., Larbig, K., Lim, R., Smith, K., Spencer, D., Wong, G.K.-S., Wu, Z., Paulsen, I.T., Reizer, J., Saler, M.H., Hancock, R.E.W., Lory, S. and Olson, M.V. (2000). Complete genome sequence of *Pseudomonas aeruginosa* PA01, an opportunistic pathogen. *Nature* **406**, 959–964.

710. Sun, H., Zusman, D.R. and Shi, W. (2000). Type IV pilus of *Myxococcus xanthus* is a motility apparatus controlled by the *frz* chemosensory system. *Curr. Biol.* **10**, 1143–1146.

711. Surette, M.G., Levit, M., Liu, Y., Lukat, G., Ninfa, E.G., Ninfa, A. and Stock, J.B. (1996). Dimerization is required for the activity of the protein histidine kinase

CheA that mediates signal transduction in bacterial chemotaxis. *J. Biol. Chem.* **271**, 939–945.

712. Surette, M.G. and Stock, J.B. (1996). Role of α-helical coiled-coil interactions in receptor dimerization, signaling, and adaptation during bacterial chemotaxis. *J. Biol. Chem.* **271**, 17966–17973.

713. Suzuki, T., Iino, T., Horiguchi, T. and Yamaguchi, S. (1978). Incomplete flagellar structures in nonflagellate mutants of *Salmonella typhimurium*. *J. Bacteriol.* **133**, 904–915.

714. Suzuki, T. and Komeda, Y. (1981). Incomplete flagellar structure in *Escherichia coli* mutants. *J. Bacteriol.* **145**, 1036–1041.

715. Swanson, R.V., Bourret, R.B. and Simon, M.I. (1993). Intermolecular complementation of the kinase activity of CheA. *Mol. Microbiol.* **8**, 435–441.

716. Swanson, R.V., Schuster, S.C. and Simon, M.I. (1993). Expression of CheA fragments which define domains encoding kinase, phosphotransfer, and CheY binding activities. *Biochemistry* **32**, 7623–7629.

717. Swanson, R.V., Lowry, D.F., Matsumura, P., McEvoy, M.M., Simon, M.I. and Dahlquist, F.W. (1995). Localized perturbations in CheY structure monitored by NMR identify a CheA-binding interface. *Nature Struct. Biol.* **2**, 906–910.

718. Tang, H., Braun, T.F. and Blair, D.F. (1996). Motility protein complexes in the bacterial flagellar motor. *J. Mol. Biol.* **261**, 209–221.

719. Tatsuno, I., Homma, M., Oosawa, K. and Kawagishi, I. (1996). Signaling by the *Escherichia coli* aspartate chemoreceptor Tar with a single cytoplasmic domain per dimer. *Science* **274**, 423–425.

720. Tawa, P. and Stewart, R.C. (1994). Kinetics of CheA autophosphorylation and dephosphorylation reactions. *Biochemistry* **33**, 7917–7924.

721. Taylor, B.L., Tribhuwan, R.C., Rowsell, E.H., Smith, J.M. and Shioi, J. (1985). Role of ATP and cyclic nucleotides in bacterial chemotaxis, in *Sensing and Response in Microorganisms* (M. Eisenbach and M. Balaban, eds.), pp. 63–71. Elsevier, Amsterdam.

722. Taylor, B.L. and Zhulin, I.B. (1999). PAS domains: Internal sensors of oxygen, redox potential, and light. *Microbiol. Mol. Biol. Rev.* **63**, 479+.

723. Taylor, B.L., Zhulin, I.B. and Johnson, M.S. (1999). Aerotaxis and other energy-sensing behavior in bacteria. *Annu. Rev. Microbiol.* **53**, 103–128.

724. Teeter, J.H. and Brand, J.G. (1987). Peripheral mechanisms of gustation: physiology and biochemistry, in *Neurobiology of Taste and Smell* (T.E.F.a.W.L. Silver, ed.) pp. 299–329. John Wiley & Sons, New York.

725. Terwilliger, T.C. and Koshland, D.E., Jr. (1984). Sites of methyl esterification and deamination on the aspartate receptor involved in chemotaxis. *J. Biol. Chem.* **259**, 7719–7725.

726. Thipayathasana, P. and Valentine, R.C. (1974). The requirement for energy transducing ATPase for anaerobic motility in *Escherichia coli*. *Biochim. Biophys. Acta* **347**, 464–468.

727. Thoelke, M.S., Kirby, J.R. and Ordal, G.W. (1989). Novel methyl transfer during chemotaxis in *Bacillus subtilis*. *Biochemistry* **28**, 5585–5589.

728. Thoelke, M.S., Casper, J.M. and Ordal, G.W. (1990). Methyl group turnover on methyl-accepting chemotaxis proteins during chemotaxis by *Bacillus subtilis*. *J. Biol. Chem.* **265**, 1928–1932.

729. Thoelke, M.S., Casper, J.M. and Ordal, G.W. (1990). Methyl transfer in chemotaxis toward sugars by *Bacillus subtilis*. *J. Bacteriol.* **172**, 1148–1150.

730. Thomas, D.R., Morgan, D.G. and DeRosier, D.J. (1999). Rotational symmetry of the C ring and a mechanism for the flagellar rotary motor. *Proc. Natl. Acad. Sci. U.S.A.* **96**, 10 134–10 139.

731. Tieman, S., Koch, A. and White, D. (1996). Gliding motility in slide cultures of *Myxococcus xanthus* in stable and steep chemical gradients. *J. Bacteriol.* **178**, 3480–3485.

732. Tisa, L.S. and Adler, J. (1992). Calcium ions are involved in *Escherichia coli* chemotaxis. *Proc. Natl. Acad. Sci. U.S.A.* **89**, 11 804–11 808.

733. Tisa, L.S., Olivera, B.M. and Adler, J. (1993). Inhibition of *Escherichia coli* chemotaxis by ω-conotoxin, a calcium ion channel blocker. *J. Bacteriol.* **175**, 1235–1238.

734. Tisa, L.S. and Adler, J. (1995). Cytoplasmic free-Ca^{2+} level rises with repellents and falls with attractants in *Escherichia coli* chemotaxis. *Proc. Natl. Acad. Sci. U.S.A.* **92**, 10 777–10 781.

735. Tisa, L.S. and Adler, J. (1995). Chemotactic properties of *Escherichia coli* mutants having abnormal Ca^{2+} content. *J. Bacteriol.* **177**, 7112–7118.

736. Tisa, L.S., Sekelsky, J.J. and Adler, J. (2000). Effects of organic antagonists of Ca^{2+}, Na^{+}, and K^{+} on chemotaxis and motility of *Escherichia coli*. *J. Bacteriol.* **182**, 4856–4861.

737. Toews, M.L. and Adler, J. (1979). Methanol formation *in vivo* from methylated chemotaxis proteins in *Escherichia coli*. *J. Biol. Chem.* **254**, 1761–1764.

738. Togashi, F., Yamaguchi, S., Kihara, M., Aizawa, S.-I. and Macnab, R.M. (1997). An extreme clockwise switch bias mutation in *fliG* of *Salmonella typhimurium* and its suppression by slow-motile mutations in *motA* and *motB*. *J. Bacteriol.* **179**, 2994–3003.

739. Toguchi, A., Siano, M., Burkart, M. and Harshey, R.M. (2000). Genetics of swarming motility in *Salmonella enterica* serovar Typhimurium: critical role for lipopolysaccharide. *J. Bacteriol.* **182**, 6308–6321.

740. Toker, A.S., Kihara, M. and Macnab, R.M. (1996). Deletion analysis of the FliM flagellar switch protein of *Salmonella typhimurium*. *J. Bacteriol.* **178**, 7069–7079.

741. Toker, A.S. and Macnab, R.M. (1997). Distinct regions of bacterial flagellar switch protein FliM interact with FliG, FliN, and CheY. *J. Mol. Biol.* **273**, 623–634.

742. Trachtenberg, S. and DeRosier, D.J. (1987). Three-dimensional structure of the frozen-hydrated flagellar filament. The left-handed filament of *Salmonella typhimurium*. *J. Mol. Biol.* **195**, 581–601.

743. Trachtenberg, S. and Gilad, R. (2001). A bacterial linear motor: cellular and molecular organization of the contractile cytoskeleton of the helical bacterium *Spiroplasma melliferum* BC3. *Mol. Microbiol.* **41**, 827–848.

744. Trachtenberg, S., Gilad, R. and Geffen, N. (2003). The bacterial linear motor of *Spiroplasma melliferum* BC3: from single molecules to swimming cells. *Mol. Microbiol.* **47**, 671–697.

745. Trudeau, K.G., Ward, M.J. and Zusman, D.R. (1996). Identification and characterization of FrzZ, a novel response regulator necessary for swarming and fruiting-body formation in *Myxococcus xanthus*. *Mol. Microbiol.* **20**, 645–655.

746. Tsang, N., Macnab, R.M. and Koshland, D.E., Jr. (1973). Common mechanism for repellents and attractants in bacterial chemotaxis. *Science* **181**, 60–63.

747. Tso, W.-W. and Adler, J. (1974). Negative chemotaxis in *Escherichia coli*. *J. Bacteriol.* **118**, 560–576.

748. Turner, L., Caplan, S.R. and Berg, H.C. (1996). Temperature-induced switching of the bacterial flagellar motor. *Biophys. J.* **71**, 2227–2233.

749. Turner, L., Samuel, A.D.T., Stern, A.S. and Berg, H.C. (1999). Temperature dependence of switching of the bacterial flagellar motor by the protein CheY(13DK106YW). *Biophys. J.* **77**, 597–603.

750. Turner, L., Ryu, W.S. and Berg, H.C. (2000). Real-time imaging of fluorescent flagellar filaments. *J. Bacteriol.* **182**, 2793–2801.

751. Ueno, T., Oosawa, K. and Aizawa, S.-I. (1992). M ring, S ring and proximal rod of the flagellar basal body of *Salmonella typhimurium* are composed of subunits of a single protein, *FliF*. *J. Mol. Biol.* **227**, 672–677.

752. Vaituzis, Z. and Doetsch, R.N. (1969). Motility tracks: technique for quantitative study of bacterial movement. *Appl. Microbiol.* **17**, 584–588.

753. van der Drift, C. and Jong, M.H.d. (1974). Chemotaxis toward amino acids in *Bacillus subtilis*. *Arch. Microbiol.* **96**, 83–92.

754. Van-der-Werf, P. and Koshland, D.E., Jr. (1977). Identification of a γ-glutamyl methyl ester in bacterial membrane protein involved in chemotaxis. *J. Biol. Chem.* **252**, 2793–2795.

755. Volz, K. (1993). Structural conservation in the CheY superfamily. *Biochemistry* **32**, 11 741–11 753.

756. Wadhams, G.H., Martin, A.C., Porter, S.L., Maddock, J.R., Mantotta, J.C., King, H.M. and Armitage, J.P. (2002). TlpC, a novel chemotaxis protein in *Rhodobacter sphaeroides*, localizes to a discrete region in the cytoplasm. *Mol. Microbiol.* **46**, 1211–1221.

757. Walz, D. and Caplan, S.R. (1998). An electrostatic model of the bacterial flagellar motor. *Bioelectrochem. Bioenerg.* **47**, 19–24.

758. Walz, D. and Caplan, S.R. (2000). An electrostatic mechanism closely reproducing observed behavior in the bacterial flagellar motor. *Biophys. J.* **78**, 626–651.

759. Wang, H. and Matsumura, P. (1996). Characterization of the $CheA_s$/CheZ complex: a specific interaction resulting in enhanced dephosphorylating activity on CheY-phosphate. *Mol. Microbiol.* **19**, 695–703.

760. Wang, H. and Matsumura, P. (1997). Phosphorylating and dephosphorylating protein complexes in bacterial chemotaxis. *J. Bacteriol.* **179**, 287–289.

761. Wang, J.X., Balazs, Y.S. and Thompson, L.K. (1997). Solid-state REDOR NMR distance measurements at the ligand site of a bacterial chemotaxis membrane receptor. *Biochemistry* **36**, 1699–1703.

762. Wanner, B.L. and Wilmes-Riesenberg, M.R. (1992). Involvement of phospho-transacetylase, acetate kinase, and acetyl phosphate synthesis in control of the phosphate regulon in *Escherichia coli*. *J. Bacteriol.* **174**, 2124–2130.

763. Wanner, B.L. (1995). Signal transduction and cross regulation in the *Escherichia coli* phosphate regulon by PhoR, CreC, and acetyl phosphate, in *Two-Component Signal Transduction* (J.A. Hoch and T.J. Silhavy, eds.), pp. 203–221. American Society for Microbiology, Washington, DC.

764. Ward, M.J., Bell, A.W., Hamblin, P.A., Packer, H.L. and Armitage, J.P. (1995). Identification of chemotaxis operon with two *cheY* genes in *Rhodobacter sphaeroides*. *Mol. Microbiol.* **17**, 357–366.

765. Ward, M.J. and Zusman, D.R. (1997). Regulation of directed motility in *Myxococcus xanthus*. *Mol. Microbiol.* **24**, 885–893.

766. Washizu, M., Kurahashi, Y., Iochi, H., Kurosawa, O., Aizawa, S.-I., Kudo, S., Magariyama, Y. and Hotani, H. (1993). Dielectrophoretic measurement of bacterial motor characteristics. *IEEE Trans. Industry Appl.* **29**, 286–294.

767. Waterbury, J.B., Willey, J.M., Franks, D.G., Valois, F.W. and Watson, S.W. (1985). A cyanobacterium capable of swimming motility. *Science* **230**, 74–76.

768. Watkins, N.J., Knight, M.R., Trewavas, A.J. and Campbell, A.K. (1995). Free calcium transients in chemotactic and non-chemotactic strains of *Escherichia coli* determined by using recombinant aequorin. *Biochem. J.* **306**, 865–869.

769. Watnick, P. and Kolter, R. (2000). Biofilm, city of microbes. *J. Bacteriol.* **182**, 2675–2679.

770. Weerasuriya, S., Schneider, B.M. and Manson, M.D. (1998). Chimeric chemoreceptors in *Escherichia coli*: signaling properties of Tar-Tap and Tap-Tar hybrids. *J. Bacteriol.* **180**, 914–920.

771. Wei, B.L., Brun-Zinkernagel, A.-M., Simecka, J.W., Prüss, B.M., Babitzke, P. and Romeo, T. (2001). Positive regulation of motility and *flhDC* expression by the RNA-binding protein CsrA of *Escherichia coli*. *Mol. Microbiol.* **40**, 245–256.

772. Weis, R.M. and Koshland, D.E., Jr. (1988). Reversible receptor methylation is essential for normal chemotaxis of *Escherichia coli* in gradients of aspartic acid. *Proc. Natl. Acad. Sci. U.S.A.* **85**, 83–87.

773. Welch, M., Oosawa, K., Aizawa, S.-I. and Eisenbach, M. (1993). Phosphorylation-dependent binding of a signal molecule to the flagellar switch of bacteria. *Proc. Natl. Acad. Sci. U.S.A.* **90**, 8787–8791.

774. Welch, M., Oosawa, K., Aizawa, S.-I. and Eisenbach, M. (1994). Effects of phosphorylation, Mg^{2+}, and conformation of the chemotaxis protein CheY on its binding to the flagellar switch protein FliM. *Biochemistry* **33**, 10470–10476.

775. Welch, M., Margolin, Y., Caplan, S.R. and Eisenbach, M. (1995). Rotational asymmetry of *Escherichia coli* flagellar motor in the presence of arsenate. *Biochim. Biophys. Acta* **1268**, 81–87.

776. Welch, M., Chinardet, N., Mourey, L., Birck, C. and Samama, J.-P. (1998). Structure of the CheY-binding domain of histidine kinase CheA in complex with CheY. *Nature Struct. Biol.* **5**, 25–29.

777. Wellman, A.M. and Paerl, H.W. (1981). Rapid chemotaxis assay using radioactively labeled bacterial cells. *Appl. Environ. Microbiol.* **42**, 216–221.

778. West, A.H., Martinez-Hackert, E. and Stock, A.M. (1995). Crystal structure of the catalytic domain of the chemotaxis receptor methylesterase, CheB. *J. Mol. Biol.* **250**, 276–290.

779. Widman, M.T., Emerson, D., Chiu, C.C. and Worden, R.M. (1997). Modeling microbial chemotaxis in a diffusion gradient chamber. *Biotechnol. Bioengineer.* **55**, 191–205.

780. Williams, A.W., Yamaguchi, S., Togashi, F., Aizawa, S.-I., Kawagishi, I. and Macnab, R.M. (1996). Mutations in *fliK* and *flhB* affecting flagellar hook and filament assembly in *Salmonella typhimurium*. *J. Bacteriol.* **178**, 2960–2970.

781. Wilson, M.L. and Macnab, R.M. (1990). Co-overproduction and localization of the *Escherichia coli* motility proteins MotA and MotB. *J. Bacteriol.* **172**, 3932–3939.

782. Wolfe, A.J., Conley, M.P., Kramer, T.J. and Berg, H.C. (1987). Reconstitution of signaling in bacterial chemotaxis. *J. Bacteriol.* **169**, 1878–1885.

783. Wolfe, A.J., Conley, M.P. and Berg, H.C. (1988). Acetyladenylate plays a role in controlling the direction of flagellar rotation. *Proc. Natl. Acad. Sci. U.S.A.* **85**, 6711–6715.

784. Wolfe, A.J. and Berg, H.C. (1989). Migration of bacteria in semisolid agar. *Proc. Natl. Acad. Sci. U.S.A.* **86**, 6973–6977.

785. Wolfe, A.J. and Stewart, R.C. (1993). The short form of the CheA protein restores kinase activity and chemotactic ability to kinase-deficient mutants. *Proc. Natl. Acad. Sci. U.S.A.* **90**, 1518–1522.

786. Wolgemuth, C., Hoiczyk, E., Kaiser, D. and Oster, G. (2002). How myxobacteria glide. *Curr. Biol.* **12**, 369–377.

787. Wright, E.L., Deakin, W.J. and Shaw, C.H. (1998). A chemotaxis cluster from agrobacterium tumefaciens. *Gene* **220**, 83–89.

788. Wu, J.G., Li, J.Y., Li, G.Y., Long, D.G. and Weis, R.M. (1996). The receptor binding site for the methyltransferase of bacterial chemotaxis is distinct from the sites of methylation. *Biochemistry* **35**, 4984–4993.

789. Wu, J.G. and Newton, A. (1997). Regulation of the *Caulobacter* flagellar gene hierarchy; not just for motility. *Mol. Microbiol.* **24**, 233–239.

790. Wu, J.R., Long, D.G. and Weis, R.M. (1995). Reversible dissociation and unfolding of the *Escherichia coli* aspartate receptor cytoplasmic fragment. *Biochemistry* **34**, 3056–3065.

791. Wu, S.S. and Kaiser, D. (1995). Genetic and functional evidence that type IV pili are required for social gliding motility in *Myxococcus xanthus*. *Mol. Microbiol.* **18**, 547–558.

792. Wylie, D., Stock, A., Wong, C.-Y. and Stock, J. (1988). Sensory transduction in bacterial chemotaxis involves phosphotransfer between Che proteins. *Biochem. Biophys. Res. Commun.* **151**, 891–896.

793. Yamaguchi, S., Fujita, H., Kuroiwa, T. and Iino, T. (1977). Sensitivity of non-flagellate *Salmonella* mutants to the flagellotropic bacteriophage χ. *J. Gen. Microbiol.* **99**, 209–214.

794. Yamaguchi, S., Aizawa, S.-I., Kihara, M., Isomura, M., Jones, C.J. and Macnab, R.M. (1986). Genetic evidence for a switching and energy-transducing complex in the flagellar motor of *Salmonella typhimurium*. *J. Bacteriol.* **168**, 1172–1179.

795. Yamamoto, K., Macnab, R.M. and Imae, Y. (1990). Repellent-response functions of the Trg and Tap chemoreceptors of *Escherichia coli*. *J. Bacteriol.* **172**, 383–388.

796. Yamamoto, K. and Imae, Y. (1993). Cloning and characterization of the *Salmonella typhimurium*-specific chemoreceptor Tcp for taxis to citrate and from phenol. *Proc. Natl. Acad. Sci. U.S.A.* **90**, 217–221.

797. Yan, D., Cho, H.S., Hastings, C.A., Igo, M.M., Lee, S.-Y., Pelton, J.G., Stewart, V., Wemmer, D.E. and Kustu, S. (1999). Beryllofluoride mimics phosphorylation of NtrC and other bacterial response regulators. *Proc. Natl. Acad. Sci. U.S.A.* **96**, 14789–14794.

798. Yang, Z.M., Geng, Y.Z., Xu, D., Kaplan, H.B. and Shi, W.Y. (1998). A new set of chemotaxis homologues is essential for *Myxococcus xanthus* social motility. *Mol. Microbiol.* **30**, 1123–1130.

799. Yao, V.J. and Spudich, J.L. (1992). Primary structure of an archaebacterial transducer, a methyl-accepting protein associated with sensory rhodopsin I. *Proc. Natl. Acad. Sci. U.S.A.* **89**, 11915–11919.

800. Yarden, Y. and Schlessinger, J. (1987). Epidermal growth factor induces rapid reversible aggregation of the purified epidermal growth factor receptor. *Biochemistry* **26**, 1443–1451.

801. Yeh, J.I., Biemann, H.-P., Privè, G.G., Pandit, J., Koshland, D.E., Jr. and Kim, S.-H. (1996). High-resolution structures of the ligand binding domain of the wild-type bacterial aspartate receptor. *J. Mol. Biol.* **262**, 186–201.

802. Yokoseki, T., Iino, T. and Kutsukake, K. (1996). Negative regulation by FliD, FliS, and FliT of the export of the flagellum-specific anti-sigma factor, FlgM, in *Salmonella typhimurium*. *J. Bacteriol.* **178**, 899–901.

803. Yonekawa, H., Hayashi, H. and Parkinson, J.S. (1983). Requirement of the cheB function for sensory adaptation in *Escherichia coli*. *J. Bacteriol.* **156**, 1228–1235.

804. Yonekura, K., Maki, S., Morgan, D.G., DeRosier, D.J., Vonderviszt, F., Imada, K. and Namba, K. (2000). The bacterial flagellar cap as the rotary promoter of flagellin self-assembly. *Science* **290**, 2148–2152.

805. Yorimitsu, T. and Homma, M. (2001). Na$^+$-driven flagellar motor of *Vibrio*. *Biochim. Biophys. Acta* **1505**, 82–93.

806. Youderian, P. (1998). Bacterial motility: Secretory secrets of gliding bacteria. *Curr. Biol.* **8**, R408–R411.

807. Young, G.M., Schmiel, D.H. and Miller, V.L. (1999). A new pathway for the secretion of virulence factors by bacteria: The flagellar export apparatus functions as a protein-secretion system. *Proc. Natl. Acad. Sci. U.S.A.* **96**, 6456–6461.

808. Yu, H.S. and Alam, M. (1997). An agarose-in-plug bridge method to study chemotaxis in the Archaeon *Halobacterium salinarum*. *FEMS Microbiol. Lett.* **156**, 265–269.

809. Zhang, W.S., Brooun, A., McCandless, J., Banda, P. and Alam, M. (1996). Signal transduction in the Archaeon *Halobacterium salinarium* is processed through three subfamilies of 13 soluble and membrane-bound transducer proteins. *Proc. Natl. Acad. Sci. U.S.A.* **93**, 4649–4654.

810. Zhang, Y.H., Conway, C., Rosato, M., Suh, Y. and Manson, M.D. (1992). Maltose chemotaxis involves residues in the N-terminal and C-terminal domains on the same face of maltose-binding protein. *J. Biol. Chem.* **267**, 22 813–22 820.

811. Zhang, Y.H., Gardina, P.J., Kuebler, A.S., Kang, H.S., Christopher, J.A. and Manson, M.D. (1999). Model of maltose-binding protein/chemoreceptor complex supports intrasubunit signaling mechanism. *Proc. Natl. Acad. Sci. U.S.A.* **96**, 939–944.

812. Zhao, R., Collins, E.J., Bourret, R.B. and Silversmith, R.E. (2002). Structure and catalytic mechanism of the *E. coli* chemotaxis phosphatase CheZ. *Nature Struct. Biol.* **9**, 570–575.

813. Zhao, R.B., Schuster, S.C. and Khan, S. (1995). Structural effects of mutations in *Salmonella typhimurium* flagellar switch complex. *J. Mol. Biol.* **251**, 400–412.

814. Zhao, R.H., Pathak, N., Jaffe, H., Reese, T.S. and Khan, S. (1996). FliN is a major structural protein of the c-ring in the *Salmonella typhimurium* flagellar basal body. *J. Mol. Biol.* **261**, 195–208.

815. Zheng, X.Y. and Sinclair, J.B. (1996). Chemotactic response of *Bacillus megaterium* strain B153–2–2 to soybean root and seed exudates. *Physiol. Mol. Plant Pathol.* **48**, 21–35.

816. Zhou, H.J., Lowry, D.F., Swanson, R.V., Simon, M.I. and Dahlquist, F.W. (1995). NMR studies of the phosphotransfer domain of the histidine kinase CheA from

Escherichia coli: assignments, secondary structure, general fold, and backbone dynamics. *Biochemistry* **34**, 13 858–13 870.

817. Zhou, H.J., McEvoy, M.M., Lowry, D.F., Swanson, R.V., Simon, M.I. and Dahlquist, F.W. (1996). Phosphotransfer and CheY-binding domains of the histidine autokinase CheA are joined by a flexible linker. *Biochemistry* **35**, 433–443.

818. Zhou, H.J. and Dahlquist, F.W. (1997). Phosphotransfer site of the chemotaxis-specific protein kinase CheA as revealed by NMR. *Biochemistry* **36**, 699–710.

819. Zhou, J.D., Fazzio, R.T. and Blair, D.F. (1995). Membrane topology of the MotA protein of *Escherichia coli*. *J. Mol. Biol.* **251**, 237–242.

820. Zhou, J.D. and Blair, D.F. (1997). Residues of the cytoplasmic domain of MotA essential for torque generation in the bacterial flagellar motor. *J. Mol. Biol.* **273**, 428–439.

821. Zhou, J.D., Lloyd, S.A. and Blair, D.F. (1998). Electrostatic interactions between rotor and stator in the bacterial flagellar motor. *Proc. Natl. Acad. Sci. U.S.A.* **95**, 6436–6441.

822. Zhou, J.D., Sharp, L.L., Tang, H.L., Lloyd, S.A., Billings, S., Braun, T.F. and Blair, D.F. (1998). Function of protonatable residues in the flagellar motor of *Escherichia coli*: a critical role for Asp 32 of MotB. *J. Bacteriol.* **180**, 2729–2735.

823. Zhu, X.Y., Amsler, C.D., Volz, K. and Matsumura, P. (1996). Tyrosine 106 of CheY plays an important role in chemotaxis signal transduction in *Escherichia coli*. *J. Bacteriol.* **178**, 4208–4215.

824. Zhu, X.Y., Rebello, J., Matsumura, P. and Volz, K. (1997). Crystal structures of CheY mutants Y106W and T87I/Y106W: CheY activation correlates with movement of residue 106. *J. Biol. Chem.* **272**, 5000–5006.

825. Zhu, X.Y., Volz, K. and Matsumura, P. (1997). The CheZ-binding surface of CheY overlaps the CheA- and FliM-binding surfaces. *J. Biol. Chem.* **272**, 23 758–23 764.

826. Zhulin, I.B., Gibel, I.B. and Ignatov, V.V. (1991). A rapid method for the measurement of bacterial chemotaxis. *Curr. Microbiol.* **22**, 307–310.

827. Zhulin, I.B. and Armitage, J.P. (1993). Motility, chemokinesis, and methylation-independent chemotaxis in *Azospirillum brasilense*. *J. Bacteriol.* **175**, 952–958.

828. Zhulin, I.B. and Taylor, B.L. (1995). Chemotaxis in plant-associated bacteria: the search for the ecological niche, in *Azospirillum VI and Related Microorganisms* (I. Fendrik, ed.), pp. 451–459. Springer-Verlag, Berlin, Heidelberg.

829. Zhulin, I.B., Bespalov, V.A., Johnson, M.S. and Taylor, B.L. (1996). Oxygen taxis and proton motive force in *Azospirillum brasilense*. *J. Bacteriol.* **178**, 5199–5204.

830. Zhulin, I.B., Rowsell, E.H., Johnson, M.S. and Taylor, B.L. (1997). Glycerol elicits energy taxis of *Escherichia coli* and *Salmonella typhimurium*. *J. Bacteriol.* **179**, 3196–3201.

831. Zusman, D.R. (1982). "Frizzy" mutants: a new class of aggregation-defective developmental mutants of *Myxococcus xanthus*. *J. Bacteriol.* **150**, 1430–1437.

Chemotaxis as a Means of Cell–Cell Communication in Bacteria *

1. Introduction

To survive in nature and to compete successfully in environments that are often hostile, bacteria have employed a variety of sophisticated strategies often involving multicellular behavior [107]. Examples of such systems include biofilm formation [117], sporulation and genetic competence [96], conjugation [92], production of virulence factors [70], swarming motility [101], complex colonial pattern formation [10], and, arguably the most striking example, the developmental program of *Myxococcus xanthus* [53, 114]. These processes require coordination involving various modes of cell-to-cell signaling. Bacteria produce and excrete compounds which, when present during appropriate conditions and in sufficient quantities, trigger specific responses at many levels from gene expression to the control of finely tuned motors (see Chapter 3). These compounds are often small diffusible molecules, which perform a role in monitoring cell density (quorum sensing), e.g., N-acyl homoserine lactones or small peptides, or they can be special "dedicated" signals such as specific oligopeptides or proteins, e.g., the C signal in

* Written by Mazal Varon and David Gutnick, Department of Molecular Microbiology and Biotechnology, George S. Wise, Faculty of Life Sciences, Tel-Aviv University, 69978 Tel-Aviv, Israel.

M. xanthus development. The key to bacterial coordinated behavior resides in the ability of a cell to receive, interpret, and respond to these signals. Relaying the information may require direct contact between the donor and the recipient cells or it may be carried out from a distance, i.e., by means of diffusible molecules, which are detected through their interaction with specific receptors. As discussed in Chapter 3, bacterial chemotaxis is the most completely understood of the bacterial sensory transduction systems. While the role of chemotaxis in intercellular communication is still not clear, in some cases, mutants deleted for chemotaxis genes fail to carry out one or another of the processes mentioned above. Components of the chemotaxis system seem to play a role in swarming motility, pattern formation, and myxobacterial development. In fact, it appears that certain chemotaxis functions have been recruited by certain species to mediate intercellular communication. In this chapter, we review some examples, and provide evidence as well as hypotheses concerning a role for proteins and systems involved in chemotaxis in multicellular behavior.

2. Pattern Formation by *E. coli* and *Salmonella*

2.1. *Pattern formation in minimal medium and the role of the Tar receptor*

In 1966, Adler showed that motile and chemotactic cells of *E. coli* spotted on a Petri dish containing agar that was soft enough to allow the cells to move through it, in the presence of an attractant that was consumed, grew and spread across the plate in a symmetric ring. The inclusion of multiple attractants in the medium resulted in the formation of multiple rings [1]. The cells generated, through consumption of the attractants, concentration gradients, and then they migrated down the concentration gradient forming a continuous ring. Budrene and Berg [18, 19] discovered that much more complex patterns were formed when the carbon sources in the growth medium were weak attractants. When *E. coli* was spotted onto a soft agar plate containing intermediates of the tricarboxylic acid cycle such as succinate or fumarate as carbon sources, the following sequence of events was observed: After a period of several hours (depending on the bacterial strain and substrate concentration) at $25°C$, a swarm ring appeared at the periphery of the swarm. The ring migrated outwards at a constant speed (depending on conditions), carrying with it cells from the original inoculation site. As the swarm ring attained a certain cell density, numerous bacterial aggregates in the form of spots or stripes appeared in its wake. Every few hours, new sets of spots or stripes appeared. Complex patterns were generated, their geometry depending on the nature of the carbon source and on its concentration. A host of patterns were observed including sunflower-like, radial arrays of spots, stripes, and chevrons (Figure 1). In the presence of multiple attractants, patterns that were even more intricate were displayed, apparently owing to the complex interactions between cells moving towards (and possibly away from) various attractants of different affinities.

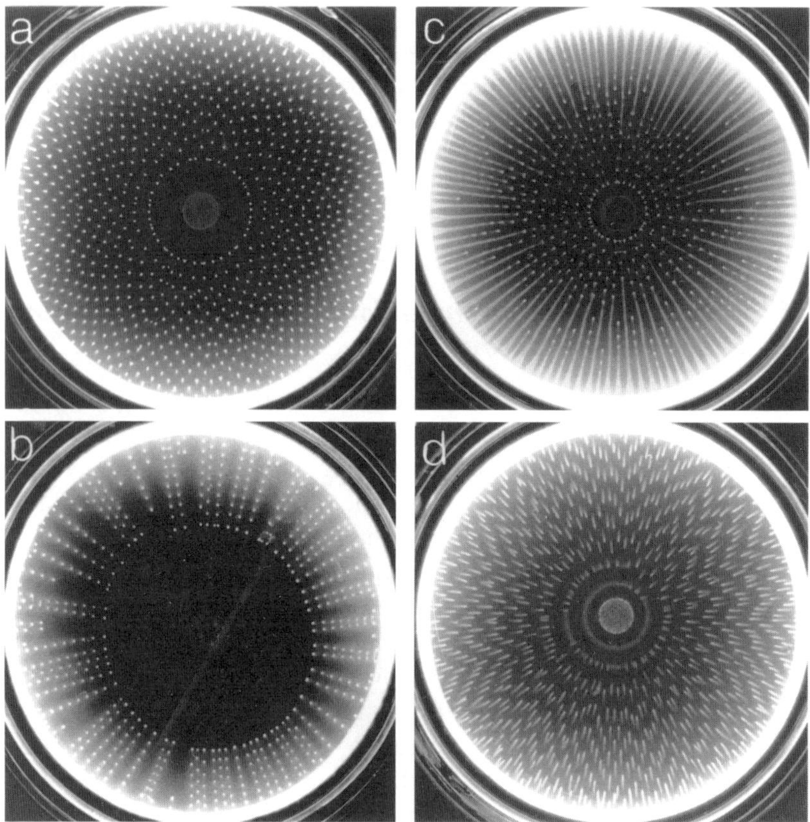

Figure 1. Patterns formed by *E. coli* in semisolid agar. (a) Sunflower-like arrays of spots; (b) radial arrays of spots; (c) radial arrays of spots and stripes; (d) spots with radial tails arranged in chevrons. Provided by H. Berg and reproduced (with permission) from *Nature* [18].

A somewhat different picture was obtained for *Salmonella* [16, 132]. (In keeping with the style introduced in Chapter 3, we refer to *Salmonella enterica* serovar Typhimurium as *Salmonella*). In this case, the cells first spread out into an unstructured lawn of low cell density and only after about 40 h, a ring containing a dense collection of cells appeared around the central part of the colony. Subsequently, sequential concentric rings formed at a constant distance (about 2 mm) from each other. The rings were continuous at high succinate concentration and broken into discrete spots or arcs at lower concentrations of the carbon source [132]. The patterns formed on succinate or on fumarate were stable for days. On the other hand, patterns formed by *Salmonella* on sugar carbon sources such as glucose or mannitol, were transient though increased buffering could stabilize them [16] (Figure 2a).

Patterning, under all the different conditions mentioned above, depended on active cellular motility as well as on chemotaxis. Pattern formation was suppressed in the presence of substances sensed by the aspartate (Tar) receptor (such as the nonmetabolizable analogue alpha methyl, D, L, aspartate [16, 18]. In addition, no patterning was observed in mutants lacking the Tar receptor. Budrene and Berg [19] proposed that patterning resulted from the secretion of a chemoattractant leading to the formation of local cellular aggregates. They identified the attractant as aspartate and concluded that the cells aggregated, eventually developing a pattern, in response to gradients of this compound, which they produced themselves from succinate or fumarate, and excreted into the medium.

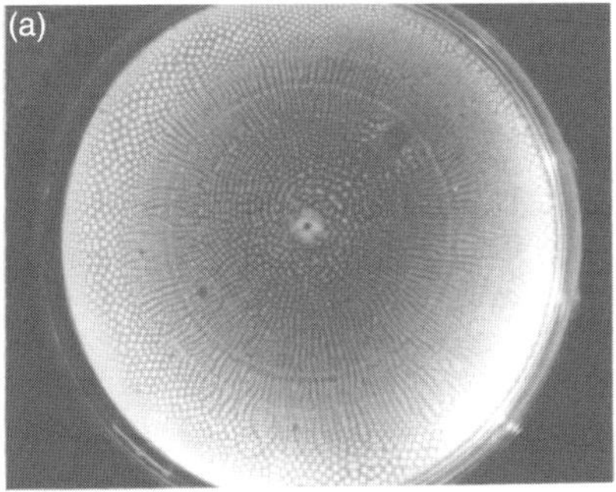
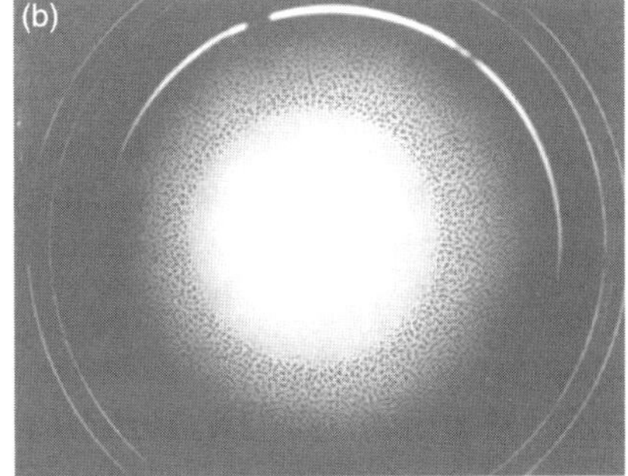

Figure 2. Patterns formed by *Salmonella* [16]. (a) Pattern formation on minimal medium (mannitol plates supplemented with MOPS buffer, pH 7). (b) Pattern of spots formed on tryptone agar (0.3%) photographed 10.5 h after inoculation in the center of the plate. Provided by M. Eisenbach and reproduced with the permission of The American Society For Microbiology.

Berg [13] suggested that when the gradient results from metabolic breakdown of the attractant, migrating cells swim outwards in circular rings. In contrast, production of the attractant by the cells themselves causes the rings to break up into discrete aggregates or a ring of spots. Especially elaborate patterns were observed upon growth of *Salmonella* on a mixture of citrate and succinate: The citrate, being a potent attractant, forms a spatial gradient due to its uptake by the cells, whereas succinate, a weak attractant, must be metabolized first in order to produce the attractant, aspartate. Together, they cause the formation of complex and beautiful patterns [13].

2.2. *Pattern formation by* Salmonella *in complex medium*

If patterning in minimal medium is mediated by a cell-produced attractant that binds to the Tar receptor, pattern formation in a complex rich medium appears to follow a different pathway(s). On soft tryptone agar (0.20–0.35%), *Salmonella* generated patterns distinct from those produced on succinate-minimal medium in several respects (Figure 2) [16]. They exhibited different geometry, were unstable, and their formation was independent of the Tar receptor or any other chemotaxis receptor. Nevertheless, nonchemotactic (motile) mutants did not form patterns. Several questions emerge from these results. In particular, the nature of the signal that gives rise to the formation of the cellular aggregates is apparently peculiar to *Salmonella* and is not generated by its close relative *E. coli* K-12. In this regard, it would be interesting to test other strains of *E. coli* and *Salmonella* for patterning properties. If the *Salmonella* cells are not using the chemotaxis receptors for sensing an attractant on rich media, are they sensing a gradient? It will be very interesting to investigate the properties of mutants of *Salmonella* that form patterns on minimal media, but which are defective in patterning on complex energy sources.

2.3. *A potential biological role for pattern formation in* E. coli *and* Salmonella

To date there is no conclusive evidence pointing to a biological role for pattern formation in *E. coli* and *Salmonella*. In the studies reported by Budrene and Berg [18] and in the investigations of Blat and Eisenbach [16], patterning could be triggered by hydrogen peroxide, whereas reducing agents inhibited it. Budrene and Berg suggested that following chemotaxis down the concentration gradient of the carbon source, a rapid oxidation ensues which potentially leads to the formation of toxic oxygen radicals. Cellular aggregation could create local anaerobic conditions owing to enhanced rates of respiration, thereby minimizing the damaging effect of such toxic radicals. Aggregation would result from the elaboration of the chemoattractant, aspartate, or another extracellular metabolite sensed by the Tar receptor. Indeed, hydrogen peroxide triggered the secretion of attractant by the cells [18].

A number of mathematical models have been constructed to simulate the formation of patterns by *E. coli* and *Salmonella* [17, 19, 123, 132]. All of these models were based on chemotaxis of the bacteria towards metabolizable substrates as well as to chemoattractants they both produce and consume. Ben-Jacob *et al.* [8] argued that in order to account for all of the patterns reported by Budrene and Berg [18, 19], the terms for chemotaxis towards a metabolizable substrate and chemotaxis towards a cell-generated chemoattractant were insufficient. They introduced a third chemotactic force, which involved chemotaxis *away* from a cell-generated chemorepellent. According to the model, chemoattraction is a short-range force operating over short distances, while chemorepulsion is long range. The authors predicted that the geometry of the patterns was governed by the sensitive interplay between these three chemotactic forces. A more precise understanding of the biology of this remarkable system awaits a detailed description of the genetic basis of pattern formation.

3. Chemotaxis Systems in Bacterial Swarming

3.1. *Swarming motility*

Swarming has been defined as an organized form of multicellular translocation of bacteria across solid surfaces [45]. An extensive treatment of swarming motility is presented in Chapter 3, and will be only briefly described here. Swarming has been observed in gram-negative and gram-positive phylogenetically diverse bacterial species. In liquid, or in a very soft agar (0.3% or less) these organisms swim, as individuals, by means of one polar flagellum (e.g., *Vibrio parahaemolyticus*, *Pseudomonas aeruginosa*) or a relatively few peritrichous flagella (e.g., *Proteus mirabilis*, *Bacillus subtilis*). On semisolid agar (0.4–1.0%), or even higher (up to 2%) for some bacteria, cells may differentiate from swimming into swarming cells. This differentiation process involves cell elongation (in *Proteus* up to 40–80 times the cell length of the swimming cells [5]). In most cases, differentiation also involves the overproduction of peritrichous flagella. In *E. coli* or *Salmonella* the number of flagella is twice as high as in the swimming cells; while in *P. mirabilis* or *V. parahaemolyticus* this number may be 50 times higher [42]. *P. aeruginosa* appears to be an exception, since its swarmer cells have no peritrichous flagella. Differentiation, in this case, involves the synthesis of type IV pili, which are required for swarming [69]. This is reminiscent of the social (S), gliding motility in myxobacteria (see below), which is also driven by type IV pili. Another parallel between the two types of motility is their dependence on high cell density. Isolated swarmer cells barely move, until a sufficiently high cell density is obtained. The cells align closely along their long axis, forming rafts that move as a unit [41]. A *P. mirabilis* mutant with a deformed shape was found to be incapable of proper alignment, and did not swarm [44].

3.2. *Intercellular signaling and swarming*

The requirement for cell alignment and the cell-density dependence of the process suggest that swarming is a "social" activity. As such, it is likely to require cell-to-cell communication and intercellular signaling. Various signaling molecules have been found to mediate this. Eberl and his colleagues [30, 31] and Lindum *et al.* [74] found that N-acyl homoserine lactones promote swarming in *S. liquefaciens*. The homoserine lactones control, through quorum sensing, the synthesis of a biosurfactant thought to be required as a surface lubricant. Similar findings were reported for *P. aeruginosa* [69], and a number of other bacteria (e.g., *B. subtilis and P. mirabilis* [31, 97]). Toguchi *et al.* [121] found that many mutants of *Salmonella* that were unable to swarm were also defective in lipopolysaccharide (LPS) synthesis. While no direct role for LPS was demonstrated, some of the mutants were spared for swarming deficiency by the addition of a biosurfactant (surfactin) from *B. subtilis* to the medium. If the generic model for swarming and gliding [9, 12] is applicable in the case of *Salmonella*, it may be that LPS or a similar biopolymer may serve as the boundary layer for raft mobility, enclosing the cells within an amphipathic polymer layer, thereby ensuring intercellular communication and interaction.

In addition to the homoserine lactones, it has been suggested that other molecules such as peptides and fatty acids act as signals for swarming motility [34, 47, 67, 72]. The branching factor, an extracellular protein required for pattern formation during colonial development of the *Paenibacilli* (see below), might also play such a role. Interestingly, the signals triggering swarming do not appear to be species-specific since signals produced by one species can affect the swarming behavior of other species as well.

The chemotaxis system is essential for swarming motility. This has been demonstrated for a number of different, unrelated bacteria, such as *V. parahaemolyticus* [105], *P. mirabilis* [5], *S. marcescens* [93], *P. aeruginosa* [79], and *B. cereus* [106]. In *E. coli* and *Salmonella*, defects in all known *che* genes completely abolish swarming [42]. However, chemotaxis itself is not needed. For example, bacteria that—for a variety of reasons—are nonchemotactic, can swarm provided they still possess an intact chemotaxis system. Thus, *E. coli* mutants carrying a flagellar switch mutation rendering the chemotactic response inactive, swarmed normally [20]. Similar behavior was seen with mutants lacking any one of the intermembrane chemoreceptors (Trg, Tsr, Tar, Tap) [42]. Likewise, saturation of chemotaxis receptors by their specific ligands inhibited chemotaxis but had no effect on the swarming motility of *B. cereus* [106] or *E. coli* [20]. Harshey and colleagues [20, 41, 42] concluded that the "chemotaxis pathway" was necessary for inducing hyperflagellation and, consequently, swarming in *E. coli* and *Salmonella*, but swarming was not dependent on the presence of any specific chemoreceptor. Interestingly, mutants defective in any of the genes associated with the signal transduction pathway such as *cheA*, *cheB*, *cheR*, and *cheY* were defective in swarming. A compensatory mutation in *cheY* (*cheY**) restored swarming, suggesting that the key protein involved in swarming is CheY-P. Moreover, the swarming defect in a strain deleted for both CheY and CheZ was suppressed by a mutation in FliM, the switch protein with which CheY normally interacts. The authors have proposed a model in which the set point governing clockwise/counterclockwise rotation is different for swarming and swimming motility, perhaps owing to the increased load on the motor [78]. Further genetic and biochemical analysis should be forthcoming in order to shed additional light on this interesting system.

Thus, chemotaxis gene products are involved in two very different modes of multicellular behavior in *E. coli* and *Salmonella*: swarming motility and the complex pattern formation discussed above.

4. Pattern Formation in *Paenibacilli*

4.1. Paenibacillus dendritiformis *morphotypes*

Another example of highly organized multicellular behavior is the formation of spectacular patterns by colonies of some strains of the motile, spore-forming *Paenibacillus* [7–12, 38] (Figure 3). Phylogenetic analysis has demonstrated that members of this genus are tightly grouped within a cluster consisting primarily of species previously assigned to the genus *Bacillus* [119]. Spotted on medium containing relatively high agar concentration (up to 2.0%), morphotypes of *P. dendritiformis*, and *P. thiaminolyticus* form branched, tip-splitting patterns (T morphotype). When cells isolated from such colonies are spot-inoculated on a plate containing 1.4% or less agar, a colony with T morphology develops for about 36 h, but then the branches begin to curl at the tips, all with the same handedness, ultimately giving rise to a chiral pattern (C morphotype). Cells from these regions transferred to plates containing the lower concentration of agar form curled branches at the outset, giving

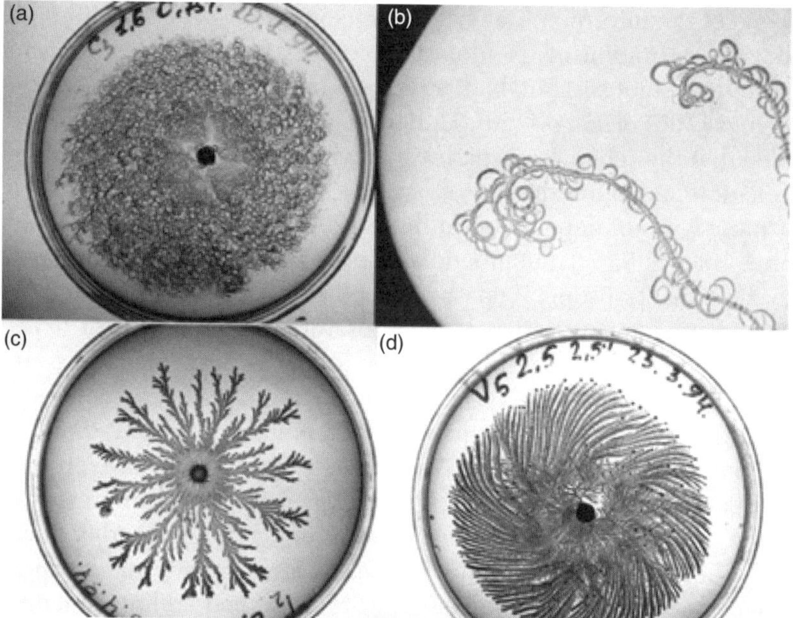

Figure 3. Pattern formation of three *Paenibacillus* morphotypes during colonial development showing different modes of patterning. (a) *P. dendritiformis* C (chiral); (b) a higher magnification of one of the chiral branches from (a); (c) *P. dendritiformis* T (tip splitting); and (d) *P. vortex* (V).

rise to the C morphotype, in which the branches all twist in a counterclockwise direction. Members of *P. vortex*, *P. glucanolyticus*, and *P. lautus* form, on harder (up to 2.5%) agar, unique, highly complex vortex patterns in which the tips of the branches contain a high density of cells all rotating around a common center at a real-time speed of several cell lengths per second [9]. Light microscopy and time-lapse video photography have been used to characterize the dynamics of patterning for each of the morphotypes [9, 10, 12]. In all cases, the collective movement of the bacteria involves the formation of a viscous boundary layer (wetting fluid). Virtually all the cells are present within this envelope. In the case of T cells, the collective movement is accompanied by a random walk like motion of the individual cells enclosed within the envelope. Cells nearest the boundary occasionally collide with it, and repeated collisions are thought to give rise to the emergence of new branches. The C cells are relatively elongated (1.5–2 microns), they are enveloped in thinner branches than the T cells, and they move only forward and backwards. As noted above, the branches of the C morphotype all twist in a counterclockwise direction even though the individual cells retain their rod shape.

Patterns formed by the different morphotypes were described by Ben-Jacob *et al.*, by a morphology diagram in which slight but distinctive changes in the patterns occur as the concentration of energy source and agar hardness are varied [7, 12, 21]. The authors devised a generic modeling approach that incorporates the general features of patterning as inferred from close inspection of colonial morphology and dynamics. These parameters were incorporated into a computer model, which also included terms describing the contribution of specific, hypothetical biological forces that might be involved, in order to account for the unique patterning observed under all conditions [21]. According to the models, the delicate changes in patterning arise as a result of a sensitive interplay between an environmental force (nutritional chemotaxis towards a food source) and two additional extracellular signals generated by the cells themselves: a short-range chemoattractant system, which regulates the intercellular dynamics within the branches, and a long-range chemorepulsive system, which controls the organization of the branches themselves. The length scale of each signal is determined by the diffusion constant of the particular cellular product and the rate of its decomposition [11, 12]. Using this model, the authors have been able to reproduce all of the patterns formed by *P. dendritiformis*. It is remarkable that the same interplay of predicted forces was operative in the simulation of

E. coli and *Salmonella* patterning (see Section 2). What is different about the two models are the special features of the motility, the collective organization of the cells, and the ability of the *P. dendritiformis* cells to produce a wetting fluid which both ensures the collective behavior of the cells as well as permits the movement across the hard surfaces.

4.2. *Pattern formation by* **P. vortex**

Patterning by *P. vortex* is substantially different from that observed with C and T morphotypes. In this system, the cells appear to move by swarming without tumbling, while *P. dendritiformis* appears to tumble. V cells are much longer than the others (2.5–3.5 microns). In addition, patterning takes place on hard agar (up to 3%). The cells are usually aligned and appear only to move forward. However, collective movement is generated by rotating groups of cells, moving at a speed of several cell lengths per second [9]. The rotating cells are often several layers thick, and intact vortices are seen to migrate and fuse to others. Moreover, as the vortices move, they leave what appear to be tracks where slower moving groups of cells are seen. These cells are often seen to be "swept up" by a migrating vortex. Modeling of this very complex system predicts a fourth cell-generated signal, a specialized chemoattractant, which attracts the V cells towards the center as they move forward, causing them to rotate around the common center and generating the vortex [9]. The dynamics and the complexity can be seen in the model which incorporates the features of nutritional and rotational chemotaxis as well as the short- and long-range chemoattractant and chemorepellent, wetting fluid, and features accounting for the strong alignment of the cells [9]. Interestingly, when observed under the microscope, the dynamics of vortex behavior are reminiscent of *M. xanthus* (http://www.microbiology.med.umn.edu/faculty/myxobacteria/movie. html). The model for *P. vortex* is thought to be applicable to the collective movement of other swarming or gliding organisms [9], although this has yet to be demonstrated.

4.3. *Morphotype "nebula"*

As mentioned above, pattern formation has been observed for a number of species of *Paenibacillus* [119]. In almost all the cases where it was investigated the patterning in these species resembled that of morphotypes T, C, and V. In addition, a new morphotype referred to as "nebula" was described for *P. alvei* [22]. The pattern formed consisted of diffuse clusters of cells distributed more or less symmetrically throughout the plate. When the morphology diagram was examined, colonial morphology appeared to vary between T, C, and V, depending on both agar hardness and nutrient concentration. The authors pointed out thus far they have not succeeded in formulating a generic model to account for this complex behavior.

4.4. *Mutants of* P. dendritiformis *defective in pattern formation: phenotypes and reconstitution*

Biological studies characterizing the patterning process have been largely descriptive. Direct microscopic observations show that patterning actually begins at a very early stage following inoculation. This stage leads to the formation of a highly organized "mother colony" from which the branches emerge (Figure 4). Mother colony formation apparently involves several stages. Once inoculated, the cells divide and multiply at the inoculation site. Growth is not uniform: shortly after inoculation, a ring consisting of numerous cells appears at the drop periphery and small aggregates appear throughout the inoculation spot. The aggregates consist of elongated cells, positioned adjacent to each other along their long axis. They grow, leaving between them zones containing only a few cells. Within 24 h, the colony matures displaying a three-dimensional surface structure of crests and troughs with a densely

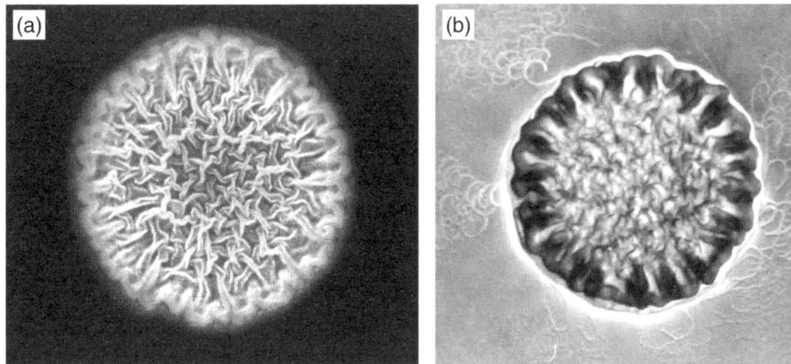

Figure 4. "Mother colony" organization in *P. dendritiformis* morphotype C. (a) 24 h after inoculation; (b) 48 h after inoculation, showing branching. The diameter of the mother colony remained unchanged at 4.5–5 mm.

packed, elevated outer ring of cells (Figure 4). The diameter of this colony is only slightly larger than the diameter of the original inoculated drop. At this stage, growth at the site of the drop appears to be arrested. There is no expansion of colony diameter or increase in height. Once formed, the organized mother colony persists throughout the subsequent stages of colony development.

To date there is little information regarding the genes and/or gene functions involved in pattern formation. It is likely that pattern formation is a complex process involving many genes and gene products which might be selected against when the cells are passed through hundreds of generations of growth under nonselective conditions such as rapid aeration in liquid culture on very rich medium. Using such a procedure, evolved populations of *P. dendritiformis* were obtained after continuous growth for 800 generations (Varon and Gutnick, in preparation). In the evolved cultures about 5% (14/250) of the colony forming units exhibited patterning defects. Among the mutants isolated from evolved C cultures were three morphologically different classes of mutants: one class, termed "precocious" (after Belas *et al.* [6]), were impaired in mother colony organization but formed a regular branching pattern. Moreover, the branches appeared earlier than in the case of the parent. They also formed on harder agar as well (class I). The second class (II) showed colony organization, but no branching even on softer agar. The third class of mutants (III), termed "crippled" (Cri), was defective in both mother colony organization and branching (Figure 5). The patterning-defective phenotype of class III mutants could be partially suppressed in the presence of an extracellular protein (10 kDa) termed "branching factor" obtained from the spent medium of the parent strain (Bach, Avigad, and Gutnick, in preparation). The reconstitution of patterning in the Cri mutants formed the basis of a bioassay for the branching factor. Using this assay, it was possible to detect branching factor in the broths of both T and C cells, in the broths of the precocious mutant and in the broth of *P. thiaminolyticus*. In contrast, the factor was not found in the spent medium of other defective mutants, or any of the *Vortex* morphotypes. Interestingly, the branching factor (which by itself lacks any proteolytic activity) could be replaced by a variety of proteases. Mutants have been isolated which failed to respond to the factor but which were rescued by proteases (Gutnick, unpublished). A hypothetical scheme for how the pathway leading to pattern formation in the C morphotype might occur is shown in Figure 6. The model predicts the production of an extracellular factor, which is thought to accumulate during the organization of

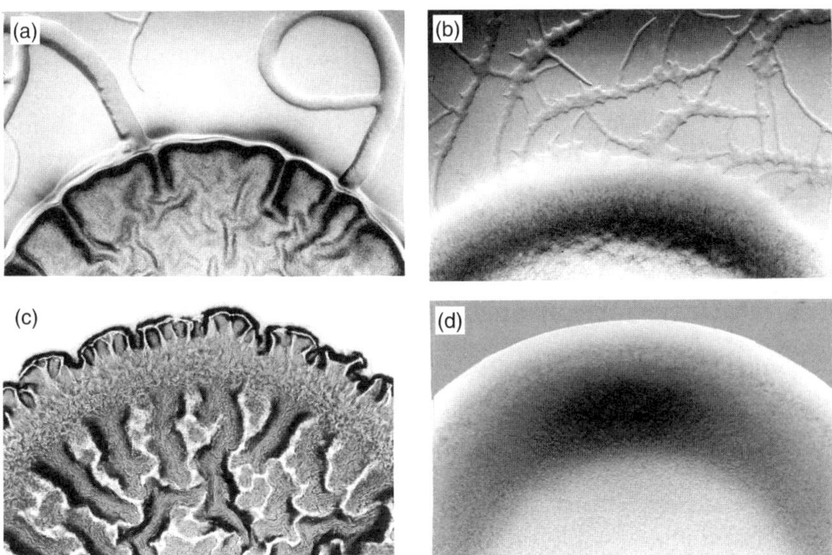

Figure 5. Phenotypes of patterning mutants of *P. dendritiformis* morphotype C. Mutants differ from their parent in self organization of the "mother colony" and/or branching. (a) The parent strain, exhibiting mother colony organization and early stages of branching. (b) Class I, "precocious" mutant. There is essentially no organization, but at a very early stage cells begin to form numerous tiny branches which surround the colony. (c) Class II mutant. The mother colony develops normally showing a complex network of ridges, similar, if not identical, to that observed in the parent strain, but no branching. (d) Class III, "crippled" mutant. Cells in the inoculated area propagate and form a colony, with growth more abundant at the periphery of the drop. There is virtually no organization and no branching even after prolonged incubation. All four colonies were photographed under the microscope after 18 h (b) or 21 h (a, c, d) of growth at 37°C. Colony diameter was 4.5–5.0 mm.

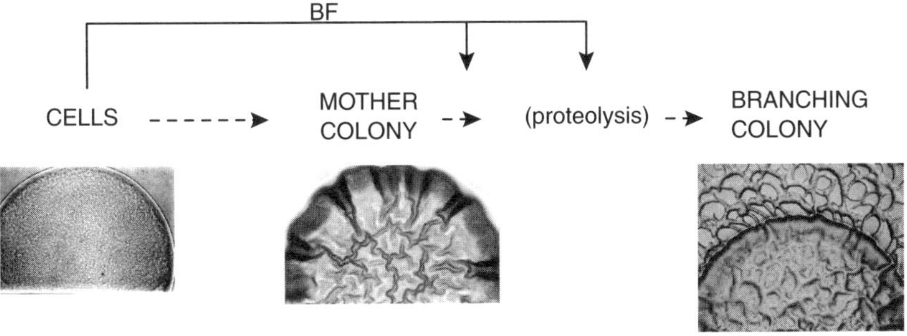

Figure 6. A hypothetical scheme for reconstitution of Class III mutants of *P. dendritiformis* morphotype C by an extracellular branching factor (BF). The scheme accounts for the fact that BF does not restore mother colony organization to Class III mutants, but does restore branching (see text for details).

the mother colony prior to branching. The branching factor might then trigger a proteolytic event(s) leading to subsequent patterning.

Clearly, the multistage process of pattern formation involves a variety of collective cellular activities, including organization and branch formation. While it is likely that chemotaxis either towards a nutrient source, or in response to cell-generated signals is involved in any or all of these stages, its role remains to be directly demonstrated.

4.5. *Pattern formation by other bacteria*

A unique 3-D pattern formed by *B. subtilis* has been described by Mendelson [90]. On fresh, moist and nutrient-rich agar surfaces, motile strains grew into colonies with finger-like projections extending from the colony periphery. Colony development occurred in sequential phases of finger expansion. Once a particular finger reached its full length, new fingers were initiated while the cell mass in the "old" finger continued to increase within the boundaries, and continued to expand in width.

Fractal pattern reminiscent of that formed by *P. dendritiformis* was observed in other bacteria that exhibit collective motility such as *S. marcescens* and *P. mirabilis* [80], and some mutants of *M. xanthus* defective in S motility (see Section 5) [75]. This latter case represents a new class of pattern-forming strains that seem to be "unmasked" by mutation.

5. Chemotaxis in the Life Cycle of the Myxobacteria

5.1. *Introduction to the myxobacteria*

The myxobacteria are gram-negative, rod-shaped, soil-dwelling bacteria (The Myxobacteria Web Page, http://www.microbiology.med.umn.edu/myxobacteria/index.html). They feed on organic matter or on prey organisms, utilizing for this purpose an array of hydrolytic enzymes (proteases, nucleases, lipases, glycanases, cell wall lytic enzymes), which efficiently degrade bacteria and yeast cells [98]. Under starvation conditions, the myxobacteria undergo a spectacular morphogenesis in the course of which tens of thousands of cells interact and form complex aggregates [27]. The aggregates eventually mature into fruiting bodies of different shapes, colors, and complexity, depending on the species (Figure 7). Inside the fruiting bodies, vegetative cells differentiate into resistant myxospores. Under favorable conditions, the myxospores germinate and start a new cycle. Throughout their life cycle, the myxobacteria display cooperative behavior. They feed, move, and develop according to a carefully

Figure 7. Fruiting bodies of *Myxobacteria*. (a) *Chondromyces pediculatus*, (b) *C. crocatus*, (c) *C. apiculatus*, (d) *Melittangium boletus*, (e) *Myxococcus fulvus*, (f) *C. lanuginosus*. Courtesy of Hans Reichenbach.

coordinated program. This coordination relies on cell-to-cell signaling. The cells transmit, receive, interpret, and respond to these signals. Relaying the information requires, in some instances, direct contact between the donor and the recipient cells. In others, it is achieved through the secretion of diffusible molecules accumulating in the growth medium.

Motility is a crucial factor in the development of the myxobacteria. The cells, which are nonflagellated, move by gliding over a solid surface. They advance in the direction of their long axis, periodically stopping and/or reversing directions so that the leading end of the cell becomes the lagging one (reviewed in [81, 116]). Gliding is very slow compared with swimming and the cells advance only several micrometers per minute. Two types of gliding motility have been described for *M. xanthus*. Adventurous (A) and social (S) motility, each coded by its own set of genes [see Hans Reichenbach's time-lapse motion picture films at http://www.microbiology.med.umn.edu/myxobacteria/movie.html]. Wild-type colonies spread over agar, with cells moving in groups, and individual cells migrating "adventurously" beyond the confines of these "social" groups. Cells with mutations in A-motility genes (A^-S^+) move well in groups but are unable to glide as isolated single cells [55]. They form spreading colonies with no individual cells beyond the edges of the colony. Strains with mutations in S-motility genes (A^+S^-) move well as individual cells but less efficiently in groups and the development of fruiting bodies is usually defective. Their motility is facilitated at high cell density [55], indicating some sort of cell interaction. The dependence on high cell density is absolute in the case of S-motility. Such motility does not occur unless the cells are within a few micrometers of one another [55]. Thus, both types of motility appear to involve cell–cell interaction and coordination. Disruption of both motility systems (A^-S^-) results in complete loss of motility [46].

The two types of motility are driven by different mechanisms. Adventurous gliding appears to rely on the flux of large volumes of highly viscous biopolymer through nozzle-like structures located in the cell membrane and clustered mainly at the two poles [131] (Figure 8).

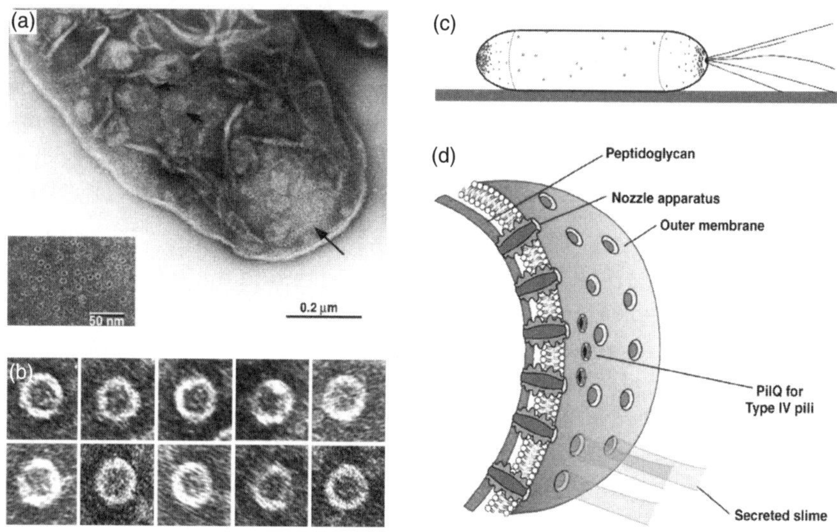

Figure 8. Slime nozzles postulated to be responsible for adventurous gliding motility in *M. xanthus*. (a) An electron micrograph of an isolated cell envelope of *M. xanthus* showing a collection of nozzles at one of the poles (long arrows). The insert shows this region at higher magnification. (b) A gallery of isolated nozzles. (c) A cartoon illustrating the presence of pili on one of the poles and nozzles at both poles. (d) Cartoon illustrating a distribution of the nozzles in the polar region shown in A. See text for further discussion of A- and S-motility in *M. xanthus*. Reproduced from *Current Biology*, Vol. 12, Wolgemuth *et al.*, "How Myxobacteria glide," p. 370, 2002 [131], with permission from the author and from Elsevier Science.

According to a model proposed by Wolgemuth *et al.* [131], the propulsive force is generated by hydration of polyelectrolyte slime in the nozzle. The slime is hydrated and consequently swells. The swollen polymer is extruded through the narrow opening of the nozzle pushing against the substratum.

As described in Chapter 3, social motility is powered by the retraction of type IV pili [118]. The pili are polar, thin fibers, up to 6 nm in diameter and several microns long, with very strong tension (reviewed in [126]. They are located at one cell pole, usually in a tuft of two to eight fibers. Their removal by genetic mutation or by mechanical shearing results in impaired S-motility [51, 102, 133]. Sun *et al.* [118] showed that when wild-type *M. xanthus* cells glide on a polystyrene surface, they occasionally become tethered to the surface and stand on end. They also appear to be pulled toward the surface. Cells of a *pil*A mutant lacking pili did not become tethered, suggesting that pili are involved in tethering. A model proposed by Sun *et al.* suggests that S-motility is achieved as a gliding cell repeatedly extends pili from its leading pole. Upon attaching to a surface, the pili retract, resulting in cell movement (Figure 9). Cells could reverse their direction of movement by extending and retracting pili from the opposite pole. To explain the smooth gliding motion of *M. xanthus*, McBride [81] suggested that some pili extend from the leading pole at the same time that others, having come into contact with a surface, retract.

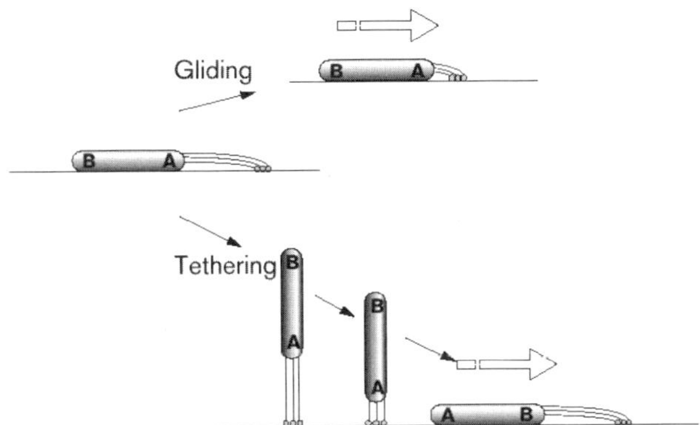

Figure 9. Model of motility mediated by the type IV pilus. Reproduced from *Current Biology*, Vol. 10, Sun *et al.*, "Type IV pilus of *Myxococcus xanthus* is a motility apparatus controlled by the *frz* chemosensory system," p. 1146, 2000 [118], with permission from the author and from Elsevier Science.

When their food supply drops below some threshold, and provided the myxobacteria are both at a sufficiently high density and on a solid substratum, they initiate a developmental program that culminates in the formation of multicellular fruiting bodies [27]. Initially, small, asymmetric aggregates appear. Some of the aggregates expand into relatively evenly spaced hemispheres, each containing about 10^5 cells. The remaining aggregates disappear. At the end of the process macroscopic, multicellular structures are formed, varying in complexity between simple mounds and elaborate, treelike forms, depending on species (Figure 7). The simplest form is that of *M. xanthus*, which is also the most extensively studied species of the group. Ultimately, the cells inside the fruiting body differentiate into metabolically dormant, resistant spores. During the aggregation phase, the cells may pass through a period in which the surface is swept by a complex pattern of waves and ripples consisting of bacteria moving in concert [99–100, 111]. The developmental program of *M. xanthus* requires cell-to-cell communication and collaboration via chemical signals. Five such signals have been identified for fruiting-body development in *M. xanthus*: A, B, C, D [39], and E [24].

5.2. *Direct evidence for chemotaxis*

The notion that chemotaxis may be involved in the myxobacterial life cycle is an appealing explanation both for feeding and for fruiting body formation [85]. Some indication to that effect comes from experiments done forty and fifty years ago [73, 88, 89]. When *M. xanthus* cells were separated from preformed fruiting bodies by a dialysis membrane, the cells aggregated and formed fruiting bodies directly over the preformed ones, suggesting that the preformed fruiting bodies lying beneath the membrane secreted a compound(s) capable of passing through the membrane and initiating the formation of new fruiting bodies on top of the membrane [88]. McVittie *et al.* [89] coined the term "developmental chemotaxis."

Twenty years later Dworkin [26] demonstrated that a flare of *M. xanthus* moved towards clumps of potential prey bacteria (*Micrococcus luteus*). However, the same response was obtained with polystyrene latex beads or glass beads, ruling out the possibility that the cells were responding to a concentration gradient of some chemoattractant. Indeed, Dworkin and Eide [29] failed to detect any chemotactic response when various chemicals were examined as potential attractants or repellents. They postulated that since the motility of *M. xanthus* is very slow, the time required for a small soluble molecule to diffuse the length of the bacterium is short compared with the time required for a cell to glide the same distance, thereby suggesting that chemotaxis was an unlikely mechanism.

In an attempt to create conditions more favorable for chemotaxis, Shi *et al.* [108] devised a way to establish steep chemical gradients. They used for this purpose compartmentalized petri dishes. The selected chemical was introduced into one compartment and it diffused into the compartment on the other side of the barrier via a thin overlay of agar, generating steep gradients that were stable for many hours. In addition, the authors used culture conditions that increased the rate of gliding (low agar concentration). They showed that under such conditions a colony expanded primarily in the direction of increasing concentrations of nutrients such as yeast extract or Casitone and away from repellents such as DMSO or short-chain alcohols. Whether these findings reflected chemotaxis was called into question by Tieman *et al.* [120]. These authors constructed chemical gradients by placing two relatively large blocks of agar 2 mm apart on top of a very thin agar (1%) layer. The selected chemical diffused very slowly from one block of agar to the other through the thin agar layer, resulting in the establishment of a steep and very stable gradient in the 2 mm bridge. Cells were placed on this bridge before or after the establishment of the gradient, and their movement was studied. There was no response to either Casitone or to yeast extract. The authors concluded that *M. xanthus* does not respond to chemical gradients per se but rather to the presence or absence of certain chemicals, e.g., nutrients.

Koch [68] suggested that chemotactic motility was not necessary for group movement of *M. xanthus*. He proposed a model according to which the cells move forward and backwards. A frequent (and reversible) mutation brings forward a cell that moves unidirectionally and "leads" a cohort of cells away from the bulk of the population.

Despite this model and the mounting body of evidence to the contrary, the intense search to establish a role for chemotaxis in myxobacteria persisted. Zusman and his colleagues contended [108, 127] that, although there may be no chemotaxis toward added chemicals, the bacteria could respond to self-generated attractant(s) produced by the cells themselves. They assumed that this attractant is produced in a cell-density-dependent manner, and is located on the cell surface.

Finally, in 1998 Kearns and Shimkets clearly demonstrated that *M. xanthus* responds chemotactically to specific stimuli [57]. The chemoattractant was phosphatidylethanolamine (PE), a phospholipid that was either extracted from the myxobacterial cell membrane or produced synthetically [60]. PE was applied to an agar surface which had been inoculated with cells of *M. xanthus*, and after evaporation of the delivery solvent the cells were examined using time-lapse video microscopy. PE, due to its hydrophobicity, diffuses very slowly in aqueous environments, thus offsetting the effect of the slow movement of the cells. Ordinarily, isolated cells move along their long axis in a manner analogous to a "run" in swimming bacteria, reversing their direction of movement ("tumble") every 5–7 min [15]. In the presence of PE, the time between reversal events was increased, resulting in directed motility. The response depended on the fatty acid composition of the PE. PE-diC16:1 ω5c elicited a response at 2 ng, the response dropping sharply at concentrations exceeding 5 ng [60]. Dilauroyl PE (diC12:0) and dioleoyl (diC18:1 ω9c) PE were active at much higher amounts (2 μg) while other PE molecules (dimyristoyl, dipalmitoyl, diheptadecanoyl and distearoyl) were inactive [57, 58, 60].

What role might chemotaxis toward PE play in the life cycle of myxobacteria? Thus far, there is no definite answer, but it is tempting to suggest that PE plays a role both in feeding as well as in fruiting body formation. Phospholipids are an important component of the bacterial membrane. The *E. coli* outer membrane consists of about 25% phospholipids, and the inner membrane may consist of up to 40%. This represents almost 75% of the total phospholipids [23]. A similar value (76%) was obtained for *M. xanthus* [94]. When bacteria, potential prey for myxobacteria, lyse, as the result of autolysis or some other cause, PE would be released. This PE might serve to attract nearby myxobacteria to their prey. As the cell density around the food source increases, there will be more PE and faster aggregation. Rosenberg *et al.* [103] have shown that high cell density is necessary for efficient utilization of macromolecules by myxobacteria. Chemotaxis to PE might attract nearby cells and consequently establish the required cell density.

It has been reported that during fruiting-body formation, a large part of the *M. xanthus* population undergoes autolysis [130]. Under conditions where this occurs nutrients and/or signals would be available for the remaining population to complete development [130]. PE appears to play a role in this process. Gelvan *et al.* [35] showed that under starvation conditions PE accumulates in the growth medium. This PE is toxic to myxobacteria and its lytic activity is due to the fatty acid moiety, released by extracellular phospholipases. The PE cleavage rate is determined by the phospholipase level, which, in turn, depends on cell density. Thus, attraction to PE would cause local aggregation, increased levels of phospholipases and a higher concentration of free fatty acids, sufficient to allow developmental lysis.

Fontes and Kaiser [33] described a special case of tactic behavior in *M. xanthus*. They showed that the cells respond to stress in the agar. This response (elasticotaxis) depends on A-, but not on S-motility. According to Wolgemuth *et al.* [131], elasticotaxis arises from the tendency of the polyelectrolyte chains of extruding slime to align with polymer chains in the agar.

5.3. *Sensory transduction systems*

Whereas direct evidence for chemotaxis in myxobacteria was slow to come, indirect evidence has been accumulating for over two decades, based on genetic studies of *M. xanthus*. As discussed in Chapter 3, the enteric bacteria possess a remarkably sophisticated system of signal transduction catalyzed and controlled by a group of tightly coordinated proteins. *M. xanthus* contains a complete set of sensory components homologous to those found in the enteric bacteria. A large number of operons encoding the Che proteins and MCP homologues have been identified, and the role of some of them has been elucidated. Two sets of genes, the *Frz* genes and the *Dif* genes, have a direct effect on the motility of *M. xanthus*.

5.3.1. *The Frz system*

In 1982, Zusman [136] (reviewed in [128, 129] described a new class of developmental mutants of *M. xanthus*, "frizzy" (*frz*) mutants. Under conditions conducive for fruiting body formation, the mutants failed to aggregate into the regular dome-shaped mounds, forming, instead, tangled, "frizzy" filaments. The mutants were capable of both A- and S-motility but were affected in a chemosensory system. Analysis of single cell movement in these mutants suggested that they were unable to regulate the reversal frequency of cell gliding [15]. The reversal rate of the mutants was approximately once an hour (compared with one reversal every 5–7 min in the wild type), suggesting that the process controlling this behavior is nonfunctional. One group of mutants carrying a gain-of-function mutation in the *frz*CD gene showed a very high frequency of reversals. In these mutants, the cells reversed their direction every 1–2 min. These differences in the reversal frequency are likely to be meaningful for directed motility and for the net advancement of the cells. Several studies indicate that the *frz* genes are developmentally regulated and respond to cell-to-cell signaling [49, 115].

Sequence analysis of the proteins encoded by the *frz* genes revealed that most of them are homologous to the major chemotactic proteins of the enteric bacteria: CheA, CheY, CheW, CheR, CheB, and the methyl-accepting chemotaxis protein (MCP) [83, 86]. However, there are also significant differences. For example, FrzE is a fusion protein composed of both a CheA and a CheY domain [86] while FrzZ appears to be a fusion of two CheY-like domains [122]. The MCP homologue of *M. xanthus*, FrzCD, is a cytoplasmic protein, lacking the two transmembrane domains and the periplasmic sensory domain typical of the enteric MCPs. Nevertheless, FrzCD appears to function similarly to the enteric MCPs: Zusman and his colleagues [82, 84–87] showed that some compounds (e.g., phospholipids containing lauric acid, [attractants?]) stimulated methylation of FrzCD, whereas others (e.g., isopropanol, [repellents?]) caused demethylation. This response is similar to that found in enteric bacteria where methylation and demethylation of MCPs mediate adaptation to chemoeffectors (Chapter 3). This, and the homology between the components of the Frz and Che systems, suggest parallels between the chemotactic behavior of the swimming bacteria and the directed motility of the gliding *M. xanthus*. Table 1 lists the *frz*

Table 1. *M. xanthus frz* genes show similarity to enteric chemotaxis genes (after Ward and Zusman [129]).

frz gene	Homologue in the Che system	Proposed function
A	CheW	Adapter
B	None	Unknown
CD	Tar (MCP)	Receptor
E (N terminal)	CheA	Histidine kinase
(C terminal)	CheY	Response regulator
F (N terminal)	None	Unknown
(C terminal)	CheR	Methyltransferase
G	CheB	Methylesterase
S	None	S-motility regulator
Z (N terminal)	CheY	Response regulator
(C terminal)	CheY	Response regulator

genes, their homology to the Che proteins, and their function as proposed by Ward and Zusman [129].

5.3.2. *The Dif system*

A second operon encoding proteins homologous to enteric chemotaxis proteins was discovered by Yang *et al.* in 1998 [134]. These investigators isolated mutants, *defective in fruiting* (*dif*), which were also defective in social motility, and did not aggregate under conditions suitable for fruiting body formation. One of the Dif proteins (DifA) is a MCP homologue, with two potential transmembrane domains and, possibly, a small periplasmic domain. Other Dif proteins are homologous to CheW, CheA, and CheY. One protein (DifB) shows no homology to known chemotaxis proteins.

The Dif system is required for production of peritrichous cell-surface appendages called fibrils [134, 135]. These are flexible filaments, 10–30 nm in diameter and up to many times the length of the cell [28, 65], composed of approximately equal amounts of polysaccharide and protein [3, 4]. A fibril protein was recently identified as a zinc metalloprotease [61]. Under the microscope, the fibrils appear to coat the cells and form a cohesive interconnective matrix ([58], Figure 10). Chemical disruption of the fibrils, like that of the pili, results in defective S-motility [2]. In

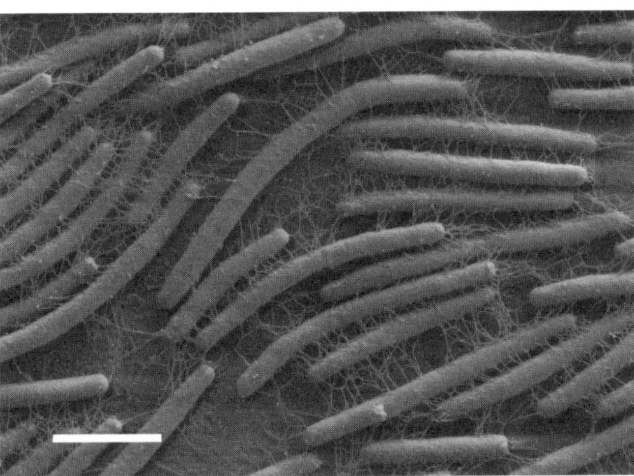

Figure 10. Scanning electron micrograph of *M. xanthus* cells showing interconnecting fibrils. Scale bar = 2 μm. Provided by L. Shimkets and reproduced from *Trends in Microbiology*, Vol. 9, Kearns and Shimkets, "Lipid chemotaxis and signal transduction in *Myxococcus xanthus*," p. 127, 2001 [58], with permission from Elsevier Science.

their study of lipid chemotaxis, Kearns *et al.* found that the fibrils behaved as chemoreceptors [59, 61]. Mutants either lacking fibrils or those defective in the zinc metalloprotease did not undergo excitation in response to dilauroyl PE (they showed excitation to dioleoyl PE). However, beyond the requirement for fibrils, PE signal transduction also required Dif proteins [58]. Therefore, Kearns and Shimkets proposed [58] a model according to which the fibrils contain chemoreceptors, possibly at their distal end, and ligand interaction is communicated to the base of the cell by an unknown mechanism. The fibrils then interact with the DifA methyl-accepting chemotaxis protein and other components of the Dif system. This location of the chemoreceptor would provide the cells with a means of interacting with insoluble signals at some distance, at their source in the environment. On the other hand, it would prevent autostimulation by cellular membrane lipids.

5.3.3. *Other chemotaxis gene clusters*

Two additional chemotaxis gene clusters have recently been identified. One of them, Che3 [66], consists of 2 methyl-accepting proteins, homologues of CheW, CheR, CheB, a hybrid CheA, but no CheY. Mutants defective in these proteins were still able to carry out chemotaxis suggesting an alternate role for the signal transduction pathway. The cluster was found to encode a divergently transcribed putative signal response regulator, termed CrdA (*chemosensory regulator of development*), which is a transcriptional activator of (σ^{54}-dependent promoters. CrdA was found, by yeast two-hybrid analysis to interact with the CheA homologue, suggesting that the two proteins might comprise a two-component regulatory system for regulating gene expression during development. Interestingly, a similar but less well-defined system has recently been described for *Rhodospirillum centenum* [14], which exhibits three distinct cellular morphologies depending on environmental conditions. Cells swim in liquid media, but in response to surface contact, they differentiate into swarm cells. Upon starvation, cells enter a pathway leading to the formation of heat-resistant cysts. *R. centenum* was found to contain three distinct chemotaxis operons. Mutants in Che2 and Che3 were isolated in which cyst development was accelerated even on rich media. It has been postulated that the developmentally regulated signaling system in *R. centenum* is mediated by a two-component regulatory pathway using histidine kinases and signal response regulators from Che2 and Che3 operons.

Another chemotaxis gene cluster in *M. xanthus*, Che4, is involved in pilus-mediated motility [125]. It encodes a membrane-bound, methyl-accepting chemotaxis protein (Mcp4), two CheW homologues, a CheA fusion, a CheY, and a CheR. Deleting Mcp4 from a A^-S^+ strain (whose motility is mediated by type IV pili) resulted in diminished swarming. In contrast, deletion of either the response regulator CheY or the entire operon resulted in enhanced swarming, suggesting an inhibitory role for these proteins in the control of pilus-mediated motility.

In total, *M. xanthus* appears to have 9 chemotaxis operons and 19 chemoreceptors (D. Zusman, personal communication). The biological role of the majority of all these systems in the myxobacteria remains to be elucidated.

Another system in which chemotaxis gene products are involved has been found in the biofilm-producing organism *Pseudomonas aeruginosa* [43]. This organism contains five distinct chemotaxis gene clusters one of which is involved in chemotaxis itself (Cluster V) and another in pilus-mediated twitching motility (Cluster IV). Cluster II contains a complete set of classical chemotaxis genes including CheA, CheB, CheY, and CheW as well as two MCPs which do not appear to be directly involved in the chemotaxis process itself. Mutants defective in *cheA* and *cheB* genes were found to be impaired in initial adherence to the surface of microtiter plates and subsequently gave rise to unstructured and unor-ganized biofilms. Transcriptome analysis of CheA, CheB, and CheW mutants showed altered transcription of 115 genes when compared with the wild-type strain, suggesting a role for these chemotaxis genes in reg-ulating gene expression, perhaps through the chemotaxis signal trans-duction system. The microarray analysis comparing mutants and wild type revealed that the chemotaxis genes were upregulated about tenfold in stationary phase and additional threefold in response to quorum-sensing signaling which also plays a role in biofilm organization. Some of the target genes included genes associated with anaerobic respiration, genes for stress response and some genes that may be involved in poly-saccharide production.

5.4. *The role of intercellular signaling in development*

As mentioned above, the formation of fruiting bodies in *M. xanthus* requires five cell-to-cell signaling factors: A, B, C, D, and E. Mutants that are unable to produce any one of these signals are incapable of fruiting body formation. Signals A, C, and E have been identified chemically.

Signal A is a mixture of amino acids and peptides; signal C is a protein; signal E is a mixture of branched-chain fatty acids. The mode of action is different for each signal but none seems to involve chemotaxis (reviewed in [110]).

5.4.1. *A-signal*

Initiation of fruiting body formation depends on continuing starvation and a high cell density. In response to starvation, the cells accumulate the highly phosphorylated, tetra- and penta-phosphate guanosine nucleotides, (p)ppGpp [76, 77]. These nucleotides induce production of the A-factor [40], apparently by proteolysis of cell surface proteins [71]. Each starved cell releases a fixed amount of the A-factor, which accumulates in the medium. As its external concentration reaches a certain level, the A-factor, together with newly-formed (p)ppGpp, induce the expression of genes responsible for the early aggregation step in the developmental program [52]. It has been proposed that the A-signal serves a quorum-sensing function in determining whether the population has reached a cell density sufficient for fruiting body formation [56].

5.4.2. *E-signal*

E-signal is a mixture of branched-chain fatty acids [25]. Downard and Toal [25] suggested a model in which branched-chain fatty acids, synthesized during vegetative growth, are released from cellular phospholipids by a developmentally regulated phospholipase during fruiting-body formation. The function of the E-signal has not been fully elucidated. Varon *et al.* showed [124] that a mixture of fatty acids (AMI) produced by the myxobacteria and excreted into the medium is autocidal and acts specifically against the myxobacteria. Autocide AMI is produced under conditions promoting fruiting body formation, apparently by enzymatic cleavage of PE [35]. Mutants resistant to AMI fail to develop normally. AMI is active at low cell densities, which tends to rule out a possible role in quorum sensing.

5.4.3. *C-signal*

The C-signal is a 17 kDa, cell surface–associated protein encoded by the *csgA* gene [113]. It controls rippling [104], the formation of the symmetric, large aggregates that will develop into mature fruiting bodies [63], and the differentiation of myxospores, which occurs several hours later [112]. Each of these three processes requires specific signal intensity. Rippling requires the lowest intensity, aggregation—higher, and sporulation—the highest signal intensity [36]. Gronewold and Kaiser found that the amount of CsgA as well as its signaling activity increases during the course of development in accordance with the timing of the morphological events it controls [36]. The rise in C-factor concentration is a consequence of a positive feedback in the C-signaling circuit: one of the genes whose expression is controlled by C-factor is the *csgA* gene itself. Thus, C-factor induces more production of itself [64].

Studies conducted primarily in the laboratory of Dale Kaiser indicate that unlike other signals that are exchanged between widely separated cells, C-factor is cell-bound and is directly transmitted from one cell to another. Each cell serves simultaneously as a transmitter and a receiver of the signal [53, 54]. Kaiser and his colleagues [37, 53, 115] proposed the following model for the C-signal transduction pathway: Triggered by starvation, *csgA* is first expressed at a low level. Within the nascent fruiting body (and following the early A-signal-dependent aggregation), cells align via end-to-end contact. Such alignment enables C-signal transmission [62], thereby enhancing *csgA* expression in the recipient cell. Thus, starting from an initially low level of C-factor, *csgA* expression rises steeply as more and more cells transmit the C-signal to one another.

The CsgA protein inside the cell fulfills two functions. As CsgA attains a certain level, it activates the Frz phosphorelay system leading to a decrease in cell reversal frequency and stopping time. This in turn leads to formation of cell chains, aggregates, and ultimately to fruiting bodies [37, 48, 49, 115]. As the CsgA level in the cells within the aggregate increases, it eventually reaches the concentration required for sporulation. The *dev* operon is then activated, triggering the differentiation of vegetative rods into spherical spores. The activation of both the Frz and Dev systems is mediated through a transcription factor, FruA.

The following scheme summarizes the pathway for C-signal transduction as conceived by Kaiser *et al.* [37, 53] (Figure 11). The CsgA signal emanating from cell A enhances *csgA* gene expression in cell B, and vice versa, creating positive feedback loops. As CsgA reaches a certain threshold, it activates (by phosphorylation) transcription factor FruA [32, 91, 115]. FruA, once activated, has two targets: Frz and Dev. The Frz phosphorelay system responds to low-level C-signaling and is activated early in the developmental program, inducing cell aggregation. The Dev system is activated at high-level C-signaling. It is induced at a later stage and leads to differentiation of the vegetative cells inside the fruiting body into myxospores. Cells outside the aggregates, peripheral cells, remain rod-shaped,

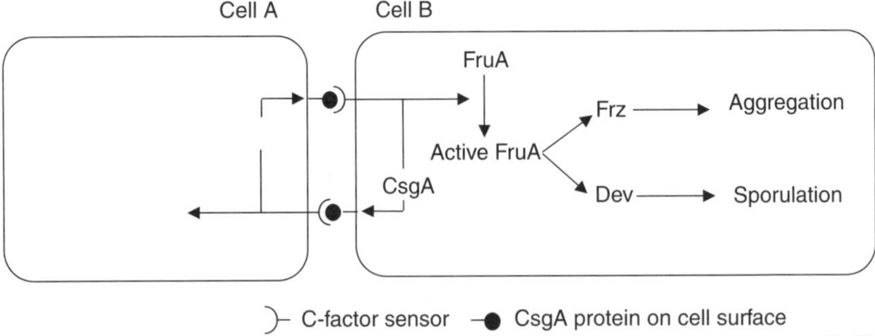

Figure 11. The C-signal transduction pathway. Modified (with permission) from [37].

apparently because they do not reach the cell density and spatial arrangement required for the production of C-factor at the level necessary for activation of the Dev system [50]. Thus, the effect of C-factor on sporulation, like its effect on aggregation, is cell-contact-dependent.

6. Concluding Remarks

The advent of the genomic and post genomic era revealed the wide distribution of large gene clusters encoding chemotaxis and motility genes throughout nature. In many cases, these genes may even be present within the prokaryotic genome in many copies within multiple regulons. It is clear that these systems must have survived evolution by providing the organism with a competitive advantage(s) in natural and often harsh environments. Only in a few cases have these genes been shown to play key roles in processes involving multicellular behavior and tightly controlled chemosensing. If the few examples presented in this Chapter are any indication, it is likely that as new processes and modes of collective behavior are discovered, new sensory transduction pathways will be revealed along with new receptors controlled through the chemotaxis system. It is interesting, in this regard, that even the most extensively studied organisms such as *E. coli* and *Salmonella* have been shown to possess modes of collective behavior, recruiting chemotaxis gene products for this process. In the case of *Salmonella*, growth of the organism on complex media seems to involve new and heretofore undiscovered pathways of intercellular communication. In addition, as new technologies are developed to assess the temporal expression of genetic networks and transient production of gene products, a more precise picture of how chemotaxis systems are recruited during complex multicellular behavior is likely to emerge.

Note Added in Proof

Very recently Park *et al.* [137, 138] described another remarkable example by which chemotaxis is involved in the collective behavior of *E. coli*. It was found that when log phase cells were grown in a confining environment poor in nutrients, they rapidly aggregated and formed dense concentrations of cells. Mutant cells deleted for chemotaxis genes did not form such dense aggregates, suggesting that the cells tended to move towards a chemoattractant subsequently shown to be produced by the cells themselves. The chemoreceptor was shown to be the Tsr receptor, and the cell-generated chemoattractant was found to be glycine. Using suitable mutants the collective behavior was shown not to be dependent on quorum sensing. In moving towards the "trap," the cells moved collectively in the form of waves which broke down as the cells entered the topologically confining space. The authors suggest that the capacity of the waves to break up when they encounter a small opening into a confining space may reflect a special strategy for organisms to seek out a unique environment using the chemotaxis system to attract a sufficient number of organisms to promote essential cell-density-dependent processes such as quorum-sensing-dependent biofilm formation. Interestingly, similar results were found for the highly motile luminescent species *Vibrio harveyi*. Collective movement of the waves is thus a stage of chemoattraction preceding the quorum sensing stage in multicellular organization.

References

1. Adler, J. (1966). Chemotaxis in bacteria. *Science* **153**, 708–716.
2. Arnold, J.W. and Shimkets, L.J. (1988). Cell surface properties correlated with cohesion in *Myxococcus xanthus*. *J. Bacteriol.* **170**, 5771–5777.
3. Behmlander, R.M. and Dworkin, M. (1994). Biochemical and structural analyses of the extracellular matrix fibrils of *Myxococcus xanthus*. *J. Bacteriol.* **176**, 6295–6303.
4. Behmlander, R.M. and Dworkin, M. (1994). Integral proteins of the extracellular matrix fibrils of *Myxococcus xanthus*. *J. Bacteriol.* **176**, 6304–6311.
5. Belas, R., Erskine, D. and Flaherty, D. (1991). *Proteus mirabilis* mutants defective in swarmer cell differentiation and multicellular behavior. *J. Bacteriol.* **173**, 6279–6288.
6. Belas, R., Schneider, R. and Melch, M. (1998). Characterization of *Proteus mirabilis* precocious swarming mutants: identification of *rsbA*, encoding a regulator of swarming behavior. *J. Bacteriol.* **180**, 6126–6139.
7. Ben-Jacob, E., Schochet, O., Tenenbaum, A., Cohen, I., Czirók, A. and Vicsek, T. (1994). Generic modeling of cooperative growth patterns in bacterial colonies. *Nature* **368**, 46–49.
8. Ben-Jacob, E., Cohen, I., Shochet, O., Aronson, I., Levine, H. and Tsimering, L. (1995). Complex bacterial patterns. *Nature* **373**, 566–567.

9. Ben-Jacob, E., Cohen, I., Czirók, A., Vicsek, T. and Gutnick, D.L. (1997). Chemo-modulation of cellular movement and collective formation of vortices by swarming bacteria and colonial development. *Physica A* **238**, 181–197.

10. Ben-Jacob, E., Cohen, I. and Gutnick, D. (1998). Cooperative organization of bacterial colonies: from genotype to morphotype. *Ann. Rev. Microbiol.* **52**, 779–806.

11. Ben-Jacob, E., Cohen, I., Golding, I. and Kozlovsky, Y. (2000). Modeling branching and chiral colonial patterning of lubricating bacteria, in *Mathematical Models for Biological Pattern Formation* (P.K. Maini, H.G. Othmer, eds.), Springer, New York.

12. Ben-Jacob, E., Cohen, I. and Levine, H. (2000). Cooperative self-organization of microorganisms. *Advances in Physics* **49**, 395–554.

13. Berg, H.C. (1996). Symmetries in bacterial motility. *Proc. Natl. Acad. Sci. U.S.A.* **93**, 14 225–14 228.

14. Berleman, J. and Bauer, C. (2003). Analysis of Che-like signal transduction components that affect cyst cell development in *Rhodospirillum centenum*. Abstracts of BLAST VII Symposium, 13.

15. Blackhart, B.D. and Zusman, D.R. (1985). Frizzy! genes of *Myxococcus xanthus* are involved in control of frequency of reversal of gliding motility. *Proc. Natl. Acad. Sci. U.S.A.* **82**, 8767–8770.

16. Blat, Y. and Eisenbach, M. (1995). Tar-dependent and -independent pattern formation by *Salmonella typhimurium*. *J. Bacteriol.* **177**, 1683–1691.

17. Brenner, M.P., Levitov, L.S. and Budrene, E.O. (1998). Physical mechanisms for chemotactic pattern formation by bacteria. *Biophys. J.* **74**, 1677–1693.

18. Budrene, E.O. and Berg, H.C. (1991). Complex patterns formed by motile cells of *E. coli*. *Nature* **349**, 630–633.

19. Budrene, E.O. and Berg, H.C. (1995). Dynamics of formation of symmetrical patterns by chemotactic bacteria. *Nature* **376**, 49–53.

20. Burkart, M., Toguchi, A. and Harshey, R.M. (1998). The chemotaxis system, but not chemotaxis, is essential for swarming motility in *Escherichia coli*. *Proc. Natl. Acad. Sci. U.S.A.* **95**, 2568–2573.

21. Cohen, I., Czirók, A. and Ben-Jacob, E. (1996). Chemotactic-based adaptive self organization during colonial development. *Physica A* **233**, 678–698.

22. Cohen, I., Ron, I.G. and Ben-Jacob, E. (2000). From branching to nebula patterning during colonial development of the *Paenibacillus alvei* bacteria. *Physica A* **286**, 321–336.

23. DiRusso, C.C., Black, P.N. and Weimar, J.D. (1999). Molecular inroads into the regulation and metabolism of fatty acids, lessons from bacteria. *Progress in Lipid Research* **38**, 129–197.

24. Downard, J., Ramaswamy, S.V. and Kil, K-S. (1993). Identification of *esg*, a genetic locus involved in cell-cell signaling during *Myxococcus xanthus* development. *J. Bacteriol.* **175**, 7762–7770.

25. Downard, J. and Toal, D. (1995). Branched chain fatty acids: the case for a novel form of cell-cell signaling during *Myxococcus xanthus* development. *Mol. Microbiol.* **16**,171–175.

26. Dworkin, M. (1983). Tactic behavior of *Myxococcus xanthus*. *J. Bacteriol.* **154**, 452–459.

27. Dworkin, M. (1996). Recent advances in the social and developmental biology of the myxobacteria. *Microbiol. Rev.* **60**, 70–102.

28. Dworkin, M. (1999). Fibrils as extracellular appendages of bacteria: their role in contact-mediated cell-cell interactions in *Myxococcus xanthus*. *Bioessays* **21**, 590–595.

29. Dworkin, M. and Eide, D. (1983). *Myxococcus xanthus* does not respond chemotactically to moderate concentration gradients. *J. Bacteriol.* **154**, 437Ø442.

30. Eberl, L., Winson, M.K., Sternberg, C., Stewart, G.S.A.B., Christiansen, G., Chabra, S.R., Daykin, M., Williams, P., Molin, S. and Givskov, M. (1996). Involvement of *N*- acyl-L-homoserine lactone autoinducers in controlling the multicellular behavior of *Serratia liquefaciens*. *Mol. Microbiol.* **20**, 127–136.

31. Eberl, L., Molin, S. and Givskov, M. (1999). Surface motility of *Serratia liquefaciens* MG1. *J. Bacteriol.* **181**, 1703–1712.

32. Ellehauge, E., Norregaard-Madsen, M. and Sogaard-Andersen, L. (1998). The FruA signal transduction protein provides a checkpoint for the temporal coordination of intercellular signals in *M. xanthus* development. *Mol. Microbiol.* **30**, 807–817.

33. Fontes, M. and Kaiser, D. (1999). *Myxococcus* cells respond to elastic forces in their substrate. *Proc. Natl. Acad. Sci. U.S.A.* **96**, 8052–8057.

34. Fraser, G.M. and Hughes, C. (1999). Swarming motility. *Current Opinion in Microbiology* **2**, 630–635.

35. Gelvan, I., Varon, M. and Rosenberg, E. (1987). Cell-density-dependent killing of *Myxococcus xanthus* by autocide AMl. *J. Bacteriol.* **169**, 844–848.

36. Gronewold, T.M.A. and Kaiser, D. (2001). The *act* operon controls the level and time of C-signal production for *M. xanthus* development. *Mol. Microbiol.* **40**, 744–756.

37. Gronewold, T.M.A. and Kaiser, D. (2002). *act* operon control of developmental gene expression in *Myxococcus xanthus*. *J. Bacteriol.* **184**, 1172–1179.

38. Gutnick, D. and Ben-Jacob, E. (1999). Complex pattern formation and cooperative organization of bacterial colonies, in *Microbial Ecology and Infectious Disease* (E. Rosenberg, ed.), Am. Soc. Microbiol., Washington, D.C.

39. Hagen, D.C., Bretscher, A.P. and Kaiser, D. (1978). Synergism between morphogenetic mutants of *Myxococcus xanthus*. *Dev. Biol.* **64**, 284–296.

40. Harris, B.Z., Kaiser, D. and Singer, M. (1998). The guanosine nucleotide (p)ppGpp initiates development and A-factor production in *Myxococcus xanthus*. *Genes Dev.* **12**, 1022–1035.

41. Harshey, R.M. (1994). Bees aren't the only ones: swarming in gram-negative bacteria. *Mol. Microbiol.* **13**, 389–394.

42. Harshey, R.M. and Matsuyama, T. (1994). Dimorphic transition in *Escherichia coli* and *Salmonella typhimurium*: surface-induced differentiation into hyperflagellate swarmer cells. *Proc. Natl. Acad. Sci. U.S.A.* **91**, 8631–8635.

43. Hawkins, A.C., Schuster, M. and Harwood, C.S. (2003). An alternative set of *Pseudomonas aeruginosa* chemotaxis genes is preferentially expressed in the stationary phase of growth and further up-regulated by quorum sensing. *Abstracts of BLAST VII Symposium*, **36**.

44. Hay, N.A., Tipper, D.J., Gygi, D. and Hughes, C. (1999). A novel membrane protein influencing cell shape and multicellular swarming of *Proteus mirabilis*. *J. Bacteriol.* **181**, 2008–2016.

45. Henrichsen, J. (1972) Bacterial surface translocation: a survey and a classification. *Bacteriol. Rev.* **36**, 478–503.

46. Hodgkin, J. and Kaiser, D. (1979). Genetics of gliding motility in *Myxococcus xanthus* (Myxobacterales): two gene systems control movement. *Mol. Gen. Genet.* **171**, 177–191.

47. Holden, M.T., Ram Chhabra, S., de Nys, R., Stead, P., Bainton, N.J., Hill, P.J., Manefield, M., Kumar, N., Labatte, M., England, D., Rice, S., Givskov, M., Salmond, G.P., Stewart, G.S., Bycroft, B.W., Kjelleberg, S. and Williams, P. (1999). Quorum-sensing cross talk: isolation and chemical characterization of cyclic dipeptides from *Pseudomonas aeruginosa* and other gram-negative bacteria. *Mol. Microbiol.* **33**, 1254–1266.

48. Jelsbak, L. and Sogaard-Andersen, L. (1999). The cell surface-associated intercellular C-signal induces behavioral changes in individual *Myxococcus xanthus* cells during fruiting body morphogenesis. *Proc. Natl. Acad. Sci. U.S.A.* **96**, 5031–5036.

49. Jelsbak, L. and Sogaard-Andersen, L. (2000). Pattern formation: fruiting body morphogenesis in *Myxococcus xanthus. Curr. Opinion in Microbiol.* **3**, 637–642.

50. Julien, B., Kaiser, D. and Garza, A. (2000). Spatial control of cell differentiation in *Myxococcus xanthus. Proc. Natl. Acad. Sci. U.S.A.* **97**, 9098–9103.

51. Kaiser, D. (1979). Social gliding is correlated with the presence of pili in *Myxococcus xanthus. Proc. Natl. Acad. Sci. U.S.A.* **76**, 5952–5956.

52. Kaiser, D. (1998). How and why myxobacteria talk to each other? *Curr. Opinion Microbiol.* **1**, 663–668.

53. Kaiser, D. (1999). Intercellular signaling for multicellular morphogenesis, in *Microbial Signaling and Communication* (R. England, G. Hobbs, N. Bainton and D. McL. Roberts, eds.). Cambridge University Press.

54. Kaiser, D. (2000). Bacterial motility: How do pili pull? *Curr. Biol.* **10**, R777–R780.

55. Kaiser, D. and Crosby, C. (1983). Cell movement and its coordination in swarms of Myxococcus xanthus. *Cell Motility* **3**, 227–224.

56. Kaiser, D. and Kroos, L. (1993). Intercellular signaling, in *Myxobacteria II* (M. Dworkin and D. Kaiser, eds.). American Society for Microbiology Press.

57. Kearns, D.B. and Shimkets, L.J. (1998). Chemotaxis in a gliding bacterium. *Proc. Natl. Acad. Sci. U.S.A.* **95**, 11957–11962.

58. Kearns, D.B. and Shimkets, L.J. (2001). Lipid chemotaxis and signal transduction in *Myxococcus xanthus. Trends Microbiol.* **9**, 126–129.

59. Kearns, D.B., Campbell, B.D. and Shimkets, L.J. (2000). *Myxococcus xanthus* fibril appendages are essential for excitation by a phospholipid attractant. *Proc. Natl. Acad. Sci. U.S.A.* **97**, 11 505–11 510.

60. Kearns, D.B., Venot, A., Bonner, P.J., Stevens, B., Boons, G.-J. and Shimkets, L.J. (2001). Identification of a developmental chemoattractant in *Myxococcus xanthus* through metabolic engineering. *Proc. Natl. Acad. Sci. U.S.A.* **98**, 13990–13994.

61. Kearns, D.B., Bonner, P.J., Smith, D.R. and Shimkets, L.J. (2002). An extracellular matrix-associated zinc metalloprotease is required for dilauroyl phosphatidylethanolamine chemotactic excitation in *Myxococcus xanthus. J. Bacteriol.* **184**, 1678–1684.

62. Kim, S.K. and Kaiser, D. (1990). Cell alignment required in differentiation of *Myxococcus xanthus. Science* **249**, 926–928.

63. Kim, S.K. and Kaiser, D. (1990). C-factor: a cell-cell signaling protein required for fruiting body morphogenesis of *M. xanthus. Cell* **61**, 19–26.

64. Kim, S.K. and Kaiser, D. (1991). C-factor has distinct aggregation and sporulation thresholds during *Myxococcus* development. *J. Bacteriol.* **173**, 1722–1728.

65. Kim, S.H., Ramaswamy, S. and Downard, J. (1999). Regulated exopolysaccharide production in *Myxococcus xanthus. J. Bacteriol.* **181**, 1496–1507.

66. Kirby, J.R. and Zusman, D. (2003). Chemosensory regulation of developmental gene expression in *Myxococcus xanthus*. *Proc. Natl. Acad. Sci. U.S.A.* **100**, 2008–2013.

67. Kleerebezem, M. and Quadric, L.E. (2001). Peptide pheromone-dependent regulation of antimicrobial peptide production in gram-positive bacteria: a case of multicellular behavior. *Peptides* **22**, 1579–1596.

68. Koch, A.L. (1998). The strategy of *Myxococcus xanthus* for group cooperative behavior. Antonie van Leeuwenhoek *J. Gen. Mol. Microbiol.* **73**, 299–313.

69. Köhler, T., Curty, L.K., Barja, F., van Delden, C. and Pechère, J.C. (2000). Swarming of *Pseudomonas aeruginosa* is dependent on cell-to-cell signaling and requires flagella and pili. *J. Bacteriol.* **182**, 5990–5996.

70. Kovacikova, G. and Skorupski, K. (2002). Regulation of virulence gene expression in *Vibrio cholerae* by quorum sensing: HapR functions at the *aphA* promoter. *Mol. Microbiol.* **46**, 1135–1147.

71. Kuspa, A., Plamann, L. and Kaiser, D. (1992). A-Signaling and the cell density requirement for *Myxococcus xanthus* development. *J. Bacteriol.* **174**, 7360–7369.

72. Lazazzera, B.A. (2001). The intracellular function of extracellular signaling peptides. *Peptides* **22**, 1519–1527.

73. Lev, M. (1954). Demonstration of a diffusible fruiting factor in myxobacteria. *Nature* **173**, 501.

74. Lindum, P.W., Anthoni, U., Christoffersen, C., Eberl, L., Molin, S. and Givskov, M. (1998). *N*-acyl-L-homoserine lactone autoinducers control production of an extracellular surface active lipopeptide required for swarming motility of *Serratia liquefaciens* MG1. *J. Bacteriol.* **180**, 6384–6388.

75. MacNeil, S.D., Mouzeyan, A. and Hartzell, P.L. (1994). Genes required for both gliding motility and development in *Myxococcus xanthus*. *Mol. Microbiol.* **14**, 785–795.

76. Manoil, C. and Kaiser, D. (1980). Accumulation of guanosine tetraphosphate and guanosine pentaphosphate in *Myxococcus xanthus* during starvation and myxospore formation. *J. Bacteriol.* **141**, 297–304.

77. Manoil, C. and Kaiser, D. (1980). Guanosine pentaphosphate and guanosine tetraphosphate accumulation and induction of *Myxococcus xanthus* fruiting body development. *J. Bacteriol.* **141**, 305–315.

78. Mariconda, S., Wang, T., Stell, A., Toguchi, A., and Harshey, R.M. (2003). Why the chemotaxis system (but not chemotaxis) is essential for swarming motility in *Salmonella typhimurium*: a model. *Abstracts of BLAST VII Symposium*, **114**.

79. Masduki, A., Nakamura, J., Ohga, T., Umezaki, R., Kato, J. and Ohtake, H. (1995). Isolation and characterization of chemotaxis mutants and genes of *Pseudomonas aeruginosa*. *J. Bacteriol.* **177**, 948–952.

80. Matsuyama, T. and Matsushita, M. (2001). Population morphogenesis by cooperative bacteria. *Forma* **16**, 307–326.

81. McBride, M.J. (2001). Bacterial gliding motility: multiple mechanisms for cell movement over surfaces. *Ann. Rev. Microbiol.* **55**, 49–75.

82. McBride, M.J. and Zusman, D.R. (1993). FrzCD, a methyl-accepting taxis protein from *Myxococcus xanthus* shows modulated methylation during fruiting body formation. *J. Bacteriol.* **175**, 4936–4940.

83. McBride, M.J., Weinberg, R.A. and Zusman, D.R. (1989). Frizzy aggregation genes of the gliding bacterium *Myxococcus xanthus* show sequence similarities to the chemotaxis genes of enteric bacteria. *Proc. Natl. Acad. Sci. U.S.A.* **86**, 424–428.

84. McBride, M.J., Kohler, T. and Zusman, D.R. (1992). Methylation of FrzCD, a methyl-accepting taxis protein of *Myxococcus xanthus*, is correlated with factors affecting cell behavior. *J. Bacteriol.* **174**, 4246–4257.

85. McBride, M.J., Hartzell, P. and Zusman, D.R. (1993). Motility and tactic behavior of *Myxococcus xanthus*, in *Myxobacteria II* (M. Dworkin and D. Kaiser, eds.). American Society for Microbiology Press.

86. McCleary, W. and Zusman, D. (1990). FrzE of *M. xanthus* is homologous to both CheA and CheY of *Salmonella typhimurium*. *Proc. Natl. Acad. Sci. U.S.A.* **87**, 5898–5902.

87. McCleary, W., McBride, M. and Zusman, D. (1990). Developmental sensor transduction in *M. xanthus* involves methylation and demethylation of FrzCD. *J. Bacteriol.* **172**, 4877–4887.

88. McVittie, A. and Zahler, S.A. (1962). Chemotaxis in *Myxococcus*. *Nature* **194**, 1299–1300.

89. McVittie, A., Messik, F. and Zahler, S.A. (1962). Developmental biology of *Myxococcus*. *J. Bacteriol.* **84**, 546–551.

90. Mendelson, N.H. (1999). *Bacillus subtilis* microfebres, colonies and bioconvection patterns use different strategies to achieve multicellular organization. *Environmental Microbiology* **1**, 471–477.

91. Ogawa, M., Fujitani, S., Mao, X., Inouye, S. and Komano, T. (1996). FruA, a putative transcription factor essential for the development of *Myxococcus xanthus*. *Mol. Microbiol.* **22**, 757–767.

92. Oger, P., Stephen K. and Farrand, S.K. (2002). Two opines control conjugal transfer of an *Agrobacterium* plasmid by regulating expression of separate copies of the quorum-sensing activator gene *traR*. *J. Bacteriol.* **184**, 1121–1131.

93. O'Rear, J., Alberti, L. and Harshey, R.M. (1992). Mutations that impair swarming motility in *Serratia marcescens* 274 include but are not limited to those affecting chemotaxis or flagellar functions. *J. Bacteriol.* **174**, 6125–6137.

94. Orndorff, P.E. and Dworkin, M. (1980). Separation and properties of the cytoplasmic and outer membrane of vegetative cells of *Myxococcus xanthus*. *J. Bacteriol.* **141**, 914–927.

95. Parangan, A.B., Ingmire, P. and Márquez-Magaña, L. (2003). A *Paenibacillus* isolate from a *B. subtilis* stock exhibits multiple motility behaviors. *Abstracts of BLAST VII Symposium*, **70**.

96. Pottathil, M. and Lazazzera, B.A. (2003). The extracellular phr Peptide-rap phosphatase signaling circuit of *bacillus subtilis*. *Front Biosci.* **1**, 8:D32–45.

97. Rahman, M.M., Guard-Petter, J., Asokan, K., Hughes, C. and Carlson, R.W. (1999). The structure of the colony migration factor from pathogenic *Proteus mirabilis*: a capsular polysaccharide that facilitates swarming. *J. Biol. Chem.* **274**, 22993–22998.

98. Reichenbach, H. and Dworkin, M. (1999). The myxobacteria, in *The Prokaryotes, electronic edition*, chief editor M. Dworkin.

99. Reichenbach, H., Heunert, H.H. and Kuczka, H. (1965). *Myxococcus spp.* (Myxobacterales)—Schwarmentwicklung und Bildung von Protocysten. Encyclopaedia Cinematographica E. 778, film of the Institut für den wissenschaftlichen Film Göttingen, Germany.

100. Reichenbach, H., Heunert, H.H. and Kuczka, H. (1965). Schwarmentwicklung und Morphogenese bei Myxobakterien—*Archangium, Myxococcus, Chondrococ-*

cus, Chondromyces. Film C 893 of the Institut für den wissenschaftlichen Film Göttingen, Germany.

101. Reimmann, C., Ginet, N., Michel, L., Keel, C., Michaux, P., Krishnapillai, V., Zala, M., Heurlier, K., Triandafillu, K., Harms, H., Défago, G. and Haas, D. (2002). Genetically programmed autoinducer destruction reduces virulence gene expression and swarming motility in *Pseudomonas aeruginosa. PAO1 Microbiology* **148**, 923–932.

102. Rosenbluh, A. and Eisenbach, M. (1992). Effect of mechanical removal of pili on gliding motility of *Myxococcus xanthus. J. Bacteriol.* **174**, 5406–5413.

103. Rosenberg, E., Keller, K.H. and Dworkin, M. (1977). Cell-density-dependent growth of *Myxococcus xanthus* on casein. *J. Bacteriol.* **129**, 770–777.

104. Sager, B. and Kaiser, D. (1994). Intercellular C-signaling and the traveling waves of *Myxococcus. Genes Dev.* **8**, 2793–2804.

105. Sar, N., McCarter, L., Simon, M. and Silverman, M. (1990). Chemotactic control of the two flagellar systems of *Vibrio parahaemolyticus. J. Bacteriol.* **172**, 334–341.

106. Senesi, S., Celandroni, F., Salvetti, S., Beecher, D.J., Wong, A.C.L. and Ghelardi, E. (2002). Swarming motility in *Bacillus cereus* and characterization of a *fliY* mutant impaired in swarm cell differentiation. *Microbiology* **148**, 1785–1794.

107. Shapiro, J.A. (1998). Thinking about bacterial populations as multicellular organisms. *Ann. Rev. Microbiol.* **52**, 81–104.

108. Shi, W.Y., Kohler, T. and Zusman, D.R. (1993). Chemotaxis plays a role in the social behaviour of *Myxococcus xanthus. Mol. Microbiol.* **9**, 601–611.

109. Shi, W., Ngok, F. and Zusman, D.R. (1996). Cell density regulates cellular reversal frequency in *Myxococcus xanthus. Proc. Natl. Acad. Sci. U.S.A.* **93**, 4142–4146.

110. Shimkets, L.J. (1999). Intercellular signaling during fruiting-body development of *Myxococcus xanthus. Ann. Rev. Microbiol.* **53**, 525–549.

111. Shimkets, L.J. and Kaiser, D. (1982). Induction of coordinated movement in *Myxococcus xanthus* cells. *J. Bacteriol.* **152**, 451–461.

112. Shimkets, L.J., Gill, R.E. and Kaiser, D. (1983). Developmental cell interactions in *Myxococcus xanthus* and the *spoC* locus. *Proc. Natl. Acad. Sci. U.S.A.* **80**, 1406–1410.

113. Shimkets, L.J. and Rafiee, H. (1990). CsgA, an extracellular protein essential for *Myxococcus xanthus* development. *J. Bacteriol.* **172**, 5299–5306.

114. Shimkets, L.J. and Dworkin, M. (1997). Myxobacterial multicellularity, in *Bacteria as Multicellular Organisms.* (J.A. Shapiro and M. Dworkin, eds.), 220–244. New York: Oxford Univ. Press.

115. Sogaard-Andersen, L. and Kaiser, D. (1996). C factor, a cell-surface-associated intercellular signaling protein, stimulates the cytoplasmic Frz signal transduction system in *Myxococcus xanthus. Proc. Natl. Acad. Sci. U.S.A.* **93**, 2675–2679.

116. Spormann, A.M. (1999). Gliding motility in bacteria: insights from studies of *Myxococcus xanthus. Microbiol. Mol. Biol. Rev.* **63**, 621–641.

117. Stoodley, P., Sauer, K., Davies, D.G. and Costerton, J.W. (2002). Biofilms as complex differentiated communities. *Ann. Rev. Microbiol.* **56**, 187–209.

118. Sun, H., Zusman, D.R. and Shi, W. (2000). Type IV pilus of *Myxococcus xanthus* is a motility apparatus controlled by the *frz* chemosensory system. *Curr. Biol.* **10**, 1143–1146.

119. Tcherpakov, M., Ben-Jacob, E. and Gutnick, D.L. (1999). *Paenibacillus dendritiformis* sp. nov., proposal for a new pattern-forming species and its localization within a phylogenetic cluster. *Int. J. Syst. Bacteriol.* **49**, 239–246.

120. Tieman, S., Koch, A. and White, D. (1996). Gliding motility in slide cultures of *Myxococcus xanthus* in stable and steep chemical gradients. *J. Bacteriol.* **178**, 3480–3485.

121. Toguchi, A., Siano, M., Burkart, M. and Harshey, R.M. (2000). Genetics of swarming motility in *Salmonella enterica* serovar typhimurium: critical role for lipopolysaccharide. *J. Bacteriol.* **182**, 6308–6321.

122. Trudeau, K.G., Ward, M.J. and Zusman, D.R. (1996). Identification and characterization of FrzZ, a novel response regulator necessary for swarming and fruiting body formation in *Myxococcus xanthus*. *Mol. Microbiol.* **20**, 645–655.

123. Tyson, R., Lubkin, S.R. and Murray, J.D. (1999). A minimal mechanism for bacterial pattern formation. *Proc. R. Soc. Lond. B. Biol. Sci.* **266**, 299–304. *Abstracts of BLAST VII Symposium*, **109**.

124. Varon, M., Tietz, A. and Rosenberg, E. (1986). *Myxococcus xanthus* Autocide AMl. *J. Bacteriol.* **167**, 356–361.

125. Vlamakis, H., Kirby, J. and Zusman, D. (2003). The Che4 operon of *Myxococcus xanthus* is involved in pilus-mediated motility. *Abstracts of BLAST VII Symposium*, **109**.

126. Wall, D. and Kaiser, D. (1999). Type IV pili and cell motility. *Mol. Microbiol.* **32**, 1–10.

127. Ward, M.J. and Zusman, D.R. (1997). Regulation of directed motility in *Myxococcus xanthus*. *Mol. Microbiol.* **24**, 885–893.

128. Ward, M.J. and David R. Zusman. (1999). Motility in *Myxococcus xanthus* and its role in developmental aggregation. *Curr. Opinion Microbiol.* **2**, 624–629.

129. Ward, M.J. and Zusman, D.R. (2000). Developmental aggregation and fruiting body formation in the gliding bacterium *Myxococcus xanthus*, in *Prokaryotic Development* (Y.V. Brun and L.J. Shimkets, eds.), 243–262. ASM Press, Washington, D.C.

130. Wireman, J.W. and Dworkin, M. (1977). Developmentally induced autolysis during fruiting body formation by *Myxococcus xanthus*. *J. Bacteriol.* **129**, 796–802.

131. Wolgemuth, C., Hoiczyk, E., Kaiser, D. and Oster, G. (2002). How *Myxobacteria* glide. *Curr. Biol.* **12**, 369–377.

132. Woodward, D.E., Tyson, R., Myerscough, M.R., Murray, J.D., Budrene, E.O. and Berg, H.C. (1995). Spatial temporal patterns generated by *Salmonella typhimurium*. *Biophysical J.* **68**, 2181–2189.

133. Wu, S.S. and Kaiser, D. (1995). Genetic and functional evidence that type IV pili are required for social gliding motility in *Myxococcus xanthus*. *Mol. Microbiol.* **18**, 547–558.

134. Yang, Z.M., Geng, Y.Z., Xu, D., Kaplan, H.B. and Shi, W.Y. (1998). A new set of chemotaxis homologues is essential for *Myxococcus xanthus* social motility. *Mol. Microbiol.* **30**, 1123–1130.

135. Yang, Z., Ma, X., Tong, L., Kaplan, H.B., Shimkets, L.J. and Shi, W. (2000). *Myxococcus xanthus dif* genes are required for biogenesis of cell surface fibrils essential for social gliding motility. *J. Bacteriol.* **182**, 5793–5798.

136. Zusman, D.R. (1982). "Frizzy" mutants: a new class of aggregation-defective developmental mutants of *Myxococcus xanthus*. *J. Bacteriol.* **150**, 1430–1437.

137. Park, S., Wolanin, P.M., Yuzbashyan, E.A., Silberzan, P., Stock, J.B. and Austin, R.H. (2003). Motion to form a quorum. *Science* **301**, 188.

138. Park, S., Wolanin, P.M., Yuzbashyan, E.A., Lin, H., Darnton, N.C., Stock, J.B. and Austin, R.H. (2003). Influence of topology on bacterial social interaction. *Proc. Natl. Acad. Sci. U.S.A.* **100**, 13910–13915.

Molecular Mechanisms of Chemotaxis in Amoebae*

1. Introduction

Amoeboid cell motility and chemotaxis are found in eukaryotes ranging from single-celled organisms to man. This chapter will focus on two cell types as examples of the range of amoeboid cell motility. *Dictyostelium* cells live as individual amoebae feeding on bacteria, show high rates of cell translocation, and are highly chemotactic to extracellular adenosine-3', 5'- cyclic monophosphate (cAMP) via a G-protein-coupled receptor [33, 121, 138]. A significant contribution to our understanding of the pathways that control the chemotaxis of amoeboid cells including neutrophils and macrophages stems from genetic, biochemical, and cell biological studies of model systems such as *Dictyostelium*. These studies have provided insights into the machinery and mechanisms necessary to remodel the cellular actin and myosin cytoskeletons and respond with directional movement to a chemoattractant. They also reveal a remarkable conservation between *Dictyostelium* and higher eukaryotes

* Written by Ruedi Meili[a], Richard A. Firtel[a], and Jeffrey E. Segall[b], [a]Section of Cell and Developmental Biology, Center for Molecular Genetics, Rm. 220, University of California, San Diego, 9500 Gilman Drive, La Jolla, CA 92093-0634; [b]Department of Anatomy and Structural Biology, Albert Einstein College of Medicine, 1300 Morris Park Avenue, New York, NY 10461.

on all levels, including regulatory pathways that control cytoskeletal reorganization, changes in subcellular localization of signaling and cytoskeletal proteins, and the gene products essential for cell motility and directional movement [73].

For multicellular eukaryotes, there is a wide range of amoeboid motility and chemotaxis behaviors, depending on cell type. Immune system cells such as neutrophils tend to have properties similar to *Dictyostelium*—high speed and exquisite sensitivity to chemotactic stimuli using G-protein-coupled receptors ([232] discussed in Chapter 6). Mesenchymally derived cells such as fibroblasts, endothelial cells, and smooth muscle cells are motile in the adult under certain circumstances but move more slowly and have much higher adhesiveness to the extracellular matrix [230, 235]. Epithelial cells that form the parenchyma of particular organs, such as mammary ductal epithelial cells, are relatively nonmotile in the adult but are motile during development and can become motile when they become cancer cells. In this chapter we will also examine the motility of fibroblasts and MTLn3 tumor cells, referring to their motility generally as mesenchymal cell motility. MTLn3 cells are tumor cells derived from a rat mammary tumor [158, 234] that are metastatic to the lungs when injected into the mammary fat pad to form a primary tumor. They move more slowly than *Dictyostelium* cells or neutrophils, are tightly attached to the substratum, and are chemotactic to the growth factor EGF (epidermal growth factor) via a receptor tyrosine kinase (Figure 1) [43]. The differences in cell motility and mechanisms of chemotactic response exemplified by *Dictyostelium* and mesenchymal cells demonstrate the variety of signaling pathways and behavioral strategies that can be used to generate chemotactic responses.

1.1. *Functions of amoeboid chemotaxis*

Dictyostelium is a social soil amoeba that feeds on bacteria in the wild. Individual cells find food by detecting bacterial metabolites such as folic acid. If food is depleted locally, starved amoebae start to emit a pulsatile cAMP signal that functions as a chemoattractant, attracting nearby cells and resulting in the formation of multicellular aggregates of up to 10^5 cells. The aggregates subsequently undergo a complex developmental program involving cell-type differentiation and morphogenetic cell migration to produce a mature organism with a spore-containing fruiting body on top of a slender stalk. Both foraging and development depend on chemotaxis (or directional cell movement guided by soluble

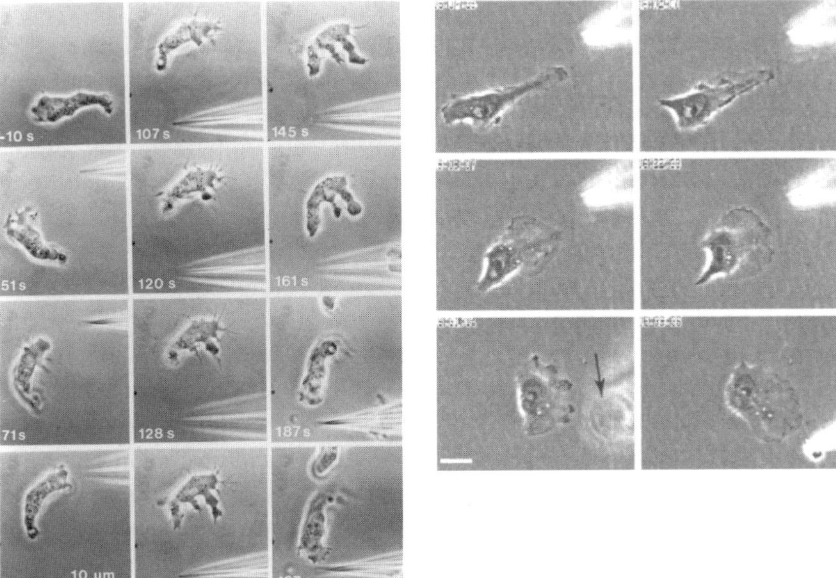

Figure 1. Chemotactic responses of *Dictyostelium* (*left*) and MTLn3 (*right*) cells. *Left image:* Typical chemotactic responses of a *Dictyostelium* cell to a pipet filled with cAMP. The pipet was introduced at 0 s, and the cell turned towards the tip maintaining its initial front. At 100 s the pipet was moved to a new position, and the cell now responded by producing multiple pseudopods. After a phase of competition, the original rear end became established as the new front. Taken with permission from Segall and Gerisch [201]. *Right image:* Chemotactic response of MTLn3 cells to EGF. A pipet containing EGF was placed at the rear of a cell that was originally moving towards the lower left. The cell reversed direction, extending a broad, flat lamellipod towards the pipet. When the pipet was moved, the cell reoriented to follow the pipet. Scale bar 20 μm. Taken with permission from Bailly *et al.* [14].

molecules) mediated by G-protein-coupled receptors. *Dictyostelium* has long been valued as a powerful experimental system with well-developed genetic methods and outstanding amenability to biochemical and cell biological methods. The nearly completed genome sequence [85] demonstrates the high degree of homology between corresponding genes of *Dictyostelium* and higher eukaryotes and corroborates the applicability of *Dictyostelium* as a model system for the study of evolutionarily conserved processes like cytokinesis, phagocytosis, general locomotion, and chemotaxis [140].

In higher eukaryotes, G-protein-coupled receptors can also mediate chemotactic responses—chemotaxis to bacteria is performed by neutrophils using FMLP (the tripeptide formyl-methionine leucine phenylalanine) as the chemoattractant secreted by bacteria (covered in Chapter 6). However, receptor tyrosine kinases (cell surface proteins that can phosphorylate other proteins on tyrosine residues) also provide major contributions to cell coordination during embryogenesis, development, and wound healing. Chemotaxis to EGF is important for cell migration of border and distal tip cells in invertebrates [135] and chemotaxis to fibroblast growth factor is important in branching morphogenesis in the lung [228]. In wound healing, chemotaxis to EGF receptor family ligands and platelet-derived growth factor is likely to aid in wound tensile strength, reepithelialization, and neovascularization [93, 99]. Blood vessel formation, important both during development and wound healing, is critically dependent upon endothelial cell migration, which is stimulated by angiogenic factors such as vascular endothelial growth factor and its corresponding receptors [52, 115, 220]. During metastasis, chemotaxis may contribute to movement of tumor cells out of the primary tumor to invade the surrounding connective tissue and blood vessels [43, 68, 235].

2. Amoeboid Motility

The fundamental feature of amoeboid motility of *Dictyostelium* and tumor cells is a change of shape that reflects the process of translocation. These shape changes depend upon the coordinated interaction of the cytoskeleton with the membrane and with adhesion receptors to form a motility cycle that results in net movement of the amoeboid cell. Cell movement of the amoeboid type is characterized by a polarized cell shape with both a leading edge or pseudopod (a structure similar to a fibroblast lamellipod) that is extended in the direction of cell movement, and a posterior end or uropod that is retracted as the cell moves [159, 163]. The processes at the front are very dynamic, with protruding and retracting pseudopods and microspikes. As a consequence, the front of the cell is frequently lifted off the substratum while the back tends to follow the front. However, the back is important for anchoring the cell to the substratum, and in later development for *Dictyostelium* it may attach to the fronts of other cells, enabling chains of cells to move in concert and to coalesce into so-called aggregation streams. Polarized localization and activity of the cytoskeletal components actin and myosin as well as their regulators underlies this process [39], giving rise to alternating cycles of extension and contraction and net cell movement which can be either directional or random depending on conditions. The sides of the cell are stiffened by thick myosin II filaments that interact with actin filaments to provide mechanical resilience or cortical tension [67]. The forces driving cell movement are generated by actin polymerization at the leading edge and actin-myosin-based contractility at the rear. *Dictyostelium* cells move quickly (approx. 10 µm/min for wild-type cells) and the cytoskeleton can react within seconds to changes in external signals. Mesenchymally derived cells move more slowly (approx. 1 µm/min or less). The interplay of chemoattractant sensing, orientation of the cellular axis, and motile response which together constitute chemotaxis will be described in more detail later.

2.1. *The motility cycle*

The fundamental motility cycle as reflected in the cell shape changes described above encompasses five steps (Figure 2): (1) extension of a protrusion, (2) attachment of the protrusion via adhesion to the substratum, (3) contraction of the cell body to generate tension on the adhesion sites at the rear of the cell, (4) detachment of the adhesion sites at the rear of the cell, and (5) net movement of cell mass towards the front of the cell [1, 43, 131, 152, 204, 211, 239]. These processes may be occurring simultaneously in different regions of the cell. The spatial and temporal regulation of their efficiency determines the direction of movement. In the absence of an imposed chemoattractant gradient, cells can undergo spontaneous, random motility. This is especially true of starved *Dictyostelium* cells. It is not yet known how the spontaneous motility system intersects with chemotactic responses—whether spontaneous motility is due to random activation of the chemotactic signaling pathway or whether there is a distinct signaling pathway for spontaneous motility. To a first approximation, the motility cycles present for spontaneous and chemotactic motility seem similar.

2.1.1. *Extension*

Extensions formed by *Dictyostelium* [41, 73, 236] and mesenchymal cells [28, 44, 202] are bounded by the plasma membrane and show a high concentration of F-actin (Figure 3). The nature of the actin network determines the type of extension that is formed. Thin, elongated filopodia (Figure 4) are formed by a square-packed parallel array of actin filaments [25, 57, 66]. Filopodia are speculated to provide sensing

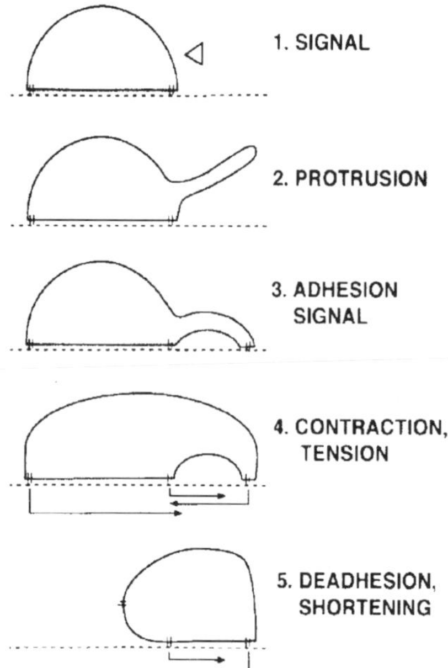

Figure 2. Amoeboid motility cycle. (1) A resting cell is stimulated by an external signal (possibly a gradient of chemoattractant). (2) The cell protrudes a lamellipod towards the source of the stimulus. (3) New focal contacts are formed under the extended lamellipod. (4) The cell exerts force on the cytoskeleton via contraction. (5) The old focal contacts detach under the exerted force, resulting in net translocation of the cell. Taken with permission from Condeelis *et al.* [43].

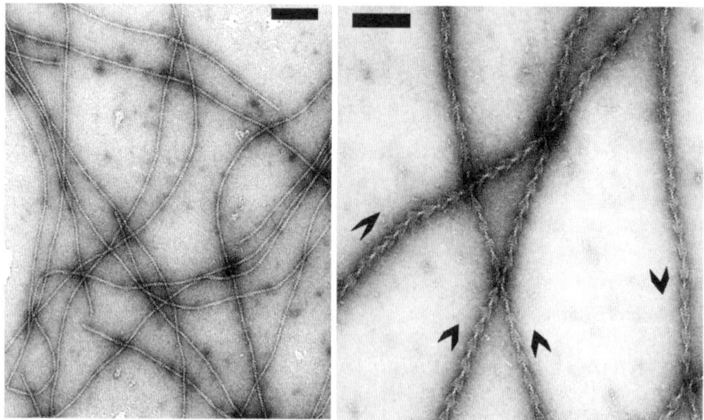

Figure 3. Actin filaments. Electron microscope images of individual negatively stained actin filaments (*left*) and actin filaments decorated by myosin S1 fragments (*right*). The black V shapes indicate the pointed and barbed end directions of the filaments as marked by labeling with myosin subfragment 1. Scale bars: 100 nm. Images provided courtesy of Dr Roger Craig.

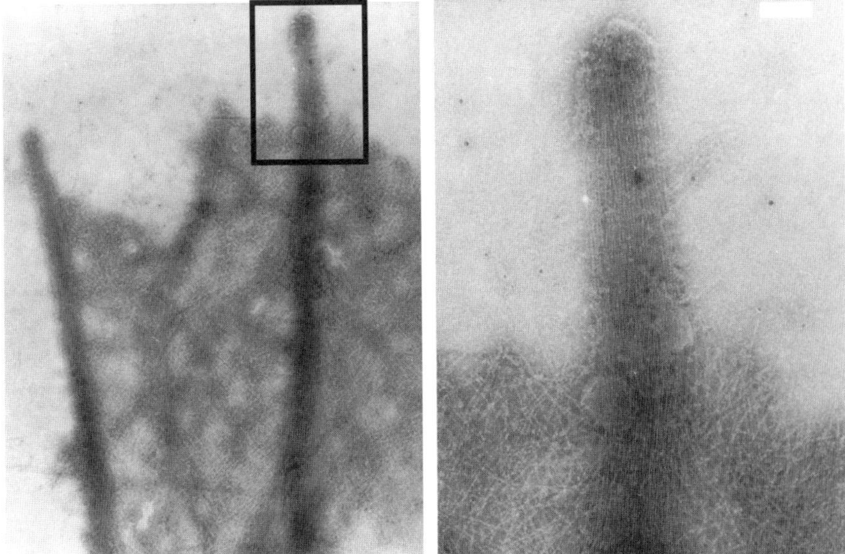

Figure 4. Actin structure in a filopod. Image of a lamellipod and filopods extending out of it. The filopod boxed in the left image is shown at higher magnification in the right image. Note the parallel array of actin filaments in the filopod. From a chicken heart fibroblast, extracted, fixed, and negatively stained with sodium silicotungstate Scale bars 100 nm. Images provided courtesy of Guenter Resch and J. Victor Small.

of spatial gradients, and can be induced by constitutively active forms of the small G-protein cdc42 [90, 144].

The major large extensions formed by the actin cytoskeletons during amoeboid motility of *Dictyostelium* and mesenchymal cells are pseudopods and lamellipods (Figure 5). They are formed from a diagonal array of actin filaments. Pseudopods are cylindrical in shape while lamellipods are flatter and more like sheets of actin meshwork covered by the cell membrane. *Dictyostelium* cells tend to form pseudopods [132, 239]. Mesenchymal cells mainly form lamellipods in tissue culture [11, 49, 209]. On a flat surface such as in tissue culture, the tips of lamellipods which do not form stable adhesions tend to rise vertically and move back over the dorsal edge of the cells [2]. Structures of this type are termed ruffles [3]. The ruffles are then resorbed into the cell body, often with formation of endocytic or phagocytic vesicles of varying sizes. *In vivo*, the balance of filopodial to pseudopodial to lamellipodial structures may shift to mainly pseudopods, reflecting a more irregularly organized extracellular environment. Extension of pseudopods and

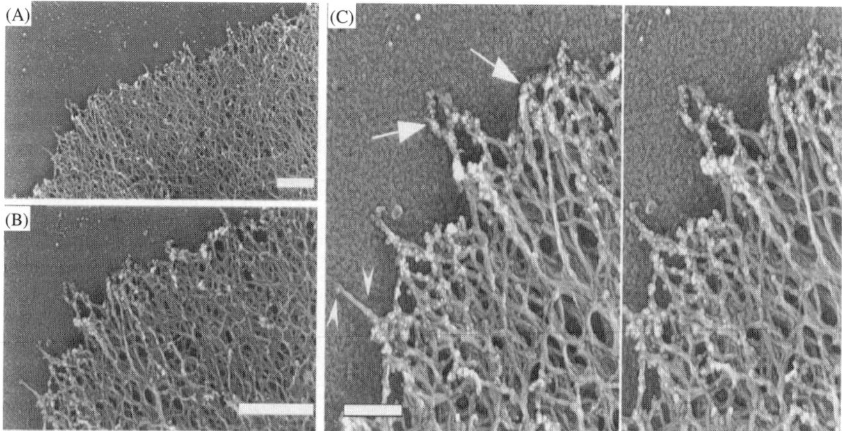

Figure 5. Actin organization in a lamellipod. Electron micrograph of a rotary shadowed, platinum coated lamellipod of an MTLn3 cell. (A) Leading edge of a lamellipod in a cell 1 min after EGF stimulation. (B) A higher magnification of A is shown and (C) a higher magnification of B (stereo view). The gold particles decorate the newly synthesized filaments at the extreme edge of the lamellipod (arrows). Scale bars A and B, 0.5 μm; C, 0.2 μm. Taken with permission from Bailly *et al.* [13].

lamellipods can be induced by constitutively active forms of the small G-protein rac [90].

For both filopods and lamellipods/pseudopods, stimulated actin polymerization directly under the membrane is thought to provide the driving force producing the protrusion [21, 40, 169, 173]. Thus the regulation of the location of actin polymerization and the cross-linking between actin filaments is critical for coordinating chemotactic responses.

2.1.2. *Adhesion*

The strength of the adhesive contacts formed by isolated *Dictyostelium* cells is significantly weaker than the contacts formed by mesenchymal cells. This may reflect a general requirement for high speed of movement on the variety of possible organic and nonorganic constituents of the soil. Indeed, *Dictyostelium* cells in suspension can have morphologies similar to cells on a surface and undergo pseudopod extension that appears quite similar to their motility on a surface, suggesting that adhesion may be less important for *Dictyostelium* motility. *Dictyostelium* pseudopods that are in contact with the substratum are more likely to remain stable [239]. The molecules mediating *Dictyostelium* attachment to the substratum are now being characterized [69], but may overlap with molecules mediating cell–cell adhesion during development. Contacts with the substratum have been identified, termed eupodia [75], and may be similar to contacts seen in macrophages termed podosomes [62]. Loss of the *Dictyostelium* homolog of the talin protein results in reduced cell adhesion, possibly reflecting a mechanism in common with mesenchymal cells described below [193, 208].

Mesenchymal cells are spherical when in suspension and form stable protrusions only upon attachment to a surface [15]. The cells can adhere tightly to the substratum in tissue culture utilizing structures termed focal adhesions [26]. Upon extension of a lamellipod, close contacts to the substratum form, which then convert into localized adhesive clusters that can be detected using interference reflection microscopy [107, 166]. These focal contacts have clusters of transmembrane proteins termed integrins that bind to extracellular matrix molecules and link intracellularly to the actin cytoskeleton to form stress fibers [251]. Adhesion to the substratum is necessary for stabilization of the lamellipod: if a lamellipod extends over a nonadhesive surface, it then retracts [14].

2.1.3. *Contraction*

Contraction in amoeboid cells makes use of nonmuscle forms of myosin type II which form bipolar thick filaments in the cytoplasm and in association with the actin filaments [23, 54]. In *Dictyostelium*, actin filaments form a cortical shell directly under the plasma membrane, with random orientation of the filaments [132]. Activation of myosin contractile activity by phosphorylation of the myosin light chain protein results in contraction of the cortical network [54, 110]. It is unclear whether this contraction is uniform throughout the cell or whether there is spatial regulation of the activity. In polarized motile *Dictyostelium* cells, myosin is concentrated at the rear of the cell [31], which could maintain the polarization, while a myosin heavy chain kinase (which phosphorylates the myosin heavy chain protein and inhibits thick filament formation), is localized at the front of the cell.

In mesenchymal cells, in addition to the actin cortical shell, myosin can be found in elongated actin structures termed stress fibers. Stress fibers typically are attached at one end to a focal adhesion site, and the other end may terminate in the cortical shell or another focal contact [26, 80]. Myosin is also present in stress fibers, and upon stimulation induces contraction of the stress fibers [5, 78, 117, 127].

2.1.4. *Detachment*

The net effect of the contractions that occur in both *Dictyostelium* and MTLn3 cells is to exert stress on the contacts with the substratum. In the case of *Dictyostelium*, when there is an increased concentration of myosin II at the rear of the cell, the net effect is to produce localized detachment of the contacts at the rear. Overall, the force generated by the contraction is likely to be much greater than the strength of the attachments for *Dictyostelium*.

In the case of mesenchymal cells, focal adhesions can be stronger than the contractile forces. In the case of fibroblasts, this enables transmission of force to the extracellular matrix, resulting in contraction of wound tissue [5, 84, 86]. As discussed in Section 2.3, there is likely to be specific control of the adhesive capability of the focal adhesions in different regions of the cell. Thus, during cell motility the focal adhesions that are weakest will detach.

2.1.5. *Translocation*

For both *Dictyostelium* and MTLn3 cells, as specific areas detach from the substratum in the presence of the contraction force, those areas will then be drawn towards the remaining regions of the cell that are held fixed by maintaining their contacts with the substratum. For *Dictyostelium*, because there is less myosin II at sites of new pseudopod extension, the rest of the cell is drawn towards those sites. For mesenchymal cells, the newer focal contacts that are formed in the newer lamellipods appear to be more stable while the older focal contacts are more likely to detach. The rest of the cell is therefore brought towards the newly extended lamellipods. In both cases, the net result is translocation of cell mass towards the most recently formed extensions.

2.2. *The cytoskeleton*

Although there is evidence that both the actin and microtubule cytoskeletal systems contribute to amoeboid chemotaxis, the evidence is strongest for the importance of the actin cytoskeleton. Drugs that inhibit the actin cytoskeleton, such as cytochalasins, have dramatic effects on chemotaxis, at least in part due to an inhibition of cell motility [202]. Assays of actin localization and polymerization indicate that chemoattractant stimulation has strong effects on the actin cytoskeleton

[33, 43, 133]. On the other hand, microtubule-depolymerizing agents such as nocodazole have generated less consistent results [202]. Nevertheless, overexpression of a tubulin deacetylase affects chemotactic cell movement, indicating possible contributions of microtubules to chemotaxis [98]. In addition, microtubules may contribute to the regulation of cell polarity and adhesion stability through interactions with the actin cytoskeleton.

2.2.1. *Actin and myosin*

The actin cytoskeleton is complex, with over 100 different actin-interacting proteins. This section will focus on aspects of the actin cytoskeleton that are specifically related to cell motility. The critical features of the actin cytoskeleton that contribute to cell motility are: polymerization/depolymerization of actin filaments, cross-linking of filaments, and tension generation (through myosin-induced contractile activity) [229].

Actin filaments are polar double helical polymers of actin protein molecules with a barbed and a pointed end (so named for their appearance in electron microscope images when decorated with myosin fragments) (Figure 3). The barbed end has a higher affinity for actin monomers (also referred to as G actin) than the pointed end. Actin polymerization occurs through addition of ATP-actin monomers to the barbed end of a pre-existing actin filament, or the association of three monomers to form a new filament. Depolymerization of actin filaments occurs when loss of actin monomers from the pointed end exceeds the rate of addition of G actin to the barbed end or if the concentration of G actin is so low that loss of actin from both ends of the filament occurs. In general, by regulating the dynamics of the barbed and pointed ends, the cell can regulate its actin filament population. The regulation occurs through proteins that can nucleate new actin seeds, bind to the barbed or pointed ends, or sever preexisting actin filaments.

Ultrastructural analysis of the leading edge of a polarized amoeboid cell shows a zone with a meshwork of actin filaments directly behind the front (Figure 5) [217]. It has long been proposed that turnover or "treadmilling" of this zone provides the driving force for locomotion [95, 123]. Only recently, a more detailed mechanistic understanding has been achieved by integrating experimental evidence from multiple systems. The actin filament meshwork is either stationary or slowly moving backwards in respect to the substratum; net forward movement is achieved by polymerization of actin monomers close to the membrane and their recycling by depolymerization at the other ends of the actin filaments in the cell interior. This dynamic behavior of actin is modulated by a large number of proteins: capping proteins which stabilize filament ends by blocking polymerization and depolymerization, cross-linking proteins which are important for the mechanical properties of the meshwork, severing proteins which regulate turnover and recycling of the monomers, etc.

At the core of the model for leading edge protrusion is dendritic nucleation of actin polymerization [155]. The barbed ends of the F-actin filaments face the plasma membrane where elongation by addition of monomers is initiated and focused. New branches sprout from the sides of existing filaments, a process which together with cross-linking of two existing filaments keeps unanchored filaments short and minimizes energy losses due to bending. Branching and nucleation can be mediated by the seven-subunit Arp2/3 complex [139], which contains the actin-related proteins 2 and 3.

The Arp2/3 complex is a major contributor to nucleation of actin filaments [94, 143]. It is regulated by members of the Wiskott–Aldrich syndrome protein family, which can in turn be activated by the Rho family small G proteins cdc42 and rac [27, 89, 100, 151, 219, 250]. Activation of cdc42 stimulates filopod formation, while rac activation is associated with lamellipod formation [90]. For filopod formation, a series of parallel actin filaments form, which are cross-linked in a square-packed array by cross-linking proteins such as EF1alpha [65, 165]. Binding of Arp2/3 to the sides of actin filaments can enhance its nucleation ability [12]. Arp2/3 can stabilize the pointed ends (where it nucleates the filament) and simultaneously cross-link the pointed end to the side of another filament at an angle of roughly 70°. The side-binding function of Arp2/3 may be particularly important in lamellipod formation, where filaments are relatively short (.2 μm) and form angles of roughly 70° relative to one another.

The capping of the barbed ends of actin filaments regulates filament length [63, 192, 229]. For formation of filopods, formation of a square-packed array together with an inhibition of barbed-end capping can lead to the typical elongated structure of filopods. For lamellipods, on the other hand, capping of barbed ends occurs relatively rapidly, resulting in short (and therefore relatively stiff) filaments in a diagonally cross-linked meshwork. One of the major actin capping activities in cells is mediated by the cap32/34 protein. It appears to provide a constitutive capping activity that regulates filament length. Proteins of the ezrin/radixin/moesin family can protect barbed ends from being capped [24]. Such protection may be important in the formation of filopods, but can be detrimental to the formation of lamellipods [18].

Severing of actin filaments provides another mechanism for generating new filaments independent of nucleation [40]. Extension of pseudopods in *Dictyostelium* can involve formation of new filaments coupled with solation of the older ones [134]. Severing of a filament by the actin-depolymerizing factor family member cofilin results in a new barbed and pointed end [17]. Depending on the lengths of the resulting filaments and the local capping protein activity, this severing can result in either generation of new filaments through polymerization from the barbed ends or increased rates of depolymerization through the pointed ends that are formed. Both processes contribute to cell motility [48]. Gelsolin is another severing protein that can be utilized to generate new barbed ends for filament growth [9, 241]. Gelsolin binds to the barbed ends of filaments that it severs, but can be released by hydrolysis of $PI(4, 5)P_2$ (phosphatidylinositol 4, 5-bisphosphate) to which it is bound.

After polymerization of an actin filament, its contribution to the actin cytoskeleton depends in part on its cross-linking to other filaments. One interaction for filaments nucleated by the Arp2/3 complex is due to the filament side-binding activity of Arp2/3. However, further stability is provided by actin-binding proteins such as filamin that cross-link the sides of filaments [216]. Cells lacking filamin show reduced chemotactic responses which are restored by expression of filamin [46, 71, 174]. EF1alpha stabilizes filopods by providing a high-density cross-linking function.

Myosin isoforms have been proposed to play multiple roles in amoeboid chemotaxis and motility. The original myosin isoforms identified, termed type II myosins, form bipolar thick filaments, which can generate contraction forces between clusters of actin filaments. Loss of myosin II in *Dictyostelium* results in reduced (but not complete loss of) cell polarity and chemotaxis [37, 238]. As noted above, one contribution of type II myosin is to generation of contraction at the rear of the cell, and in polarized *Dictyostelium* cells this myosin isoform is concentrated there [31]. In the rear of the cell, myosin II is assembled into filaments and localizes to a cap-shaped domain. In *Dictyostelium*, myosin II assembly is negatively regulated by phosphorylation of the heavy chain [128, 177] at sites in the tail that are likely be phosphorylated by several different kinases. Among the kinases are a family of myosin heavy chain kinases, which are recruited to the leading edge [215], where their activity presumably prevents the formation of myosin II filaments [22, 183, 214, 215]. The regulation of type II myosin is multiple: phosphorylation of the regulatory light chain stimulates the actin-associated ATPase activity, while phosphorylation of the heavy chain regulates thick filament formation. It seems that the light chain phosphorylation affects mainly the ability to respond to temporally changing chemoattractant gradients, while chemotaxis in stable gradients does not require light chain phosphorylation [111, 237, 253]. The regulation of thick filament formation by heavy chain kinases is important for localization of the thick filaments to the rear of the cell—the kinases are localized at the anterior side of the cell. Myosin II may also be involved in lamellipod/pseudopod dynamics. In *Dictyostelium*, tracking of green fluorescent protein-tagged myosin reveals transient increases in myosin in the bases of retracting pseudopods [154].

The functions of myosin II in mesenchymal cells expand upon the functions seen in *Dictyostelium*. There are two forms of nonmuscle myosin II—A and B [23]. Myosin IIA is found in stress fibers as well as in fibers parallel to the leading edge and in the rear of migrating fibroblasts [126, 186]. Myosin IIB is found mainly in the rear of migrating cells. In nonmigrating cells, both isoforms are found in stress fibers, reflecting a possible difference in the character of stress fibers in motile and nonmotile cells. Regulation of myosin II in mesenchymal cells, as in *Dictyostelium*, can be due to light chain or heavy chain phosphorylation. Calmodulin can regulate myosin light chain kinases, providing a calcium-mediated regulation pathway. Alternatively, or perhaps in synergy with calmodulin, kinases activated by the small G protein RhoA [72, 240], such as Rho kinase and p21-activated kinase (PAK1) can phosphorylate myosin light chain and/or a light chain phosphatase [7, 189, 203, 212]. Inhibition of myosin activity affects lamellipod extension [47], stress fiber stability [7], and cell motility. Myosin activity is important for formation of stress fibers and focal contacts, as described below in Section 2.3.2. However, the mechanism by which myosin activity contributes to lamellipod extension is unclear.

Other myosins may play supporting roles in regulating protrusion. Cells lacking single-headed myosin I isoforms show increased generation of side protrusions in chemotaxing *Dictyostelium* [38, 149, 197, 245]. Myosin I phosphorylation is increased by chemotactic stimulation [84]. Thus, one function of type I myosins may be to suppress inappropriate extensions from regions of the cell exposed to lower concentrations of chemoattractant. Myosin I has also been shown to interact with actin polymerization factors, and might selectively transport them to the leading edges of cells [114]. In particular, the isoforms myosin IB and IC have been shown to interact with the Arp2/3 complex as well as capping proteins [114]. Myosin VII has been reported to be important for adhesion of the leading edge to the substratum [221], presumably by delivering adhesion molecules to the correct subcellular location. Myosins also play roles in the transport of other components that contribute to cell motility. For example, localization of beta-actin messenger RNA via myosin-based transport to the front of the cell has been proposed to be important for mesenchymal cell polarity [205]. The localized translation of actin messenger RNA into protein at sites of lamellipod extension may enhance lamellipod extension, which in turn enhance localized focal contact formation [157].

2.2.2. *Microtubules*

As noted previously, the evidence for the contribution of microtubules to amoeboid cell motility and chemotaxis is mixed [242]. The microtubule organizing center has been reported to be localized to either the front or rear side of the nucleus, depending upon the cell type [167, 187, 188]. Alterations in microtubules can affect fibroblast lamellipod extension and motility [150], but in some assays, chemotactic responses may be unaffected [202]. Alterations in acetylation enhance chemotactic ability [98]. Microtubules have been proposed to alter the stability of adhesion sites, enhancing their disassembly [119]. In sum, microtubules are likely to be permissive for amoeboid motility and chemotaxis, and respond to polarization of the cell generated by the actin system with polarization of the microtubule system. This may in turn stabilize cell polarity and enhance overall chemotactic efficiency. In the absence of a strong external stimulus, or in cases in which autocrine secretion influences cell polarization, the microtubule apparatus may provide critical signals for cell polarity [164].

2.3. *Adhesion*

This section will focus on cell-substratum adhesion, although in multicellular organisms, cell–cell adhesion may be an important process for cell motility in some circumstances [60, 243]. For *Dictyostelium*, the major contacts appear to be broad close contacts with the substratum together with dynamic eupodia that enable relatively rapid motility. This is consistent with the actomyosin system structure in *Dictyostelium*—a highly dynamic actin cortex with no stress fibers or focal contacts. On the other hand, mesenchymal cells can form much tighter contacts, termed focal contacts, with the substratum *in vitro*. Such contacts may be useful in generating traction in the *in vivo* environment consisting of randomly oriented matrix fibers that are irregularly spaced. The regulated control of adhesion plays a role in both the rate of amoeboid motility as well as in the generation of cell polarity [4, 120]. Although additional types of contacts are present between mesenchymal cells and the substratum, the focus will be on focal contacts since they are the best-characterized contacts.

2.3.1. *Close contacts*

Dictyostelium has been shown thus far to only have relatively weak contacts with the substratum. Observations by interference reflection microscopy show that highly chemotactic *Dictyostelium* cells have relatively small areas of contact with the substratum [193, 231]. They do not demonstrate the focal contacts that are seen with mesenchymal cells. The contact areas with the substratum do not show focal accumulations of cytoskeletal proteins, consistent with the lack of focal contacts. Flow force measurements of cell adhesion and detachment show that mutations in cytoskeletal proteins can reduce the adhesion strength [55, 161, 208, 221]. *Dictyostelium* cells can form "eupodia" on their dorsal surfaces when in contact with agarose on that surface [74, 75]. These structures may represent more localized adhesion sites, which are present when the appropriate substratum is available. In addition, during aggregation, chemotactic responses may be modified by cell–cell adhesion.

Mesenchymal cells show close contacts at the leading edges prior to formation of focal contacts, as discussed below [58, 107, 108].

2.3.2. *Focal contacts*

Focal contacts are regions of close contact and tight adhesion to the substratum [50, 190, 230, 251]. Interference reflection microscopy was originally used to identify these structures [1, 109], and recently immunofluorescent or live imaging (using green fluorescent protein fusions [116, 142]) of the localization of specific proteins has become more common. The best-characterized transmembrane proteins that generate focal contacts are the integrins [206]. These molecules are heterodimers (composed of alpha and beta subunits) that bind to specific extracellular matrix molecules outside of the cell, and then couple to signaling and cytoskeletal proteins inside the cell. There are multiple alpha and beta isoforms, and different combinations of isoforms result in both different binding specificities for extracellular matrix ligands as well as different intracellular signaling events. In general, integrin signaling is stimulated by aggregation of the integrins and binding to their ligands. In addition, "inside out" signaling refers to signal transduction events inside the cell that can alter the affinity of integrins for ligand [196].

Upon extension of a lamellipod, structures termed focal complexes form at the tip of the lamellipod [162]. Focal complexes can be induced by activation of rac. Although the components of focal complexes have not been exhaustively elucidated, they contain adhesion-related proteins such as integrins, vinculin, paxillin and high levels of tyrosine phosphorylation [80]. Formation of mature focal contacts requires both RhoA activity and aggregation of the integrins with binding to their extracellular matrix ligands. Binding of integrins to matrix-coated beads triggers association of intracellular signaling molecules such as talin, tensin, and focal adhesion kinase [153]. Integrins do not have intrinsic kinase activities, but the association of focal adhesion kinase with integrin multimers results in increased focal adhesion kinase activity [171]. In combination with the tyrosine kinase src, focal adhesion kinase itself shows increased tyrosine phosphorylation and becomes a scaffold protein for assembling a number of typical focal adhesion proteins such as paxillin and vinculin. Focal adhesion kinase is most important for focal adhesion dynamics—fibroblasts lacking focal adhesion kinase show extremely slow turnover of focal adhesions, resulting in low motility [168].

As noted above, part of the process of formation of focal adhesions requires the small G protein RhoA [120]. Constitutively active RhoA stimulates, while dominant negative RhoA inhibits, focal adhesion formation. RhoA activity can itself be regulated by adhesion [8, 113, 195], G protein-coupled receptors [185], and receptor tyrosine kinases [124]. RhoA has multiple effectors, including the mammalian diaphanous protein and regulators of myosin II function such as Rho-associated kinase. Upon activation, RhoA stimulates myosin II activity by inhibiting the myosin light chain phosphatase and stimulating a myosin light chain kinase [212]. It is thought that this contractile activity brings actin filaments into roughly parallel bundles for stress fibers. At the cell membrane, the stress fibers attach to integrins via actin-binding proteins such as alpha-actinin, vinculin, and talin. Thus the contractile activity may contribute to focal adhesion assembly by stimulating integrin association.

The role of the diaphanous protein appears to be in orientation of the stress fibers, potentially coordinating them with the orientation of the microtubule cytoskeleton [106, 167, 191].

Removal of focal contacts is as critical for amoeboid motility as formation of focal contacts. The speed of motility has been shown to correlate with the degree of adhesiveness [166], and spatial regulation of disassembly of focal contacts could aid both in generating cell polarity and in enhancing motility. As noted above, focal adhesion kinase appears to be important for focal adhesion disassembly. One possibility is that focal adhesion kinase activates the extracellular signal regulated kinase (ERK/MAP kinase), which in turn activates calpains [83]. Calpains are calcium-dependent proteases, which can cleave talin and other proteins that are part of focal adhesions. Receptor tyrosine kinases such as the EGF receptor also trigger ERK activation to stimulate calpain [82, 207]. Since ERK can be activated by a variety of important signaling pathways, this could be a general mechanism for growth-factor-stimulated deadhesion during motility. This mechanism of disassembly may be important for amoeboid chemotactic responses as will be discussed later. Reduction in focal adhesion kinase levels by antisense RNA can reduce growth-factor-stimulated signaling and chemotaxis, indicating that another way that focal adhesion kinase may contribute to chemotaxis of mesenchymal cells could be through enhancement of chemotaxis receptor signaling [92].

2.4. *The gradient sensed by amoebae: temporal vs. spatial sensing*

Amoebae are able to orient and chemotax in spatial gradients of chemoattractants. For example, placing a micropipet close to one side of a cell can cause the cell to directly extend a pseudopod towards the pipet (Figure 1) [14, 43]. Thus amoeboid cells can perform spatial comparisons in that they can determine which side of the cell should extend a lamellipod—a process that is not thought to occur during bacterial chemotaxis. However, given the example of how temporal sensing enables bacterial chemotaxis, a long-standing issue has been whether temporal sensing mechanisms contribute to amoeboid chemotaxis. In the pipet example given above, the placement of the pipet near the cell generates a sudden, temporal increase in chemoattractant concentration that is greatest on the side of the cell next to the pipet, and the combination of temporal increase and spatial localization could

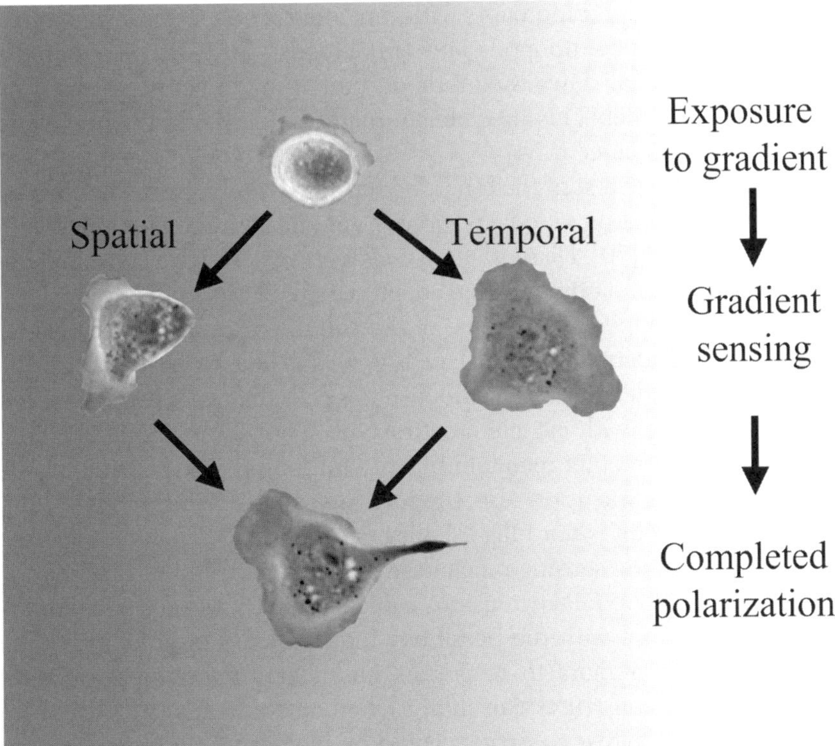

Figure 6. Idealized discrimination between spatial vs. temporal sensing models for amoeboid chemotaxis. An unpolarized cell present in a spatial gradient of chemoattractant (*top*) could determine the gradient by either spatial (*left, middle*) or temporal (*right, middle*) methods. In the spatial comparison case, the gradient determination is achieved in the absence of cell protrusion and the cell immediately extends in the direction of the gradient. In the temporal extension case, the cell extends in all directions, determines the gradient direction, and polarizes in the direction of the gradient. The difficulty in performing this experiment is due to the temporal stimulus provided upon initially imposing the gradient.

be the stimulus for inducing the extension of a lamellipod towards higher concentrations of chemoattractant (Figure 6).

There is an additional possible temporal stimulus that could be occurring in amoebae migrating in stable spatial gradients as well. As noted above, amoeboid cells produce both filopodia and pseudopodia/ lamellipodia. Filopodia typically protrude out beyond lamellipodia and pseudopodia, and can be rapidly extended and retracted. Filopodia could behave as analogs to bacteria: if the filopod is extended up the gradient, the fraction of receptors occupied by chemoattractant on the surface of the filopod increases with time, and it generates a signal that locally stimulates pseudopods and lamellipods [148]. If it is extended perpendicular to or down the gradient, then it might generate no signal or a negative signal. This mechanism might be particularly effective in growth cone guidance [254].

The most direct test of the temporal sensing hypothesis was performed using a system that provided a negative temporal stimulus combined with a spatial gradient [70]. *Dictyostelium* cells were exposed to a constant concentration of chemoattractant and adapted to that level, and then the level of attractant dropped on one side of the cell to create a spatial gradient. After an initial adaptation period, cells were still able to chemotax up the gradient, although not quite as effectively as cells that were exposed to the gradient starting with no chemoattractant. The observation that cells treated with latrunculin maintain polarity for long periods of time in the absence of any protrusion also supports a spatial sensing model [170]. However, this point is not fully established (see [225]).

Even if temporal sensing mechanisms do not contribute to chemotactic orientation, filopodial extensions could still be providing spatial information. Consistent with the possibility that filopods could be providing information for chemotactic responses, under some conditions dominant negative cdc42 constructs can inhibit chemotactic responses [6]. Since the dominant negative constructs could be blocking filopod production, that result is consistent with filopods contributing to chemotactic responses. However, the correlation between filopod extension and chemotactic ability is not perfect. For *Dictyostelium*, although growth phase cells can show dramatic filopod extension to folate (the growth phase chemoattractant) [179], the chemotactic response to folate is relatively weak. Starved *Dictyostelium* cells show much stronger chemotactic responses to cAMP (the aggregation phase chemoattractant) but much less filopod formation. Thus, amoeboid chemotactic responses may not absolutely require either filopods or a temporal sensing mechanism, but such factors are likely to enhance chemotactic responses.

In summary, it is clear that amoebae can spatially discriminate receptor binding of chemoattractant. However, the importance of temporal concentration changes for chemotaxis is still unresolved.

2.5. *Excitation and adaptation*

The terms excitation and adaptation attempt to define chemotactic responses in terms of a systems analysis approach in which excitation refers to signaling pathways that are directly leading to motility changes and adaptation refers to pathways that suppress the excitation pathways. This terminology and approach to analyzing chemotaxis is based upon the bacterial chemotaxis pathways, in which the emphasis is on temporal responses to temporal changes in chemoattractant concentration. As described in Chapter 3, feedback pathways such as methylation of the methyl-accepting chemotaxis proteins occur more slowly than the excitation pathways involving CheY (indeed are accelerated by CheY-P), and thus adaptation events tend in bacterial chemotaxis to act directly on receptors and because of the reduced rate of adaptation, enable detection of a temporal stimulus.

The issues regarding excitation and adaptation for spatial sensing of amoeboid cells are more complex. A general model that is used to described spatial chemotactic signaling includes the spatial range of the signal: excitation signals are likely to be short-range, affecting the local actin polymerization while adaptation signals could include longer range signals that may act directly on the receptors but may also affect the cytoskeletal responses directly that the excitation signals are also stimulating [33, 103, 199]. For example, activation of phosphoinositide 3-kinase may locally increase the concentration of phosphatidylinositol (3, 4, 5) trisphosphate (PIP_3), resulting in stimulation of cdc42 and rac, actin polymerization, and generation of filopods and lamellipods. However, in addition, the extracellular signal-regulated kinase (ERK) signaling pathway may be activated. ERKs travel throughout the cell, and through phosphorylation of calpains can cause reduction in focal contacts and cell detachment, providing an example of long-range spatial suppression of adhesion [82]. In addition, ERKs may also provide a more traditional adaptation pathway through phosphorylation of receptors or components of the signaling pathway stimulating actin polymerization.

Thus, for amoeboid chemotaxis, the spatial aspects of the signaling pathways need to be examined as well when considering excitation and adaptation pathways. These issues will be discussed in more detail in the section on signal transduction.

3. Techniques to Measure Motility and Chemotaxis

A variety of assays are available to quantitate amoeboid motility and chemotaxis. More detailed discussions of the strategic and practical aspects of deciding on which chemotaxis assay to use are given in [257]. What follows is a brief overview. Because the assays are of varying complexity and difficulty, the choice of the specific assay to be used depends upon the question that is asked. For measurement of the complete chemotactic response: directional cell translocation in a spatial gradient, most assays utilize spatial gradients of chemoattractant generated by diffusion (Section 3.1). However, detailed dissection of specific responses in a temporal fashion can utilize rapid, uniform increases in concentration (Section 3.2). Finally, tracking of amoeboid motility *in situ* enables analysis of motility responses under physiological conditions (Section 3.3), typically at the expense of control and/or knowledge of the actual gradients that are present.

3.1. *Assays in which a gradient of the stimulant is established by diffusion*

These assays are used to quantitate the complete chemotactic response. The micropipet assay can provide steep gradient stimuli with direct visualization of chemotaxis at the price of low numbers of cells analyzed and less precise knowledge of the absolute concentration levels. The semistable diffusion gradient methods provide the ability to visualize chemotactic motility in response to more precisely defined gradients than the micropipet assay. Other assays such as the agar well assay and Boyden chamber assay are more rapid and amenable to high throughput, but with limitations in terms of dissecting individual behaviors.

3.1.1. *Micropipet assay*

Micropipets with tip diameters $<.5\,\mu m$ containing chemoattractant provide a point-source stimulation method. The concentration gradient of diffusion out of the pipet varies with the inverse of the distance from the tip, but the absolute release rate from the tip is variable and difficult to quantitate. By placing the pipet within a cell diameter from a cell, one can generate gradients of 50% across the cell—extremely steep gradients. In addition, one can rapidly move the pipet, enabling the tracking of cell responses to reorientation of the gradient. However for such steep gradients only one or two cells can be stimulated at a time. Mesenchymal cells using receptor tyrosine kinases may require steep gradients [10] either using micropipets or the Boyden chamber assay, possibly because of a less efficient chemotactic sensing mechanism. For cells that respond more sensitively, such as *Dictyostelium*, the pipet can be placed many cell diameters away, enabling the simultaneous stimulation of many cells.

3.1.2. *Semistable diffusion gradient assays*

Using semistable diffusion gradient assays such as the Zigmond and Dunn chemotaxis chambers [256, 257], linear gradients of chemoattractant are produced on glass slides that enable observation of cells moving in response to the gradients. The gradients are not as steep as in the micropipet assay, but are more uniform and the absolute concentrations are known, allowing careful comparisons of specific cell behaviors under different gradient conditions. Technically, these chambers can be challenging to work with, and the number of cells analyzed per experiment is still small—typically 10–20.

3.1.3. *Boyden chambers*

Boyden chambers [20] utilize a filter with holes large enough (on the order of 3–10 µm) to allow cells to migrate through the filter. Typically the cells are placed on the top of the filter and chemoattractant is placed on the bottom. Enough cells are deposited to cover the surface of the filter, and thus cells sitting over the pores in the filter are exposed to extremely steep gradients—high concentration in the pore, and lower concentrations on the surface of the cell away from the pore. After a period of time (typically several hours) to allow a significant number of cells to cross the membrane, the cells that did not cross are wiped off and the cells that did cross are fixed, stained, and counted. This general method has been adapted to enable multiple simultaneous assays in the microchemotaxis chamber [91] and Transwell formats. Thus dose-response and kinetic curves can be rapidly generated, but the actual behaviors of the cells during the response cannot be observed. A variant of this assay uses gradients of chemoattractants across nitrocellulose filters, and can evaluate the depth to which cells penetrate into the filter.

3.1.4. *Agar assays*

The advantages of agarose-based assays are that longer-term (several hours to days) experiments can be performed due to stabilization of diffusion gradients against convection. These assays typically utilize large numbers of cells and track the leading edge of the cell population. Typical measurements are then looking at the maximal distance traveled from the original site of inoculation. Cells can be placed on the surface (and then crawl on the surface) of the agarose or in a well in the agarose (and then crawl under or through the agarose).

3.2. *Assays utilizing global temporal increases (upshift)*

Although chemotactic behaviors occur in spatial gradients of chemoattractant with the corresponding spatial segregation of specific cytoskeletal responses, it is difficult to compare biochemical experiments with the full chemotactic behavior. Normal biochemical experiments involve sudden addition of a uniform concentration of chemoattractant to a population of cells, followed by a biochemical analysis that measures that average response. To understand what the amoeboid behaviors are that correlate with the temporal sequence of biochemical responses, the behavioral responses to rapid, uniform changes in chemoattractant concentration are also often measured. Changes in actin polymerization, cell translocation, adhesion, and shape change have been measured both on surfaces and in suspension in response to a temporal increase in concentration.

This stimulation regime could in principle enable a temporal dissection of responses that normally occur in spatially segregated regions of the cell. For example, local excitation responses such as actin polymerization tend to occur relatively rapidly, while responses that are likely to be global, suppressive responses may have slower kinetics. These kinetic differences most likely reflect time lags required for diffusion and coupling to downstream signaling molecules for the two types of behaviors.

3.2.1. *Actin polymerization assays*

Filopod, pseudopod, and lamellipod extension are absolutely dependent upon actin polymerization. Thus assays of total polymerization of actin as well as barbed end number as a function of time after stimulation have been heavily utilized. Studies of cells in suspension have often been used in order to measure total changes in these parameters. However, because the spatial localization of the sites of actin polymerization is critical to proper chemotactic responses, studies then have included analyses of individual cells attached to substrata [40]. In general, barbed ends increase near the plasma membrane and increases in polymerized actin follow in the same sites, producing protrusion of the plasma membrane, as discussed before. The spatial localization of specific signaling molecules and their signaling state can be examined to determine whether the localization is consistent with a role in the location and dynamics of actin polymerization.

3.2.2. *Adhesion assays*

Because adhesion can play a role in the polarization of mesenchymal cells, examination of the dynamics of focal contacts and cell/substratum adhesion as a function of time after chemoattractant stimulation has been performed. Using interference reflection microscopy [15] or green fluorescent protein-tagged adhesion proteins [252], the dynamics and location of specific cell substratum contacts can be followed during stimulation. In general, there is stimulation of new adhesion sites in newly extended lamellipods and retraction of the older adhesion sites, including those at the rear of the cell.

The strength of cell-substratum adhesion sites per cell can be measured using physical assays to determine the force required to detach cells from a substratum. Relatively slow measurements can be performed using centrifugation assays, while assays utilizing fluid flow enable a more rapid kinetic analysis. On the single cell level, flexible substrata provide a mechanism for following traction forces that are generated around individual cells [16, 19].

3.2.3. *Behavioral assays*

Behavioral assays for following responses to sudden increases in chemoattractant have included light scattering assays of cell populations as well as the tracking of individual cells. Light scattering assays were initially performed because of the ease and rapidity of analysis where an average over a large number of cells can be made [81]. More recently, the focus has been in following individual cell shape changes and motility changes in order to more precisely correlate specific behaviors with biochemical changes. For example, stimulation of *Dictyostelium* with a sudden increase in chemoattractant produces a decrease in net translocation together with a brief contraction, followed by an area increase corresponding to extending pseudopods [198, 224]. Loss of a cGMP-specific phosphodiesterase results in prolonged increases in intracellular cGMP in response to chemoattractant, and correlated with that is a prolonged decrease in translocation [30, 147, 160, 200].

3.3. *Tracking amoebae* in situ

With the identification of green fluorescent protein as an easy method for stably marking cell populations, the ability to track individual green fluorescent protein-expressing cells in a population of unlabeled cells has become possible. This enables the analysis of cell behavior *"in vivo"* under physiologically relevant conditions. Thus, with *Dictyostelium*, tracking of a labeled subpopulation enables characterization of the cell behaviors that occur during aggregation and formation of fruiting bodies. Similarly, for mesenchymal cells, the behavior of cells during development [125, 129], wound healing, and of tumor cells in the primary tumor is accessible [42, 97]. The limitation with these methods is that the size and temporal dynamics of the chemoattractant gradients driving the cells are not known.

4. Signal Transduction During Chemotaxis

The following sections integrate the material described above to provide our current understanding of the mechanisms involved in *Dictyostelium* and mesenchymal cell chemotaxis. *Dictyostelium* will be covered first, and provides specific paradigms for spatial discrimination that are relatively well worked out. How these paradigms could apply to mesenchymal cell chemotaxis is then discussed in the subsequent section along with additional pathways.

4.1. *Dictyostelium*

4.1.1. *Molecular mechanisms of* Dictyostelium *chemotaxis*

Chemotaxis to extracellular cAMP will be the focus of this section because it is the best characterized chemotactic response in *Dictyostelium*, but chemotaxis at other times of the *Dictyostelium* life cycle relies on the same principles and uses closely related pathways which are based on the same or homologous molecules. This is illustrated by the observation that ablation of the single gene coding for the alpha subunit of the heterotrimeric G-protein working downstream of the membrane receptors completely abolishes chemotaxis [172, 244] in *Dictyostelium*.

At the onset of development, *Dictyostelium* cells start to produce and respond to the small molecular chemoattractant cAMP [146]. The details of the inner workings of the oscillatory cAMP-signaling network are not directly relevant to the description of the chemotactic response and will only be touched on briefly. In a field of cells, periodic cAMP waves emerge which coordinate the formation of centers towards which the cells move by chemotaxis. On the level of an individual cell, the amazing feat is the detection and response to small external concentration differences between the front and the back of the cell (2–10%, depending on the shape of the gradient) with a very strong polarization of the cytoskeleton. Movement in the direction of the source of the signal is driven by this highly polarized cytoskeleton. The front is extruded as a pseudopod by actin polymerization, while the back of the cell or uropod is retracted by actomyosin-based contractions.

4.1.2. *Receptor*

Unlike higher eukaryotes, *Dictyostelium* possesses no receptor tyrosine kinases and chemotaxis is mediated by G-protein-coupled receptors, as in human neutrophils. During aggregation, the main cAMP sensor is the heterotrimeric G-protein-coupled serpentine receptor cAMP receptor 1. Binding of cAMP leads to the dissociation of the G-protein into the alpha subunit and the beta-gamma heterodimer, both of which remain membrane-associated. The way these subunits couple to downstream events is currently unclear. It is possible that free beta-gamma heterodimer is the major signaling element, similar to yeast orientation mechanisms [59]. In an external cAMP gradient, the receptor maintains uniform membrane localization [248], while the released beta-gamma heterodimer shows a slight anterior–posterior distribution gradient, essentially mirroring the external gradient [112]. This is how the cell initially receives the signal, but for efficient chemotaxis the anterior–posterior signal differences need to be amplified. It is not known how amplification of a presumably slight anterior–posterior gradient occurs, as the next known downstream events already exhibit strong polarization (as described in the next section). Phosphorylation of the receptor reduces the affinity of the receptor for chemoattractant threefold [122, 247].

4.1.3. *The phosphoinositide 3-kinase pathway*

Progress made in recent years has revealed the central role the phosphoinositide 3-kinase pathway plays in the establishment and the maintenance of cell polarity (Figure 7) [246]. The substrate for the Class I phosphoinositide 3-kinase [223] involved in chemotaxis is the membrane lipid phosphatidylinositol-(4, 5) bisphosphate, or $PI(4,5)P_2$, which becomes phosphorylated at its $3'$ position to phosphatidylinositol-(3, 4, 5) trisphosphate, or $PI(3, 4, 5)P_3$. A second 3-phosphoinositide, $PI(3, 4)P_2$, can be formed by the dephosphorylation of $PI(3, 4, 5)P_3$ at

Activation of the PI 3-kinase pathway

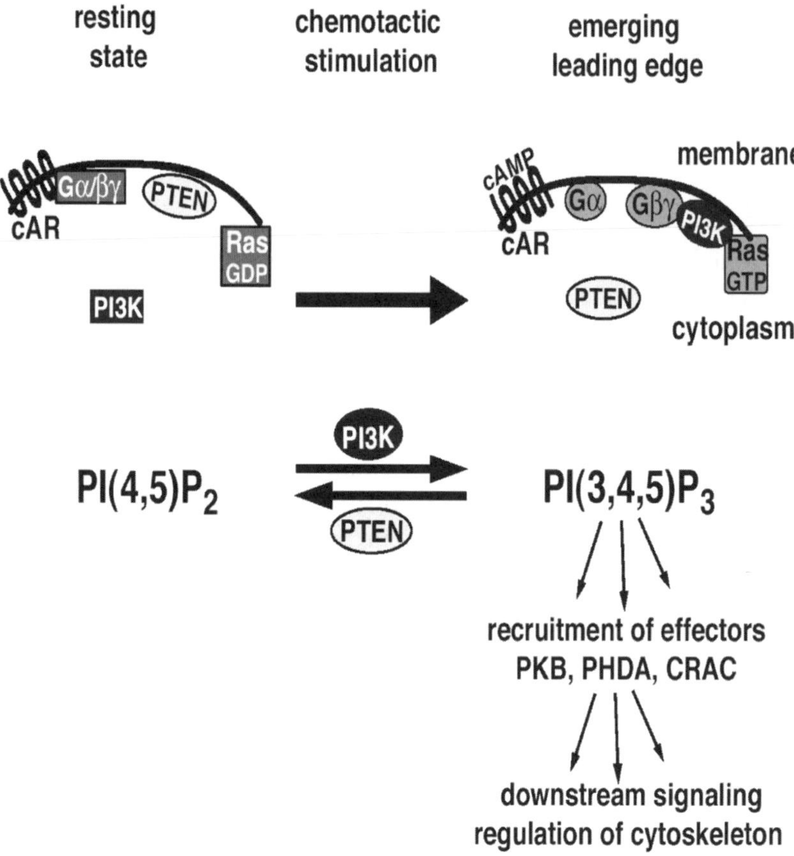

Figure 7. Activation of the phosphoinositide 3-kinase pathway. Activation of the cAMP receptor leads to dissociation of the heterotrimeric G-protein, which by an unknown mechanism generates binding sites for the recruitment of phosphoinositide 3-kinase at the membrane facing the source of cAMP. Phosphoinositide 3-kinase is further activated by interaction with Ras^{GTP}. The 3-phosphoinositide phosphatase PTEN is released from the same region of the membrane. The net result is localized generation of phosphatidylinositol-(3, 4, 5) trisphosphate. In response, multiple pleckstrin homology-domain containing proteins, which bind phosphatidylinositol-(3, 4, 5) trisphosphate with a high affinity, are recruited to this site. These proteins mediate a variety of downstream processes, chief among them regulation of the cytoskeleton and cAMP production.

the 5′ position or by phosphorylation of PI(4)P by phosphoinositide 3-kinase. The lipids serve as specific docking sites for a subfamily of pleckstrin homology domain-containing proteins that either contain or bind to downstream effectors that are recruited to the membrane when the pathway is activated. An important negative regulator is the 3-phosphoinositide phosphatase PTEN (for Phosphatase and TENsin homology), a tumor suppressor, which removes the 3-phosphate group from these lipids [141, 156]. Other potential negative regulators degrading PI(3, 4, 5)P$_3$ are phospholipase C, which has been described in *Dictyostelium* [61] but does not seem to be involved, and 5-phosphate specific phosphatases of which family members are present in the *Dictyostelium* genome but not yet completely characterized. Inhibition of the phosphoinositide 3-kinase pathway results in reduced polarity, formation of multiple weak pseudopodia, reduced cytoskeletal activity, and generally inefficient chemotaxis, whereas hyperactivation disrupts polarity and results in excessive multifocal actin polymerization, with protrusion of multiple simultaneous pseudopodia.

Two Class I phosphoinositide 3-kinases, PIK1 and PIK2 [255], are involved in *Dictyostelium* chemotaxis. A double knockout cell line exhibits severe cell polarity defects and the cells fail to become elongated in the direction of movement [35, 77, 145]. Even though they are still capable of slow cell movement towards cAMP, the knockout cells have trouble maintaining their direction. Actin polymerization, as an indicator of cytoskeletal response to cAMP stimulation, is significantly reduced. Treatment of wild-type cells with the phosphoinositide 3-kinase inhibitor LY294002 closely mimics the phenotype of the double knockout cells. A third Class I phosphoinositide 3-kinase, PIK3, is present in *Dictyostelium*. A PIK3 single knockout has no obvious phenotype and it has not been possible to obtain any multiple knockout cell lines in combination with the other two phosphoinositide 3-kinases. It is therefore currently unclear whether PIK3 mediates the residual chemotactic behavior of the double knockout cells.

The *Dictyostelium* phosphoinositide 3-kinases have a domain organization similar to mammalian Class I phosphoinositide 3-kinases. They have an N-terminal domain that does not show much conservation, followed by a Ras-binding domain, a C2 domain, and a bipartite kinase domain. The involvement of phosphoinositide 3-kinases in polarized cellular responses suggested that their subcellular localization during chemotaxis might be important. Deletion analysis revealed that the N-terminal domain alone is required and sufficient for cAMP-induced membrane recruitment of both PIK1 and PIK2 [76]. The mechanism of this recruitment is currently unknown. It does not require phosphoinositide 3-kinase activity, as it occurs both in the phosphoinositide 3-kinase double knockout cells and in the presence of the phosphoinositide 3-kinase inhibitor LY294002. Interestingly, phosphoinositide 3-kinase localization is already highly polarized relative to the external gradient, suggesting that part of the polarization increase occurs upstream of phosphoinositide 3-kinase. Neither the Ras-binding domain nor the C2 domain are required for membrane association. It has been shown for mammalian cells that the Ras-binding domain is required for phosphoinositide 3-kinase function [180]. A point mutation in the *Dictyostelium* phosphoinositide 3-kinase 1 or phosphoinositide 3-kinase 2 Ras-binding domain abolishing interaction of this domain with RasGTP revealed that Ras binding is required for *Dictyostelium* phosphoinositide 3-kinase function. In addition, this mutant failed to complement the *pik1/pik2* double knockout. Overexpression of a permanently membrane-targeted phosphoinositide 3-kinase leads to the formation of multiple pseudopods and an augmented and prolonged response to cAMP stimulation. This phenotype is closely mirrored by the phenotype of cells lacking the 3-phosphoinositide phosphatase PTEN [102]. If the Ras-binding site of the membrane-targeted phosphoinositide 3-kinase is mutated, overexpression has no effect on cell polarity, suggesting that membrane localized, active phosphoinositide 3-kinase is a limiting mediator of the chemotactic response. The multifocal activation of the membrane-targeted phosphoinositide 3-kinase also leads to the expectation that the Ras activation pattern does not have a steep anterior–posterior gradient, although this remains to be shown. It is important to note that membrane targeting of phosphoinositide 3-kinase does not lead to ubiquitous and persistent activation of downstream effector pathways. There must be an additional layer of regulation that could involve spatially restricted activation of Ras or additional activating and inhibitory molecules, or which might depend on dynamic processes involving negative and positive feedback loops yet to be defined. The identity of the Ras involved in activating phosphoinositide 3-kinase has not been finally resolved. The Ras isoforms that probably activate phosphoinositide 3-kinase are RasG and RasC, based on the phenotypes of their null mutants and affinity studies. Because *Dictyostelium* has many Ras genes [32], it is reasonable to assume that multiple Ras proteins may contribute to phosphoinositide 3-kinase activation. *Dictyostelium* also contains many regulators of Ras activity (Ras guanine nucleotide exchange factors and Ras guanine nucleotide phosphatase-activating proteins), of which only a few have been characterized. One of these, the Ras guanine nucleotide exchange factor Aimless [105], is a potential candidate for an upstream activator, since *aimless* null cells are deficient in chemotaxis. The identification of the binding partner for the phosphoinositide 3-kinase N-terminal domain is of paramount interest, as it will provide a link to the upstream activating processes.

PTEN (3-phosphoinositide phosphatase), the negative regulator of the phosphoinositide 3-kinase pathway, has a distribution complementary to phosphoinositide 3-kinase [76, 102]. It shows a uniform membrane localization in resting cells and delocalizes from the leading edges of chemotaxing cells. This is a very functional subcellular localization for a negative regulator of the phosphoinositide 3-kinase pathway. The regulation of PTEN localization is not dependent on phosphoinositide 3-kinase, as localization and release from the membrane are indistinguishable in wild-type cells and *pik1/pik2* double null cells. The N-terminus of PTEN contains a putative phosphatidylinositol-(4, 5) bisphosphate $(PI(4, 5)P_2)$ binding motif that might be involved in localizing PTEN to the membrane. This motif alone is insufficient to explain the observed release from the membrane, and additional work will be needed to understand the mechanism. It will be interesting to see whether there is a single component directly upstream which regulates the localization of both phosphoinositide 3-kinase and PTEN or whether the regulation occurs via parallel pathways originating from the membrane receptor. Ablation of PTEN expression results in cells that exhibit elevated and prolonged responses to cAMP stimulation, including membrane recruitment of downstream effectors, adenylyl cyclase activity, protein kinase B (Akt/PKB) activity, and actin polymerization [76, 102]. These studies provide additional evidence for a regulatory role of phosphoinositide 3-kinase in mediating localized F-actin polymerization at the leading edge.

The downstream events of the phosphoinositide 3-kinase pathway are generally activated via phosphatidylinositol-(3, 4, 5) trisphosphate $(PI(3, 4, 5)P_3)$. A subset of pleckstrin homology domain-containing proteins binds this lipid or its derivative phosphatidylinositol-(3, 4) bisphosphate $(PI(3, 4)P_2)$ with high affinity and is recruited to the leading edge by this interaction. So far, only three such targets have been described in the literature, but the *Dictyostelium* genome contains many more pleckstrin homology domain proteins that may bind phosphatidylinositol lipids and are thus potential downstream effectors. One of the known targets is the cytosolic regulator of adenylyl cyclase [104, 136]. Its function is probably not directly related to cytoskeletal regulation. Another target is the proto-oncogene protein kinase B (Akt/PKB), a pleckstrin homology domain-containing kinase which is activated at the leading edge in response to chemoattractant stimulation in *Dictyostelium* [145] and mammalian [100] cells, suggesting that the role of phosphoinositide 3-kinase in chemotaxis is very ancient. *Akt/pkb* null cells exhibit strongly reduced polarization and chemotactic efficiency. One of the protein kinase B substrates, the p21-regulated kinase homolog PAKa, requires phosphorylation for activation and localization to the back of the cell [35]. PAKa or PAK-related kinases also influence actin dynamics at the front of the cell by directly regulating the motor activity of myosin I family members. It is of interest that key regulators of mammalian cell growth, such as phosphoinositide 3-kinase, PTEN, and protein kinase B, are also regulators of chemotaxis in *Dictyostelium*.

Another regulator of actin polymerization is PhdA [77]. This is a protein with a pleckstrin homology domain closely related to the cytosolic regulator of adenylyl cyclase and a divergent C-terminus. *PhdA* null cells exhibit a severe chemotaxis defect. Analysis of the motion of individual cells shows that the defect might be due to reduced pseudopod extension and impaired pseudopod retraction, and a consequence of this might be the observed difficulties in following a chemoattractant gradient. The mechanistic link to the actin cytoskeleton is not yet known.

Our current understanding of how the phosphoinositide 3-kinase pathway is linked to the cytoskeleton is incomplete. Most likely, a subset of the many pleckstrin homology domain-containing proteins in the genome is important. It also seems likely that it will involve multiple members of the rac family of small GTPases and their regulatory proteins like guanine nucleotide exchange proteins and guanine nucleotide phosphatase-activating proteins. A regulatory connection with the phosphoinositide 3-kinase pathway could be provided by rac guanine nucleotide exchange proteins which contain pleckstrin homology domain as direct targets of phosphatidylinositol-(3, 4, 5) trisphosphate (PI(3, 4, 5)P$_3$).

For Rac1B and its associated guanine nucleotide phosphatase-activating protein RacGap1, it has been shown that the activation state of Rac correlates with actin polymerization [34]. Either an increase or a reduction in the level of activated Rac is detrimental to chemotactic efficiency. The link to the cytoskeleton most likely involves p21-activated protein kinases and Wiskott–Aldrich syndrome protein family members that directly interact with activated racs.

The results acquired so far have been described as a pathway, which rationalizes a subset of the experimental data but is necessarily incomplete. It is clear that the reality will be better modeled by a regulatory network involving feedback, crosstalk, and other pathways as described in the next section.

4.1.4. *Other pathways*

Other pathways also clearly contribute to chemotactic orientation (Figure 8). Mutants lacking the guanylyl cyclases show reduced chemotactic responses [181, 222]. Loss of the cGMP phosphodiesterases results in prolonged cGMP signaling and prolonged cell motility responses to sudden changes in chemoattractant concentration [22, 159]. The primary target of the cGMP pathway appears to be myosin function. However, the regulation is complex because there is phosphorylation of both the myosin regulatory light chains and the myosin heavy chains, which can have opposing effects on myosin function. Phosphorylation of the regulatory light chains via the myosin light chain kinase MLCK-A is regulated by cGMP, possibly via phosphorylation of MLCK-A. A strong candidate for a cGMP-regulated kinase that could phosphorylate MLCK-A is the cGMP-binding protein GbpC [22]. Phosphorylation of the regulatory light chains activates myosin contractile function, and may contribute to formation of thick filaments which are important for contraction at the rear of the cell. Phosphorylation of the heavy chains inhibits contraction, and is likely to inhibit formation of thick filaments at the front of the cell. Myosin heavy chain kinases which have been localized to the front of the cell [184, 214] could provide a mechanism for disassembling thick filaments at the front of the cell, resulting in enhanced formation of thick filaments at the rear.

Additional pathways clearly contribute to chemotactic function but need further study. Mutants lacking the extracellular signal-regulated kinases ERK1 or ERK2 show different defects in chemotaxis [79, 210, 227], which may reflect different regulatory pathways. In addition, permissive regulation of chemotactic signaling occurs via other extracellular factors such as conditioned medium factor [56]. Myosin I phosphorylation is stimulated by chemoattractant stimulation [84] and also can contribute to cell polarization. Some of these pathways are likely to contribute to longer range spatial signaling since the dispersion ranges for small soluble molecules such as cGMP and even soluble proteins such as extracellular signal regulated kinases (ERKs) can be on the order of a cell diameter [175, 178].

Localized Activation Leading to Cell Polarity and Movement

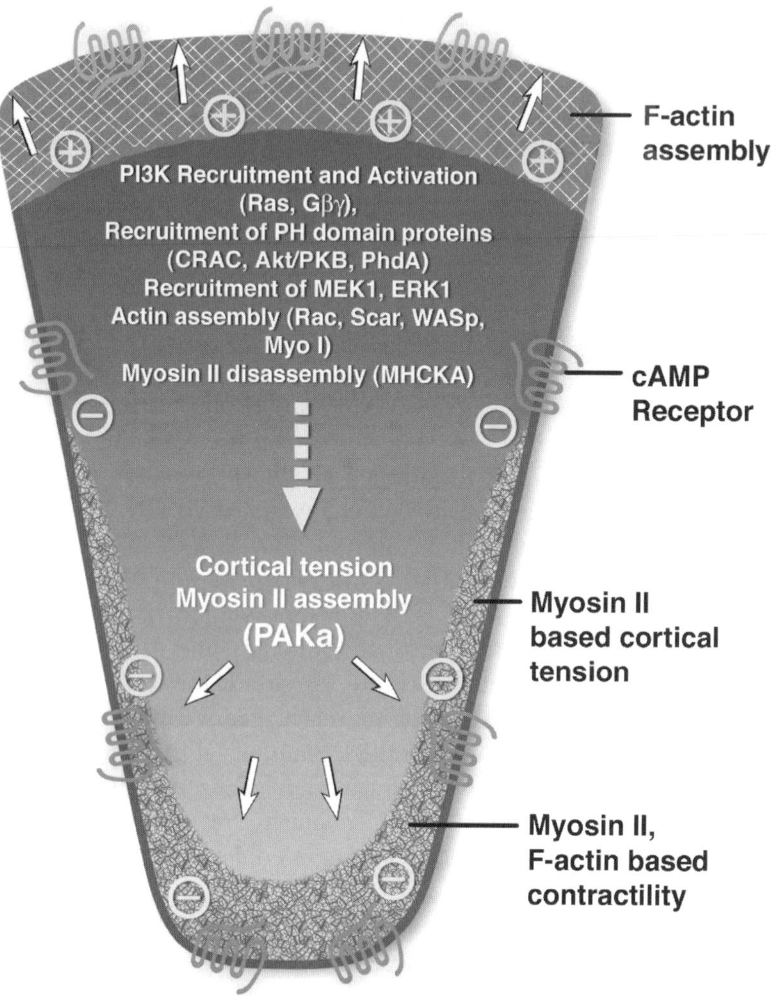

Figure 8. Localized activation leading to cell polarity and movement. A chemotaxing *Dictyostelium* amoeboid cell is characterized by a polarized morphology with a leading edge where filamentous actin assembly drives membrane protrusion (+) and a posterior or uropod that is retracted by actin/myosin II-mediated contraction (−). The establishment and regulation of the polar cytoskeleton depends on polarized regulatory signals. While the cAMP receptor is uniformly distributed along the membrane, 3-phosphinositide lipids such as phosphatidylinositol-(3, 4, 5) trisphosphate, the products of phosphoinositide 3-kinase, are highly enriched at the front of the cell. The reason for this is the selective recruitment and activation of a phosphatidylinositol

4.2. *Mesenchymal cells*

This section will focus on receptor tyrosine kinase-mediated chemotactic responses of mesenchymal cells. G-protein-coupled receptor-mediated chemotaxis mechanisms were discussed above in the *Dictyostelium* section.

4.2.1. *Receptors*

Receptors such as the fibroblast growth factor, hepatocyte growth factor, platelet-derived growth factor, and EGF receptors (among others) are type I receptor tyrosine kinases, typically containing an extracellular ligand-binding domain, a transmembrane domain, and cytoplasmic tyrosine kinase and signaling domains [194]. Binding of ligand stimulates multimerization of the receptors, which then cross-phosphorylate each other, stimulating the tyrosine kinase activity and generating phosphorylated tyrosine residues in the signaling domains. The phosphorylated tyrosine residues then bind to downstream signaling molecules, producing a cluster of receptors and signaling molecules. This cluster can then be internalized with potentially a change in signaling. Coupled to internalization quite often is ubiquitination, which leads eventually to degradation of the receptor. Alternatively, after removal of the ligand in an intracellular compartment, the receptors may recycle to the cell surface. During internalization, the signaling pathways activated by the receptor may shift in relative amplitude prior to being shut off.

3-kinase at the leading edge and the release of its antagonistic 3-phosphoinositide phosphatase PTEN from the front in response receptor activation. PTEN (not shown) remains associated with the lateral sides and posterior of chemotaxing cells, sharpening the anterior phosphatidylinositol-(3, 4, 5) trisphosphate gradient. Currently, the mechanism for this localization process is unknown but the resulting polarity in lipid distribution seems to govern much of the downstream processes. A critical link to downstream signaling is the recruitment of pleckstrin homology-domain containing proteins (which bind phosphatidylinositol-(3, 4, 5) trisphosphate or phosphatidylinositol-(3, 4) bisphosphate) to the front. These proteins, in turn, initiate a multitude of polarized signaling events that are only partially understood. Some of these, like the regulation of p21 activated kinase (PAKa) activity and localization, provide a first link to the genesis, maintenance, and regulation of a polarized cytoskeleton. Other signaling pathways, including cGMP, also play essential roles, with cGMP along with PAKa being linked to myosin II assembly and contraction of the uropod. Taken with permission from Chung *et al.* [33].

4.2.2. *Receptor coupling to intracellular pathways*

The multiple functions that are played by receptor tyrosine kinases—regulation of growth and survival as well as chemotaxis—lead to additional signaling pathways that do not directly affect chemotactic responses. The result is a web of interacting proteins, with perturbation of a particular protein usually having effects on multiple functions [101, 176].

The first layer of coupling to receptor tyrosine kinases involves direct protein–protein interactions mediated by phosphotyrosines on the C terminus of the receptor tyrosine kinase. The platelet-derived growth factor receptor is a particularly instructive example in which specific phosphotyrosines can be shown to mediate the activation of specific downstream pathways [182]. For example, direct activation of phospho-inositide 3-kinase and phospholipase C gamma is mediated by tyrosines 740/751 and 1021 respectively of the platelet-derived growth factor receptor—mutation of these tyrosines to phenylalanines blocks activation of these molecules by the platelet-derived growth factor receptor and has specific effects on chemotaxis. The initial activation of these molecules may be stabilized by tyrosine phosphorylation, as in the case of phospholipase C gamma, in which tyrosine phosphorylation by the receptor tyrosine kinase then enables dissociation of phospholipase C gamma from the receptor tyrosine kinase and diffusion to find and hydrolyze substrate. Phosphoinositide 3-kinase may also be activated in this way [29, 51], or through other protein–protein interactions, as described below.

Additional signaling pathways are activated by formation of a complex of proteins. Such complexes tend to make use of adaptor proteins. Adaptors typically contain combinations of protein–protein interaction domains such as src homology domain 2 (SH2) or phosphotyrosine-binding domains (PTB) for binding to receptor tyrosine kinases together with src homology domain 3 (SH3) or other domains for interacting with other proteins. For example, activation of Ras involves binding of a growth factor receptor-bound protein-2/son of sevenless (Grb2/sos) complex to the receptor tyrosine kinase, followed by activation of Ras by the guanine nucleotide exchange activity of the son of sevenless protein. Growth factor receptor-bound protein-2 (Grb2) is an adaptor protein containing a src homology domain 2 and 2 src homology domain 3 sequences. Sometimes an additional layer of protein–protein interactions occurs with the src homology domain 2 containing proto-oncogene shc binding to the receptor tyrosine kinase, becoming phosphorylated on tyrosine, and then in turn binding the Grb2/sos complex. Variations on this theme include proteins with direct enzymatic activity, protein–protein interaction domains as well as multiple proteins that bind to a particular adaptor.

4.2.3. *Stimulation of extension*

Several pathways lead to stimulation of the extension phase in the amoeboid chemotaxis response. One pathway for activation of actin polymerization involves double homology domain proteins, rho family G proteins, and the Arp2/3 complex. As noted above, receptor tyrosine kinases stimulate double homology domain-containing proteins such as son of sevenless (sos). In parallel, phosphoinositide 3-kinase can be activated by binding of its 85 kilo Dalton subunit to phosphorylated receptor or phosphorylated docking proteins such as the Grb2-associated binding proteins [137]. The product of phosphoinositide 3-kinase can activate double homology domain-containing proteins which contain pleckstrin homology domains (which bind phosphatidylinositol lipids). The double homology domains then stimulate the GDP/GTP exchange of small rho family proteins such as cdc42 and rac, and thus such proteins are guanine nucleotide exchange factors. Cdc42 can directly bind Wiskott–Aldrich syndrome proteins to activate Arp2/3, while rac interactions with the WAVE subfamily of Wiskott–Aldrich syndrome proteins to activate the Arp2/3 complex appear to involve additional proteins such as IRSp53 [64]. Upon activation of the Arp2/3 complex, new actin nuclei are formed together with a branched network.

A second pathway for stimulation of actin polymerization involves phospholipase C gamma. Phospholipase C gamma cleaves phosphatidylinositol 4, 5 bisphosphate ($PI(4, 5)P_2$) to produce inositol trisphosphate (IP_3) and diacylglycerol. Actin-severing proteins such as cofilin and gelsolin bind phosphatidylinositol 4, 5 bisphosphate ($PI(4, 5)P_2$) and are inhibited by this binding [45, 218]. Cleavage of phosphatidylinositol 4, 5 bisphosphate ($PI(4, 5)P_2$) will release the bound cofilin and gelsolin, enhancing their severing activity. Cofilin severing of actin filaments will then generate free barbed ends that will polymerize longer actin filaments. Gelsolin severing of actin filaments requires high calcium or low pH as well [130]. The IP_3 released from phospholipase C gamma hydrolysis of phosphatidylinositol 4, 5 bisphosphate ($PI(4, 5)P_2$) can also stimulate release of calcium from intracellular stores to produce the required increase in calcium to enable severing of actin filaments by gelsolin. However, the gelsolin remains bound to the barbed ends of the severed filaments. Resynthesis of phosphatidylinositol 4, 5 bisphosphate (PI(4, 5)P$_2$) will rebind gelsolin, releasing it from the barbed ends and thus stimulating barbed end growth.

Both mechanisms for stimulation of actin polymerization mentioned above are likely to be relatively short range—diffusion of activated G-proteins or phospholipase C gamma is likely to be relatively limited. Thus, the polymerization of actin stimulated by receptor tyrosine kinases can be localized, consistent with generation of a localized extension. In addition, there are several possible mechanisms for positive feedback stimulation of these localized pathways. As noted above, some guanine nucleotide exchange factors for rac and cdc42 have pleckstrin homology domains, which can bind phospholipids such as the lipid product of phosphoinositide 3-kinase, PIP$_3$. Binding of PIP$_3$ can enhance guanine nucleotide exchange factors, producing enhanced activation of rac and cdc42. Furthermore, rac can bind to and activate phosphoinositide 3-kinase. The result is the potential for a mutual positive feedback process to amplify a localized signal via phosphoinositide 3-kinase activity [33, 103, 233].

4.2.4. *Adhesion*

Generation of adhesive contacts appears to be a natural consequence of extending the plasma membrane over an adhesive substrate. Thus, plating cells from suspension onto an adhesive surface stimulates formation of focal adhesions. In a similar fashion, extension of cell membranes as a consequence of localized actin polymerization will generate new sites of attachment between integrins and appropriate extracellular matrix molecules. Thus, upon stimulation with chemoattractant, adhesion becomes stimulated in two ways, which may interact synergistically. First, activation of intracellular signaling pathways triggers activation of Rho signaling proteins, resulting in increased focal contact formation. Second, extension of lamellipods generates new interaction sites with the extracellular matrix. The actual formation of new focal contacts appears to be concentrated in these new extension sites, possibly because of the generation of new integrin—extracellular matrix interactions [14].

However, there is also stimulation of turnover of old adhesion sites. One possible mechanism for stimulated removal of old adhesion sites is via extracellular signal-regulated kinase (ERK) activation of calpain [249], followed by calpain-mediated cleavage of adhesion proteins such as talin [83]. Activation of Ras by receptor tyrosine kinases results in activation of extracellular signal-regulated kinase (ERK). ERK is likely able to spread throughout the cell, as evidenced by its transport into the nucleus after stimulation. Since ERK activation is relatively rapid compared to lamellipod extension and formation of new adhesion sites, it appears that upon attractant stimulation, there is first destabilization of adhesion sites present during the initial stimulus, and then new adhesion sites form after the lamellipod has extended. In addition, there may be biochemical distinctions between new and old adhesion sites that enable targeting of old adhesion sites to occur.

4.2.5. *Contraction*

Mechanisms of contraction for mesenchymal cells may be more dependent on formation of stress fibers and focal contacts than on the cortical contraction mechanisms seen in *Dictyostelium* [195]. Receptor tyrosine kinases can stimulate increased myosin light chain phosphorylation via rho activation as a mechanism for contraction. Alternatively, increased contraction may follow formation of new focal contacts, possibly via focal adhesion kinase and src-based mechanisms [212]. This would be consistent with the temporal sequence of events that are observed, with contraction being a late step that should follow formation of new focal contacts.

4.2.6. *Negative feedback*

Negative feedback mechanisms are important for providing adaptation mechanisms for spatial sensing as well as providing responsiveness to changing stimuli. At the receptor level, phosphorylation, internalization and degradation can occur. Phosphorylation of the EGF receptor by extracellular signal-regulated kinase (ERK) and protein kinase C has been reported to reduce signaling efficiency [53]. Internalization of the EGF receptor in mesenchymal cells occurs via clathrin-coated pits as well as larger pinocytic structures [36, 118]. After endocytosis, acidification of the endosomes leads to release of the ligand and then either recycling of the receptor to the cell surface or degradation. Binding of the ubiquitin ligase cbl to phosphotyrosines on the receptor can lead to ubiquitinization and bias the receptor fate towards degradation.

The removal of the receptor from the plasma membrane is coupled with a reduction in signaling to some pathways such as the phosphoinositide

3-kinase pathway [226]. However, activation of Ras may still continue during the initial stages of endocytosis [213]. The continued activation of Ras could contribute to chemotaxis via extracellular signal-regulated kinase (ERK) activation in two ways. First, as mentioned above, ERK could continue to phosphorylate and activate calpain to weaken old focal adhesions. Second, ERK can phosphorylate the receptor itself as well as other signaling pathway components. The net effect would be a general suppression of signaling, corresponding to a global inhibitory signal.

Other negative feedback mechanisms could include activation of phosphatases or proteins which accelerate GTP hydrolysis on small G-proteins (so-called GAP proteins).

4.2.7. *Integration mechanisms for spatial gradient sensing*

Direct demonstrations of specific mechanisms for spatial sensing have yet to be made for mesenchymal cells. Using the basic scheme of localized positive signals and global negative signals for generating cell polarity, candidates for both can be proposed. For localized positive signals, following models based on studies in *Dictyostelium* and neutrophils, the positive feedback between local rac activation by phosphatidylinositol-(3, 4, 5) trisphosphate-induced activation of rac guanine nucleotide exchange proteins, and rac activation of phosphoinositide 3-kinase has the potential to generate a strong localized increase in phosphatidylinositol-(3, 4, 5) trisphosphate. A number of other factors may also contribute to highly localized phosphatidylinositol-(3, 4, 5) trisphosphate production, including reduced 3-phosphoinositide phosphatase (PTEN) amounts at the leading edge and enhancement of phosphoinositide 3-kinase activity by increased localized filamentous actin. Localized activation of phospholipase C gamma may also contribute to cell polarity. For long-range signals, extracellular signal-regulated kinase (ERK) can distribute throughout the cell and suppress signaling by phosphorylating the receptor as well as downstream targets, together with destabilization of focal adhesions. A downstream target of phosphatidylinositol-(3, 4, 5) trisphosphate, protein kinase B (Akt/PKB) may also distribute throughout the cell, as has been seen in *Dictyostelium*. Mechanical integration over longer distances could be important for mesenchymal cells due to the strong interactions with matrix produced by focal contacts. Generation of new focal contacts at the front of the cell and loss of older ones from the rest of the cell may reinforce strong signal generation from the cell due to synergistic interactions between pathways activated by receptor tyrosine kinases and focal adhesions.

4.2.8. *Coupling between receptor tyrosine kinase and G-protein-coupled receptor signaling pathways*

Intriguing connections between some receptor tyrosine kinases and G-protein-coupled receptors have been described. For the platelet-derived growth factor receptor, it appears that the G-protein-coupled receptor Edg-1 may mediate some of the responses that are observed. Inhibition of Edg-1 has effects on platelet-derived growth factor–mediated chemotactic responses, as does G-protein inhibition [96]. Conversely, the EGF receptor has been implicated in downstream signaling from other G-protein-coupled receptors, especially the lysophosphatidic acid receptor [87, 88]. A speculative possibility is that coupling between G-protein-coupled receptors and receptor tyrosine kinases may help to integrate activation of different downstream signaling systems—G-protein-coupled receptors may particularly effectively activate a phosphoinositide 3-kinase/rac positive feedback loop, while receptor tyrosine kinases may more effectively activate extracellular signal-regulated kinases (ERKs).

5. Conclusion

The most important concept developed in the last couple of years is the idea of an internal compass that aligns itself with the external gradient. This concept is most fully realized thus far in the polarized distribution of 3-phosphoinositides, which in turn are hypothesized to guide the polarization of cytoskeleton assembly. Our picture of *Dictyostelium* chemotaxis is incomplete, but the chemotactic response to cAMP is one of the best-characterized chemotactic responses and provides a framework for further studies. A strong amplification of the chemotactic gradient occurs at the level of phosphoinositide 3-kinase activation, resulting in identification of the side of the cell closest to the source of chemoattractant. This marking probably leads to increased local actin polymerization and myosin heavy chain kinase localization. The myosin heavy chain kinase delocalizes myosin II away from the front of the cell, resulting in a cell that has increased actin polymerization at the front, and increased myosin contractility at the rear. It is clear that the chemotactic machinery consists of proteins numbering in the tens or even hundreds. Working out the detailed mechanistic interactions and filling in the gaps will provide an exciting challenge for the years to come.

References

1. Abercrombie, M., Dunn, G.A. and Heath, J.P. (1977). The shape and movement of fibroblasts in culture. *Soc. Gen. Physiol. Ser.* **32**, 57–70.
2. Abercrombie, M., Heaysman, J.E. and Pegrum, S.M. (1970). The locomotion of fibroblasts in culture. I. Movements of the leading edge. *Exp. Cell Res.* **59**, 393–398.
3. Abercrombie, M., Heaysman, J.E. and Pegrum, S.M. (1970). The locomotion of fibroblasts in culture. II. "Ruffling." *Exp. Cell Res.* **60**, 437–444.
4. Adams, J.C. (2002). Regulation of protrusive and contractile cell-matrix contacts. *J. Cell Sci.* **115**, 257–265.
5. Allen, F.D., Asnes, C.F., Chang, P. *et al.* (2002). Epidermal growth factor induces acute matrix contraction and subsequent calpain-modulated relaxation. *Wound Repair Regen.* **10**, 67–76.
6. Allen, W.E., Zicha, D., Ridley, A.J. *et al.* (1998). A role for Cdc42 in macrophage chemotaxis. *J. Cell Biol.* **141**, 1147–1157.
7. Amano, M., Fukata, Y. and Kaibuchi, K. (2000). Regulation and functions of Rho-associated kinase. *Exp. Cell Res.* **261**, 44–51.
8. Arthur, W.T. and Burridge, K. (2001). RhoA inactivation by p190RhoGAP regulates cell spreading and migration by promoting membrane protrusion and polarity. *Mol. Biol. Cell* **12**, 2711–2720.
9. Azuma, T., Witke, W., Stossel, T.P. *et al.* (1998). Gelsolin is a downstream effector of rac for fibroblast motility. *EMBO J.* **17**, 1362–1370.
10. Bailly, M., Condeelis, J.S. and Segall, J.E. (1998). Chemoattractant-induced lamellipod extension. *Microsc. Res. Tech.* **43**, 433–443.
11. Bailly, M., Condeelis, J.S. and Segall, J.E. (1998). Chemoattractant-induced lamellipod extension. *Microscopy Research & Technique* **43**, 433–443.
12. Bailly, M., Ichetovkin, I., Grant, W. *et al.* (2001). The F-actin side-binding activity of the Arp2/3 complex is essential for actin nucleation and lamellipod extension. *Curr. Biol.* **11**, 620–625.
13. Bailly, M., Macaluso, F., Cammer, M. *et al.* (1999). Relationship between Arp2/3 complex and the barbed ends of actin filaments at the leading edge of carcinoma cells after epidermal growth factor stimulation. *J. Cell Biol.* **145**, 331–345.
14. Bailly, M., Yan, L., Whitesides, G.M. *et al.* (1998). Regulation of protrusion shape and adhesion to the substratum during chemotactic responses of mammalian carcinoma cells. *Exp. Cell Res.* **241**, 285–299.
15. Bailly, M., Yan, L., Whitesides, G.M. *et al.* (1998). Regulation of protrusion shape and adhesion to the substratum during chemotactic responses of mammalian carcinoma cells. *Exp. Cell Res.* **241**, 285–299.
16. Balaban, N.Q., Schwarz, U.S., Riveline, D. *et al.* (2001). Force and focal adhesion assembly: a close relationship studied using elastic micropatterned substrates. *Nat. Cell Biol.* **3**, 466–472.

17. Bamburg, J.R. (1999). Proteins of the ADF/cofilin family: essential regulators of actin dynamics. *Ann. Rev. Cell Dev. Biol.* **15**, 185–230.

18. Bear, J.E., Svitkina, T.M., Krause, M. *et al.* (2002). Antagonism between Ena/VASP proteins and actin filament capping regulates fibroblast motility. *Cell* **109**, 509–521.

19. Beningo, K.A. and Wang, Y.L. (2002). Flexible substrata for the detection of cellular traction forces. *Trends Cell Biol.* **12**, 79–84.

20. Bonner, J.T., Hirshfield, M.F. and Hall, E.M. (1971). Comparison of a leukocyte and a cellular slime mold chemotaxis test. *Exp. Cell Res.* **68**, 61–64.

21. Borisy, G.G. and Svitkina, T.M. (2000). Actin machinery: pushing the envelope. *Curr. Opin. Cell Biol.* **12**, 104–112.

22. Bosgraaf, L., Russcher, H., Smith, J.L. *et al.* (2002). A novel cGMP signaling pathway mediating myosin phosphorylation and chemotaxis in *Dictyostelium*. *EMBO J.* **21**, 4560–4570.

23. Bresnick, A.R. (1999). Molecular mechanisms of nonmuscle myosin-II regulation. *Curr. Opin. Cell Biol.* **11**, 26–33.

24. Bretscher, A., Chambers, D., Nguyen, R. *et al.* (2000). ERM-Merlin and EBP50 protein families in plasma membrane organization and function. *Ann. Rev. Cell Dev. Biol.* **16**, 113–143.

25. Bryant, P.J. (1999). Filopodia: fickle fingers of cell fate? *Curr. Biol.* **9**, R655–R657.

26. Burridge, K. and Chrzanowska-Wodnicka, M. (1996). Focal adhesions, contractility, and signaling. *Ann. Rev. Cell Dev. Biol.* **12**, 463–518.

27. Caron, E. (2002). Regulation of Wiskott-Aldrich syndrome protein and related molecules. *Curr. Opin. Cell Biol.* **14**, 82–87.

28. Chan, A.Y., Raft, S., Bailly, M. *et al.* (1998). EGF stimulates an increase in actin nucleation and filament number at the leading edge of the lamellipod in mammary adenocarcinoma cells. *J. Cell Sci.* **111**, 199–211.

29. Chan, T.O., Rodeck, U., Chan, A.M. *et al.* (2002). Small GTPases and tyrosine kinases coregulate a molecular switch in the phosphoinositide 3-kinase regulatory subunit. *Cancer Cell* **1**, 181–191.

30. Chandrasekhar, A., Wessels, D. and Soll, D.R. (1995). A mutation that depresses cGMP phosphodiesterase activity in Dictyostelium affects cell motility through an altered chemotactic signal. *Developmental Biology* **169**, 109–122.

31. Chu, Q. and Fukui, Y. (1996). *In vivo* dynamics of myosin II in *Dictyostelium* by fluorescent analogue cytochemistry. *Cell Motil. Cytoskeleton* **35**, 254–268.

32. Chubb, J.R. and Insall, R.H. (2001). *Dictyostelium*: an ideal organism for genetic dissection of Ras signaling networks. *Biochim. Biophys. Acta* **1525**, 262–271.

33. Chung, C.Y., Funamoto, S., and Firtel, R.A. (2001). Signaling pathways controlling cell polarity and chemotaxis. *Trends Biochem. Sci.* **26**, 557–566.

34. Chung, C.Y., Lee, S., Briscoe, C. *et al.* (2000). Role of Rac in controlling the actin cytoskeleton and chemotaxis in motile cells. *Proc. Natl. Acad. Sci. U.S.A.* **97**, 5225–5230.

35. Chung, C.Y., Potikyan, G. and Firtel, R.A. (2001). Control of cell polarity and chemotaxis by Akt/PKB and PI3 kinase through the regulation of PAKa. *Mol. Cell* **7**, 937–947.

36. Clague, M.J. and Urbe, S. (2001). The interface of receptor trafficking and signaling. *J. Cell Sci.* **114**, 3075–3081.

37. Clow, P.A. and McNally, J.G. (1999). *In vivo* observations of myosin II dynamics support a role in rear retraction. *Mol. Biol. Cell* **10**, 1309–1323.
38. Coluccio, L.M. (1997). Myosin I. *Am. J. Physiol.* **273**, C347–C359.
39. Comer, F.I. and Parent, C.A. (2002). PI 3-kinases and PTEN: how opposites chemoattract. *Cell* **109**, 541–544.
40. Condeelis, J. (2001). How is actin polymerization nucleated *in vivo*? *Trends Cell Biol.* **11**, 288–293.
41. Condeelis, J. and Hall, A.L. (1991). Measurement of actin polymerization and cross-linking in agonist- stimulated cells. *Methods Enzymol.* **196**, 486–496.
42. Condeelis, J.S., Wyckoff, J. and Segall, J.E. (2000). Imaging of cancer invasion and metastasis using green fluorescent protein. *Eur. J. Cancer* **36**, 1671–1680.
43. Condeelis, J.S., Wyckoff, J.B., Bailly, M. *et al.* (2001). Lamellipodia in invasion. *Semin. Cancer Biol.* **11**, 119–128.
44. Conrad, P.A., Nederlof, M.A., Herman, I.M. *et al.* (1989). Correlated distribution of actin, myosin, and microtubules at the leading edge of migrating Swiss 3T3 fibroblasts. *Cell Motil. Cytoskeleton* **14**, 527–543.
45. Cooper, J.A. and Schafer, D.A. (2000). Control of actin assembly and disassembly at filament ends. *Curr. Opin. Cell Biol.* **12**, 97–103.
46. Cox, D., Wessels, D., Soll, D.R. *et al.* (1996). Re-expression of ABP-120 rescues cytoskeletal, motility, and phagocytosis defects of ABP-120-*Dictyostelium* mutants. *Mol. Biol. Cell* **7**, 803–823.
47. Cramer, L.P. (1999). Organization and polarity of actin filament networks in cells: implications for the mechanism of myosin-based cell motility. *Biochem. Soc. Symp.* **65**, 173–205.
48. Cramer, L.P. (1999). Role of actin-filament disassembly in lamellipodium protrusion in motile cells revealed using the drug jasplakinolide. *Curr. Biol.* **9**, 1095–1105.
49. Cramer, L.P., Siebert, M. and Mitchison, T.J. (1997). Identification of novel graded polarity actin filament bundles in locomoting heart fibroblasts: implications for the generation of motile force. *J. Cell Biol.* **136**, 1287–1305.
50. Critchley, D.R. (2000). Focal adhesions—the cytoskeletal connection. *Curr. Opin. Cell Biol.* **12**, 133–139.
51. Cuevas, B.D., Lu, Y., Mao, M. *et al.* (2001). Tyrosine phosphorylation of p85 relieves its inhibitory activity on phosphatidylinositol 3-kinase. *J. Biol. Chem.* **276**, 27 455–27 461.
52. Daniel, T.O. and Abrahamson, D. (2000). Endothelial signal integration in vascular assembly. *Ann. Rev. Physiol.* **62**, 649–671.
53. Davis, R.J. (1993). The mitogen-activated protein kinase signal transduction pathway. *J. Biol. Chem.* **268**, 14 553–14 556.
54. de la Roche, M.A. and Cote, G.P. (2001). Regulation of *Dictyostelium* myosin I and II. *Biochim. Biophys. Acta* **1525**, 245–261.
55. Decave, E., Garrivier, D., Brechet, Y. *et al.* (2002). Shear flow-induced detachment kinetics of *Dictyostelium discoideum* cells from solid substrate. *Biophys. J.* **82**, 2383–2395.
56. Deery, W.J., Gao, T., Ammann, R. *et al.* (2002). A single cell density-sensing factor stimulates distinct signal transduction pathways through two different receptors. *J. Biol. Chem.* **277**, 31 972–31 979.

57. Demma, M., Warren, V., Hock, R. *et al.* (1990). Isolation of an abundant 50,000-dalton actin filament bundling protein from *Dictyostelium* amoebae. *J. Biol. Chem.* **265**, 2286–2291.

58. DePasquale, J.A. (1998). Cell matrix adhesions and localization of the vitronectin receptor in MCF-7 human mammary carcinoma cells. *Histochem. Cell Biol.* **110**, 485–494.

59. Dohlman, H.G. (2002). G proteins and pheromone signaling. *Ann. Rev. Physiol.* **64**, 129–152.

60. Dormann, D., Vasiev, B. and Weijer, C.J. (2000). The control of chemotactic cell movement during *Dictyostelium* morphogenesis. *Philos. Trans. R. Soc. Lond B Biol. Sci.* **355**, 983–991.

61. Drayer, A.L. and Van Haastert, P.J. (1992). Molecular cloning and expression of a phosphoinositide-specific phospholipase C of *Dictyostelium discoideum. J. Biol. Chem.* **267**, 18 387–18 392.

62. Duong, L.T. and Rodan, G.A. (2000). PYK2 is an adhesion kinase in macrophages, localized in podosomes and activated by beta(2)-integrin ligation. *Cell Motil. Cytoskeleton* **47**, 174–188.

63. Eddy, R.J., Han, J. and Condeelis, J.S. (1997). Capping protein terminates but does not initiate chemoattractant-induced actin assembly in *Dictyostelium. J. Cell Biol.* **139**, 1243–1253.

64. Eden, S., Rohatgi, R., Podtelejnikov, A.V. *et al.* (2002). Mechanism of regulation of WAVE1-induced actin nucleation by Rac1 and Nck. *Nature* **418**, 790–793.

65. Edmonds, B.T., Murray, J. and Condeelis, J. (1995). pH regulation of the F-actin binding properties of *Dictyostelium* elongation factor 1 alpha. *J. Biol. Chem.* **270**, 15 222–15 230.

66. Edwards, R.A. and Bryan, J. (1995). Fascins, a family of actin bundling proteins. *Cell Motil. Cytoskeleton* **32**, 1–9.

67. Egelhoff, T.T., Naismith, T.V. and Brozovich, F.V. (1996). Myosin-based cortical tension in *Dictyostelium* resolved into heavy and light chain-regulated components. *J. Muscle Res. Cell Motil.* **17**, 269–274.

68. Feldner, J.C. and Brandt, B.H. (2002). Cancer cell motility—on the road from c-erbB-2 receptor steered signaling to actin reorganization. *Exp. Cell Res.* **272**, 93–108.

69. Fey, P., Stephens, S., Titus, M.A. *et al.* (2002). SadA, a novel adhesion receptor in *Dictyostelium. J. Cell Biol.* **159**, 1109–1119.

70. Fisher, P.R., Merkl, R. and Gerisch, G. (1989). Quantitative analysis of cell motility and chemotaxis in *Dictyostelium* discoideum by using an image processing system and a novel chemotaxis chamber providing stationary chemical gradients. *J. Cell Biol.* **108**, 973–984.

71. Flanagan, L.A., Chou, J., Falet, H. *et al.* (2001). Filamin A, the Arp2/3 complex, and the morphology and function of cortical actin filaments in human melanoma cells. *J. Cell Biol.* **155**, 511–517.

72. Frame, M.C. and Brunton, V.G. (2002). Advances in Rho-dependent actin regulation and oncogenic transformation. *Curr. Opin. Genet. Dev.* **12**, 36–43.

73. Friedl, P., Borgmann, S. and Brocker, E.B. (2001). Amoeboid leukocyte crawling through extracellular matrix: lessons from the *Dictyostelium* paradigm of cell movement. *J. Leukoc. Biol.* **70**, 491–509.

74. Fukui, Y., De Hostos, E., Yumura, S. *et al.* (1999). Architectural dynamics of F-actin in eupodia suggests their role in invasive locomotion in *Dictyostelium. Exp. Cell Res.* **249**, 33–45.

75. Fukui, Y. and Inoue, S. (1997). Amoeboid movement anchored by eupodia, new actin-rich knobby feet in *Dictyostelium. Cell Motil. Cytoskeleton* **36**, 339–354.

76. Funamoto, S., Meili, R., Lee, S. *et al.* (2002). Spatial and temporal regulation of 3-phosphoinositides by PI 3-kinase and PTEN mediates chemotaxis. *Cell* **109**, 611–623.

77. Funamoto, S., Milan, K., Meili, R. *et al.* (2001). Role of phosphatidylinositol 3′ kinase and a downstream pleckstrin homology domain-containing protein in controlling chemotaxis in *Dictyostelium. J. Cell Biol.* **153**, 795–810.

78. Galbraith, C.G. and Sheetz, M.P. (1998). Forces on adhesive contacts affect cell function. *Curr. Opin. Cell Biol.* **10**, 566–571.

79. Gaskins, C., Clark, A.M., Aubry, L. *et al.* (1996). The *Dictyostelium* MAP kinase ERK2 regulates multiple, independent developmental pathways. *Genes & Development* **10**, 118–128.

80. Geiger, B., Bershadsky, A., Pankov, R. *et al.* (2001). Transmembrane crosstalk between the extracellular matrix—cytoskeleton crosstalk. *Nat. Rev. Mol. Cell Biol.* **2**, 793–805.

81. Gerisch, G. (1987). Cyclic AMP and other signals controlling cell development and differentiation in *Dictyostelium. Ann. Rev. Biochem.* **56**, 853–879.

82. Glading, A., Chang, P., Lauffenburger, D.A. *et al.* (2000). Epidermal growth factor receptor activation of calpain is required for fibroblast motility and occurs via an ERK/MAP kinase signaling pathway. *J. Biol. Chem.* **275**, 2390–2398.

83. Glading, A., Lauffenburger, D.A. and Wells, A. (2002). Cutting to the chase: calpain proteases in cell motility. *Trends Cell Biol.* **12**, 46–54.

84. Gliksman, N.R., Santoyo, G., Novak, K.D. *et al.* (2001). Myosin I phosphorylation is increased by chemotactic stimulation. *J. Biol. Chem.* **276**, 5235–5239.

85. Glockner, G., Eichinger, L., Szafranski, K. *et al.* (2002). Sequence and analysis of chromosome 2 of *Dictyostelium discoideum. Nature* **418**, 79–85.

86. Grinnell, F. (2000). Fibroblast-collagen-matrix contraction: growth-factor signaling and mechanical loading. *Trends Cell Biol.* **10**, 362–365.

87. Gschwind, A., Prenzel, N. and Ullrich, A. (2002). Lysophosphatidic acid-induced squamous cell carcinoma cell proliferation and motility involves epidermal growth factor receptor signal transactivation. *Cancer Res.* **62**, 6329–6336.

88. Gschwind, A., Zwick, E., Prenzel, N. *et al.* (2001). Cell communication networks: epidermal growth factor receptor transactivation as the paradigm for interreceptor signal transmission. *Oncogene* **20**, 1594–1600.

89. Hall, A. (1998). Rho GTPases and the actin cytoskeleton. *Science* **279**, 509–514.

90. Hall, A. and Nobes, C.D. (2000). Rho GTPases: molecular switches that control the organization and dynamics of the actin cytoskeleton. *Philos. Trans. R. Soc. Lond. B Biol. Sci.* **355**, 965–970.

91. Harvath, L., Falk, W. and Leonard, E.J. (1980). Rapid quantitation of neutrophil chemotaxis: use of a polyvinylpyrrolidone-free polycarbonate membrane in a multiwell assembly. *J. Immunol. Methods* **37**, 39–45.

92. Hauck, C.R., Hsia, D.A. and Schlaepfer, D.D. (2002). The focal adhesion kinase—a regulator of cell migration and invasion. *IUBMB. Life* **53**, 115–119.

93. Heldin, C.H. and Westermark, B. (1999). Mechanism of action and *in vivo* role of platelet-derived growth factor. *Physiol. Rev.* **79**, 1283–1316.

94. Higgs, H.N. and Pollard, T.D. (2001). Regulation of actin filament network formation through ARP2/3 complex: activation by a diverse array of proteins. *Ann. Rev. Biochem.* **70**, 649–676.

95. Hill, T.L. and Kirschner, M.W. (1982). Subunit treadmilling of microtubules or actin in the presence of cellular barriers: possible conversion of chemical free energy into mechanical work. *Proc. Natl. Acad. Sci. U.S.A.* **79**, 490–494.

96. Hobson, J.P., Rosenfeldt, H.M., Barak, L.S. *et al.* (2001). Role of the sphingosine-1-phosphate receptor EDG-1 in PDGF-induced cell motility. *Science* **291**, 1800–1803.

97. Hoffman, R.M. (2002). Green fluorescent protein imaging of tumor cells in mice. *Lab. Anim. (N.Y.)* **31**, 34–41.

98. Hubbert, C., Guardiola, A., Shao, R. *et al.* (2002). HDAC6 is a microtubule-associated deacetylase. *Nature* **417**, 455–458.

99. Hudson, L.G. and McCawley, L.J. (1998). Contributions of the epidermal growth factor receptor to keratinocyte motility. *Microsc. Res. Tech.* **43**, 444–455.

100. Hurley, J.H. and Meyer, T. (2001). Subcellular targeting by membrane lipids. *Curr. Opin. Cell Biol.* **13**, 146–152.

101. Hynes, N.E., Horsch, K., Olayioye, M.A. *et al.* (2001). The ErbB receptor tyrosine family as signal integrators. *Endocr. Relat. Cancer* **8**, 151–159.

102. Iijima, M. and Devreotes, P. (2002). Tumor suppressor PTEN mediates sensing of chemoattractant gradients. *Cell* **109**, 599–610.

103. Iijima, M., Huang, Y.E. and Devreotes, P. (2002). Temporal and spatial regulation of chemotaxis. *Dev. Cell* **3**, 469–478.

104. Insall, R., Kuspa, A., Lilly, P.J. *et al.* (1994). CRAC, a cytosolic protein containing a pleckstrin homology domain, is required for receptor and G protein-mediated activation of adenylyl cyclase in *Dictyostelium*. *J. Cell Biol.* **126**, 1537–1545.

105. Insall, R.H., Borleis, J. and Devreotes, P.N. (1996). The aimless RasGEF is required for processing of chemotactic signals through G-protein-coupled receptors in *Dictyostelium*. *Curr. Biol.* **6**, 719–729.

106. Ishizaki, T., Morishima, Y., Okamoto, M. *et al.* (2001). Coordination of microtubules and the actin cytoskeleton by the Rho effector mDia1. *Nat. Cell Biol.* **3**, 8–14.

107. Izzard, C.S. (1988). A precursor of the focal contact in cultured fibroblasts. *Cell Motil. Cytoskeleton* **10**, 137–142.

108. Izzard, C.S. and Lochner, L.R. (1976). Cell-to-substrate contacts in living fibroblasts: an interference reflexion study with an evaluation of the technique. *J. Cell Sci.* **21**, 129–159.

109. Izzard, C.S. and Lochner, L.R. (1980). Formation of cell-to-substrate contacts during fibroblast motility: an interference-reflexion study. *J. Cell Sci.* **42**, 81–116.

110. Jay, P.Y., Pasternak, C. and Elson, E.L. (1993). Studies of mechanical aspects of amoeboid locomotion. *Blood Cells* **19**, 375–386.

111. Jay, P.Y., Pham, P.A., Wong, S.A. *et al.* (1995). A mechanical function of myosin II in cell motility. *J. Cell Sci.* **108**, 387–393.

112. Jin, T., Zhang, N., Long, Y. *et al.* (2000). Localization of the G protein beta-gamma complex in living cells during chemotaxis. *Science* **287**, 1034–1036.

113. Juliano, R.L. (2002). Signal transduction by cell adhesion receptors and the cytoskeleton: functions of integrins, cadherins, selectins, and immunoglobulin-superfamily members. *Ann. Rev. Pharmacol. Toxicol.* **42**, 283–323.

114. Jung, G., Remmert, K., Wu, X. *et al.* (2001). The *Dictyostelium* CARMIL protein links capping protein and the Arp2/3 complex to type I myosins through their SH3 domains. *J. Cell Biol.* **153**, 1479–1497.

115. Jussila, L. and Alitalo, K. (2002). Vascular growth factors and lymphangiogenesis. *Physiol. Rev.* **82**, 673–700.

116. Kam, Z., Zamir, E. and Geiger, B. (2001). Probing molecular processes in live cells by quantitative multidimensional microscopy. *Trends Cell Biol.* **11**, 329–334.

117. Katoh, K., Kano, Y., Amano, M. *et al.* (2001). Rho-kinase-mediated contraction of isolated stress fibers. *J. Cell Biol.* **153**, 569–584.

118. Katzmann, D.J., Odorizzi, G. and Emr, S.D. (2002). Receptor downregulation and multivesicular-body sorting. *Nat. Rev. Mol. Cell Biol.* **3**, 893–905.

119. Kaverina, I., Krylyshkina, O., Gimona, M. *et al.* (2000). Enforced polarisation and locomotion of fibroblasts lacking microtubules. *Curr. Biol.* **10**, 739–742.

120. Kaverina, I., Krylyshkina, O. and Small, J.V. (2002). Regulation of substrate adhesion dynamics during cell motility. *Int. J. Biochem. Cell Biol.* **34**, 746–761.

121. Kessin, R.H. *Dictyostelium: Evolution, Cell Biology, and the Development of Multicellularity.* 2001. Cambridge, U.K., Cambridge University Press.

122. Kim, J.H., Soede, R.D., Schaap, P. *et al.* (1997). Phosphorylation of chemoattractant receptors is not essential for chemotaxis or termination of G-protein-coupled responses. *J. Biol. Chem.* **272**, 27313–27318.

123. Kirschner, M.W. (1980). Implications of treadmilling for the stability and polarity of actin and tubulin polymers *in vivo*. *J. Cell Biol.* **86**, 330–334.

124. Kjoller, L. and Hall, A. (1999). Signaling to Rho GTPases. *Exp. Cell Res.* **253**, 166–179.

125. Knight, B., Laukaitis, C., Akhtar, N. *et al.* (2000). Visualizing muscle cell migration *in situ*. *Curr. Biol.* **10**, 576–585.

126. Kolega, J. (1997). Asymmetry in the distribution of free versus cytoskeletal myosin II in locomoting microcapillary endothelial cells. *Exp. Cell Res.* **231**, 66–82.

127. Kreis, T.E. and Birchmeier, W. (1980). Stress fiber sarcomeres of fibroblasts are contractile. *Cell* **22**, 555–561.

128. Kuczmarski, E.R. and Spudich, J.A. (1980). Regulation of myosin self-assembly: phosphorylation of *Dictyostelium* heavy chain inhibits thick filament formation. *Proc. Natl. Acad. Sci. U.S.A.* **77**, 7292–7296.

129. Kulesa, P.M. and Fraser, S.E. (2002). Cell dynamics during somite boundary formation revealed by time-lapse analysis. *Science* **298**, 991–995.

130. Kwiatkowski, D.J. (1999). Functions of gelsolin: motility, signaling, apoptosis, cancer. *Curr. Opin. Cell Biol.* **11**, 103–108.

131. Lauffenburger, D.A. and Horwitz, A.F. (1996). Cell migration: a physically integrated molecular process. *Cell* **84**, 359–369.

132. Lee, E. and Knecht, D.A. (2001). Cytoskeletal alterations in *Dictyostelium* induced by expression of human cdc42. *Eur. J. Cell Biol.* **80**, 399–409.

133. Lee, E., Pang, K. and Knecht, D. (2001). The regulation of actin polymerization and cross-linking in *Dictyostelium*. *Biochim. Biophys. Acta.* **1525**, 217–227.

134. Lee, E., Shelden, E.A. and Knecht, D.A. (1997). Changes in actin filament organization during pseudopod formation. *Exp. Cell Res.* **235**, 295–299.

135. Lehmann, R. (2001). Cell migration in invertebrates: clues from border and distal tip cells. *Curr. Opin. Genet. Dev.* **11**, 457–463.

136. Lilly, P.J. and Devreotes, P.N. (1994). Identification of CRAC, a cytosolic regulator required for guanine nucleotide stimulation of adenylyl cyclase in *Dictyostelium*. *J. Biol. Chem.* **269**, 14123–14129.

137. Liu, Y. and Rohrschneider, L.R. (2002). The gift of Gab. FEBS Lett. **515**, 1–7.

138. Loomis, W.F. *Dictyostelium Discoideum: A Developmental System.* 1975. New York, Academic Press.

139. Machesky, L.M., Atkinson, S.J., Ampe, C. *et al.* (1994). Purification of a cortical complex containing two unconventional actins from Acanthamoeba by affinity chromatography on profilin-agarose. *J. Cell Biol.* **127**, 107–115.

140. Maeda, Y, Inouye, K. and Takeuchi, I. *Dictyostelium: A Model Organism for Cell and Developmental Biology.* 1997. Tokyo, Universal Academy Press.

141. Maehama, T. and Dixon, J.E. (1998). The tumor suppressor, PTEN/MMAC1, dephosphorylates the lipid second messenger, phosphatidylinositol 3,4,5-trisphosphate. *J. Biol. Chem.* **273**, 13375–13378.

142. Manabe, Ri R., Kovalenko, M., Webb, D.J. *et al.* (2002). GIT1 functions in a motile, multi-molecular signaling complex that regulates protrusive activity and cell migration. *J. Cell Sci.* **115**, 1497–1510.

143. May, R.C. (2001). The Arp2/3 complex: a central regulator of the actin cytoskeleton. *Cell Mol. Life Sci.* **58**, 1607–1626.

144. McClay, D.R. (1999). The role of thin filopodia in motility and morphogenesis. *Exp. Cell Res.* **253**, 296–301.

145. Meili, R., Ellsworth, C., Lee, S. *et al.* (1999). Chemoattractant-mediated transient activation and membrane localization of Akt/PKB is required for efficient chemotaxis to cAMP in *Dictyostelium*. *EMBO J.* **18**, 2092–2105.

146. Meima, M. and Schaap, P. (1999). *Dictyostelium* development-socializing through cAMP. *Semin. Cell Dev. Biol.* **10**, 567–576.

147. Meima, M.E., Biondi, R.M. and Schaap, P. (2002). Identification of a Novel Type of cGMP Phosphodiesterase That Is Defective in the Chemotactic stmF Mutants. *Mol. Biol. Cell* **13**, 3870–3877.

148. Meinhardt, H. (1999). Orientation of chemotactic cells and growth cones: models and mechanisms. *J. Cell Sci.* **112**, 2867–2874.

149. Mermall, V., Post, P.L. and Mooseker, M.S. (1998). Unconventional myosins in cell movement, membrane traffic, and signal transduction. *Science* **279**, 527–533.

150. Mikhailov, A. and Gundersen, G.G. (1998). Relationship between microtubule dynamics and lamellipodium formation revealed by direct imaging of microtubules in cells treated with nocodazole or taxol. *Cell Motil. Cytoskeleton* **41**, 325–340.

151. Millard, T.H. and Machesky, L.M. (2001). The Wiskott-Aldrich syndrome protein (WASP) family. *Trends Biochem. Sci.* **26**, 198–199.

152. Mitchison, T.J. and Cramer, L.P. (1996). Actin-based cell motility and cell locomotion. *Cell* **84**, 371–379.

153. Miyamoto, S., Teramoto, H., Coso, O.A. *et al.* (1995). Integrin function: molecular hierarchies of cytoskeletal and signaling molecules. *J. Cell Biol.* **131**, 791–805.

154. Moores, S.L., Sabry, J.H. and Spudich, J.A. (1996). Myosin dynamics in live *Dictyostelium* cells. *Proc. Natl. Acad. Sci. U.S.A.* **93**, 443–446.

155. Mullins, R.D., Heuser, J.A. and Pollard, T.D. (1998). The interaction of Arp2/3 complex with actin: nucleation, high affinity pointed end capping, and formation of branching networks of filaments. *Proc. Natl. Acad. Sci. U.S.A.* **95**, 6181–6186.

156. Myers, M.P., Pass, I., Batty, I.H. *et al.* (1998). The lipid phosphatase activity of PTEN is critical for its tumor supressor function. *Proc. Natl. Acad. Sci. U.S.A.* **95**, 13513–13518.

157. Nabi, I.R. (1999). The polarization of the motile cell. *J. Cell Sci.* **112**, 1803–1811.

158. Neri, A., Welch, D., Kawaguchi, T. *et al.* (1982). Development and biologic properties of malignant cell sublines and clones of a spontaneously metastasizing rat mammary adenocarcinoma. *J. Natl. Cancer Inst.* **68**, 507–517.

159. Newell, P.C. (1995). Signal transduction and motility of *Dictyostelium*. *Bioscience Reports* **15**, 445–462.

160. Newell, P.C., Europe Finner, G.N. and Small, N.V. (1987). Signal transduction during amoebal chemotaxis of *Dictyostelium discoideum*. *Microbiol. Sci.* **4**, 5–11.

161. Niewohner, J., Weber, I., Maniak, M. *et al.* (1997). Talin-null cells of *Dictyostelium* are strongly defective in adhesion to particle and substrate surfaces and slightly impaired in cytokinesis. *J. Cell Biol.* **138**, 349–361.

162. Nobes, C.D. and Hall, A. (1995). Rho, rac, and cdc42 GTPases regulate the assembly of multimolecular focal complexes associated with actin stress fibers, lamellipodia, and filopodia. *Cell* **81**, 53–62.

163. Noegel, A.A. and Luna, J.E. (1995). The *Dictyostelium* cytoskeleton. *Experientia.* **51**, 1135–1143.

164. Omelchenko, T., Vasiliev, J.M., Gelfand, I.M. *et al.* (2002). Mechanisms of polarization of the shape of fibroblasts and epitheliocytes: Separation of the roles of microtubules and Rho-dependent actin-myosin contractility. *Proc. Natl. Acad. Sci. U.S.A.* **99**, 10452–10457.

165. Owen, C.H., DeRosier, D.J. and Condeelis, J. (1992). Actin cross-linking protein EF-1a of *Dictyostelium discoideum* has a unique bonding rule that allows square-packed bundles. *Journal of Structural Biology* **109**, 248–254.

166. Palacek, S.P., Huttenlocher, A., Horwitz, A.F. *et al.* (1998). Physical and biochemical regulation of integrin release during rear detachment of migrating cells. *J. Cell Sci.* **111**, 929–940.

167. Palazzo, A.F., Cook, T.A., Alberts, A.S. *et al.* (2001). mDia mediates Rho-regulated formation and orientation of stable microtubules. *Nat. Cell Biol.* **3**, 723–729.

168. Palecek, S.P., Horwitz, A.F. and Lauffenburger, D.A. (1999). Kinetic model for integrin-mediated adhesion release during cell migration. *Ann. Biomed. Eng.* **27**, 219–235.

169. Pantaloni, D., Le Clainche, C. and Carlier, M.F. (2001). Mechanism of actin-based motility. *Science* **292**, 1502–1506.

170. Parent, C.A. and Devreotes, P.N. (1999). A cell's sense of direction. *Science* **284**, 765–770.

171. Parsons, J.T., Martin, K.H., Slack, J.K. *et al.* (2000). Focal adhesion kinase: a regulator of focal adhesion dynamics and cell movement. *Oncogene* **19**, 5606–5613.

172. Peracino, B., Borleis, J., Jin, T. *et al.* (1998). G protein beta subunit-null mutants are impaired in phagocytosis and chemotaxis due to inappropriate regulation of the actin cytoskeleton. *J. Cell Biol.* **141**, 1529–1537.

173. Pollard, T.D., Blanchoin, L. and Mullins, R.D. (2000). Molecular mechanisms controlling actin filament dynamics in nonmuscle cells. *Ann. Rev. Biophys. Biomol. Struct.* **29**, 545–576.

174. Ponte, E., Rivero, F., Fechheimer, M. *et al.* (2000). Severe developmental defects in *Dictyostelium* null mutants for actin-binding proteins. *Mech. Dev.* **91**, 153–161.

175. Postma, M. and Van Haastert, P.J. (2001). A diffusion-translocation model for gradient sensing by chemotactic cells. *Biophys. J.* **81**, 1314–1323.

176. Prenzel, N., Fischer, O.M., Streit, S. *et al.* (2001). The epidermal growth factor receptor family as a central element for cellular signal transduction and diversification. *Endocr. Relat. Cancer* **8**, 11–31.

177. Rahmsdorf, H.J., Malchow, D. and Gerisch, G. (1978). Cyclic AMP-induced phosphorylation in *Dictyostelium* of a polypeptide comigrating with myosin heavy chains. *FEBS Lett.* **88**, 322–326.

178. Rappel, W.J., Thomas, P.J., Levine, H. *et al.* (2002). Establishing direction during chemotaxis in eukaryotic cells. *Biophys. J.* **83**, 1361–1367.

179. Rifkin, J.L. (2001). Folate reception by vegetative *Dictyostelium discoideum* amoebae: distribution of receptors and trafficking of ligand. *Cell Motil. Cytoskeleton* **48**, 121–129.

180. Rodriguez-Viciana, P., Warne, P.H., Vanhaesebroeck, B. *et al.* (1996). Activation of phosphoinositide 3-kinase by interaction with Ras and by point mutation. *EMBO J.* **15**, 2442–2451.

181. Roelofs, J. and Van Haastert, P.J. (2002). Characterization of two unusual guanylyl cyclases from *Dictyostelium*. *J. Biol. Chem.* **277**, 9167–9174.

182. Ronnstrand, L. and Heldin, C.H. (2001). Mechanisms of platelet-derived growth factor-induced chemotaxis. *Int. J. Cancer* **91**, 757–762.

183. Rubin, H. and Ravid, S. (2002). Polarization of MHC-PKC in chemotaxing *Dictyostelium* cells. *J. Biol. Chem.* **277**, 36005–36008.

184. Rubin, H. and Ravid, S. (2002). Polarization of myosin II heavy chain-protein kinase C in chemotaxing *Dictyostelium* cells. *J. Biol. Chem.* **277**, 36005–36008.

185. Sah, V.P., Seasholtz, T.M., Sagi, S.A. *et al.* (2000). The role of Rho in G protein-coupled receptor signal transduction. *Ann. Rev. Pharmacol. Toxicol.* **40**, 459–489.

186. Saitoh, T., Takemura, S., Ueda, K. *et al.* (2001). Differential localization of nonmuscle myosin II isoforms and phosphorylated regulatory light chains in human MRC-5 fibroblasts. *FEBS Lett.* **509**, 365–369.

187. Sameshima, M., Imai, Y., Fujimoto, H. *et al.* (1992). Correlation of axenic linkage groups with the position of the microtubule-organizing center in aggregating *Dictyostelium*. *Cell Structure & Function* **17**, 417–425.

188. Sameshima, M., Imai, Y. and Hashimoto, Y. (1988). The position of the microtubule-organizing center relative to the nucleus is independent of the direction of cell migration in *Dictyostelium discoideum*. *Cell Motil. Cytoskeleton* **9**, 111–116.

189. Sanders, L.C., Matsumura, F., Bokoch, G.M. *et al.* (1999). Inhibition of myosin light chain kinase by p21-activated kinase. *Science* **283**, 2083–2085.

190. Sastry, S.K. and Burridge, K. (2000). Focal adhesions: a nexus for intracellular signaling and cytoskeletal dynamics. *Exp. Cell Res.* **261**, 25–36.

191. Satoh, S. and Tominaga, T. (2001). mDia-interacting protein acts downstream of Rho-mDia and modifies Src activation and stress fiber formation. *J. Biol. Chem.* **276**, 39 290–39 294.

192. Schafer, D.A., Jennings, P.B. and Cooper, J.A. (1996). Dynamics of capping protein and actin assembly in vitro: uncapping barbed ends by polyphosphoinositides. *J. Cell Biol.* **135**, 169–179.

193. Schindl, M., Wallraff, E., Deubzer, B. *et al.* (1995). Cell-substrate interactions and locomotion of *Dictyostelium* wild-type and mutants defective in three cytoskeletal proteins: a study using quantitative reflection interference contrast microscopy. *Biophys. J.* **68**, 1177–1190.

194. Schlessinger, J. (2000). Cell signaling by receptor tyrosine kinases. *Cell* **103**, 211–225.

195. Schoenwaelder, S.M. and Burridge, K. (1999). Bidirectional signaling between the cytoskeleton and integrins. *Curr. Opin. Cell Biol.* **11**, 274–286.

196. Schwartz, M.A. (2001). Integrin signaling revisited. *Trends Cell Biol.* **11**, 466–470.

197. Schwarz, E.C., Neuhaus, E.M., Kistler, C. *et al.* (2000). *Dictyostelium* myosin IK is involved in the maintenance of cortical tension and affects motility and phagocytosis. *J. Cell Sci.* **113** (Pt 4), 621–633.

198. Segall, J.E. (1988). Quantification of motility and area changes of *Dictyostelium discoideum* amoebae in response to chemoattractants. *J. Muscle Res. Cell Motil.* **9**, 481–490.

199. Segall, J.E. (1990). Mutational studies of amoeboid chemotaxis using *Dictyostelium discoideum*. Armitage, J.P. and Lackie, J.M. Biology of the chemotactic response. 241–272. Cambridge, Cambridge University Press.

200. Segall, J.E. (1992). Behavioral responses of streamer F mutants of *Dictyostelium discoideum*: effects of cyclic GMP on cell motility. *J. Cell Sci.* **101**, 589–597.

201. Segall, J.E. and Gerisch, G. (1989). Genetic approaches to cytoskeleton function and the control of cell motility. *Curr. Opin. Cell Biol.* **1**, 44–50.

202. Segall, J.E., Tyerech, S., Boselli, L. *et al.* (1996). EGF stimulates lamellipod extension in metastatic mammary adenocarcinoma cells by an actin-dependent mechanism. *Clin. Exp. Met.* **14**, 61–72.

203. Sells, M.A., Boyd, J.T. and Chernoff, J. (1999). p21-activated kinase 1 (Pak1) regulates cell motility in mammalian fibroblasts. *J. Cell Biol.* **145**, 837–849.

204. Sheetz, M.P., Felsenfeld, D., Galbraith, C.G. *et al.* (1999). Cell migration as a five-step cycle. *Biochem. Soc. Symp.* **65**, 233–243.

205. Shestakova, E.A., Singer, R.H. and Condeelis, J. (2001). The physiological significance of beta-actin mRNA localization in determining cell polarity and directional motility. *Proc. Natl. Acad. Sci. U.S.A.* **98**, 7045–7050.

206. Shimaoka, M., Takagi, J. and Springer, T.A. (2002). Conformational regulation of integrin structure and function. *Ann. Rev. Biophys. Biomol. Struct.* **31**, 485–516.

207. Shiraha, H., Glading, A., Chou, J. *et al.* (2002). Activation of m-calpain (calpain II) by epidermal growth factor is limited by protein kinase A phosphorylation of m-calpain. *Mol. Cell Biol.* **22**, 2716–2727.

208. Simson, R., Wallraff, E., Faix, J. *et al.* (1998). Membrane bending modulus and adhesion energy of wild-type and mutant cells of *Dictyostelium* lacking talin or cortexillins. *Biophys. J.* **74**, 514–522.

209. Small, J.V., Stradal, T., Vignal, E. *et al.* (2002). The lamellipodium: where motility begins. *Trends Cell Biol.* **12**, 112–120.
210. Sobko, A., Ma, H. and Firtel, R.A. (2002). Regulated SUMOylation and ubiquitination of DdMEK1 is required for proper chemotaxis. *Dev. Cell* **2**, 745–756.
211. Soll, D.R., Wessels, D., Voss, E. *et al.* (2001). Computer-assisted systems for the analysis of amoeboid cell motility. *Methods Mol. Biol.* **161**, 45–58.
212. Somlyo, A.P. and Somlyo, A.V. (2000). Signal transduction by G-proteins, rho-kinase and protein phosphatase to smooth muscle and non-muscle myosin II. *J. Physiol.* 522 Pt **2**, 177–185.
213. Sorkin, A. (2001). Internalization of the epidermal growth factor receptor: role in signaling. *Biochem. Soc. Trans.* **29**, 480–484.
214. Steimle, P.A., Licate, L., Cote, G.P. *et al.* (2002). Lamellipodial localization of *Dictyostelium* myosin heavy chain kinase A is mediated via F-actin binding by the coiled-coil domain. *FEBS Lett.* **516**, 58–62.
215. Steimle, P.A., Yumura, S., Cote, G.P. *et al.* (2001). Recruitment of a myosin heavy chain kinase to actin-rich protrusions in *Dictyostelium*. *Curr. Biol.* **11**, 708–713.
216. Stossel, T.P., Condeelis, J., Cooley, L. *et al.* (2001). Filamins as integrators of cell mechanics and signaling. *Nat. Rev. Mol. Cell Biol.* **2**, 138–145.
217. Svitkina, T.M., Verkhovsky, A.B., McQuade, K.M. *et al.* (1997). Analysis of the actin-myosin II system in fish epidermal keratocytes: mechanism of cell body translocation. *J. Cell Biol.* **139**, 397–415.
218. Takenawa, T. and Itoh, T. (2001). Phosphoinositides, key molecules for regulation of actin cytoskeletal organization and membrane traffic from the plasma membrane. *Biochim. Biophys. Acta.* **1533**, 190–206.
219. Takenawa, T. and Miki, H. (2001). WASP and WAVE family proteins: key molecules for rapid rearrangement of cortical actin filaments and cell movement. *J. Cell Sci.* **114**, 1801–1809.
220. Traver, D. and Zon, L.I. (2002). Walking the walk: migration and other common themes in blood and vascular development. *Cell* **108**, 731–734.
221. Tuxworth, R.I., Weber, I., Wessels, D. *et al.* (2001). A role for myosin VII in dynamic cell adhesion. *Curr. Biol.* **11**, 318–329.
222. Van Haastert, P.J. and Kuwayama, H. (1997). cGMP as second messenger during *Dictyostelium* chemotaxis. *FEBS Lett.* **410**, 25–28.
223. Vanhaesebroeck, B. and Waterfield, M.D. (1999). Signaling by distinct classes of phosphoinositide 3-kinases. *Exp. Cell Res.* **253**, 239–254.
224. Varnum Finney, B., Schroeder, N.A. and Soll, D.R. (1988). Adaptation in the motility response to cAMP in *Dictyostelium discoideum*. *Cell Motil. Cytoskeleton* **9**, 9–16.
225. Vicker, M.G. (1994). The regulation of chemotaxis and chemokinesis in *Dictyostelium* amoebae by temporal signals and spatial gradients of cyclic AMP. *J. Cell Sci.* **107**, 659–667.
226. Vieira, A.V., Lamaze, C. and Schmid, S.L. (1996). Control of EGF receptor signaling by clathrin-mediated endocytosis. *Science* **274**, 2086–2089.
227. Wang, Y., Liu, J. and Segall, J.E. (1998). MAP kinase function in amoeboid chemotaxis. *J. Cell Sci.* **111**, 373–383.
228. Warburton, D., Schwarz, M., Tefft, D. *et al.* (2000). The molecular basis of lung morphogenesis. *Mech. Dev.* **92**, 55–81.

229. Wear, M.A., Schafer, D.A. and Cooper, J.A. (2000). Actin dynamics: assembly and disassembly of actin networks. *Curr. Biol.* **10**, R891–R895.

230. Webb, D.J., Parsons, J.T. and Horwitz, A.F. (2002). Adhesion assembly, disassembly and turnover in migrating cells—over and over and over again. *Nat. Cell Biol.* **4**, E97–E100.

231. Weber, I., Wallraff, E., Albrecht, R. *et al.* (1995). Motility and substratum adhesion of *Dictyostelium* wild-type and cytoskeletal mutant cells: a study by RICM/bright-field double-view image analysis. *J. Cell Sci.* **108**, 1519–1530.

232. Weiner, O.D. (2002). Regulation of cell polarity during eukaryotic chemotaxis: the chemotactic compass. *Curr. Opin. Cell Biol.* **14**, 196–202.

233. Weiner, O.D., Neilsen, P.O., Prestwich, G.D. *et al.* (2002). A PtdInsP(3)- and Rho GTPase-mediated positive feedback loop regulates neutrophil polarity. *Nat. Cell Biol.* **4**, 509–513.

234. Welch, D.R., Neri, A. and Nicolson, G.L. (1983). Comparison of "spontaneous" and "experimental" metastasis using rat 13 762 mammary adenocarcinoma metastatic cell clones. *Inv. Met.* **3**, 65–80.

235. Wells, A. (2000). Tumor invasion: role of growth factor-induced cell motility. *Adv. Cancer Res.* **78**, 31–101.

236. Wessels, D., Schroeder, N.A., Voss, E. *et al.* (1989). cAMP-mediated inhibition of intracellular particle movement and actin reorganization in *Dictyostelium. J. Cell Biol.* **109**, 2841–2851.

237. Wessels, D. and Soll, D.R. (1990). Myosin II heavy chain null mutant of *Dictyostelium* exhibits defective intracellular particle movement. *J. Cell Biol.* **111**, 1137–1148.

238. Wessels, D., Soll, D.R., Knecht, D. *et al.* (1988). Cell motility and chemotaxis in *Dictyostelium* amebae lacking myosin heavy chain. *Dev. Biol.* **128**, 164–177.

239. Wessels, D., Vawter-Hugart, H., Murray, J. *et al.* (1994). Three-dimensional dynamics of pseudopod formation and the regulation of turning during the motility cycle of *Dictyostelium. Cell Motil. Cytoskeleton* **27**, 1–12.

240. Wherlock, M. and Mellor, H. (2002). The Rho GTPase family: a Racs to Wrchs story. *J. Cell Sci.* **115**, 239–240.

241. Witke, W., Sharpe, A.H., Hartwig, J.H. *et al.* (1995). Hemostatic, inflammatory, and fibroblast responses are blunted in mice lacking gelsolin. *Cell* **81**, 41–51.

242. Wittmann, T. and Waterman-Storer, C.M. (2001). Cell motility: can Rho GTPases and microtubules point the way? *J. Cell Sci.* **114**, 3795–3803.

243. Wong, E., Yang, C., Wang, J. *et al.* (2002). Disruption of the gene encoding the cell adhesion molecule DdCAD-1 leads to aberrant cell sorting and cell-type proportioning during *Dictyostelium* development. *Development* **129**, 3839–3850.

244. Wu, L., Valkema, R., Van Haastert, P.J. *et al.* (1995). The G protein beta subunit is essential for multiple responses to chemoattractants in *Dictyostelium. J. Cell Biol.* **129**, 1667–1675.

245. Wu, X., Jung, G. and Hammer, J.A., III (2000). Functions of unconventional myosins. *Curr. Opin. Cell Biol.* **12**, 42–51.

246. Wymann, M. and Arcaro, A. (1994). Platelet-derived growth factor-induced phosphatidylinositol 3-kinase activation mediates actin rearrangements in fibroblasts. *Biochem. J.* **298** Pt 3, 517–520.

247. Xiao, Z., Yao, Y., Long, Y. *et al.* (1999). Desensitization of G-protein-coupled receptors. agonist-induced phosphorylation of the chemoattractant receptor cAR1 lowers its intrinsic affinity for cAMP. *J. Biol. Chem.* **274**, 1440–1448.
248. Xiao, Z., Zhang, N., Murphy, D.B. *et al.* (1997). Dynamic distribution of chemoattractant receptors in living cells during chemotaxis and persistent stimulation. *J. Cell Biol.* **139**, 365–374.
249. Xie, H., Pallero, M.A., Gupta, K. *et al.* (1998). EGF receptor regulation of cell motility: EGF induces disassembly of focal adhesions independently of the motility-associated PLCgamma signaling pathway. *J. Cell Sci.* **111**, 615–624.
250. Zalevsky, J., Lempert, L., Kranitz, H. *et al.* (2001). Different WASP family proteins stimulate different Arp2/3 complex-dependent actin-nucleating activities. *Curr. Biol.* **11**, 1903–1913.
251. Zamir, E. and Geiger, B. (2001). Molecular complexity and dynamics of cell-matrix adhesions. *J. Cell Sci.* **114**, 3583–3590.
252. Zamir, E., Katz, M., Posen, Y. *et al.* (2000). Dynamics and segregation of cell-matrix adhesions in cultured fibroblasts. *Nat. Cell Biol.* **2**, 191–196.
253. Zhang, H., Wessels, D., Fey, P. *et al.* (2002). Phosphorylation of the myosin regulatory light chain plays a role in motility and polarity during *Dictyostelium* chemotaxis. *J. Cell Sci.* **115**, 1733–1747.
254. Zheng, J.Q., Wan, J.J. and Poo, M.M. (1996). Essential role of filopodia in chemotropic turning of nerve growth cone induced by a glutamate gradient. *Journal of Neuroscience* **16**, 1140–1149.
255. Zhou, K., Takegawa, K., Emr, S.D. *et al.* (1995). A phosphatidylinositol (PI) kinase gene family in *Dictyostelium discoideum*: biological roles of putative mammalian p110 and yeast Vps34p PI 3-kinase homologs during growth and development. *Mol. Biol. Cell.* **15**, 5645–5656.
256. Zicha, D., Dunn, G.A. and Jones, G. (1997). Analyzing chemotaxis using the Dunn direct-viewing chamber. Pollard, J.W. and Walker, J.M. *Basic Cell Culture Protocols.* **2**(36), 449–458. Totowa, N.J., Humana Press.
257. Zigmond, S.H., Foxman, E.H. and Segall, J.E. (1998). Chemotaxis Assays for Eukaryotic Cells, in *Current Protocols in Cell Biology* (J.S. Bonifacino, J.B. Harford, M. Dasso, J. Lippincott-Schwartz and K.M. Yamada, eds.), New York, John Wiley & Sons.

Physiology and Molecular Mechanisms of Chemotaxis of White Blood Cells*

1. Introduction

Migration is an essential function of cells of the immune system. White cells forming in the bone marrow migrate to sites of release into the bloodstream. Monocytes migrate from the blood into tissues where they develop into resident macrophages. Lymphocytes circulate in the bloodstream and migrate to and within the lymph nodes where further development occurs, and then re-enter the circulation. Neutrophils, monocytes, and lymphocytes migrate from the vasculature through tissues to sites of infection. Thus the migratory capabilities of the white cells are essential for their functions. All of these cells migrate by a process referred to as amoeboid chemotaxis, which means they crawl on a substratum or through a matrix.

White cells are physiologically and pathologically important as mediators of the inflammatory response (reviewed in [288]). Neutrophils are the first of the inflammatory cells to arrive at sites of injury. The neutrophils' phagocytic capabilities and the release of proteinases and oxidants are important for fighting bacterial infections as well as clearing dead cell

---

* Written by Geneva M. Omann, Surgery and Biological Chemistry, University of Michigan/VA Medical Center, 2215 Fuller Road, Ann Arbor, MI 48105, USA.

debris. Neutrophils accumulate at the site of injury within the first 30–60 min of the initial insult. Thus neutrophils are important for initiating and augmenting the inflammatory cascade. Within 4–5 h, monocytes and lymphocytes begin to accumulate at the site of inflammation. Monocytes contribute phagocytic activity and lymphocytes mediate antibody-specific lysis of cells. All of these cell types release additional inflammatory mediators that can augment the inflammatory response. In addition, factors are released that promote the resolution of inflammation and wound healing. In this chapter, neutrophil migration will be discussed as a model of white blood cell migration. Many aspects of migration are similar for all white blood cells, although significant differences, particularly in signaling systems, are also evident. Recent reviews of migration in other leukocytes are available (e.g., [74, 147, 159, 254, 425]).

2. Neutrophil Chemoattractants

The migratory function of neutrophils is mediated by chemoattractants that are released from the site of injury. The structures of these molecules share little in common, with structures ranging from lipids to short peptides to polypeptides. Neutrophils (and monocytes) have receptors for bacterial peptides that serve as detectors for bacterial infections. These peptides, N-formyl peptides, are short peptide sequences that have a formyl group on the nitrogen of an N-terminal methionyl residue. Bacterial protein synthesis begins with N-formyl methionine, hence this is a common bacterial peptide. N-formyl-met-leu-phe is a bacterial tripeptide [231] commonly used for studies of neutrophil activation and migration. However, inflammation can occur in the absence of a bacterial infection; thus, other factors are also released from or activated by damaged tissues. The tissues themselves may release N-formyl peptide sequences, since the mitochondrial protein synthesis machinery, being evolutionarily of bacterial origin, also uses N-formyl-methionine to initiate protein synthesis [57]. Tissues and neutrophils at the inflammatory site release additional chemoattractants, including the lipids leukotriene B_4 and platelet-activating factor, and chemokines including the polypeptide interleukin-8 [21, 109]. Activation of the complement system by pathogens releases the polypeptide chemoattractant, C5a. Together these chemoattractants further recruit inflammatory cells to the injury/infection site. Each of these chemoattractant has its own receptor. Although the structures of these chemoattractants bear little similarity, the receptors they bind are structurally similar G-protein-coupled receptors [257, 258, 440]. These include N-formyl peptide receptor, leukotriene B_4 receptor, C5a receptor, interleukin-8 receptor, and platelet-activating factor receptor, all of which mediate recruitment of neutrophils to inflammatory sites. Table 1 summarizes the properties of known neutrophil chemoattractants.

In addition to migration, other responses are activated that combine to carry out the inflammatory functions of neutrophils (Figure 1). Once at the site of inflammation, neutrophils release proteinases, hydrolases, and bactericidal peptides from preformed granules and an NADPH oxidase is assembled and activated to release superoxide. These compounds function to kill invading organisms. Lipid inflammatory mediators such as leukotrienes and prostaglandins are synthesized via enzymatic reactions. In addition, production of cytokines and growth factors, stimulated at the level of gene expression, recruit additional immune cells to the inflamed site and promote healing. All of these responses must be highly coordinated for proper clearance of infection and healing to occur. Yet not all of these responses are stimulated equally by all chemoattractants. N-formyl peptides, leukotriene B_4, C5a, interleukin-8, and platelet-activating factor are all potent stimulators of chemotaxis. Yet leukotriene B_4, platelet-activating factor, and IL-8 are relatively weak activators of the NADPH oxidase and degranulation [276, 284, 403].

3. The Transition from Circulating to Migrating Cells

In the physiological setting of the blood stream, circulating neutrophils are normally unactivated and nonmotile. When viewed microscopically, they exhibit a behavior called "rolling" in which they repeatedly attach to and release from the vascular endothelium (the cells lining the blood vessels). When an injury (or some source of inflammation) occurs, the neutrophils attach tightly to the vascular endothelium near the site of inflammation, cross the endothelial cell layer without disrupting its barrier function (referred to as diapedesis), and migrate through the tissues to the site of inflammation. Rolling, tight attachment, and diapedesis are mediated by sequentially regulated specific molecules on neutrophils and endothelial cells, schematically represented in Figure 2 (reviewed in [52, 193, 366, 380, 406]).

The rolling behavior of neutrophils along the vascular endothelium is mediated by reversible binding of a constitutively expressed surface

Table 1. Neutrophil chemoattractants and their receptors.

Receptor[a]	Chemoattractant	Comments and references
		"Classic" chemoattractants
FPR	N-formyl-met-leu-phe	N-formyl-met-leu-phe binds with high affinity ($K_d \sim 1$ nM) to FPR [440]. Many N-formyl peptide sequences bind this receptor with K_ds in the nanomolar and sub-nanomolar concentration ranges [107, 408].
FPRL1	N-formyl-met-leu-phe	N-formyl-met-leu-phe binds with low affinity ($K_d \sim 400$ nM) to FPRL1 [440]. The synthetic peptide Trp-Lys-Tyr-Met-Val-Met-NH_2 is a chemoattractant that binds FPRL1 with high affinity, but not FPR [60]. FPRL1 has also been identified as the lipoxin A_4 receptor, although lipoxin A_4 is NOT a chemoattractant [103].
C5aR	C5a	C5a is a fragment from complement component C5 released during complement activation [42, 96].
C3aR	C3a	C3a is a fragment from complement component C3 released during complement activation [42, 233].
PAFR	Platelet-activating factor (PAF)	The PAFR is found in a broad range of cell types and tissues, including neutrophils and PAF binds with high affinities are in the low nanomolar range [42, 311].
BLT1	Leukotriene B_4 (LTB$_4$)	BLT1 has high affinity for LTB$_4$ ($K_d \sim 1$ nM) and is expressed mainly in leukocytes [441].
BLT2	LTB$_4$	BLT2 is 45% identical in amino acids to BLT1, but binds LTB$_4$ with lower affinity ($K_d \sim 20$ nM) and is found in a wide variety of tissues, including leukocytes [441].
		CXC chemokines
IL8RA (CXCR1)	Interleukin-8 (IL-8)	Nomenclature for chemokines is defined elsewhere [22, 460]. IL8RA is specific for IL-8 [257, 258, 440].

(Continued)

Table 1. Continued

Receptor[a]	Chemoattractant	Comments and references
IL8RB (CXCR2)	IL-8, GROα, GROβ, GROγ, NAP-2, ENA-78	IL8RB binds IL-8 and the several additional chemokines listed with binding constants typically in the region of 10 nM [22, 257, 258, 440].
LESTR/fusin (CXCR4)	Stromal cell-derived factor 1 (SDF-1)	SDF-1 is produced by bone marrow stromal cells [35, 262].
		Other
TβR type I and II	Transforming growth factor β (TGFβ) 1, 2, and 3	The known receptors for TGFβs are single membrane spanning, serine/threonine kinases [76]. TGFβs are the most potent neutrophil chemoattractants known, with maximum chemotaxis elicited by 40 fM ligand [44, 118, 294, 328].
NK1R?	Substance P (SP)	SP induces neutrophil chemotaxis under agarose and transmigration through fibroblast monolayers [118, 160]. Substance P is a neuropeptide known to have high affinity for the neurokinin 1 receptor, a G-protein coupled receptor [400]. However it is not clear if this receptor mediates its activity in neutrophils. Although NK1R antagonists block many of the effects of SP on neutrophils (e.g., [160]), there is evidence that SP binds to the N-formyl peptide receptor [230], and alone it can penetrate a membrane bilayer and activate G-proteins [252].

[a]All receptors listed are G-protein-coupled receptors except TβR type I and II.

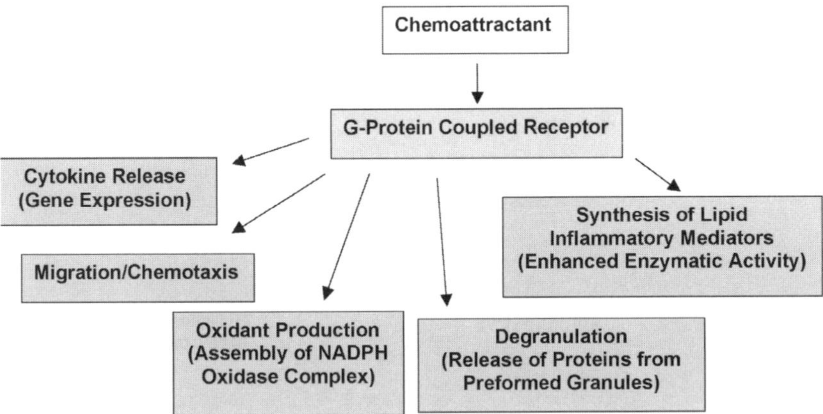

Figure 1. Inflammatory functions of neutrophils activated by G-protein-coupled chemoattractant receptors.

protein on the neutrophil, L-selectin, and its constitutively expressed ligands on the surface of the endothelial cells, sialomucins. When an injury occurs, inflammatory mediators released from the injury site (complement products, oxygen radicals, various cytokines) activate the endothelium to express higher affinity receptors (P-selectin) that bind to constitutively expressed neutrophil surface proteins, P-selectin glyco-protein ligand-1 (PSGL-1, CD162). A second endothelial cell protein, E-selectin, is also expressed upon activation and binds to E-selectin lig-and 1 (ESL-1) on neutrophils. These binding interactions result in slowed neutrophil rolling, and tethering of neutrophils to the activated endothelium. Tethering is followed by firm adhesion, either in response to inflammatory cytokines or as a result of selectin-mediated activation. Firm adhesion is a consequence of activation of a class of adhesion mol-ecules called integrins on the neutrophils. Mac-1, LFA-1, and p150,95 are neutrophil integrins of the subclass $\beta2$ (CD18), important for pro-moting neutrophil trafficking. These activated integrins bind ligands on the endothelial cells, intercellular adhesion molecules (ICAM-1, -2, and -3), which are expressed at low levels on unactivated endothelial cells, but are markedly increased upon activation by inflammatory mediators. Evidence indicates that passage of neutrophils through the endothelial cell layer occurs preferentially at tricellular junctions where the mole-cule, platelet-endothelial cell adhesion molecule-1 (PECAM-1) is con-centrated. PECAM-1 is also evenly distributed on the surface of circulating neutrophils and can bind to PECAM-1 on the endothelial cells. Thus PECAM-1 may be important for directing neutrophils to the

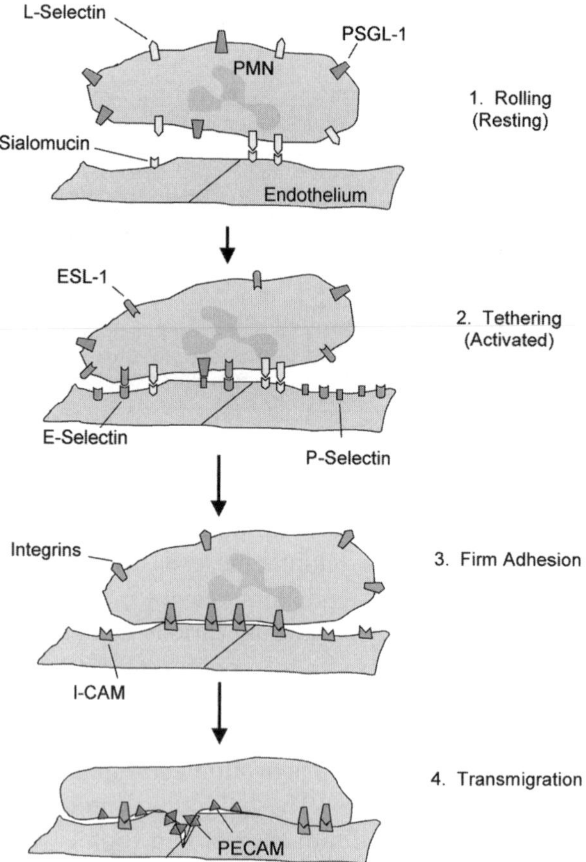

Figure 2. Sequential steps in attachment and transendothelial migration of neutrophils. (1) Rolling: circulating (resting) neutrophils transiently associate with the endothelium by reversible binding of constitutively expressed L-selectin on neutrophils and sialomucins on the endothelium. (2) Tethering: activation results in upregulation of endothelial P selectin and E selectin and neutrophil ESL-1. Binidng of P-selectin to PSGL-1 on neutrophils and E-selectin to ESL-1 results in tethering of the neutrophil to the endothelium. (3) Firm adhesion: integrins upregulated to the surface of neutrophils serve as high-affinity receptors for ICAMs upregulated to the surface of endothelial cells. (4) PECAM-1 on neutrophils binds PECAM-1 at tricellular junctions of endothelial cells mediating transmigration of neutrophils across the endothelium.

portals for egress through the endothelium. Alternatively, evidence also exists for transcellular transmigration of neutrophils, that is migration through passages that traverse endothelial cell cytoplasm, although the mechanisms and molecules involved have not been clearly delineated. Both mechanisms may be physiologically important.

4. Neutrophil Migration

4.1. *Overview*

Once neutrophils have exited the vasculature, they crawl through tissues to the site of infection. Much of what is understood about neutrophil migration comes from *in vitro* microscopy studies using purified cells migrating in two dimensions, and the main features of the process are analogous to amoeboid chemotaxis (see Chapter 4 and [207, 242]). Migrating adherent neutrophils develop a prominent polar structure with thin, veil-like protrusions (lamellipodia) at the leading edge and a tail (uropod) that is pulled along behind (Figure 3) [171, 212, 322, 449]. Movement occurs by continual extension of this lamellipodium and retraction of the uropod. The cell body appears to undulate up and down as extension of the lamellipodium causes the cell body to flatten, and uropod retraction causes the cell body to rise up [102, 259, 321]. Cytoplasmic granules can be seen moving from the rear toward the front of the cell body [102, 322, 346]. A constriction ring can be seen propagating from front to tail [346, 354]. Formation of the lamellipodium requires an increase in the surface area of the cell, and this additional membrane comes from fusion of intracellular granules with the plasma membrane [138].

The lamellipodium of the cell is high in polymerized actin and free of organelles and microtubules [64, 75, 101, 172, 209, 225, 245, 280, 281, 339]. Actin polymerization is necessary for protrusion of the lamellipodium and for migration to occur [413, 447]. Moreover, constant signaling is required to maintain the polar morphology, since removal of the chemoattractant results in the loss of polarity [452]. If while a cell is migrating the stimulus is removed, the lamellipodial F-actin quickly (within 10 s) depolymerizes and the lamellipodium retracts [58]. Likewise, if while a cell is migrating, cytochalasin is added to prevent actin polymerization, the lamellipodial F-actin depolymerizes and the lamellipodial structure retracts [58]. Thus signaling for a sustained F-actin structure in the lamellipodium is required for polarization and migration to occur.

Neutrophils become polarized in response to chemoattractants even when stimulated in suspension. Thus the mechanisms for cell polarization do not require adherence to a substratum, however migration does. In order for movement to occur, traction must be generated against the substratum. For neutrophils migrating on a two-dimensional surface, adhesion molecules called integrins are the treads for traction [121, 179].

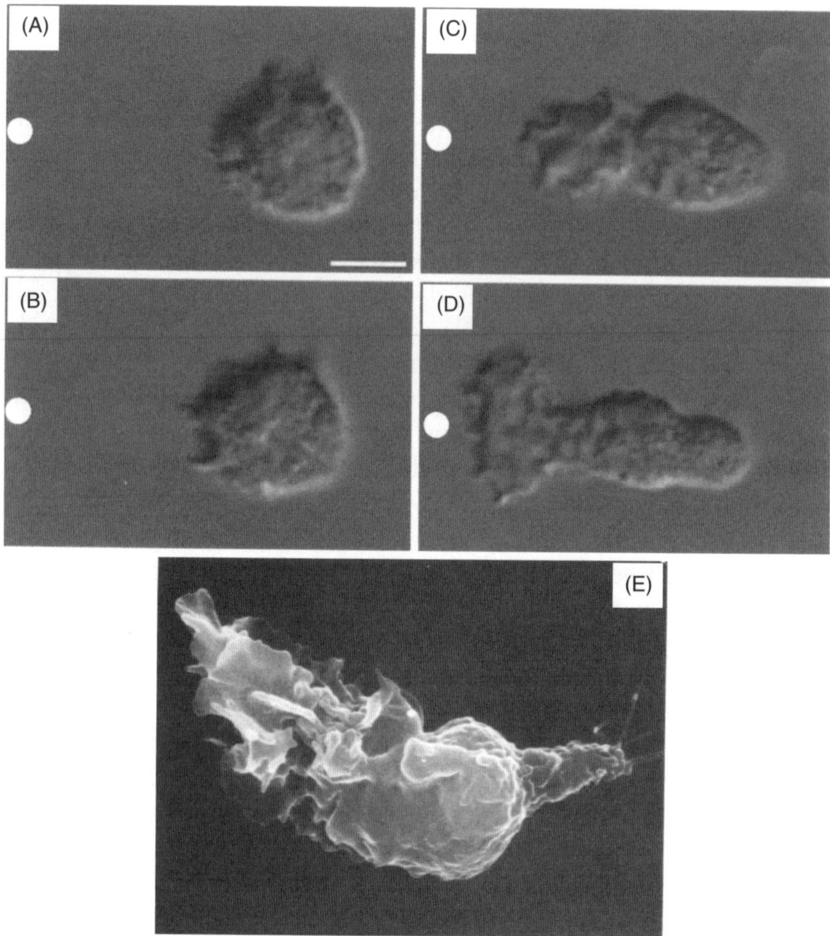

Figure 3. Morphology of polarized neutrophils. (A–D) Nomarski images of a neutrophil responding to a micropipet containing N-formyl-met-leu-phe. Images were collected 5 s (A), 30 s (B), 81 s (C) and 129 s (D) after introducing stimulus. The white dot indicates the position of the micropipet and the bar represents 5 μm. (E) Electron micrograph of a neutrophil incubated for 10 min with N-formyl-met-leu-phe. The bar equals 1 μm. Panels A–D reprinted with permission from [416]. Panel E reprinted with permission from [75].

However, migration through three-dimensional substrata occurs independently of integrins [226, 227, 342]. Thus two mechanisms for development of traction exist, one integrin-dependent for migration on a two-dimensional surface, and one integrin-independent for migration through three-dimensional matrices. This distinction is important for interpretation of migration assays and understanding signaling systems.

4.2. *Assays of neutrophil motility*

In animals, accumulation of neutrophils in the peritoneal cavity after injection of a stimulus is often used as an *in vivo* indicator of migration (e.g., [135, 213, 337, 341, 429]). *In vivo* assays for human studies utilize a skin chamber that is placed over a portion of disrupted skin; accumulation of cells in the chamber is then quantified [15, 325]. Whereas these assays are dependent upon the migratory capabilities of the cells, they also depend upon the ability of cells to adhere to the endothelium and mechanisms for clearance of cells from these compartments. Thus these assays are not uniquely migration assays and do not give detailed information about the characteristics of cell movement.

Many *in vitro* assays (Table 2) of neutrophil migration have been developed from which details of neutrophil migration have been obtained. Several indirect assays measure not migration itself, but events or processes that are known to be associated with migration. These assays in general are useful for screening of potential chemoattractants. However, little information about the process of migration itself may be obtained. Population assays measure the movement of a population of cells, and this movement can be understood in terms of the flux or dispersion of the population. This movement is characterized by parameters μ, the random motility coefficient (analogous to a molecular diffusion coefficient) and χ, a chemotaxis coefficient (analogous to a fluid flow velocity). Of course, the movement of a population of cells depends upon the movement of the individual cells in that population. Thus, single cell tracking studies have been used to measure the actual paths taken by the cells. Characteristic parameters measured are cell translocation speed, persistence time, and chemotactic index. The parameters of the single cell tracking data are related to the parameters obtained from population data.

4.2.1. *Indirect assays*

Cell polarization assays
Nonmotile neutrophils have a spherical morphology. Within a few minutes of activation with a chemoattractant (either in a gradient or in a uniform concentration), they develop the characteristic polar morphology of a lamellipodium at one end and a uropod at the other. This polarization is a prerequisite for migration of leukocytes, and thus measures of cell polarization are often used as indicators of migration [125, 170].

Table 2. *In vitro* methods for measuring neutrophil motility and chemotaxis [289, 391, 426].

Assay type	Basis for assay	Comments	Reference
		Indirect assays	
Cell polarization	Microscopic visualization of polarized morphology	Polarization is a prerequisite for neutrophil migration; thus this assay is a measure of migratory potential, but not a direct measure of motility.	[125, 170]
Actin polymerization		Neutrophil motility requires significant rearrangement of the actin cytoskeleton and chemoattractants induce a dramatic increase in cellular F-actin. Measures of chemoattractant-induced actin polymerization provide a measure of chemotactic potential, but are not a direct measure of motility.	[289, 391]
	Fluorescent phallotoxin binding to F-actin	In this assay, cells are fixed, stained with fluorescent phallotoxins that bind to F-actin, and then bound fluorescence is quantified by flow cytometry or spectrofluorimetry. Total cellular F-actin is measured. This assay is easily adapted for high throughput assays.	[142, 407]
	Mass of actin in detergent-soluble and -insoluble cytoskeleton	Triton X-100 detergent extraction of neutrophils separates detergent soluble and insoluble F-actin fractions that are typically quantified by staining proteins after separation by SDS-PAGE. This is a laborious assay, but allows the characterization of the two actin pools.	[418]
	G-actin inhibition of DNAse I	Whereas the above two assays quantify F-actin, this assay quantifies G-actin. G-actin potently inhibits DNAse I, the activity of which can be monitored spectroscopically. A cell lysate is made and assayed for its ability to inhibit DNAse I. Total actin is measured in an aliquot of the same cell lysate after treatment with guanidine hydrochloride which depolymerizes F-actin.	[101, 324]
	Right angle light scatter	Right angle light scattering measurements of neutrophils are inversely correlated with total F-actin content. This assay is easy, fast, and provides for high time resolution kinetic measurements (subsecond) of actin polymerization in living cells. The assay is easily performed in a spectrofluorometer. However, it is an indirect measure of F-actin and other cellular processes can also influence light scattering from cells. Thus results with this assay must be verified with additional methods.	[359]

Population-based assays

Cellulose filter assay	Cell migration occurs through a porous filter	Cells are placed in a chamber above a cellulose filter and chemoattractant is placed below the filter. The number of cells that move through the filter is quantified. This assay measures migration in three dimensions, but does not distinguish chemokinesis and chemotaxis.	[43]
Checkerboard assay	Cell migration occurs through a porous filter	This assay is a variation of the cellulose filter assay that uses multiple arrangements of chemoattractant above and below the filter to distinguish chemokinesis and chemotaxis. In addition, the depth to which the cells penetrate the filter is determined. These features allow calculation of population parameters μ, the random motility coefficient and χ, a chemotaxis coefficient.	[49, 448]
Polycarbonate filter assay	Cell migration occurs through a porous filter	Like the cellulose filter assay, cells migrate through a filter that separates two chambers, with the cells in the top chamber and chemoattractant in the bottom. Because polycarbonate filters are very thin, migration in two dimensions along the top surface of the filter is primarily measured. This assay does not distinguish between chemokinesis and chemotaxis.	[140]
Under agarose assay	Migration occurs between an agarose gel and the plastic surface that supports it	In this assay, agarose is layered on a plastic surface. Two circular wells or parallel troughs are cut into the solidified agarose. Chemoattractant is put into one well from which it can diffuse out into the agarose. Cells are placed in the other well, and because they cannot penetrate the agarose, they migrate on the plastic surface under the agarose in response to the chemoattractant gradient established in the agarose. The distance the cells migrate in a given amount of time is quantified. This assay can be configured to calculate population parameters μ and χ.	[73, 201, 204, 264]

(Continued)

Table 2. *Continued*

Assay type	Basis for Assay	Comments	Reference
Gel invasion assays	Migration through physiologically relevant, 3-D matrices	Defined chemoattractant gradients can be constructed in physiologically relevant three-dimensional matrices and cell migration monitored in these gels. Movement of the cell population is measured. Population-based parameters μ and χ can be determined.	[47, 79, 151, 191, 228, 247]
Zigmond chamber	Tracking the paths taken by individual cells	*Single cell tracking assays* Two-dimensional migration is observed on a specially designed microscope slide that produces a linear gradient of chemoattractant. Movements of single cells are tracked. This allows for calculation of single cell parameters S, the translational speed and P, the persistence time.	[450, 455]
Gel invasion assays	Tracking the paths taken by individual cells	Three-dimensional gels are produced with defined chemoattractant gradients, and individual cell movement is measured in all three dimensions. Single cell parameters S and P can be calculated.	[117, 148, 151, 247]

In this assay, cells in suspension are activated with chemoattractant for a designated time (on the order of 10–30 min), then fixed and viewed under the microscope. Cells exhibiting a polar morphology (clearly deviating from a circular outline) are scored as a percent of the total cells. Additionally, the long axis of each cell can be measured. This assay gives no information about the migratory properties of the cells, only that the putative chemoattractant can induce the requisite polar morphology and its dose response properties can be assessed.

Actin polymerization assay
An additional indirect assay of neutrophil motility is the actin polymerization assay. Chemoattractants induce a rapid, transient actin polymerization response in neutrophils (reviewed in [289, 391]), and since actin polymerization is required for migration to occur, it is often used as an indication of chemotactic capability. Several approaches have been developed to measure this response (reviewed in [289, 391]) (Table 2). Most commonly, cells in suspension are mixed with a putative chemoattractant then fixed and stained with a fluorescent phallotoxin that binds to polymerized actin, but not monomeric actin. The bound fluorescence is quantified using flow cytometry or spectrofluorometry. The actin polymerization response to many neutrophil chemoattractants is rapid, reaching a maximum within 10 s of stimulus addition at 37° C. As with the cell polarization assay, no information about the migratory properties of the cells is obtained with the actin polymerization assay, only the likelihood of chemotactic activity is assessed. Like the cell polarization assay, the actin polymerization assay is useful for initial screening of chemoattractants, but must be followed up with an actual measurement of motility.

4.2.2. *Population-based assays*

Filter assays
Filter assays have been very popular for identifying soluble compounds that stimulate locomotion (reviewed in [424, 426]). Two chambers are separated by a filter with pore sizes less than the diameter of the cells, such that cells must actively migrate to get through. The apparatus is oriented such that cells are put in the top chamber and can settle onto the filter. Solution with chemoattractant is put in the bottom chamber, and as it diffuses through the pores a concentration gradient is established such that the cells crawl through the filter. There are two main types of filter

assays, those using cellulose filters and those using polycarbonate filters, that differ significantly in the dimensionality of migration measured.

In 1962, Boyden [43] introduced the cellulose filter assay. These filters are made of cellulose esters that are approximately 120 μm thick. Filters are available in a variety of pores sizes, generally less than the diameter of the cell, meaning that cells squeezing through the pores migrate in three dimensions, an important point that will be discussed later (Section 6). A suspension of cells is placed in the top compartment and the chemoattractant solution placed in the bottom compartment of a chamber (often called a Boyden chamber). Cells that migrate through the pores to the other side of the filter are counted. Variations on this theme are to count the cells that fall off of the filter into the chemoattractant solution, and to count both cells adhering to the bottom of the filter and those that fall off of it [167]. By itself, this assay does not distinguish between chemokinesis (stimulated random motility) and chemotaxis (directed motility). To rectify this limitation, Zigmond and Hirsch [448] introduced a modification in the quantification of migration in these filters, the leading front assay, and a technique for distinguishing chemokinesis and chemotaxis. For the leading front assay, the incubation time of the assay is shortened such that cells are not allowed to penetrate the entire depth of the filter. Each filter is fixed and stained, and "clarified" (the filters become transparent after treatment with xylene). The filter is viewed with a microscope, and the distance between the top of the filter and the leading front (field containing the two furthest cells) is measured using the micrometer on the fine focus knob.

In the same paper, Zigmond and Hirsch introduced the so called "checkerboard assay" that allowed the distinction between chemokinesis and chemotaxis. In this assay, a range of chemoattractant concentrations is used above and below the filter to create a concentration matrix (or checkerboard). Thus the cells' abilities to migrate in a variety of concentration gradients, both positive and negative, are measured. The concentration dependence of random motility also is determined (a range of chemoattractant concentrations, the same above and below the filter). Using the leading front measurements from the uniform concentration samples, and assuming the concentration gradient across the membrane is linear, the distance traveled in a gradient due to chemokinesis (random motility) can be predicted [424, 448]. If the predicted values differ from the measured values, nonrandom motion (chemotaxis) is inferred. These calculations assume random motility is analogous to diffusion, an assumption that appears to be quite adequate given the time scale over which the measurements are performed (30–60 min). The mathematical treatment of migration will be further discussed below.

The use of a second type of filter made of polycarbonate was introduced in the 1970s [140, 368]. These filters are very thin, of the order of 10 μm, the diameter of a cell, with holes punched in them. This short distance between the top and bottom of the filter leaves little opportunity for a gradient to develop across the membrane. Thus migration on these filters occurs in two dimensions on the top of the filter as the cells migrate toward the chemoattractant diffusing from the other side of the filter through the nearest pore. As with the cellulose filter assay, cells either on the bottom of the filter or cells that have dropped off the filter into the chemoattractant solution are counted. This assay has been useful for detecting factors that stimulate movement of cells; however, chemokinesis and chemotaxis are not easily resolved, and in comparison, the cellulose filter assays provide more quantitative data. In addition, it is important to note that these two filter assays differ in dimensionality of motion that is being measured: on cellulose filters migration is in 3-D, whereas on polycarbonate filters, migration is in 2-D [426].

Under agarose assay

In the under agarose assay [73, 264, 265, 426] a layer of agarose is allowed to solidify in a plastic petri dish. Three wells in a line are punched out of it. Cells are put in the center well and chemoattractant and control solutions (no chemoattractant) placed on opposite sides. A gradient develops as the chemoattractant diffuses out from its well through the gel. Cells, which cannot penetrate the agarose, migrate *under* the agarose toward the chemoattractant well. Cells are utilized at a concentration such that the leading front of the migrating cells can be viewed with a standard microscope, either before [73] or after fixing and staining the cells [264]. The distance from the well margin to the leading front of cells in the direction of the chemoattractant well (A) is compared with the analogous distance in the direction of the control solution well (B). The directionality of migration can then be quantified as a chemotactic differential (A-B) or a chemotactic index (A/B or [A−B]/B) [264, 265]. It should be noted that the under agarose assay is an assay of 2-D migration on the plastic surface of the petri dish.

Gel invasion assays

Gel invasion assays have been developed for monitoring migration in three dimensions through physiologically relevant matrices [47, 148,

191, 228, 342, 422, 426]. Collagen gels are models for extracellular matrix proteins, and fibrin gels are models for stroma in solid tumors and blood clots in inflammation. Reconstituted basement membrane extracts that contain collagen type IV, laminin, entactin, and heparin sulfate proteoglycan (commercially available under the trademark "Matrigel") have also been used as models of the subendothelial basal lamina. Human amnion has been utilized as a heterogeneous matrix typical of the structure of loose fibro-connective tissue. For these assays, the cells may be placed on top or under gels into which chemoattractant has been incorporated, then after a set amount of time the gels are viewed with a microscope and the distance to the leading front measured analogously to measurements in the cellulose filter assay.

Quantitative analysis of population assays
Common strategies for quantifying migration in population assays, such as the leading front distance or chemotactic differential, are somewhat arbitrary in that the physical parameters of the system, such as concentration gradient of chemoattractant and time of assay, may affect the values of the parameters. What are most desirable are parameters that reflect the intrinsic migratory functions of the cells. Population assays measure the movement of a population of cells, and this movement can be understood in terms of the flux or dispersion of the population. This movement is characterized by parameters μ, the random motility coefficient (analogous to a molecular diffusion coefficient) and χ, a chemotaxis coefficient (analogous to a fluid flow velocity). We shall see that μ depends upon cell speed and direction persistence time whereas χ depends upon cell speed and directional orientation accuracy. Thus both parameters reflect intrinsic cell properties. The number density of motile cells distributed in space and time can be approximated by the following partial differential equation [12, 203, 224]:

$$\frac{\partial c}{\partial t} = \mu \frac{\partial^2 c}{\partial x^2} - \frac{\partial}{\partial x}\left\{\left[\chi - \frac{1}{2}\frac{d\mu}{da}\right]\frac{\partial a}{\partial x}\right\}c + R_c. \tag{1}$$

To the typical cellular or molecular biologist and physiologist this equation may look quite formidable. However the terms have physical meanings that can be readily understood. The equation describes the cell number density as a function of spatial position, x, and time, t, such that $\partial c/\partial t$ represents the rate at which cell number density at some position x changes with time. The first term on the right side of the equation quantifies the change in cell number density due to random migration.

μ is the random motility coefficient analogous to a molecular diffusion coefficient. The second term in the right side of the equation represents the change in cell number density due to spatial gradients of the chemotactic stimulus. The group in brackets represents the net directional flow of cells in a stimulus gradient. χ is the chemotactic coefficient characterizing the biased directional migration in response to the spatial gradient of chemotactic stimulus, $\partial a/\partial x$, where a is the stimulus concentration. The additional factor in this bracket, $(d\mu/da)(\partial a/\partial x)$, represents the net cell procession in a stimulus concentration gradient due to the effects of the stimulus on chemokinesis. Finally, the last term on the right side of the equation, R_c, is the rate of production or loss of cells at position x by mechanisms other than migration, such as proliferation or death. For most experimental systems with neutrophils, this factor is negligible.

How does this equation relate to measurable parameters? The cell number density is a measured parameter, and distance and time are measured or controlled parameters. For the cellulose filter and gel invasion assays, the cell density at any depth within the filter or gel can be quantified after a designated time interval. The leading front method of quantification measures the distance at which the cell density becomes 2 per unit area for a designated time interval. μ is first determined after measuring migration in a uniform concentration of stimulus. Then χ is determined from the measurement of migration in response to a stimulus concentration gradient, which in turn requires knowledge of the stimulus concentration profile as a function of space and time. These profiles can be estimated based on the diffusion characteristics of the chemotactic stimulus [205]. Methods for determining μ and χ have been presented for the checkerboard cellulose filter assay [49], under agarose assay [201, 204], and gel invasion assays [79, 151, 247]. The parameters μ and χ are very useful because they are independent of assay conditions such as time, length, and gradient of chemoattractant. Thus they can be used to compare motility of different cell types, or of the same cell type in response to different ligands, or the same cell type on different substrata, etc.

4.2.3. *Single cell tracking assays and analysis*

Experimental configurations
Migration of individual leukocytes *in vivo* were recorded as early as the mid-1800s; fascinating reviews of these early studies are available in which detailed accounts are given of individual leukocytes migrating in the rabbit cornea and ear, tadpole tails, and frog tongue and webs [63, 237].

Quantitative studies of neutrophil migration began in the early 1900s when Comandon [66] was the first to use time-lapse cinematography to measure directional migration of leukocytes on microscope cover slips. The time-lapse photography allows determination of the path the cell traverses. In such studies a particulate chemoattractant source, such as a bacterium, tissue fragment, or lysed red blood cell, is utilized and migration can be measured relative to that source [11, 321, 322]. However, in such studies the exact nature of the chemotactic factor may be unknown. Micropipets placed near the cell have been utilized to deliver a defined chemoattractant solution, although in this case the chemoattractant concentration the cell experiences is not clearly defined.

A special orientation chamber, often referred to as the Zigmond chamber, was developed that allows estimation of the chemoattractant gradient experienced by cells [450, 455]. For this assay, a special microscope slide was designed with two troughs on either side of a 1 mm bridge. Chemoattractant solution is placed on one side of the bridge and buffer on the other. A coverslip to which neutrophils are adherent is inverted over troughs and bridge and a gradient of chemoattractant develops over the bridge. This allows visualization of the orientation and movement of the cells in the gradient. The stimulus concentration gradient that develops across the bridge remains linear and stable for time periods that are long compared to the time over which the measurements are made [202, 205]. Thus migration parameters can be directly related to chemoattractant concentration gradient.

Like population assays, single cell movement has also been measured in three-dimensional gel invasion assays [117, 148, 247, 296]. These gels can be constructed such that the chemoattractant concentration gradients are well defined [151, 247].

Single cell tracking parameters
For single cell tracking studies, time-lapse photography or video recording is utilized to follow the cell motion with time. The path taken by a cell is marked by noting the location of the cell centroid at designated time intervals. Parameters quantified from these paths include total path (distance) traveled, net displacement of the cell after some time interval, average speed, instantaneous speed, turn angle, turn frequency, and stop frequency. A randomly migrating cell will exhibit periods of linear movement (related to a persistence time) interspersed with turns, such that a random walk occurs with no net displacement of the cell over times that are long compared to the persistence time. During

chemotaxis, fewer and smaller turns are made so the motion is directed to the chemoattractant source. The propensity for oriented migration (chemotaxis) can be quantified as the "chemotropism index" or "chemotactic index" (CI), the net cell displacement divided by the total length traveled over a certain amount of time, a concept originally introduced by Dixon and McCutcheon [84]; see also [277]. If migration is random, the total cell displacement will be zero and the chemotropism index will be zero. If migration is highly directed such that a straight path is taken toward the chemoattractant source, the chemotropism index will be 1. However, by itself this parameter does not provide information of the speed of locomotion.

Information about the average speed of locomotion is usually obtained by measuring the total path traveled and dividing by the total time of migration. However, the measured value of total path traveled may underestimate the real path if the time interval between frames is significantly larger than the persistence time. More rigorously, information about the speed of locomotion is obtained by measuring the cell centroid displacement, Δx, during a time period, Δt. Mathematically, treatment of random migration is analogous to that for Brownian diffusion; the motion is modeled as a random walk with a persistence time. For random migration, the mean value of the square of the displacement, $\langle(\Delta x)^2\rangle$, averaged over many time periods is related to the time period, Δt, as follows [89, 224]:

$$\langle(\Delta x)^2\rangle = nS^2P[(\Delta t) - P(1 - \exp\{-(\Delta t)/P\})]. \tag{2}$$

The constant, n, is the dimensionality of the system. S is referred to as the translational speed, the speed of migration over any straight-line portion of the path. P, the persistence time, reflects the tendency of the cell to keep moving in the same direction and is inversely related to the tendency of a cell to change directions while migrating. P contains information of the turn angles, and is approximately the average time it takes for a cell to turn approximately 90°. Like μ and χ, S and P are parameters that reflect the intrinsic migratory properties of the cell and are independent of assay conditions.

Using current technology, computer-based image analysis methods are utilized to analyze data collected with fast ccd cameras allowing for high time resolution of the motion both in two dimensions and three dimensions [87, 247]. From measurements of $\langle(\Delta x)^2\rangle$ and Δt the parameters S and P can be extracted. Theoretical treatment of the model and optimization of analysis methods have been reported [79, 80, 89, 224, 393].

4.2.4. *Relationship between population and cell tracking data*

Since the migratory behavior of a population of cells is determined by the behavior of the individual cells in that population, it seems reasonable that the parameters determined from population assays (μ and χ) and single cell assays (S, P, and CI) are related. For random migration, $\mu = (1/n)S^2P$, and in the presence of a gradient of chemoattractant, $\chi = (CI \times S)/(da/dx)$, where da/dx represents the concentration gradient of chemoattractant [89, 206, 224]. Values of μ and χ determined by population and single-cell assays of neutrophil motility agree well [247].

Neutrophils are among the fastest-moving mammalian cells measured, with typical cell speeds in the range of $10\,\mu$m per minute [11, 87, 321, 452, 457]. Measured persistence times are in the range of 1 min. Maximally stimulated random motility coefficients and chemotaxis coefficients are in the range of 3×10^{-9} to $8 \times 10^{-7}\,\text{cm}^2/\text{s}$ and 20–400 cm^2/sM, respectively [49, 204, 247, 393]. The large ranges are likely due to differences in the substratum on which the cells migrate and the chemoattractant used to stimulate motility. In comparison, in endothelial cells stimulated with αFGF, μ is similar to that in neutrophils, but endothelial cell speed is much slower ($0.5\,\mu$m/min) and persistence coefficient much longer (300 min) than in neutrophils [373, 374]. These differences in migratory characteristics of different cell types may arise from differences in the cell signaling systems that control migration and differences in mechanisms by which the cells interact with the substratum.

5. Cell Polarization

The process of migration is complex, involving the signaling and response mechanisms within a cell, the interaction of the cell with the matrix in which it migrates, and the generation of force for movement. The process can be divided into two components, cell polarization and translocation. Cell polarization in response to the chemoattractant requires force generation and is a process mediated by the cell itself, although the properties that develop may be influenced by the cell's interaction with the substratum. The translocation phase requires force generation and is completely dependent upon the cell's interaction with a substratum for the development of traction, and the mechanisms that mediate that traction appear to be different for migration in two dimensions and three dimensions.

5.1. *Morphological and behavioral aspects of polarization*

5.1.1. *Cells in a uniform concentration of chemoattractant*

Polarization occurs for cells in a uniform suspension of chemoattractant, indicating that the machinery for polarization is internal, i.e., independent of attachment and independent of a chemoattractant gradient [169, 365]. Unstimulated neutrophils are round and nonmotile [11, 170, 452]. Within seconds of addition of a uniform concentration of a chemoattractant many shallow ruffles form: at longer times (minutes) the cells become polarized with a lamellipodium similar to the leading edge and a uropod similar to the trailing edge of migrating cells [64, 71, 75, 138, 172, 354, 365, 433]. These polarized cells in suspension exhibit crawling motions analogous to those of adherent cells [169]. Like adherent cells, if the chemoattractant is removed from cells in suspension the cells revert to a rounded morphology and cytochalasin B pretreatment prevents chemoattractant-induced polarization [365]. Thus the development of polarity is dependent upon continual signaling for maintenance of an F-actin structure.

Adherent cells exposed to a uniform concentration of chemoattractant polarize and exhibit random motility. Thus a chemoattractant gradient is not required to obtain cell movement although it is required for directed movement to occur [11, 449]. Adherent, polarized neutrophils exhibit a behavioral polarity in their responsiveness to chemoattractant stimulation. Randomly migrating cells (i.e., those in a uniform concentration of stimulus) move by extending the lamellipodium and turn by a lateral extension of the existing lamellipodium (and withdrawal of the cytoplasm from the other side). Rarely is a new lamellipodium extended from a region of the cell surface not previously extending a lamellipodium [452]. Large turns are accomplished by a series of small turns in the same direction. This behavioral polarity is not a result of movement alone, since there is no correlation between rate of locomotion and frequency or angle of turns; the faster-moving cells simply travel further in between turns.

5.1.2. *Cells in a chemoattractant gradient*

Cells migrating in a chemoattractant gradient exhibit a biased random walk, with migration persisting in the direction of the gradient because they turn less and less often than during random migration. If a cell is migrating in a chemoattractant gradient, then the orientation of the gradient is changed, and most often the cells reorient to the new gradient by maintaining their existing polarity and laterally extending the existing lamellipodium; if the gradient is reversed by 180°, the cells make a U-turn by making a series of small turns; occasionally a new lamellipodium extends from the tail of the cell [110, 168, 321, 452]. These observations indicate that under most conditions the rear of the cell is refractory to stimulation. The same behavior occurs if cells are migrating in a gradient of N-formyl peptide, and then the peptide is removed and replaced by a gradient of C5a in the opposite direction [452]. Thus the polarization properties induced by the first stimulus are retained during activation with the second, different stimulus, implying that the polarization properties of the neutrophil are independent of the specific chemoattractant signaling the response.

The directional orientation of the cell persists as long as a uropod is present. Cells exhibiting random migration in a uniform concentration of chemoattractant will stop moving transiently when the chemoattractant concentration is increased [451, 452]. This pause is associated with the development of ruffles over most of the cell surface except for the uropod. Locomotion then resumes in the same direction the cell was moving before the increase in stimulus. If chemoattractant is removed from migrating cells for a time period sufficient to allow cell rounding (i.e., withdrawal of the uropod), then additional chemoattractant is added, the new direction of locomotion is not correlated with the direction of locomotion before rounding. Thus the uropod structure is an important component determining the orientation of the cells.

The demonstration of behavioral polarity indicates that the rear of a polarized, migrating cell is refactory to stimulation, since new lamellipodia are not formed there in the presence of ligand. In addition, both the uropod and the lamellipodium seem to be important structures determining the axis of orientation, although they appear not to be the only factors. Zigmond *et al.* [452] observed a cell migrating in response to a chemoattractant gradient for which the lamellipodium pulled off from the cell body, creating two migrating cell fragments, the forward portion containing the original lamellipodium and the cell body containing the original uropod. Both fragments continued to migrate in the same direction. Similar behavior is exhibited by enucleated "lamellipodial fragments" created by incubating neutrophils at superphysiological temperatures. These fragments, called cytokineplasts, pinch off from the main cell body and exhibit oriented migratory behavior similar to an intact cell [168, 225].

Both sets of observations indicate that the axis of orientation may be determined by an intracellular gradient throughout the entire cell, not just where the lamellipodium and/or the uropod are located.

It should be noted that there are some instances in which the uropod demonstrates an ability to respond to chemoattractant [7, 110, 321]. For example, Ramsey [321] and Gerisch and Keller [110] report if the source of stimulus is moved from the front of a migrating cell to the back, a new lamellipodium forms where the uropod was previously, and the old lamellipodium retracts. In these studies, the U-turn behavior was also seen in some cells, indicating that the two different modes of turning are not mutually exclusive. Thus the data suggest two types of polarized morphologies are possible: one in which the uropod is sensitive to chemoattractant and one in which it is not. It has been suggested the extent to which either mode of turning is manifested may depend upon the interactions of the cells with the substratum [7].

5.2. *Molecular aspects of polarization*

Besides the aforementioned asymmetry in the F-actin and organelle distributions in the cell, other molecular asymmetries have been noted. The uropod has enhanced endocytic activity and is more negatively charged than the lamellipodium [75, 138]. This enhanced negative charge is likely due to the concentration of the highly negatively charged glycoprotein, CD43, in the uropod [349, 350]. During migration the endocytotic recycling compartment retains a position just behind the base of the lamellipodium [307]. Streaming of cytoplasmic organelles from the back to the front of the cell body occurs [322, 346]. Several cell surface receptors, including Fc receptors, Concanavalin A receptors, and C3b receptors are concentrated in the lamellipodium; ligation of these receptors causes their translocation to the uropod [354, 409, 415, 421].

Cytoskeletal binding proteins also show a polarized distribution; actin-binding protein and α-actinin are located in the lamellipodium [349, 401] whereas moesin, filament-associated cyclic GMP-dependent protein kinase and its substrate, vimentin, are in the uropod [316, 349, 432]. Myosin is observed in both the uropod and lamellipodium [91, 401]. The cell signaling molecules, Rac [306], p21-activated kinase 1 (PAK1) [78], and the actin polymerization regulator Arp2/3 complex [416] are localized in the lamellipodium, a likely reflection of their roles in regulating cytoskeletal dynamics at the leading edge of the cell (further discussed in Section 9).

Polarized neutrophils also show nonuniform distribution of plasma membrane lipids. The mobility of the fluorescent lipid probe, octadecanoyl-aminofluorescein, is reduced in the rear of polarized neutrophils, indicating a difference in the physical properties of the lipids at the two poles of the cell [238]. In many cell types, specific membrane microdomains exist that are characterized by high cholesterol and sphingolipid content and an insolubility in cold Triton X-100 detergent [48]. These microdomains are referred to as detergent-resistant membranes (DRMs) or lipid rafts. These domains can be detected microscopically with fluorescent lipids probes and fluorescent antibodies to proteins that preferentially partition into them. Microscopy studies show that the uropods of migrating neutrophils contain DRMs that also contain the proteins CD44, CD43, and CD16 [351]. Moreover, fluorescence microscopy studies show that a fluorescently-labeled probe for 3′-phosphoinositides, products of the enzyme phosphatidylinositol 3-kinase (discussed further in Section 9.3.1), localize to the lamellipodia of polarized neutrophil-like HL-60 cells, implicating an important role for phosphatidylinositol 3-kinase in regulating cell polarity [348].

5.3. *Development of polarity*

Whereas there are many data demonstrating that cell polarity exists, detailed kinetic studies of the development of neutrophil polarity are not abundant. In one of the most detailed kinetic studies available, Coates and coworkers [64] studied the kinetics of the morphological and actin cytoskeletal changes that occur during the development of polarity of cells activated in suspension. When neutrophils are activated by a uniform concentration of the N-formyl peptide, N-formyl-met-leu-phe, there is an initial increase in F-actin underlying the plasma membrane that is uniformly distributed within 30 s of activation (at 25° C), and F-actin is increased throughout the cell by 45 s. These localized increases in F-actin correlate with increased total F-actin in the cell (reviewed in [289, 391]) and increased membrane ruffling. By 90 s after stimulation, total F-actin begins to decrease. At this time, as polar morphology develops, F-actin is *lowest* in the newly developing lamellipodium, but by 300 s poststimulus, when polar morphology is clearly established and sustained, F-actin is *highest* in the lamellipodium. Thus very dynamic reorganization of the actin cytoskeleton occurs during cell polarization.

The progression to developing polarity in the distribution of DRMs begins with a uniform distribution of DRM components in resting cells.

Activation with N-formyl peptides causes an initial rapid clustering of DRMs into discrete patches, followed by formation of a large cap over the cell body and uropod. This process requires myosin, since inhibitors of myosin prevent cap formation [351]. Localization of 3′-phosphoinositides in the lamellipodium may also be dependent upon the actin cytoskeleton [411]. Thus polarization of the actin cytoskeleton appears to dictate the development of polarity of the membrane and signaling components.

For cells in a uniform concentration of stimulus, polarity is thought to develop stochastically [126, 238, 354, 392], that is, as a neutrophil in suspension is exposed to a uniform concentration of stimulus, there is some probability that somewhere on that cell, one receptor will be the first to bind a ligand. It is thought that this "first hit" then determines the axis of polarization that will develop. Very locally, a signal will be generated to extend a lamellipodium. At very high concentrations of chemoattractant, a distinct bipolar structure is not observed, and this is thought to reflect a condition where the cell cannot distinguish a "first hit" [126]. This stochastic model for activation and polarization of neutrophils in a uniform concentration of stimulus extrapolates well to the condition of neutrophils exposed to a gradient of chemoattractant. In this case, there is an overwhelming probability that the "first hit" will be along an axis that is parallel to the concentration gradient, and therefore cell polarization will result in a leading edge moving up the concentration gradient.

Thus the development of polarity depends upon the ability of the cell to maintain a localized signal in the lamellipodium. Neutrophils are able to respond to as little a 1% concentration gradient across the cell body [450]. In such a situation, receptors all around the cell would be exposed to significant levels of ligand and signaling would be expected to occur all around the cell, with a slight bias at the leading edge where the concentration is highest. Yet the gradient of the signaling molecules, 3-phosphoinositides, is several-fold higher than the corresponding chemoattractant gradient that induced it [348]. There are two prominent, not mutually exclusive hypotheses of how the external gradient becomes internally amplified. One hypothesis is that there is global production of an inhibitory signal and localized production of an activation signal, the balance between the two being such that positive signaling overcomes the inhibitory signaling only at the leading edge of the cell [104, 295]. A second hypothesis is that there is a positive feedback loop that amplifies the activation signal in the lamellipodium [417]. Because understanding these hypotheses requires an understanding of the signaling systems in neutrophils, they will be further discussed in Sections 9.1.5 and 9.3.1.

6. Two-Dimensional Versus Three-Dimensional Migration

Migration of neutrophils on a two-dimensional substratum and through a three-dimensional matrix appears to be similar except for the mechanisms of developing traction against the substratum. Two-dimensional migration is mediated by integrins, whereas three-dimensional migration is integrin-independent.

6.1. *Migration on a two-dimensional surface*

Integrins are a class of transmembrane adherence and signaling molecules that were first identified and named for their abilities to integrate the extracellular environment (extracellular matrix proteins or other cells) with the intracellular cytoskeleton. They are composed of α and β subunits, and classified into groups based on the structure of the β subunit.[a] Neutrophils contain β_1, β_2, and β_3 integrins; the β_2 integrins are leukocyte specific [121, 179]. The particular integrins involved in mediating neutrophil migration depend upon the chemical characteristics of the substratum. For example, migration on glass is mediated by β_2 integrins [342], on vitronectin by $\alpha_v\beta_3$ [208], and on fibronectin by $\alpha_5\beta_1$ [307].

During migration, attachments must be made at the leading edge of the cell to tether the forming lamellipodium to the surface. These attachment sites then provide the links to the cytoskeleton (traction) required for the cell body to move forward. Finally, as the cell body is pulled up from behind, these attachments must be broken. There is a biphasic relationship between cell substratum adhesiveness and cell speed; if the adhesive forces are too small, the cells cannot efficiently attach to the substratum, if the adhesive forces are high, the cell cannot effectively release from the substratum. Thus intermediate adhesive forces lead to efficient migration [70, 292]. The integrin-mediated links between the substratum and the cell cytoskeleton are regulated by

---

[a] The nomenclature for integrins has a rich history, with the leukocyte integrins accumulating several names. The β_2 subunit is designated CD18, and the leukocyte α subunits designated CD11. Thus the following names have accumulated: CD11a/CD18 = $\alpha_L\beta_2$ = LFA-1; CD11b/CD18 = $\alpha_M\beta_2$ = Mac1 = MO1 = CR3; CD11c/CD18 = $\alpha_X\beta_2$ = p150,95. A fourth member of the β_2 integrin family has been identified—$\alpha_D\beta_2$.

complex interactions of outside-in and inside-out signaling (reviewed in [179, 217, 428]). There are three recognized mechanisms by which rear attachments of migrating cells are broken: (1) dissociation of the integrin from the matrix ligand, (2) proteinase cleavage of either the integrin or the substratum, and (3) deposition of the membrane fragment in which the integrin is imbedded. In neutrophils, release from integrin-mediated attachment proceeds predominantly by the first mechanism. For migration mediated by $\alpha_v\beta_3$ and $\alpha_5\beta_1$ integrins, little integrin remains behind on the substratum, but rather is recycled through endocytic compartments to the cell front [208, 307]. A similar fate has been observed for β_2 integrins [70].

6.2. *Migration in a three-dimensional matrix*

Migration of cells in a three-dimensional network does not require integrins. Schmalstieg *et al.* [342] carefully demonstrated this by showing that antibodies to integrins, while inhibiting migration in a two-dimensional under agarose assay, do not inhibit migration through a collagen matrix. Likewise neutrophils from patients with β_2 integrin deficiencies demonstrate restricted migration under agarose, but normal migration through collagen. This work supported earlier reports that locomotion in two dimensions is adhesion-dependent but adhesion-independent in three-dimensional matrices [47, 195]. Both *in vitro* and *in vivo* studies have further confirmed the existence of integrin-independent migration [86, 127, 250]. Moreover, when neutrophils are allowed to migrate between closely spaced, parallel glass coverslips, such that contact must be made with both surfaces, migration is integrin-independent; migration on only one surface requires integrins [226, 227].

That integrins are important for normal neutrophil function is clearly demonstrated by a class of diseases called leukocyte adherence deficiency syndromes (LAD 1) in which the β_2 integrins are deficient or impaired [15, 51]. This syndrome is characterized by recurrent infections and reduced pus formation in spite of elevated circulating neutrophil counts [15]; the neutrophils are defective in their ability to leave the blood stream and migrate to sites of infection. Neutrophils from LAD patients are normal in the ability to develop a polarized morphology in response to chemoattractant, demonstrating that β_2 integrins are not required for cell polarization [15]. Given that migration through three-dimensional matrices does not require β_2 integrins (as discussed above), it appears the role of integrins in neutrophil emigration from the vasculature into tissues may be primarily dictating adherence of neutrophils to and perhaps migration along the endothelium (as described in Section 3).

7. Force Generation During Migration

7.1. *Lamellipodium extension*

7.1.1. *The case for actin polymerization*

Most eukaryotic cells contain the major cytoskeletal protein, actin, which can exist as a monomer (G-actin) and reversibly polymerize into long filamentous structures (F-actin). The F-actin filament is polarized such that the rates of polymerization and depolymerization are different at the two ends. At the plus end (also called barbed end) monomer addition occurs readily, relative to the minus end (also called pointed end) where polymerization is less favored. In addition, the actin subunit binds adenine nucleotides and has an intrinsic ATPase activity. ATP-actin binds preferably at the plus end and, after incorporation into the filament, is converted to ADP-actin. These properties have a significant impact on the dynamics and stability of the filament [54, 293]. In the cell, polymerization and depolymerization are highly regulated reactions, this regulation being mediated by multiple actin-binding proteins.

It is clear that actin polymerization is required for lamellipodial extension, since inhibitors of actin polymerization inhibit neutrophil polarization and lamellipodial extension. The idea that actin polymerization itself could provide the force for lamellipodial extension was proposed over a decade ago [387, 446]. In neutrophils, as well as other cells [207, 242], the plus ends of F-actin filaments are located at the leading edge of the lamellipodium, and this is the region where rapid polymerization of actin occurs [326]. Models of the elastic motion of filaments (the elastic Brownian ratchet) at the leading edge of the cell predict polymerization can indeed produce forces adequate to push the cell membrane forward [2, 248, 249, 300]. Thus the fundamental motor driving lamellipodium extension appears to be actin polymerization.

Other mechanisms for producing protrusive force have been proposed, and not necessarily ruled out. Osmotic forces generated either by polymerization of actin or by activation of a sodium/proton antiport may be involved. It has been proposed that extensive cross-linking of actin filaments at the leading edge could produce osmotic forces that push the membrane forward [67, 68]. This hypothesis is consistent with the need for actin polymerization to drive movement, but is untested in neutrophils. Alternatively, it has been shown that neutrophils migrating in a three-dimensional matrix exhibit a significant volume increase (~50%) driven by activation of a sodium/proton antiport that results in water intake [338, 431]. Whereas this swelling could produce forces sufficient to cause lamellipodium protrusion, comparable swelling is not seen in neutrophils migrating on a surface [338, 431], calling into question the relevance of this mechanism for lamellipodium extension.

7.1.2. *The oscillatory motor*

Under specified conditions, the actin polymerization response oscillates, with cycles of polymerization and depolymerization occurring with a periodicity of approximately 8 s at 37°C [94, 95, 146, 285, 286, 330, 434]. The increases in F-actin correlate with the extension of lamellipodia and depolymerization correlates with lamellipodia retraction both in adherent and suspended cells [123, 146]. Thus the fundamental motor driving lamellipodium extension is oscillatory. What drives these oscillations is unknown, but it has been noted that metabolic cycles exist in neutrophils, with levels of NADPH oscillating at a comparable periodicity [reviewed in 302]. Since levels of NADPH and ATP are inexorably linked in cells [129], the metabolic oscillations could affect cell functions by modulating ATP levels. ATP, in turn, could affect actin polymerization because of its role in maintaining the G-actin-ATP pool [55, 293]. Alternatively, oscillating ATP levels could influence kinase activity because of ATP's role as a substrate for these enzymes; levels of signaling kinases in general [178] or myosin light chain kinase specifically [329, 330] could possibly influence cytoskeletal oscillations.

7.1.3. *Is myosin important for lamellipodium formation?*

The contractile protein myosin is located both in the lamellipodium and uropod of polarized neutrophils [91, 401]. The compound 2,3-butanedione monoxime, which inhibits the ATPase activity of myosins [72], inhibits chemoattractant-induced polarization and migration [91, 398]. Specifically, high concentrations of 2,3-butanedione monoxime inhibit formation of the lamellipodium [91, 398], indicating a role for myosins in lamellipodium formation. However, detailed mechanisms defining the role of myosin in lamellipodium extension remain to be elucidated.

7.2. *Traction*

For the neutrophil to migrate relative to a substratum, traction must be generated against the substratum. For neutrophils migrating on a two-dimensional surface, integrins provide the traction. However, in three dimensions, migration is independent of integrins (as noted in Section 6). Three-dimensional imaging of neutrophils migrating through a three-dimensional matrix has shown that they are able to generate traction by extending lateral projections into the matrix; these projections remain stationary relative to the matrix as the cell moves past them [228]. In addition, migration proceeds more rapidly over a two-dimensional membrane with small (0.8 μm) pores than the same membrane without pores; the migrating cell gains footholds by inserting lateral projections into the small pores. Thus the projections provide a mechanical anchor that facilitates migration. The mechanisms by which the cell pushes or pulls on these anchors have not been defined.

Migration of neutrophils between two closely spaced coverslips, such that the cell must touch both surfaces, also is independent of integrins [226, 227]. The cell appears to generate traction by exerting forces on opposing surfaces, much like a rock climber ascends a rock chimney. How these forces are generated are unknown. However, it should be noted that in a three-dimensional matrix migration of neutrophils is associated with an approximately 50% increase in cell volume, a change that does not occur for neutrophils migrating on a two-dimensional surface [338, 431]. This volume change is due to activation of a sodium/proton antiport. It is tempting to hypothesize that this volume change is capable of generating the outward forces required for traction in three-dimensional migration. Such a hypothesis awaits confirmation.

7.3. *Rear retraction*

As noted previously, myosin is located both in the lamellipodium and uropod of polarized cells [91, 401]. Whereas the myosin inhibitor, 2,3-butanedione monoxime, inhibits lamellapodial extension at high concentrations, at intermediate concentrations cell polarization and lamellipodium extension occur, but uropod retraction is inhibited, clearly indicating a role for myosin-mediated contractile forces in uropod retraction [91, 398]. The inhibitor, 2,3-butanedione monoxime, inhibits multiple forms of myosin [72], thus lamellapodial extension and uropod retraction may be mediated by different forms of myosin.

Myosin regulation by signaling mechanisms is further discussed in Section 9.

8. Migratory Responses to Multiple Stimuli

At a site of injury and inflammation, neutrophils are likely to encounter multiple stimuli, including chemoattractants released from bacteria and surrounding tissues, priming agents released from bacteria and immune cells, and integrin ligands imbedded in the extracellular matrix. Relatively little work has addressed the question of how neutrophils process the information from multiple stimuli. However, work is beginning to show that neutrophils are capable of integrating signals from multiple stimuli. We shall describe two examples that demonstrate this point.

8.1. *Contact guidance*

Contact guidance has been described as the phenomenon in which orientation of the matrix in which neutrophils are imbedded influences the direction of migration. For example, gels have been made in which the gel fibers are aligned [228, 422, 423], and glass surfaces have been micropatterned with parallel ridges/grooves [381]. Neutrophils incorporated into aligned gels show a preference for motion along the axis of the fibers; in the absence of a chemoattractant gradient this motion will be in either direction. In the presence of a chemoattractant gradient, movement is more efficient if the cells are moving along the fibers than if they have to cross the fibers. Thus both directive cues, fiber orientation and chemoattractant gradient, are processed by the cells and influence motion, although the mechanisms by which this occurs remain to be elucidated. There is intense interest in understanding contact guidance in cell migration because of the potential to engineer cell movement for the purpose of tissue engineering [381].

8.2. *Multiple chemoattractants*

Integrated migration to multiple chemoattractants has been demonstrated, and the mechanism by which this occurs likely involves receptor desensitization. Using multiple combinations of pairs of chemoattractants, it has been shown that receptor desensitization or adaptation give the cell a "memory" of prior exposure, and this may influence how the cells respond to multiple stimuli. Chemoattractants induce homologous receptor desensitization, defined as ligand-induced desensitization of the receptor to which it binds. Some chemoattractants also induce heterologous desensitization, defined as desensitization of receptors that do not bind the same chemoattractant. N-formyl peptides and C5a are known to cross-desensitize responses to IL-8, LTB_4, PAF; the converse is not true [10, 53, 81, 180]. N-formyl peptides also desensitize C5a responses.

In an under agarose assay, when neutrophils are placed in a well with one chemoattractant that induces only homologous desensitization (e.g., LTB_4 or IL-8), they will migrate up a concentration gradient of a second chemoattractant in a distant well (IL-8 or LTB_4, N-formyl peptide, C5a); to do this, they must migrate down a concentration gradient of the first chemoattractant [105]. In the same assay system, high concentrations of a chemoattractant inhibit migration, and this is thought to be due to saturation and desensitization of the receptors. These desensitized cells are then able to respond to a stimulatory gradient of a second ligand. In both experimental systems, homologous desensitization of receptors to the first stimulus occurs, and the cells are able to respond to the second chemoattractant. If competing gradients of ligand are present, neutrophils respond as if to the vector sum of the orienting signals [106]. However, if N-formyl peptide or C5a is the first stimulus, response to the second stimulus is not observed. Thus N-formyl peptides and C5a provide a chemoattractant signal that blocks the influence of other chemoattractants produced by leukocyte-recruiting tissues. However, for the chemoattractants released by the tissues (e.g., LTB_4, IL-8, PAF) summing of the signals occurs. It has been proposed that these desensitization mechanisms allow sequential chemotaxis of neutrophils to one attractant after another, thereby guiding neutrophils to their destinations within tissues by integrating competing orienting signals over time [106].

9. Signal Transduction

Many signaling events are triggered in neutrophils by activation of the chemoattractant receptors, including G-protein activation, lipid remodeling, protein kinase activation, and calcium elevation. Extensive work has been done to delineate the signaling mechanisms activated by the N-formyl peptide receptor, although the story is still incomplete. These pathways are complicated with crosstalk and feedback, and the many components of migration may require coordination of multiple arms of the signaling pathways. This section will summarize what is known about signaling events in neutrophils with a focus on signaling pathways involved in N-formyl peptide-induced migration.

9.1. *Ligand-receptor binding and processing*

The chemoattractant receptors are all seven transmembrane spanning receptors that bind to G-proteins [305]. Much work has been devoted to understanding the binding of radiolabeled N-formyl peptide to neutrophils [154, 454] as well as fluorescent N-formyl peptides utilizing high time resolution, flow cytometric and spectrofluorometric methods [136, 137, 356, 357, 360–362, 408]. Over short time periods (minutes) binding of N-formyl peptides to intact neutrophils at $37°C$ can be described by the following model [136, 360]:

$$L + R_s \overset{k_f}{\underset{k_r}{\longleftrightarrow}} LR_s \overset{k_x}{\longrightarrow} LR_x \overset{k_{r2}}{\underset{k_{f2}}{\longleftrightarrow}} L + R_x \qquad \text{Model 1}$$

$$k_{up} \uparrow \qquad k_{int} \downarrow$$

$$R_{pool} \qquad LR_{in}$$

Ligand (L) binds to surface receptors (R_s) with rate constant k_f to form low-affinity receptor-ligand complexes (LR_s), which convert to high-affinity complexes (LR_x) with rate constant k_x. High-affinity complexes are also formed by ligand binding to high-affinity state receptors (R_x) with the rate constant k_{f2}. Ligand can dissociate from low- or high-affinity receptor-ligand complexes with rate constants k_r and k_{r2}, respectively. The number of surface receptors increases as more receptors from an internal pool (R_{pool}) become upregulated to the surface with rate constant k_{up}. The high-affinity complex can be internalized with rate constant k_{int}. For the high affinity N-formyl peptide CHO-NLFNYK-fluorescein at $37°C$, the kinetic rate constants are $k_f = k_{f2} = 8.4 \times 10^7\,M^{-1}s^{-1}$, $k_r = 0.37\,s^{-1}$, $k_{up} = 8 \times 10^{-4}$, $k_{r2} = 4.6 \times 10^{-3}\,s^{-1}$, $k_x = 6.5 \times 10^{-2}\,s^{-1}$, $k_{int} = 3.3 \times 10^{-3}\,s^{-1}$, and $k_q = 3.9 \times 10^{-3}\,s^{-1}$ [136, 360]. The two receptor states differ most notably (100-fold) in their dissociation rate constants.

Over longer time periods (hours), internalized receptors recycle, bringing previously occupied receptors back to the surface. Receptor recycling appears to be important for sustained migration, since treatments that prevent receptor recycling inhibit migration [291, 298, 299, 453]. The mechanisms of receptor recycling may be important for the development of cell polarity and the cell's sense of direction, as discussed in Section 5.3.3.

9.1.1. *The nature of LR_s*

LR_s is thought to be the signaling form of the receptor, able to activate G_i protein but not precoupled to it. The ternary complex of ligand-receptor-G-protein is not detected in typical binding studies of intact cells because it is very short-lived and only detectable in rapid mixing experiments [268]. That LR_s is the signaling state is supported by the work of Jesaitis *et al.* [155] that shows the amount of oxidant response correlates with the amount of low affinity receptor on the cell surface.

9.1.2. *The nature of LR_x*

LR_x appears to be a desensitized receptor form, since high-affinity ligand binding persists well after responses have terminated [358, 361] and desensitized receptors are of high affinity for ligand [155]. Desensitization of neutrophil responses has been extensively documented [246, 278, 355, 363, 412], and desensitization appears to occur at multiple levels in the signaling cascade, including the receptor [9, 10, 81, 239, 335, 420]. The nature of the desensitized receptor has been partially characterized. Using radiolabeled N-formyl peptide, Jesaitis *et al.* [155] characterized a desensitized, high-affinity, surface receptor state that transiently associates with the detergent-insoluble cell cytoskeleton. In isolated plasma membranes, a high-affinity form of the receptor associates with actin of the membrane skeleton, an actin pool that seems to be inaccessible to cytochalasin [182]. Using protein overlay methods Jesaitis *et al.* [158] have shown that the interaction between receptor and actin is likely direct, and binding of actin to the receptor enhances its affinity for ligand. Desensitization of responses correlates with formation of this receptor-actin complex, suggesting it is a desensitized receptor state [182].

Desensitization and lateral segregation of receptor and G-protein
Subcellular fractionation studies have revealed two plasma membrane fractions in neutrophils, based on their densities: a light plasma membrane fraction and a heavy plasma membrane fraction [156]. In unstimulated cells, the light plasma membrane fraction shows concentration of the N-formyl peptide receptor and G_i-proteins. In cells desensitized by

prior exposure to N-formyl peptide, the receptor is found in the heavy plasma membrane compartment that also contains the cytoskeletal proteins, actin and fodrin; G_i-proteins remain in the light plasma membrane compartment. These results suggest the hypothesis that desensitization results from the lateral segregation of receptors from their G-protein partners into separate compartments where they cannot interact. This hypothesis is further supported by studies that measure the physical interaction between receptors and G-proteins and show that in heavy plasma membranes from desensitized cells the receptor is not coupled to endogenous G-protein, but it can couple to exogenously added G-protein [157, 181]. That actin is also found in the heavy plasma membrane fractions suggests the cytoskeleton plays a role in mediating this lateral segregation of membrane-signaling proteins.

Desensitization and receptor phosphorylation
Like other G-protein-coupled receptors, the N-formyl peptide receptor can be phosphorylated [8, 382], and neutrophils have a G-protein-coupled receptor kinase GRK2 (also known as β-adrenergic receptor kinase) [61, 312]. Phosphorylation of the N-formyl peptide receptor in HL-60 cells (a promyelocytic cell line that can be induced to differentiate into neutrophil-like cells) correlates with response desensitization [8, 382].

The N-formyl peptide receptor has been cloned, expressed, and studied in other cell systems. Analyses of mutant forms of the receptor that lack carboxy terminus phosphorylation sites provide insights into what are possible mechanisms of desensitization in intact neutrophils. In undifferentiated U937 cells transfected with the N-formyl peptide receptor, phosphorylation of the receptor is required for response desensitization [314]. The N-formyl peptide receptor can be phosphorylated at multiple sites in its carboxy terminus by G-protein-coupled receptor kinases 2 and 3 (GRK2 or β-adrenergic kinase and GRK3) [312]. Receptor phosphorylation requires an active receptor state, but not G-protein activation [313, 315]. In addition, mutant receptors with no phosphorylation sites in the carboxy terminus are deficient in receptor phosphorylation and desensitization, but are fully capable of binding ligand and migrating up a concentration gradient of N-formyl peptide [143].

Receptor phosphorylation and arrestin binding
In a reconstituted receptor system, phosphorylated wild-type N-formyl peptide receptors, but not nonphosphorylated receptors,

bind arrestin 2 and arrestin 3 [175], molecules implicated in desensitization and internalization in other G-protein-coupled systems [241]. The arrestin/receptor complex has high affinity for ligand [175]. In studies of mutant N-formyl peptide receptors [27], specific mutants partially deficient in ligand-induced receptor phosphorylation were fully capable of mounting a calcium response to ligand and internalizing ligand, although they did not desensitize. Reconstitution of the phosphorylated form of these mutant receptors with arrestin showed they were deficient in arrestin binding. Together these studies indicate that, at least in U937 cells, arrestin binding to phosphorylated N-formyl peptide receptor may mediate response desensitization, but not internalization.

9.1.3. *Relationship of receptor states in intact neutrophils and reconstituted systems*

Because formation of the high-affinity receptor state in intact neutrophils at 37°C requires ATP [361], it is possible the high-affinity form of the receptor in intact cells (LR_x in Model 1) is a phosphorylated receptor state. However, at 4°C, where receptor upregulation and internalization are inhibited, the low-affinity receptor state still converts to a higher-affinity state [137, 408]. Because at 4°C enzymatic activity is likely to be inhibited, it is possible that there is a high-affinity form of the receptor that is not phosphorylated; this may represent a conformationally distinct form of the (unphosphorylated) receptor. It is not known if this state is capable of signaling or if it exists at 37°C. Moreover, the conversion of the low- to high-affinity receptor states is not affected by cytochalasin [283], as is the case with the formation of high-affinity actin-associated receptor complexes in membranes [182]. The work of Key *et al.* [175] shows that phosphorylated wild-type receptor does not bind G-proteins, even in the absence of arrestin, whereas the work of the Jesaitis group [181, 182] shows that desensitized receptors in the heavy plasma membrane fraction are capable of binding G-proteins, suggesting this desensitized state is not phosphorylated. Taken together, the data suggest that in the intact neutrophil at physiological temperatures, LR_s (low affinity for ligand, capable of interacting with G-protein) converts to LR_x (high affinity for ligand, possibly bound to actin and laterally segregated from G-proteins) which is subsequently phosphorylated and binds arrestin, the later also being high-affinity receptor-ligand states (Figure 4).

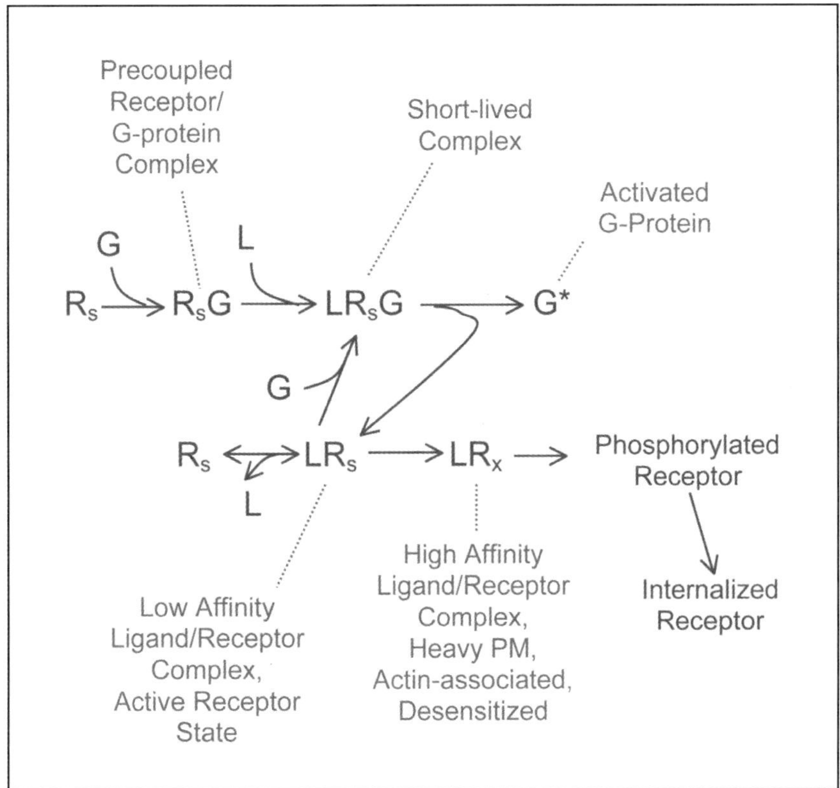

Figure 4. Ligand-receptor binding and processing for the N-formyl peptide receptor. Receptor on the cell surface (R_s) can be either free or precoupled to G-protein (R_sG). Ligand can bind to R_sG to give LR_sG and ligand can bind to R_s to give LR_s, which then binds to G-protein to give LR_sG. LR_sG is very short-lived, rapidly dissociating to give LR_s and active G-protein. LR_s can either dissociate to give free L and R or convert to a high-affinity complex, LR_x. LR_x represents a desensitized receptor state residing in a heavy plasma membrane compartment. LR_x can then be phosphorylated and internalized.

9.1.4. *Receptor internalization and recycling*

In intact neutrophils at 37° C, internalization of N-formyl peptide receptors begins within minutes of initiating ligand binding [153, 357, 376]. This internalization is independent of G-protein activation [113, 315, 361] and early internalization events occur in the presence of cytochalasins [283]; thus, the stimulated actin polymerization response is not requisite. High-affinity ligand-receptor complexes traffic through endocytotic compartments to a Golgi-enriched compartment [153]. Dissociation of ligand from the receptor in internalized compartments may be facilitated by the lower pH in those compartments and by proteases that degrade the ligand [291]. These internalized, unligated receptors can then be recycled to the cell surface [291, 298, 453]. Experimental conditions that inhibit receptor recycling also inhibit chemotaxis [290, 291, 298, 299], indicating that receptor recycling is required for sustained migration of neutrophils. Presumably, receptors must be recycled to the leading edge of the cell for migration to continue, as hypothesized in Section 5.3.3.

9.1.5. *Global inhibition and local activation hypothesis of signaling in the lamellipodium*

We return now to the question of how signals become localized to the leading edge of the cell. As indicated in Section 5.3, it has been hypothesized that the localization of signaling components in the lamellipodium results from global production of an inhibitory signal and localized production of an activation signal, the balance between the two being such that positive signaling overcomes inhibitory signaling only at the leading edge of the cell [104, 295]. Theoretical treatments of this model have been presented [211]. The molecular identities of these signaling species in the neutrophil are largely a mystery, however, the receptor itself may be a key signaling component consistent with this hypothesis.

Early work addressed the hypothesis that chemoattractant receptors were concentrated in the lamellipodium thereby amplifying the chemoattractant gradient. Several investigators have measured the distribution of N-formyl peptide receptors in polarized cells using tagged N-formyl peptides. The results have been mixed with a few reports indicating that receptors accumulate at the leading edge [238, 410], but the majority of reports indicate localization of receptors predominantly in the mid-region of the cell body [245, 343, 344, 377]. More recently, Servant *et al.* [347] have utilized green fluorescent protein (GFP)-tagged C5a receptors transfected into a leukemia cell line that polarizes and migrates in response to C5a. No active concentration of GFP-C5a receptors in the lamellipodium was observed, demonstrating that migration can occur in the absence of a receptor gradient.

It is important to note the differences in labeling methods in these studies. The ability to visualize receptors with tagged ligands may be influenced by the presence of receptors of differing binding affinities. GFP labeling technology would label all receptors irrespective of their affinities for ligand. As noted above, N-formyl peptide binding to its receptor results in the conversion from a low-affinity, signaling state to a high-affinity, desensitized state. It is possible studies utilizing tagged ligand to visualize receptor could have underestimated the levels of low-affinity receptors concentrated in the lamellipodium. In agreement with this contention, Loitto *et al.* [216] showed that N-formyl peptide receptor antibodies detect receptors that are not detected by fluorescent ligands at the extreme tip of the lamellipodium. Similarly in the amoeboid cell, *Dictyostelium discoideum*, chemoattractant cAMP receptors, and G-proteins are uniformily distributed around the cell [295], however, dissociation of ligand from receptors is faster at the anterior end of the cell [397]. These considerations lead to the modified hypothesis that concentration of *low-affinity, signaling* receptors at the leading edge of neutrophils is responsible for developing and maintaining cell polarity.

Thus development of polarity is hypothesized to occur by the following sequence of events: Binding of ligand to receptors all across the cell rapidly leads to activation of signaling pathways and global receptor desensitization. The slight excess of ligand-binding events at the side of the cell facing the gradient marks a local target for surface expression of receptors from preexisting internal stores and recycling pathways. This local target may be newly polymerized actin [245] or 3′-phosphoinositides (further discussed in Section 9.3.1). In the case of stimulation in a uniform concentration of ligand, the slight excess of binding events occurs stochastically. Desensitized receptors move to the rear of the cell, are internalized, and subsequently are recycled to the front of the cell. Thus as time progresses, activatable receptors (those capable of signaling when ligand binds) are inserted at the leading edge of the cell and desensitized receptors move back to the rear. In such a scheme, the distribution of all receptors on the cell surface could remain rather uniform. Thus the "global inhibition" component of the mechanism is receptor desensitization, and the "local activation" is a result of localized upregulation of new receptors. This mechanism is consistent with a need for a polarized actin cytoskeleton to orient receptor trafficking.

9.2. *Heterotrimeric G-proteins*

In the early 1980s, heterotrimeric G-proteins were discovered in many cell systems including the neutrophil. Binding of N-formyl peptide to neutrophil plasma membranes was shown to be sensitive to guanine nucleotides [187]. Hyslop *et al.* [145] showed that N-formylpeptides stimulated GTPase activity in neutrophil homogenates and plasma membrane fractions. That same year, Bokoch and Gilman [37] showed that N-formyl peptide-mediated release of arachidonic acid and granular enzymes was inhibited by pretreatment of neutrophils with pertussis toxin, a toxin previously identified as an enzyme that ADP-ribosylates G-proteins of the subclass, G_i [36]. It is now well established that N-formyl peptide receptors activate pertussis toxin sensitive heterotrimeric G_i-proteins of the subclasses G_{i2} and G_{i3}, with G_{i2} being the most abundant [112, 256]. Pertussis toxin-mediated ADP-riboyslation of the $G_{\alpha i}$ subunit of the G-protein makes it unable to couple to receptors. All responses and signaling events induced by N-formyl peptides, including migration, can be inhibited by pertussis toxin treatment with notable exceptions: the conversion of the ligand-receptor complex from the low- to the high-affinity state, receptor phosphorylation, and receptor internalization [136, 315, 361], as noted above.

9.2.1. *The G-protein cycle*

The ensuing G-protein signaling cycle proceeds analogous to that described in other cell systems (Figure 5) [269]. After ligand binds to receptor on the cell surface, the receptor diffuses to a heterotrimeric G_i-protein with GDP bound to the alpha subunit. Alternatively, precoupled receptor-G_i-protein complexes may exist on the cell surface. The interaction of G-protein with ligated receptor causes the dissociation of GDP from the G-protein and allows the binding of GTP, which is present at high concentrations in the cell (relative to its dissociation constant for binding to the G-protein). Binding of GTP causes dissociation of G-protein from the receptor and dissociation of the G-protein α subunit from the $\beta\gamma$ subunits. Both the GTP-bound α subunit and the $\beta\gamma$ subunits have signaling functions. The G_α subunit has an intrinsic GTPase activity that can be stimulated by regulators of G-protein signaling (RGS) proteins. In addition, while bound to the GTP-G_α, some RGS proteins are also able to stimulate guanine nucleotide exchange on small GTP-binding protein (to be discussed in Section 9.3.2), thus providing an additional mechanism for signaling downstream events [122]. The inactive GDP-bound G_α binds $G_{\beta\gamma}$ subunits, terminating their activities.

In nonneutrophil cell lines transfected with chemoattractant receptors, chemotaxis induced by leukocyte chemoattractants is mediated by the released $G_{i\beta\gamma}$ subunits, not $G_{i\alpha}$ [16, 266, 267]. Cells (human embryonic kidney cells and a lymphocyte cell line) transfected with chemoattractant receptors migrate in response to the corresponding chemoattractant,

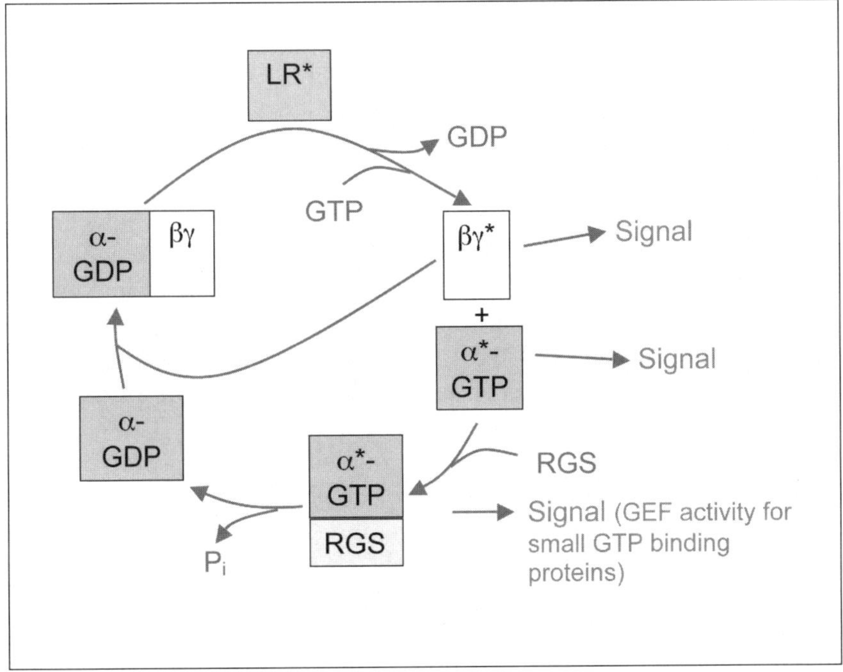

Figure 5.   G-protein activation cycle. In the resting cell, G-protein exists as a trimer of α, β, and γ subunits with GDP bound to the α subunit. Activated ligand-receptor complexes (LR*) stimulate the release of GDP from the α-subunit; GTP rapidly fills the void. Binding of GTP causes dissociation of the G-protein from the LR complex and dissociation of the G-protein α subunit from the $\beta\gamma$ subunits, both of which have signaling functions. The α subunit has an intrinsic GTPase activity that is stimulated by regulator of G-protein signaling (RGS) proteins to return the α subunit to the inactive, GDP-bound form. When coupled to α-GTP, RGS proteins may also function as guanine nucleotide exchange factors (GEF) for activation of small GTP-binding proteins. Free α-GDP binds $\beta\gamma$ subunits, terminating their activity and completing the cycle.

whereas cells transfected with chemoattractant receptors and proteins that sequester $G_{i\beta\gamma}$ subunits are defective in the chemotaxis response, but normal in responses mediated by $G_{i\alpha}$. Whether this is also true for neutrophils remains to be determined.

9.2.2. *Precoupled receptor/G-protein complexes*

The evidence that precoupled receptor/G-proteins exist in intact neutrophils is circumstantial, but likely to be important for migration. Such a state would provide rapid signaling that is not dependent upon receptor and G-protein diffusion. Successful modeling of actin polymerization response kinetics in intact cells requires the presence of a precoupled receptor/G-protein complex with faster forward rate constant than for free receptor [3]. Since actin polymerization is required for chemotaxis, it is likely that precoupled receptor/G-proteins are of particular importance for chemotaxis. Yet precoupled receptor/G-proteins have not been detected in binding studies of intact neutrophils [136]. This may be because they exist in intact cells in such a low fraction that they are not detectable with current binding technology. In contrast, in permeabilized neutrophils fifty percent of receptors are precoupled to G-proteins [100, 310], and reconstitution of detergent solubilized N-formyl peptide receptor and G-proteins in the presence of ligand results in the formation of an LRG complex [28]. These studies confirm that binding of ligand to precoupled receptor/G-proteins is of higher affinity than binding to receptor alone. However, the formation of precoupled receptor-G-protein complexes in permeabilized and/or reconstituted systems may not reflect levels of receptor-G-protein coupling in intact neutrophils for (at least) two reasons. First, in permeabililized cells or reconstituted systems, the G-protein does not have guanine nucleotide bound, and thus the precoupled LRG state is not the same as the precoupled RG complex in intact cells that presumably has GDP bound to the G-protein. Second, studies with permeabilized cells are conducted under conditions where G-protein GTPase activity does not occur, whereas in intact cells it does. Theoretical studies of the effect of G-protein GTPase activity on the level of precoupled receptor/G-proteins show that the predicted number of precoupled receptor/G-protein complexes drops when GTPase activity occurs [353]. In addition, theoretical studies of the effect of precoupled receptor/G-proteins on cellular responses indicate that even a small percentage of precoupled receptor/G-protein complexes could significantly enhance responsiveness [352]. Thus it is likely that precoupled receptor/G-protein complexes are important in signaling neutrophil migratory responses.

Reconstitution of detergent-solubilized receptors and G-proteins have been accomplished and provide interesting insights into possible receptor-G-protein interactions [28, 175]. In these studies, there is no guanine nucleotide bound to the G-protein. This nucleotide-free G-protein forms a complex only with liganded receptor; the ligand slowly dissociates from this state. Addition of GTPγS (a nonhydrolyzable analog of GTP) to this LRG complex causes the rapid dissociation

($t_{1/2} \sim 140\,$msec) [268] of the G-protein from the receptor, leaving an LR complex from which L dissociates more rapidly. Using this reconstituted system, Bennett *et al.* [28] have shown there is specificity of receptor binding to G-proteins: G_{i3} binds with higher affinity than G_{i2}; G-proteins not found in neutrophils, G_{i1} and G_o, bind poorly. In addition, the heterotrimeric form of the G-protein is required for binding to the receptor; α-subunits or βγ-subunits alone do not bind. Moreover, phosphorylated receptors do not bind to G-proteins, even in the absence of arrestins [175]. Thus N-formyl peptide receptor phosphorylation alone is enough to block receptor-G-protein interactions.

9.3. *G-protein effectors in neutrophils*

The activated neutrophil G_i-proteins then go on to initiate downstream signaling cascades. Many biochemical events are initiated in neutrophils, and in many cases the exact hierarchy of signaling events is not known. Evidence suggests that activation of phospholipase C (PLC), phosphatidylinositol 3-kinase (PI3K), tyrosine kinases, and the small GTP-binding proteins, Ras, Rac, and/or Cdc42, are directly or closely linked to activation by G-proteins. These are each discussed below, as are their downstream signaling, with an emphasis on data addressing their roles in migration.

9.3.1. *Phosphatidylinositol 3-kinase*

Enzyme characteristics
Phosphatidylinositol 3-kinase (PI3K) phosphorylates inositol phospholipids on the 3′-position of the inositol ring [232, 389, 402, 435]. Three classes of PI3Ks have been identified, of which the Class I isoforms are unique in their selectivity for phosphatidylinositol 4,5-bisphosphate (PIP_2) as a substrate to generate phosphatidylinositol 3,4,5-trisphosphate (PIP_3). In numerous cell types, PIP_3 is a signaling intermediate in pathways leading to cell adhesion, cytoskeletal rearrangement, cell growth and survival. Class I PI3Ks are dimeric proteins consisting of a 110 kDa catalytic subunit (p110) and an adaptor/regulatory subunit (of differing molecular weights). There are two members of Class I. Class I_A enzymes are activated downstream of tyrosine kinases and consist of three p110 isoforms (α, β, δ) and at least seven adaptor proteins. Class I_B enzymes are regulated by G-proteins, and only one isoform has been identified consisting of p110γ catalytic subunit and a 101 kDa regulatory protein

(also denoted PI3Kγ). The catalytic subunits of Class I isoforms bind the small GTP-binding protein, Ras, although the significance of this for PI3K activation is unclear. Class I isoforms also exhibit protein kinase activity; although the significance of this activity is, likewise, not clear, it suggests multiple signaling pathways may emerge from PI3K activation.

PI3K in neutrophils

In neutrophils, PIP_2 is a substrate for PI3K, resulting in the production of the signaling lipid PIP_3 [370, 394, 395]. Neutrophils contain Class I_A, tyrosine kinase regulated PI3Ks (p110δ catalytic subunit is leukocyte specific and the most abundant Class I_A enzyme in neutrophils coupled with an 85 kDa regulatory subunit) and a $G_{\beta\gamma}$ regulated form (p110γ catalytic subunit with 101 kDa regulatory subunit, PI3Kγ); both types can be activated by N-formyl peptides. The activation of PI3Kγ is via direct activation by $G_{\beta\gamma}$ [372]. Class I_A PI3Ks are more abundant than the PI3Kγ form [383] and are activated downstream of tyrosine kinases [317, 385]. Activation of Class I_A PI3Ks is via the Src family tyrosine kinase, Lyn kinase, and involves the adaptor protein, Shc [317]. In N-formyl peptide-activated neutrophils phosphorylated Lyn, phosphory-lated Shc, and active PI3K form a complex. Activation of these phosphorylation events and PI3K is inhibited by pretreatment of the cells with pertussis toxin. Thus it is proposed that G-protein-dependent activation of Lyn kinase results in its autophosphorylation, phosphoryla-tion of the adaptor protein Shc, and phosphorylated Shc activation of Class I_A PI3K [317]. The precise molecular interactions linking acti-vated G-protein and activated Lyn kinase are not clear. The pathways by which PIP_3 may be generated in neutrophils are summarized in Figure 6. The role played by each of the two classes of PI3Ks in neutrophil acti-vation is unclear. Studies of the sensitivity of chemoattractant-induced PIP_3 production to tyrosine kinase inhibitors (expected to inhibit the Class I_A isoforms) and measurement of the activity of PI3K immuno-precipitated from activated cells have indicated variable results. Some studies suggest as much as 70% of PIP_3 is produced through a tyrosine-kinase regulated PI3K [318, 385] and others indicate that 90% or more of PIP_3 is produced through G-protein regulated PI3K [261, 371, 404].

Most PI3Ks (including Class I_A and I_B) are inhibited by a fungal toxin, wortmannin, and the structurally unrelated inhibitor LY294002. These compounds are potent inhibitors of chemoattractant-induced degranulation and oxidant production by neutrophils although phos-pholipase C activation and the cytosolic calcium increase are not

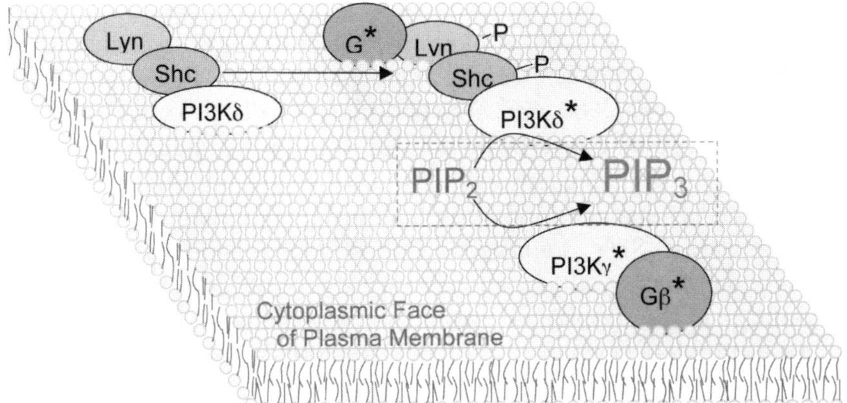

Figure 6. Pathways for stimulation of PI3K in neutrophils. PI3Kγ is activated by free Gβγ subunits released by chemoattractant-induced G-protein activation. Class I$_A$ PI3Ks (represented as PI3Kδ in the figure) are in complex with the kinase, Lyn, and the adaptor protein, Shc. Activation of G-protein results in activation of Lyn kinase, triggering autophosphorylation and phosphorylation of Shc, events that activate PI3K activity.

affected [18, 77, 275, 405]. These inhibitors also attenuate N-formyl peptide-induced development of neutrophil polarity and incorporation of F-actin into cytoskeletal actin (defined as the fraction of actin sedimenting at low g force after solubilization of cells with non-ionic detergent) but do not inhibit the increase in total cellular F-actin [18, 272, 275, 405]. However, reports of the effects of these inhibitors on N-formyl peptide-induced migration are conflicting, with some reports indicating PI3K inhibitors inhibit migration [65, 119, 184, 272] and others indicating no inhibition of migration [275, 384]. These discrepancies may indicate that PIP$_3$-mediated pathways are not required for migration, but are required for optimal migration.

PI3K inhibitor studies have been further supplemented by studies of neutrophils from mice that lack PI3Kγ [135, 213, 341]. Chemoattractants fail to induce a measurable increase in PIP$_3$ in these cells, indicating that in mice PI3Kγ is the primary form of PI3K activated by chemoattractants. Intracellular calcium elevation and actin polymerization responses are normal; migration is significantly, but not completely, inhibited. Taken together, the PI3K inhibitor studies and studies of PI3K-deficient mice suggest that while PI3K activation is not required for initiation of actin polymerization, it may be required for optimal assembly of the actin filament network and orientation for directed movement.

Signaling functions of PIP$_3$

The product of PI3K, PIP$_3$, is a known cell-signaling molecule that functions in adhesion, actin cytoskeletal rearrangement, growth and survival in a variety of cell types [389, 402]. As a phospholipid, PIP$_3$ is located in membranes and functions in part by recruiting proteins to the membrane. Many proteins have consensus sequence domains that bind PIP$_3$. Pleckstrin homology (PH) domains bind phosphoinositides including PIP$_3$, whereas Src homology-2 (SH2) domains bind phosphotyrosyl-containing proteins or PIP$_3$, the binding being mutually exclusive [389]. An example of the importance of these binding domains can be demonstrated by the ser/thr kinase Akt/protein kinase B (Akt/PKB) [163]. PIP$_3$ simultaneously recruits Akt/PKB to the plasma membrane and activates additional protein kinases (3-phosphoinositide-dependent kinases) that phosphorylate and activate Akt/PKB. The importance of PIP$_3$ in activating Akt/PKB is further supported in studies that dissect out the contributions of the lipid kinase verse protein kinase activities of PI3K; PI3Kγ mutants defective in lipid kinase activity (but retaining protein kinase activity measured as PI3Kγ autophosphorylation) did not activate Akt/PKB [39]. These same mutants retained the ability to activate the MAPK pathway, indicating that it is the protein kinase activity, not the lipid kinase activity, of PI3K that is important for mediating MAPK activity (discussed further in Section 9.3.3).

Stimulation of neutrophils with N-formyl peptide, IL-8, and the cytokine GROα results in transient activation of Akt/PKB [176, 275, 337, 341, 388]. This activation is inhibited by pertussis toxin pretreatment of neutrophils and by PI3K inhibitors, confirming G$_i$-protein-mediated activation of PI3K is required for Akt/PKB activation in neutrophils. Likewise, neutrophils from PI3Kγ null mice show no activation of Akt/PKB [341]. Direct binding of PIP$_3$ to Akt/PKB and its 3-phosphoinositide-dependent kinases may be the mechanism by which Akt/PKB is activated in neutrophils. However, an additional pathway for activation of Akt/PKB via the MAPKinase pathway has also been reported, and this pathway, along with its possible role in regulating migration, will be discussed in Section 9.3.3.

Localization of signals in the lamellipodium by 3'-phosphoinositides in a positive feedback loop

We return now to the second hypothesis of how signaling components localize in the lamellipodium (Section 5.3)—that of a positive feed forward loop of PI3K activation. Neutrophils treated with exogenous PIP$_3$ provide evidence for a role of PIP$_3$ in migration. PIP$_3$ itself is not membrane-permeant, however, it can be incorporated into cells as a membrane permeant ester (dilauroyl phosphotidylinositol 3,4,5-trisphosphate-heptakis-(acetooxy-methyl)ester) that crosses the plasma membrane by passive diffusion and is hydrolyzed by intracellular esterases, generating PIP_3 inside the neutrophils [274]. Alternatively, PIP_3 complexed with histones can be delivered to neutrophil-like HL-60 cells [417]. Incubation of neutrophils and HL-60 cells with exogenous PIP_3 causes cell polarization, accumulation of F-actin in lamellipodia, and random motility analogous to that observed in a uniform concentration of chemotactic peptide. These effects are inhibited by PI3K inhibitors—an unexpected result because addition of PIP_3 would be expected to circumvent the need for PI3K activation. These results suggest that PIP_3 induces activation of PI3K in a positive feedback loop [274, 417].

Localization of PIP_3 in the lamellipodia of neutrophil-like HL-60 cells has been demonstrated by tranfecting these cells with a GFP-tagged peptide sequence that contains a pleckstrin homology domain from Akt/PKB (PH-Akt-GFP) [348, 411, 417]. The gradient of 3-phosphoinositides in a polarized neutrophil-like HL-60 cell is much steeper than the gradient of chemoattractant to which it responds, an observation that demonstrates the ability of the cell to internally amplify the effect of the chemoattractant gradient [348]. Exogenously added PIP_3 also induces localization of the PH-Akt-GFP to the lamellipodium by a mechanism that is inhibited by PI3K inhibitors and inhibitors of Rho GTPases (further discussed in Section 9.3.2). This observation further confirms the presence of a positive feedback loop whereby PIP_3 stimulates endogenous PI3K activity and suggests the latter involves activation of a Rho GTPase [417]. Furthermore, pretreatment of these cells with disruptors of the actin cytoskeleton markedly impairs the amplification of the internal PIP_3 gradient [411]. Thus, as previously hypothesized [245], F-actin in the lamellipodium is important for concentrating signaling species there.

Partial inhibition of the positive feedback loop results in cells that migrate more slowly and wander more than control cells, suggesting the PIP_3 positive feedback loop is necessary for orientation [411]. 3'-phosphoinositides have been called the "compasses" that determine cell orientation [336], and they may be important for determining the location at which actin polymerization occurs. However, it should be noted that inhibitors of PI3K do not inhibit chemoattractant-induced actin polymerization itself [272, 405], and thus other signaling mechanism must serve this function.

Thus both mechanisms for localization of signaling in the lamellipodium, a positive feedback loop amplifying PI3K activity and local upregulation of signaling receptors coupled with global desensitization (Section 9.1.5) may function in an interdependent way. The PIP_3 positive feedback loop may be important for orientation, and additional receptor-mediated signaling events may be important for promoting actin polymerization, which in turn influences the PIP_3 amplification and signaling at the leading edge.

Downstream effectors of PI3K
Activation of PI3K is upstream of several signaling molecules including the small GTP-binding proteins Rac, Cdc42, Rho, and the MAPK cascade, each of which will be described below, and each of which has been implicated in the regulation of migration. It should also be noted that many of these pathways also can be activated by PI3K-independent pathways.

9.3.2. *Small GTP-binding proteins and regulation of neutrophil chemotaxis*

Small GTP-binding proteins provide a switch for turning on and off signaling pathways (Figure 7) [320]. Like the heterotrimeric G-proteins, the small GTP-binding proteins bind GDP in their inactive state, but binding of GTP results in activation. Likewise, the small GTP-binding proteins have an intrinsic GTPase activity. The exchange of GDP and GTP at the guanine nucleotide binding site can be regulated by guanine nucleotide exchange factors (GEFs) and guanine nucleotide dissociation inhibitors (GDIs) that promote or prevent the release of GDP. The GTPase activity can be modulated by GTPase-activating proteins (GAPs). Many small GTP-binding proteins are present in neutrophils, including Ras, Rac1, Rac2, Rho, Cdc42, Ral, Rab1A, and several putative GEFs (such as Vav and Sos), GAPs (such as p120-GAP), and GDIs (such as RhoGDI), have been identified [38, 62, 90, 445]. The following will focus on those small GTP-binding proteins whose activities appear to be important for migration.

Rac, Cdc42 and regulation of neutrophil actin polymerization
Both Rac and Cdc42, members of the Rho family of small GTP-binding proteins, are activated in neutrophils by N-formyl peptides [6, 24]. In resting cells, these proteins reside predominantly in the cytoplasm in

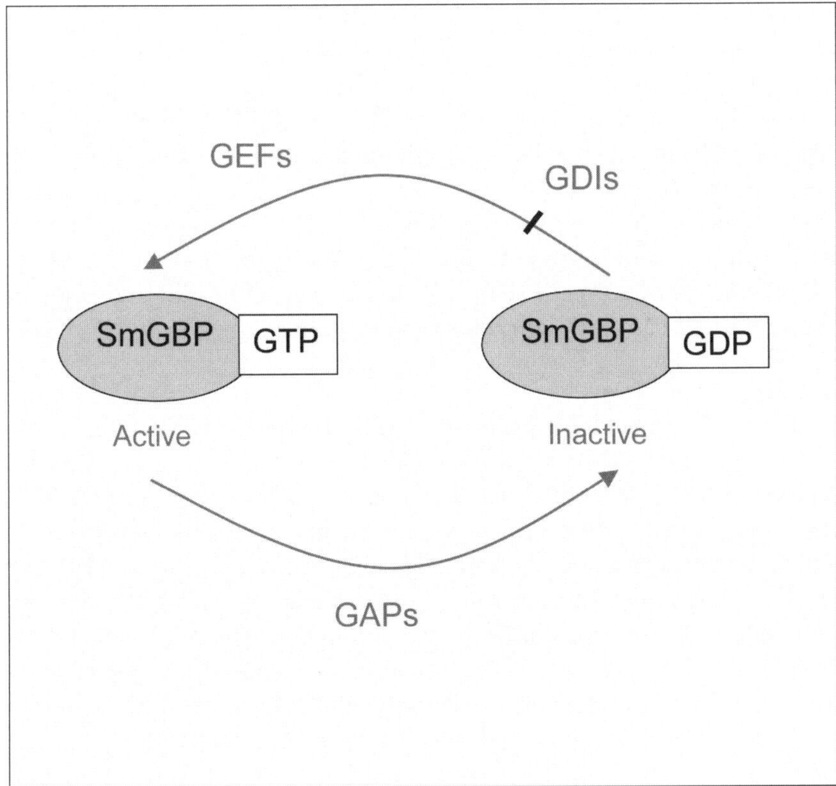

Figure 7. Regulation of small GTP-binding protein activation. Small GTP-binding proteins (SmGBP) with bound GDP are inactive. Guanine nucleotide exchange factors (GEFs) stimulate the release of GDP allowing GTP to bind and activate the protein. Guanine nucleotide dissociation inhibitors (GDIs) inhibit dissociation of GDP thereby inhibiting activation. GTPase-activating proteins (GAPs) stimulate the hydrolysis of bound GTP to GDP, returning the protein to its inactive, GDP-bound state.

complex with RhoGDI [1, 38, 62]. Activation of Rac and Cdc42 coincides with their release from RhoGDI and translocation to the plasma membrane [38, 303, 319] where GEFs and proteins that activate GEFs appear to be located [38, 458]. This activation is dependent upon G_i-protein activation but independent of PLC activation [213]. Studies of Benard *et al.* [24] show that inhibitors of PI3K substantially but not completely inhibit this activation, indicating there are PI3K-independent and -dependent pathways for N-formyl peptide–mediated activation of these proteins. Further support of PI3K-independent pathways for activating Rac is provided by studies of mice neutrophils deficient in PI3Kγ, in which N-formyl peptide–induced Rac2 activation is normal [213]. The intervening steps between G_i-protein or PI3K activation and activation of Rac and Cdc42 in neutrophils remain to be elucidated, although activation of GEFs are likely to be involved [38, 164, 458]. Recently a protein called p115RhoGEF has been identified that acts as a RGS protein to stimulate the GTPase activity of $G_{\alpha 13}$, and when bound to $G_{\alpha 13}$-GTP it acts as a RhoGEF, thus serving as a direct link between activation of heterotrimeric G-proteins and small GTP-binding proteins [122, 269]. It will be interesting to see if such a mechanism functions for activation of Rho family GTPase by G_i proteins in neutrophils.

Cdc42

Studies of the role of Cdc42 in neutrophils has been motivated in large part by the hypothesized role of Cdc42-induced activation of actin nucleation (in many cell types) via the actin-nucleating protein Arp2/3. The model developed from these studies links receptor activation of Cdc42 to activation of Wiskott–Aldrich syndrome protein (WASP), which in turn stimulates the nucleation activity of Arp2/3 [219, 220, 308]. WASP, a protein identified as containing a mutation that results in Wiskott–Aldrich syndrome [331], is expressed exclusively in hematopoietic cells, although homologs such as N-WASP have been identified in other cell types (together these proteins are referred to as the WASP/Scar family) [132]. The observation that expression of WASP in porcine aortic endothelial cells results in Cdc42-dependent increases in F-actin launched the hypothesis that WASP was an important regulator of the actin cytoskeleton [379]. Subsequent studies have shown that WASP/Scar proteins function to stimulate the actin-nucleation activity of Arp2/3 complex, a complex of seven subunits, which by itself nucleates actin polymerization with low efficiency [34, 219–221].

A role for Cdc42 in initiation of actin polymerization in neutrophils has been supported by studies of neutrophil lysates that retain the ability to polymerize actin in response to GTPγS and exogenous Cdc42 [458], although they do not retain an ability to be activated by chemoattractants. This polymerization is inhibited by cytochalasins, thus, it requires barbed-end elongation. In this system, actin nucleation and F-actin polymerization are induced by GTPγS added to cell lysates that retain particulate membrane fractions, but not lysates from which particulate membranes have been removed by centrifugation. Recombinant, activated Cdc42, but not Rac1 or Rho, added to either lysate preparation induces actin polymerization. The ability of GTPγS to induce actin polymerization in lysates depleted of membrane is restored by the addition of recombinant GEFs, Cdc24 and Dbl. These results confirm that neutrophils contain machinery for GEF-induced activation of Cdc42 that leads to actin nucleation and F-actin polymerization. Downstream targets of activated Cdc42 that lead to actin polymerization were not further identified in this study, although roles for PIP_2 and PIP_3 were ruled out. These studies were partially confirmed in studies by Katanaev and Wymann [164] who also studied GTPγS-induced actin polymerization in neutrophil cytosolic extracts. Further studies by Zigmond *et al.* [459] showed that addition of free barbed ends (either as spectrin-actin seeds or sheared F-actin filaments) to neutrophil lysates did not induce actin polymerization, indicating that an increase in free barbed ends alone cannot drive polymerization. In addition, the rate of Cdc42-induced polymerization did not increase in the presence of spectrin-actin seeds, suggesting that Cdc42 does not inhibit capping proteins or release G-actin from a sequestered pool. Rather Cdc42 appears to induce actin polymerization by both creating free barbed ends and facilitating elongation at those sites. It has been hypothesized that Cdc42 activates a complex that contains a nucleation component and an elongation component; the nucleation component creates free barbed ends (either by uncapping existing filaments or creating new filaments), then remains bound at the pointed end of the filament whereas the elongation factor binds near the barbed end and both facilitates elongation and prevents capping proteins from binding [459].

Rac

The importance of Rac in N-formyl peptide–induced neutrophil migration is demonstrated by organisms in which Rac is deficient or mutated. Neutrophils contain two isoforms of Rac, Rac1 and Rac2. Rac1 is found in many cell types, whereas Rac2 is found only in hematopoietic cells, and in neutrophils it is the more abundant of the two forms (>96%) [130, 319]. In neutrophils from Rac2-deficient mice [337] migration is significantly inhibited (although not completely) as measured by *in vitro* (migration through polycarbonate filters) and *in vivo* (accumulation of neutrophils in peritoneal exudates) assays. In addition, a Rac2 dominant negative mutation has been identified in a human [14, 427]. The neutrophils from this patient demonstrate a dramatic inability to migrate on polycarbonate filters—both random migration and chemotaxis to N-formyl peptide and IL-8 are inhibited [14, 427]. Rac2 defects are also associated with reduced tethering and rolling of cells on the L-selectin ligand, GlyCAM-1, adhesion to fibronectin, and spreading after integrin ligation [427, 337]. Resting levels of total cellular F-actin are approximately one quarter less in Rac2-deficient neutrophils than in wild-type neutrophils, and the ability of N-formyl peptide, IL-8, and LTB$_4$ to induce an increase in total cellular F-actin is dramatically reduced in Rac2-deficient neutrophils [337]. Thus Rac2 is clearly involved in signaling mechanisms that regulate the initiation of actin polymerization.

Further elucidation of the roles of Rac and Cdc42 in signaling for actin nucleation (generation of free barbed ends) comes from studies with permeabilized neutrophils that can be penetrated by putative activators and inhibitors of actin nucleation [114]. Neutrophils permeabilized with octyl glucoside retain the ability to be activated by N-formyl peptides and actin-nucleation activity can be measured with an assay that utilizes pyrene-labeled exogenous actin. Under the conditions utilized, the rate of polymerization of the exogenous pyrene-actin is proportional to the number of actin-nucleation sites in the permeabilized cells. PIP$_2$ and PIP$_3$ promote the formation of actin-nucleation sites, whereas a PIP$_2$-binding peptide derived from the gelsolin phosphoinositide-binding site partially inhibit N-formyl peptide–induced nucleation, suggesting that polyphosphoinositides are important signals for actin nucleation. Constitutively active mutants of Rac and Cdc42 stimulate actin nucleation whereas inactive mutants inhibit N-formyl peptide–induced actin nucleation. A peptide derived from N-WASP that activates Arp2/3 complex *in vitro* increases nucleation activity of permeabilized neutrophils, whereas a peptide derived from N-WASP that inhibits Arp2/3 complex *in vitro* partially inhibits N-formyl peptide–induced nucleation. Various combinations of these activators and inhibitors identify a sequence of events in which Cdc42 activates actin nucleation by two pathways, one involving Rac-dependent polyphosphoinositide synthesis presumed to induce uncapping of barbed ends, and a second N-WASP mediated pathway apparently involving Arp2/3 (Figure 8). This schema is consistent with the results from studies of lysates [458] in that a Cdc42-dependent, phosphoinositide-independent pathway for inducing actin polymerization was identified. And although no Rac-induced actin polymerization was noted, Rac-induced polyphosphoinositide production was detected. The apparent discrepancies between the two studies may lie in the likely circumstance that both pathways (perhaps as well as others) are present in neutrophils and particular experimental protocols may bias the functioning of either one. *In vivo* these pathways may also function in a compensatory fashion;

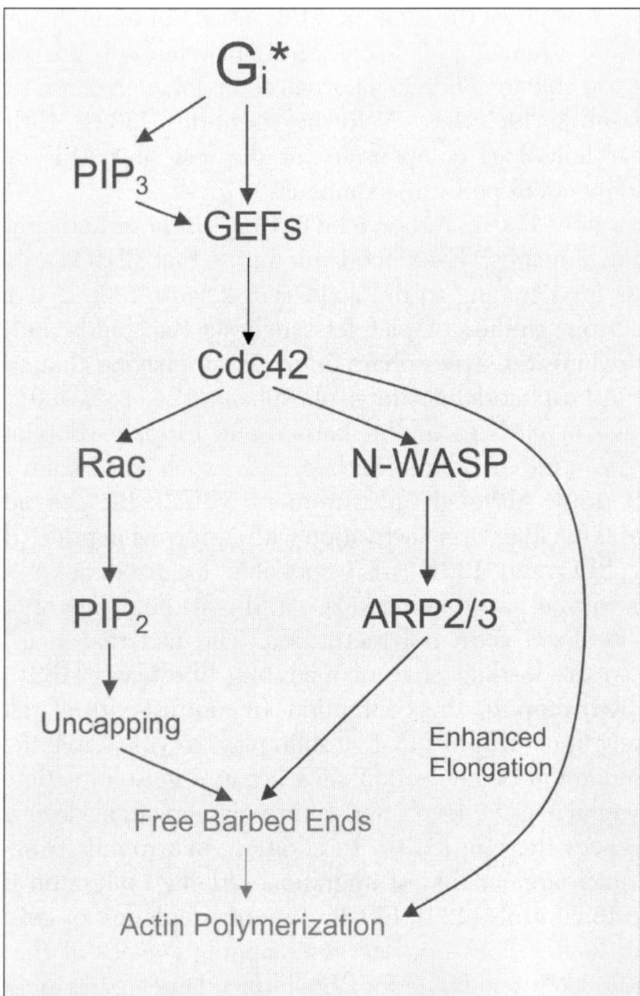

Figure 8. Schematic of proposed pathways leading to chemoattractant-induced actin polymerization in neutrophils. Activated G-protein (G_i^*) stimulates guanine nucleotide exchange factors (GEFs) by PIP_3-dependent and -independent pathways. Cdc42 stimulates actin polymerization via two pathways. In one pathway, Cdc42 activates a protein of the WASP/Scar family (N-WASP), which in turn activates Arp2/3 complex to create new barbed ends at which actin polymerization occurs. Cdc42 also stimulates Rac, which in turn stimulates PIP_2 production via phosphatidyl-4-inositol 5-kinase. PIP_2 inhibits barbed-end capping proteins thereby releasing free barbed ends at which polymerization occurs. Cdc42 also stimulates enhanced elongation of F-actin at the free barbed ends.

if one is not functional the other may take over. For example, neutrophils (as well as other hematopoietic cells) from patients with Wiskott–Aldrich syndrome are able to generate a normal actin polymerization response to several stimuli, including N-formyl peptides [332]. Either other WASP/Scar homologs compensate for the loss of WASP, or perhaps WASP-independent pathways compensate.

Just how does Rac activation lead to PIP_2 synthesis and generation of actin-nucleation sites? Evidence is mounting that PIP_2 is an important messenger for signaling to the actin cytoskeleton [33, 124, 133, 345]. Evidence from studies of platelets suggests Rac binds and activates phosphatidylinositol-4-phosphate 5-kinase, an enzyme that synthesizes PIP_2 from phosphatidylinositol-4-phosphate [124, 232, 390]. PIP_2 has been shown *in vitro* to inhibit barbed-end capping proteins (i.e., to increase the number of free barbed ends) such as gelsolin and capZ [128, 149, 150]. Although measurement of bulk PIP_2 in neutrophils shows that it declines upon activation with N-formyl peptide (due to the action of PLC and PI3K), it is possible localized action of phosphatidylinositol-4-phosphate 5-kinase at the leading edge of cells could facilitate localized actin polymerization. The fact that Rac has been localized to the leading edge of migrating fibroblasts [189] and neutrophils [306] supports this contention. In contrast with platelets [124], it is unlikely in neutrophils that gelsolin plays a critical role in N-formyl peptide induction of nucleation sites because gelsolin action requires μM intracellular Ca^{2+} levels and actin polymerization does not ([289] and references therein; [359]). In addition, neutrophils from gelsolin-deficient mice are capable of migration, although migraton is delayed compared to controls [429]. Finally, immunodepletion of gelsolin from neutrophil lysates does not decrease capping activity of these lysates [83]. A more likely candidate for PIP_2-induced uncapping in neutrophils is a Ca^{2+}-independent capping protein, capping protein-β_2, a homolog of capZ that confers capping activity upon neutrophil lysates and is inhibited by PIP_2 [83, 235].

p21$^{Rac/Cdc42}$-activated kinases (PAKs)

Additional roles for Rac and Cdc42 in regulating neutrophil migration appear to be mediated by p21-activated kinases (PAKs), serine/threonine protein kinases that are activated by Rac and Cdc42 [186]. Activation of PAKs is implicated in actin remodeling, the stress response, activation of the NADPH oxidase, and initiation of apoptosis. When GTP-bound forms of Rac or Cdc42 associate with a characteristic

GTPase-binding domain (GBD) of PAK, PAK autophosphorylation occurs and phosphotransferase activity towards substrates increases as much as 300-fold [444]. In human neutrophils two isoforms, PAK1 and PAK2, have been identified [82, 185]. N-formyl peptide induces PAK activation rapidly (within 30 s) and transiently with a time course that parallels Rac activation [144, 185]. In unstimulated neutrophils, PAK1 is located in the cytosol; PAK1 colocalizes with F-actin in membrane ruffles and lamellipodia in N-formyl peptide–stimulated neutrophils [78].

Potential substrates for PAKs that may regulate cytoskeletal function have been identified. PAK1 has been shown to phosphorylate myosin light chain kinase (MLCK) thereby decreasing its activity [340]. Contractile events in nonmuscle cells are primarily modulated by actin and myosin II [45], and actin-myosin interactions are regulated (stimulated) by the phosphorylation of the regulatory subunit of myosin, myosin light chain (MLC) by MLCK (Figure 9). Thus activation of PAKs would be expected to inhibit myosin II-mediated contraction. Studies in BHK-21 cells show that activation of PAK1 inhibits MLCK activity, MLC phosphorylation, and cell spreading. Although direct phosphorylation of MLCK by PAKs has not been demonstrated in neutrophils, N-formyl peptides are known to cause myosin phosphorylation [101, 273, 309, 330]. Given the known importance of myosin II for retraction of the neutrophil uropod during migration on fibronectin and vitronectin [91], further investigation into this pathway is warranted.

Cofilin is an actin-binding protein that binds to F-actin filaments, severs them and promotes actin depolymerization; activity of cofilin is inhibited by phosphorylation [222, 333]. In nonneutrophil cell lines PAK1 phosphorylates and activates LIM-kinase which in turn phosphorylates and inactivates cofilin [17, 92, 223, 436]. The net effect of PAK1 activation of LIM-kinase would then be to promote the accumulation of F-actin by inhibiting depolymerization. However, this mechanism does not appear to function in neutrophils. A significant fraction of cofilin in resting neutrophils is phosphorylated, and stimulation with N-formyl peptides results in dephosphorylation (and hence activation) of cofilin [85, 131, 214, 279, 378]. Thus in contrast to non-neutrophil cell lines, in neutrophils regulation of cofilin appears to be independent of LIM kinase and cell activation coincides with the increased actin depolymerization activity of cofilin. The different mechanisms may be related to differences in cell speed, with rapidly migrating cells requiring faster filament turnover than slowly mirating cells [457]. A complete understanding of the role of cofilin in neutrophil migration remains to be elucidated.

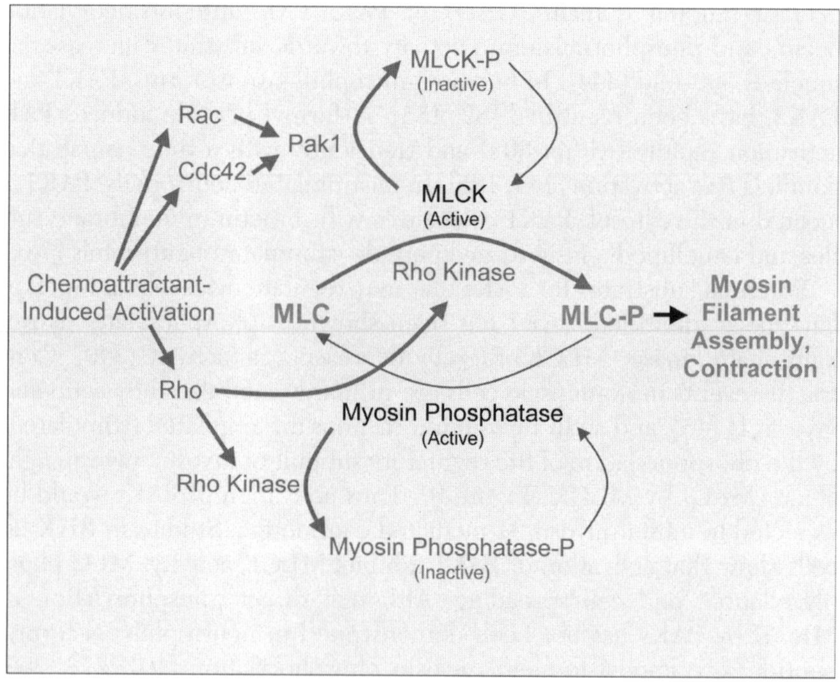

Figure 9. Schematic of proposed pathways for chemoattractant regulation of myosin light chain (MLC)–mediated contraction in neutrophils. Chemoattractant-induced activation of Rho activates Rho kinase, which catalyzes MLC phosphorylation leading to assembly of actin-myosin filaments and contraction. Rho kinase may also inactivate myosin phosphatase, further promoting the accumulation of MLC in the active phosphorylated state. PAK activation by Rac and Cdc42 inactivates myosin light chain kinase (MLCK) inhibiting its regulation of MLC activity.

Rho

Whereas Rac and Cdc42 are important for mediating the initiation of actin polymerization, Rho is important for uropod formation and release from the substratum during migration. Like Rac and Cdc42, Rho translocates from the cytosol to the plasma membrane and is activated upon stimulation with N-formyl peptide [38, 199].

Inactivation of Rho with *Clostridium botulinum* C3 ADP-ribosyl-transferase (C3 exoenzyme, which specifically ADP-ribosylates and inactivates Rho) inhibits neutrophil chemotaxis to N-formyl peptides [369, 443] without inhibiting the initiation of actin polymerization [94]. Microscopic examination of these cells reveals that they are unaffected in their abilities to polarize and extend pseudopodia, rather C3 exoenzyme inhibition of migration is associated with inhibition of rear release, which causes the cells to elongate, in some cases to the point of breaking in two [443]. This effect appears to be mediated by Rho influences on myosin contraction.

Rho regulation of myosin II

GTP-Rho is known to activate several downstream kinases. One class of Rho-kinases, Rho-associated coiled-coil forming kinases (ROCKs), regulate formation of focal contacts and stress fibers in several cell types by affecting myosin light chain [263].

N-formyl peptides are known to increase phosphorylation of MLC [101, 273, 309, 329, 330]. Closely related isoforms ROCK I and ROCK II have been identified in neutrophils [273, 443] and have been shown to translocate from the cytosol to a membrane fraction upon N-formyl peptide activation [419]. Y-27632, an inhibitor of Rho kinases including ROCK I and ROCK II, inhibits N-formyl peptide–induced neutrophil MLC phosphorylation (specifically on ser19) and chemokinesis without affecting F-actin assembly, the formation of membrane ruffles [273], intracellular Ca^{2+} elevation, or oxidant production [165]. Morphologically, inhibitor-treated cells appear able to extend F-actin-rich lamellapodia, but cannot form a distinct, contracted uropod, consistent with an effect on myosin-based contraction [273]. Serine-19 of MLC can be phosphorylated directly by MLC kinase and ROCK [13]. In neutrophils Y-27632 almost completely inhibits MLC phosphorylation, indicating that ROCK may be the primary mechanism for phosphorylating MLC at serine-19 [273]. In other cell systems, Rho kinases have been shown to phosphorylate and inactivate myosin phosphatase, an enzyme that dephosphorylates MLC [177]. Thus it is also possible in neutrophils that Rho-kinases contribute to enhance MLC phosphorylation by inhibiting its dephosphorylation.

Other possible Rho effectors—atypical ζ PKC and phosphatidylinositol-4-phosphate 5-kinase

The protein kinase C family of ser/thr kinases is divided into three subfamilies, classical, novel, and atypical, with each class containing several isozymes. Classical PKCs are activated by Ca^{2+} and diacylglycerol and thus are typically activated downstream of phospholipase C activation. Novel PKCs are activated by diacylglycerol, and atypical PKCs are not activated by Ca^{2+} or diacylglycerol and thus are not downstream of PLC [414]. Neutrophils contain classical isozymes α, β_I, and β_{II}, novel isozyme δ, and atypcial isozyme, ζ [174, 200].

In neutrophils, ζ PKC is activated and translocated to the plasma membrane upon N-formyl peptide activation [200]. Specific inhibitors of this enzyme inhibit neutrophil adherance, migration, and F-actin polymerization in response to N-formyl peptide and IL-8, suggesting this enzyme is a key step in the signal transduction pathways controlling migration. Chemoattractant-induced translocation of ζ PKC to the plasma membrane is inhibited by C3 exoenzyme treatment suggesting a role for Rho as an upstream regulator of ζ PKC. However, treatment of neutrophils with C3 exoenzyme does not inhibit actin polymerization [93], calling into question the role of Rho in regulating ζ PKC. The role of ζPKC and its possible regulation by Rho represents an exciting area in neutrophil migration research and requires further delineation.

Regulation of cytoskeletal events by Rac/Cdc42 and Rho
The actions of Rho (through Rho kinases) and Rac/Cdc42 (through PAKs), on MLC (summarized in Figure 9) are particularly fascinating, since they appear to have opposing affects. Rac/Cdc42 would be expected to promote relaxation of myosin II–based contraction since they mediate PAK-induced inactivation of MLCK. In contrast, RhoA activation of ROCKs would promote contraction. Perhaps the two pathways oscillate out of phase such that relaxation occurs during lamellipodium extension followed by a phase of uropod retraction. Such cycles in cell morphology have been noted [102, 123, 259], although there are no data in the literature indicating oscillations in Rac, Cdc42, or Rho activation occur. Alternatively, simultaneous activation of all three small GTP-binding proteins would have the net effect of shifting the regulation of MLC phosphorylation and activation from MLCK to Rho-kinases (ROCKs). This is indeed observed in neutrophils, where the Rho kinase inhibitor, Y-27632, nearly completely inhibits N-formyl peptide–induced MLC phosphorylation on serine-19 [273]. This may be a mechanism that allows N-formyl peptide activation to overtake control of myosin II contractile events while suppressing influences (other signaling pathways) that mediate contraction through MLCK.

Regulation of neutrophil migration by Rac/Cdc42 and Rho is summarized in Figure 10. Actin polymerization for extension of the lamellipodium is mediated by RAC/Cdc42 activation (outlined in Figure 8). Rho activation mediates myosin-based contraction in the uropod, while Rac/Cdc42-stimulated PAK activity inhibits myosin contraction mediated by MLCK (outlined in Figure 9).

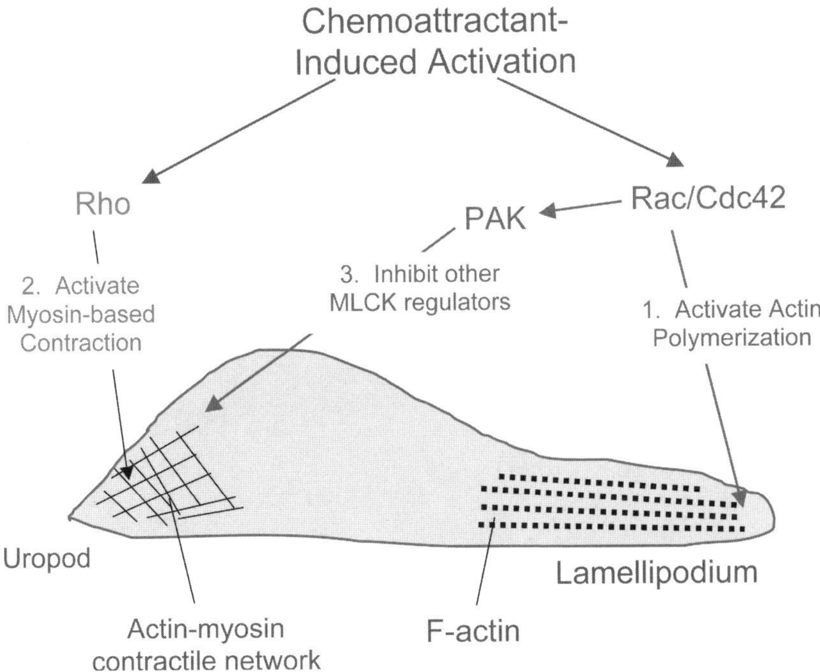

Figure 10. Regulation of neutrophil migration by Rac/Cdc42 and Rho. Chemoattractant-induced G-protein activation stimulates Rac/Cdc42 and Rho. (1) Rac/Cdc42 activates actin polymerization in the lamellipodium by mechanisms proposed in Figure 8. (2) Rho activates myosin-based contraction in the uropod, while (3) Rac/Cdc42 activation of PAK inhibits myosin contraction by MLCK as described in Figure 9.

9.3.3. *Ras and the mitogen-activated protein kinase (MAPK) cascades*

The small GTP-binding protein, Ras, was originally identified as a key signaling component regulating cell growth in a variety of cell types. Ras is activated in N-formylpeptide-stimulated neutrophils [183, 253, 430, 445]. This activation is inhibited by pertussis toxin, indicating a requirement for activated G_i-protein, but is not affected by inhibitors of tyrosine kinases, PI3K, PKC, and the intracellular Ca^{2+} increase, ruling out these signaling events as upstream requirements. The lack of effect of tyrosine kinase inhibitors on Ras activation rules out a role for Vav and Sos, since the activities of these guanine nucleotide exchange factors are regulated by tyrosine kinases [445]. Rather, activation of Ras appears to be mediated by inhibition of p120-GAP activity [445] (Figure 11). The mechanism by which this occurs is not definitively delineated, although

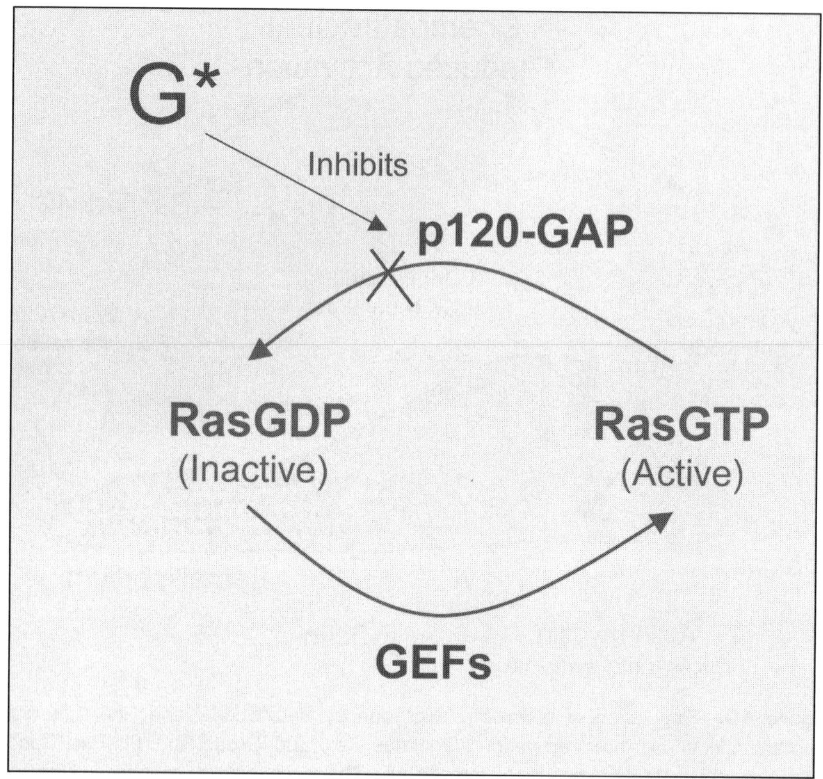

Figure 11. Activation of Ras in neutrophils. Active G-protein (G*) inhibits p120-GAP, which functions to return Ras-GTP to the inactive form. The net result (in the presence of a presumed constitutive GEF activity or spontaneous release of GDP from Ras-GDP) is accumulation of Ras in the active GTP-bound form.

it is likely to involve translocation of p120-GAP from the cytosol to the plasma membrane [90]. In many cell systems, Ras is important in activating the mitogen-activated protein kinase (MAPK) cascade, and so it is thought to do the same in neutrophils [183, 430].

The MAPK cascades

The MAPK cascades were originally identified as pathways triggered by tyrosine kinase receptors that lead to activation of gene expression and cell division. MAPK cascades are also triggered by G-protein coupled receptors [116, 399]. The MAP kinase cascades can be thought of as modules composed of three kinases that function in a series (Figure 12). Both serine/threonine and tyrosine kinase activities are involved, although the MAP kinases are serine/threonine kinases. Three such

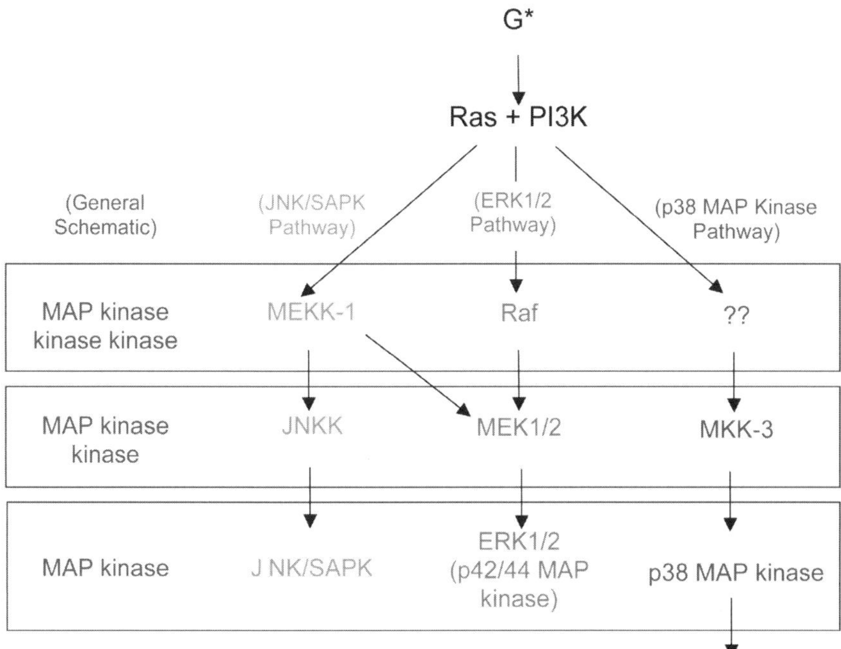

Figure 12.   MAP kinase cascade in neutrophils. The MAP kinase cascade is a series of three protein kinases represented schematically in the left column. Active G-protein (G*) stimulates Ras and PI3K activity, which in turn mediate activation of three MAP kinase cascades, the JNK/SAPK pathway, the ERK1/2 pathway, and the p38 MAP kinase pathway. p38 MAP kinase phosphorylates MK2 and mediates activation of Akt/PKB and Hsp27 phosphorylation. Hsp27 phosphorylation may stimulate actin polymerization and/or myosin contraction.

modules have been characterized, the ERK-1/2 module (extracellular signal-regulated kinases, also known as p42/44 MAPK), the p38MAPK module, and the JNK/SAPK module (c-Jun N-terminal kinase/stress-activated protein kinase) [98, 399].

Studies of the β_2-adrenergic receptor system demonstrate that activation of the MAPK cascade results from assembly of MAPK enzymes on phosphorylated receptor-β-arrestin scaffolds [304, 305]. However, this does not appear to be the case for the N-formyl peptide receptor, since MAPK activation is G-protein-dependent [19, 59] whereas receptor phosphorylation is G-protein-independent [315].

It is hypothesized that Ras activation is an initiator of the MAPK cascade in neutrophils [386, 430, 445] and this in conjunction with PI3K activity [19, 183] then activates the first kinase in the cascade

(Figure 12). Whereas PI3K inhibitors usually inhibit the MAPK cascades, these same inhibitors do not inhibit Ras activation [445]. Thus Ras activation is not dependent upon PI3K activity. PI3Ks are known to have a binding site for Ras and a close functional association between Ras and PI3K in other G-protein-coupled receptor systems is well documented [116]. In the simultaneous activation of Ras and PI3K, Ras may provide the important function of coupling PI3K activity to the MAPK pathways. Of further note is evidence that suggests activation of the MAPK cascade by PI3K is a result of its protein kinase activity rather than its lipid kinase activity [39].

Although, in general, inhibitors of PI3K inhibit N-formyl peptide–induced MAPK activation [19, 183, 323], in mice lacking PI3Kγ, JNK activation is normal and ERK activation is only partially decreased [213]. Thus the requirement for PI3K activity in MAPK activation may not be absolute.

All three of the MAP kinase modules have been demonstrated to exist in neutrophils (Figure 12), with the ERK-1/2 and p38MAP kinase modules being the predominant pathways activated by N-formyl peptides. Although phosphorylation of JNK is induced by N-formyl peptide in mouse neutrophils [135, 176], such phosphorylation has not been detected in human neutrophils [184, 270]. N-formyl peptide–induced activation of all three kinases in the ERK cascade (Raf, MEK1/2, and ERK1/2) has been demonstrated [19, 88, 183, 270, 386, 430, 445]. In addition, activation of MEKK-1, the MAP kinase kinase kinase usually associated with activation of JNK, is also induced by N-formyl peptide, and *in vitro* it phosphorylates MEK1/2 [19, 323]. Thus MEKK-1 may provide a Raf-independent pathway for activation of MEK1/2. Inhibition of MEK1/2 by a highly specific inhibitor does not inhibit N-formyl peptide–induced degranulation of primary and secondary granules, or F-actin polymerization. MEK1/2 inhibition does not inhibit chemotaxis to N-formyl peptide or IL-8 in *in vitro* polycarbonate filter assays [65, 88, 243, 461], although one report, utilizing an under agarose assay, contradicts this result [192]. Thus migratory functions appear not to be dependent upon activation of this pathway, but under some conditions this pathway may affect migration.

p38 MAP kinase and regulation of the actin cytoskeleton
N-formyl peptides also cause the activation of the p38MAP kinase and its kinase, MKK-3 [59, 190, 270, 323]. p38MAP kinase may play a role in migration, although it may not to be required. Studies with a p38MAP

kinase inhibitor indicate substantial inhibition of N-formyl peptide–induced migration [270, 461], although migration to IL-8 is not inhibited [184]. The mechanism by which p38 MAP kinase modulates neutrophil migration may involve the heat shock protein Hsp27 in a mechanism partly analogous to that occurring in smooth muscle cells [111, 197, 198]. In smooth muscle cells p38 MAP kinase phosphorylates and actives a target kinase, MAP kinase-activated protein kinase 2 (MK2). Phosphorylation of Hsp27 by MK2 results in increased actin polymerization *in vitro*, apparently by relieving inhibition of actin polymerization caused by unphosphorylated Hsp27. Hsp27 is also implicated in the regulation of myosin-based contraction [111]. In neutrophils p38 MAP kinase activates MK2 and phosphorylates Hsp27 in a mechanism that also involves MK2-dependent phosphorylation of akt/PKB [176, 190, 323]. In unstimulated neutrophils, p38 MAP kinase, MK2, Akt/PKB and Hsp27 exist as a complex; activation by N-formyl peptide causes phosphorylation of each of these components and release of Hsp27 from this complex [323]. Whether Hsp27 influences neutrophil migration by altering actin polymerization, myosin contraction, or both remains to be elucidated.

9.3.4. *Serine/threonine protein phosphatases and ezrin/radixin/moesin proteins*

Phosphorylation of proteins requires ATP, and the resulting phosphoprotein bond is very stable. Cells contain phosphatases that catalyze the removal of phosphates from the proteins. It is now recognized that phosphatases are regulated by cell signaling systems just like kinases, and the balance between kinase and phosphatase activities is highly regulated and coordinated in cells [173, 236]. Several classes of serine/threonine protein phosphatases have been identified based on similarity of amino acid sequences and functional characteristics. Protein phosphatase 1 is usually associated with cAMP-mediated signaling pathways. Protein phosphatase 2A is generally considered a broad-specificity ser/thr phosphatase. Protein phosphatase 2B is also called calcineurin, a Ca^{2+}-calmodulin sensitive phosphatase originally identified in the brain. Other phosphatases have been characterized [236] but will not be discussed here because of a lack of knowledge of their roles in neutrophil function.

Inhibitors of ser/thr protein phosphatases 1 and 2A, calyculin A and okadaic acid, which each inhibit both phosphatases [236], have been shown to inhibit N-formyl peptide–induced neutrophil migration without inhibiting actin polymerization [240, 443]. Extension of lamellipodia is not inhibited, rather release of the uropod does not occur. This effect has been linked to ERM proteins that function to link the plasma membrane with the cell cytoskeleton.

ERM proteins are a family of proteins including ezrin, radixin, and moesin that link actin filaments to the plasma membrane [46, 396]. The carboxy-terminal tail of ERM binds to F-actin, while the amino-terminal domain binds to integral membrane proteins with adhesion functions such as CD44, CD43, ICAMs 1, 2 and 3, and P-selectin glycoprotein ligand 1. ERM proteins exist in an inactive form in which the biologically active domains are masked. Activation results in unmasking of these domains. The mechanisms by which linking functions of ERM proteins are activated are not clear, but *in vitro* experiments suggest two mechanisms: phosphorylation of C-terminal threonine residues and binding of PIP_2 to their N-terminal domains. Several studies have indicated a role for Rho in the activation of ERM proteins, and this could result from Rho activation of protein kinases or Rho-dependent and -independent activation of phosphatidyl-4-phosphate 5-kinase [234, 442], although these signaling pathways need to be further characterized.

Neutrophils contain ERM proteins with moesin being the predominant form [301, 443]. In resting neutrophils moesin is phosphorylated on the carboxy terminus (measured with an antibody to moesin phosphorylated on Thr558). Microscopic localization of phosphorylated moesin in resting neutrophils shows a peripheral distribution near the plasma membrane. N-formyl peptide causes a rapid (within 10 sec) decrease in moesin Thr558 phosphorylation; phosphorylated moesin is excluded from lamellapodia and appears to be retained in the forming uropod, but then disappears as the uropod retracts.

Treatment of cells with inhibitors of ser/thr phosphatases 1 and 2A (calyculin A and okadaic acid) but not 2B (cypermethrin) prevent N-formyl peptide–induced moesin dephosphorylation, rear retraction, and cell migration, suggesting that activation of phosphatases is stimulated by N-formyl peptides and calyculin A-sensitive phosphatases play a crucial role in rear release by dephosphorylating ERM.

Evidence regarding a role for Rho in regulating ERM proteins in neutrophils is complex [443]. Treatment of neutrophils with C3 exotoxin (which inhibits Rho) reduces the resting level of phosphorylated moesin. However, inhibition of the downstream effector kinase of Rho, ROCK, does not decrease phosphorylated moesin in resting cells.

Furthermore, treatment of unstimulated cells with inhibitors of ser/thr phosphatases 1 and 2A (calyculin A and okadaic acid) results in an increase of phosphorylated moesin. Together, these results suggest in the resting cell constitutive activation of Rho maintains a phosphorylated and active pool of moesin via a ROCK-independent mechanism that is balanced by constitutive phosphatase activity. Subsequent activation of neutrophils, which is known to activate Rho, does not result in an increase in phosphorylated moesin, but rather a decrease. Phosphatase inhibitors cause an increase in phosphorylated moesin that is larger in N-formyl pepide–stimulated cells than in controls, suggesting that kinase activity is indeed stimulated. Overall the data suggest that the relative contributions of Rho activation of moesin phosphorylation are overwhelmed by a greater activation of phosphatase activity. However, the details of how ERM proteins are regulated and influence neutrophil migration remain to be defined.

9.3.5. *Other signaling events in neutrophils*

Several additional enzymes that regulate the production of lipid signaling molecules are activated in chemoattractant-stimulated neutrophils. These include phospholipase C (PLC), phospholipase A_2 (PLA$_2$), and phospholipase D (PLD). Additional types of kinases are activated, and multiple Ca^{2+} pools appear to be uniquely regulated in neutrophils. These will be discussed here for completeness and because many of these may prove to be important for regulation of migration.

Phospholipase C
The PLC$_\beta$ family of isozymes is regulated by $G_{\beta\gamma}$ subunits [334]. PLC$_{\beta2}$ is found primarily in hematopoietic cells, including neutrophils, and PLC$_{\beta3}$ is more widely distributed in cells, including neutrophils [334]. The phospholipase C enzymes cleave phosphatidylinositol 4,5-bisphosphate (PIP$_2$) and produce diacylglycerol (DAG) and inositol 1,4,5-trisphosphate (IP$_3$). IP$_3$ causes the release of Ca^{2+} from a subset of intracellular stores containing IP$_3$ receptors. Both DAG and Ca^{2+} contribute to the activation of protein kinase C, which induces phosphorylation of many proteins. In neutrophils, the best-characterized substrates for PKC are two proteins involved in activating the NADPH oxidase that produces superoxide [20].

The phospholipase C–mediated pathway appears to diverge from the pathway leading to migration. In human neutrophils, inhibition of phospholipase C does not inhibit the N-formyl peptide–induced actin polymerization response, an essential feature of the migratory response [25]. In murine neutrophils lacking $PLC_{\beta2}$ and $PLC_{\beta3}$ [213], IP_3 production and Ca^{2+} elevation in response to N-formyl peptide or interleukin-8 are completely inhibited. In contrast, chemotaxis assessed with both *in vitro* and *in vivo* assays shows no impairment, nor is F-actin polymerization in response to either chemoattractant affected. Thus phospholipase C activation and the ensuing signaling cascade it mediates, does not appear to be responsible for mediating migration. Consistent with this conclusion, a specific peptide inhibitor of αPKC and βPKC, PKC isoforms regulated by Ca^{2+} and diacylglycerol, had no effect on neutrophil migration to N-formyl peptide and IL-8 [200]. Likewise, N-formyl peptide–induced migration occurs in the absence of a notable cytosolic Ca^{2+} increase ([196, 299, 456]; see also Section 9.3.5 for further discussion of Ca^{2+}).

Phospholipase D
Phospholipase D catalyzes the hydrolysis of phospholipids to produce phosphatidic acid and the corresponding polar head group [215]. In most systems, including neutrophils, phosphotidylcholine is the preferred substrate. Two mammalian isoforms of the enzyme have been cloned, but PLD purified from human neutrophils displays different biochemical characteristics, suggesting it represents a unique isoform [139, 215]. In the presence of primary alcohols, PLD catalyzes a transphosphatidylation reaction that produces the corresponding phosphatidylalcohol. This transphosphatidylation reaction effectively competes with hydrolysis, and thus alcohols such as ethanol and 1-butanol are frequently used experimentally as inhibitors of PLD-catalysed PA production.

Chemoattractant-induced activation of PLD involves the small GTP-binding proteins RhoA and ADP-ribosylating factor (ADF) and an unidentified cytosolic protein [141, 282]. Chemoattractant-induced PLD activation is also blocked by inhibition of PLC, PI3K, and tyrosine phosphorylation [41]; thus PLD appears to be activated downstream of these signaling events and its regulation is complex.

In neutrophils, phosphatidic acid produced by PLD is then subsequently cleaved by phosphatidic acid phosphohydrolase [32]. In neutropohils, choline-containing phosphoglycerides contain approximately equal amounts of alkylacyl- and diacyl-glycerols. Thus the PA is dephosphorylated to give both DAG and alkylacylglycerol. The DAG produced by this pathway has been shown to be important for sustaining the oxidant response by activating PKC, which in turn phosphorylates components of the NADPH oxidase [4, 5, 23, 255, 439]. In addition, phosphatidic acid itself is a signaling molecule [97] and it appears to directly activate a kinase that phosphorylates components of the NADPH oxidase [327].

Primary alcohols such as ethanol and 1-butanol at concentrations that inhibit PA formation, but not PLC activation, inhibit chemoattractant-induced oxidant production and degranulation, but not actin polymerization or migration [40, 161, 162, 289, 375, 438, 439], suggesting this enzyme plays no role in signaling for chemotaxis.

Phospholipase A$_2$
Neutrophil cytosolic phospholipase A$_2$ hydrolyses phospholipids containing the arachidonyl moiety at the *sn*-2 position to liberate arachidonic acid and a lysophospholipid, with both products serving as precursors for additional inflammatory mediators [288]. Cytosolic PLA$_2$ is known to require Ca^{2+} for binding to the plasma membrane, and phosphorylation results in enhanced enzymatic activity [210]. This enzyme is activated in chemoattractant-stimulated neutrophils, and inhibitors of PKC, tyrosine phosphorylation, and PLD have all been shown to inhibit PLA$_2$, thus its activation appears to be mediated downstream of these signaling events [108]. Inhibitors of PLA$_2$ have been shown to inhibit oxidant production and degranulation. However, inhibitor effects on migration are variable [99, 108, 134], thus the role of PLA$_2$ in the migratory response remains to be clarified.

Additional nontransmembrane tyrosine kinases
In addition to Lyn kinase discussed above, two additional members of the Src family of nontransmembrane tyrosine kinases, Hck and Fgr, have been detected and shown to be activated by N-formyl peptide in neutrophils [243, 275]. A protein kinase inhibitor, PP1, shown to have high specificity of Src kinases, does not affect neutrophil migration through cellulose nitrate filters in response to N-formyl peptide [275], calling into question a requirement for any of these kinases in integrin-independent neutrophil migration. However, these kinases are important in integrin signaling [217], thus a role for Src kinases in integrin-mediated migration cannot be ruled out.

The Tec family of nontransmembrane tyrosine kinases are unique in that they contain pleckstrin homology domains, the domains that bind

PIP$_3$, and thus their functions are thought to be regulated by PI3K activity [437]. The Tec family members Tec, Bmx, and Btk have been identified in neutrophils. N-formyl peptide–induced phosphorylation and translocation of these proteins from the cytosol to membrane fractions are sensitive to pertussis toxin and PI3K inhibitors [194]. A possible role of these kinases in migration remain to be explored.

Calcium

It is well documented that N-formyl peptides induce a rapid, transient rise in intracellular calcium in neutrophils. It is also well established that inositol 1,4,5-trisphosphate released by the action of PLC triggers the release of Ca^{2+} from intracellular stores. With the exception of the specific condition of neutrophils migrating on fibronectin and vitronectin [91], migration appears to occur in the absence of the intracellular calcium transient [196, 299, 456]. However, the studies that demonstrated this conclusion utilized indicators and buffers of cytosolic calcium that may not probe the entire calcium pool. Evidence suggests two additional mechanisms for regulating calcium that may have an impact on cell migration.

Before the introduction of the cytosolic calcium probes in the early 1980s, chlortetracycline was used as a probe of membrane calcium ([287] and references therein). Chlortetracycline partitions into membranes, and binding of Ca^{2+} increases its fluorescence. Stimulation of neutrophils labeled with chlortetracycline causes a decrease in fluorescence that has been interpreted as a release of Ca^{2+} from the membrane that would increase the free Ca^{2+} proximal to the membrane [26, 260, 367]. Changes in N-formyl peptide–induced chlortetracycline fluorescence correlate with changes in F-actin content [26]. Conditions that prevent the increase in fluorescence of bulk cytosolic Ca^{2+} indicators do not alter the chlortetracycline response [26], and the dose-response characteristics of the chlortetracycline response differ from those of cytosolic calcium indicators [188]. More recently, the technique of total internal reflection fluorescence microscopy has been employed to measure membrane-proximal calcium levels in stimulated neutrophils; this approach combines the use of cytosolic calcium indicators with a technique for measuring fluorescence within 100 nm of the plasma membrane [287]. This work confirms an N-formyl peptide–induced membrane-proximal calcium increase that is regulated differently than the bulk cytosolic Ca^{2+} pool. Taken together, these results indicate that (1) there is a membrane-bound pool of calcium that is regulated differently than the cytosolic pool, (2) the calcium released from the membrane-bound pool is a small fraction of the total released into the cytosol, and (3) this pool of Ca^{2+} may affect the actin cytoskeleton. Futher studies are required to determine the significance of membrane-bound calcium for regulation of migration.

Immune cells, including neutrophils, and many other cells contain CD38, a transmembrane glycoprotein that catalyzes the production of cyclic ADP-ribose from NAD^+ [218, 297]. Cyclic ADP-ribose is a second messenger that stimulates the release of Ca^{2+} from ryanodine receptor-regulated stores that are distinctly different from the IP_3-regulated stores [115]. In permeabilized neutrophils, both cyclic ADP-ribose and ryanodine induce comparable increases in intracellular calcium (detected with cytosolic indicators), indicating that neutrophils contain ryanodine-sensitive stores that are also triggered by cyclic ADP-ribose [297]. Neutrophils from CD38-deficient mice are defective in their abilities to migrate through polycarbonate filters in response to N-formyl peptide, suggesting a role for CD38, ADP-ribose, and ryanodine-sensitive Ca^{2+} stores in neutrophil migration. However, in the same assay, CD38 is not needed for chemotaxis to IL-8. When mice are infected with *S. pneumoniae*, neutrophil accumulation in the airways after 6 hours is the same in wild-type and CD38-deficient mice, suggesting no impairment of neutrophil migration. After 12 hours of infection, airway neutrophils are significantly less in the CD38-deficient mice [297]. Further studies will be needed to sort out these apparent contradictions and elucidate the roles CD38 and ryanodine-sensitive calcium stores may play in neutrophil migration.

9.4. *Regulation of G-protein-effector coupling*

Just as the coupling of receptors and G-proteins in plasma membrane microdomains is regulated (see Section 9.1.2), evidence indicates that co-localization or segregation of G-proteins and effectors in membrane microdomains is a mechanism by which cell responsiveness to chemoattractant is regulated. Resting, unactivated neutrophils respond little to N-formyl peptides, however, when "primed" by a variety of chemical and physical conditions, these cells become very responsive [69, 166]. Studies show [166] that in unprimed cells N-formyl peptide receptors, phospholipase $C_{\beta 2}$, and Lyn kinase predominate in the heavy plasma membrane fractions, whereas G_{i2} and G_{i3} are in the light plasma membrane fractions. As a consequence of priming, G_{i2} (but not G_{i3}) moves to the heavy plasma membrane fraction and a significant amount of Lyn kinase moves to the light plasma membrane fraction. These studies suggest that regulation of the localization of G-proteins and effectors in plasma membrane microdomains is an important mechanism by which G-protein-coupled signaling pathways are modulated [166].

9.5. *Integration of the signaling pathways*

By now, the reader is likely convinced that signal transduction in neutrophils is a complex, complicated network. The above review of signaling has focused on N-formyl peptide–induced responses, and while many of the same signaling pathways are initiated by other chemoattractant receptors, such as the IL-8, C5a, platelet-activating factor, and LTB_4 receptors, not all of the same pathways are triggered [50, 183, 270, 284] although all of these receptors mediate chemotaxis. In addition, some of these pathways are triggered by integrin ligation and priming factors under conditions where chemotaxis is not stimulated [217, 271]. Finally, the signaling systems outlined above are an oversimplification in that they describe primarily activation events and ignore most of the abundant literature detailing mechanisms for turning off responses [9, 10, 81, 120, 229, 239, 244, 245, 335, 420].

Many aspects of the signaling pathways appear redundant. For example, PI3K activity can be activated by tyrosine kinase-dependent and -independent enzymes, Rac2 can be activated by PI3K-dependent and -independent mechanisms, Akt/PKB can be activated by MAPK-dependent and -independent mechanisms, and actin polymerization can be initiated by Rac-dependent and Cdc42-dependent pathways. The extent to which either pathway in a redundant pair is activated may vary with experimental conditions; apparent contradictions in the literature (for example, PIP_3 is produced by PI3Kγ in some studies and by Class I_A PI3K in others) may arise from differences in the initial state of the cells, that state being effected by cell purification protocols, priming state, incubation conditions, etc.

The complicated interconnected and overlapping nature of the pathways may reflect the evolution of a signaling system that is robust, that is, one in which some desired system characteristics are maintained despite fluctuations in its component parts or its environment [56, 251]. In the case of neutrophils, the desired system characteristic is the ability to migrate, whether the chemoattractant be N-formyl peptide, IL-8, C5a, or LTB_4, whether adherence be mediated by β_1, β_2, β_3, or no integrins, whether cells have been primed or not. The system must do complicated things, i.e., temporally and spacially coordinate protrusive forces, tractional forces, and contractile forces for the cell to move. Advanced mathematical analyses are being applied to models of cell metabolism and signaling systems and uncovering emergent properties that coincide with the experientially observed behaviors of biological systems [29–31, 56, 152]. The application of such approaches to signaling systems in neutrophils will be required to understand the significance of each of the numerous signaling components. Such complexity will likely have significant implications for the design of drugs to control neutrophil functions. It is unlikely that targeting one specific component in the signal transduction pathway will be effective. Mathematical modeling approaches will help identify which combinations of components are likely to be the most effective targets for therapeutic manipulation.

References

1. Abo, A., Webb, M.R., Grogan, A. and Segal, A.W. (1994). Activation of NADPH oxidase involves the dissociation of p21rac from its inhibitory GDP/GTP exchange protein (rhoGDI) followed by its translocation to the plasma membrane. *Biochem. J.* **298**, 585–591.
2. Abraham, V.C., Krishnamurthi, V., Taylor, D.L. and Lanni, F. (1999). The actin-based nanomachine at the leading edge of migrating cells. *Biophys. J.* **77**, 1721–1732.
3. Adams, J.A., Omann, G.M. and Linderman, J.J. (1998). A mathematical model for ligand/receptor dynamics and actin polymerization in human neutrophils. *J. Theor. Biol.* **193**, 543–560.
4. Agwu, D.E., McPhail, L.C., Chabot, M.C., Daniels, L.W., Wykle, R.L. and McCall, C.E. (1989). Choline-linked phosphoglycerides. A source of phosphatidic acid and diglycerides in stimulated neutrophils. *J. Biol. Chem.* **264**, 1405–1413.
5. Agwu, D.E., McPhail, L.C., Sozzani, S., Bass, D.A. and McCall, C.E. (1991). Phosphatidic acid as a second messenger in human polymorphonuclear leukocytes. Effects on activation of NADPH oxidase. *J. Clin. Invest.* **88**, 531–539.
6. Akasaki, T., Koga, H. and Sumimoto, H. (1999). Phosphoinositide 3-kinase-dependent and -independent activation of the small GTPase Rac2 in human neutrophils. *J. Biol. Chem.* **274**, 18055–18059.
7. Albrecht, E., and Petty, H.R. (1998). Cellular memory: Neutrophil orientation reverses during temporally decreasing chemoattractant concentrations. *Proc. Natl. Acad. Sci. U.S.A.* **95**, 5039–5044.
8. Ali, H., Richardson, R.M., Tomhave, E.D., Didsbury, J.R. and Snyderman, R. (1993). Differences in phosphorylation of formyl peptide and C5a chemoattractant receptors correlates with differences in desensitization. *J. Biol. Chem.* **268**, 24247–24254.

9. Ali, H., Fisher, I., Haribabu, B., Richardson, R.M. and Snyderman, R. (1997). Role of phospholipase Cb3 phosphorylation in the desensitization of cellular responses to platelet-activating factor. *J. Biol. Chem.* **272**, 11 706–11 709.

10. Ali, H., Richardson, R.M., Haribabu, B. and Snyderman, R. (1999). Chemoattractant receptor cross-desensitization. *J. Biol. Chem.* **274**, 6027–6030.

11. Allan, R.B. and Wilkinson, P.C. (1978). A visual analysis of chemotactic and chemkinetic locomotion of human neutrophil leucocytes. *Exp. Cell Res.* **111**, 191–203.

12. Alt, W. (1980). Biased random walk models for chemotaxis and related diffusion approximations. *J. Math. Biol.* **9**, 147–177.

13. Amano, M., Ito, M., Kimura, K., Fukata, Y., Chihara, K., Nakano, T., Matsuura, Y., Kaibuchi, K. (1996). Phosphorylation and activation of myosin by Rho-associated kinase (Rho-kinase). *J. Biol. Chem.* **271**, 20 246–20 249.

14. Ambruso, D.R., Knall, C., Abell, A.N., Panepinto, J., Kurkchebasche, A., Thurman, G., Gonzalez-Aller, C., Hiester, A., DeBoer, M., Harbeck, R.J., Oyer, R., Johnson, G.L. and Roos, D. (2000). Human neutrophil immunodeficiency syndrome is associated with an inhibitory Rac2 mutation. *Proc. Natl. Acad. Sci. U.S.A.* **97**, 4654–4659.

15. Anderson, D.C., Schmalsteig, F.C., Finegold, M.J., Hughes, B.J., Rothlein, R., Miller, L.J., Kohl, S., Tosi, M.F., Jacobs, R.L., Waldrop, T.C., Goldman, A.S., Sheare, W.T. and Springer, T.A. (1985). The severe and moderate phenotypes of heritable Mac-1, LFA-1 deficiency: their quantitative definition and relation to leukocyte dysfunction and clinical features. *J. Infect. Dis.* **152**, 668–689.

16. Arai, H., Tsou, C.-L. and Charo, I.F. (1997). Chemotaxis in a lymphocyte cell line transfected with C-C chemokine receptor 2B: Evidence that directed migration is mediated by $\beta\gamma$ dimers released by activation of $G_{\alpha i}$-coupled receptors. *Proc. Natl. Acad. Sci. U.S.A.* **94**, 14 495–14 499.

17. Arber, S., Barbayannis, F.A., Hanser, H., Schneider, C., Stanyon, C.A., Bernard, O. and Caroni, P. (1998). Regulation of actin dynamics through phosphorylation of cofilin by LIM-kinase. *Nature* **393**, 805–809.

18. Arcaro, A., and Wymann, M.P. (1993). Wortmannin is a potent phosphatidylinositol 3-kinase inhibitor: the role of phosphatidylinositol 3,4,5-trisphosphate in neutrophil responses. *Biochem. J.* **296** (Pt 2), 297–301.

19. Avdi, N.J., Winston, B.W., Russel, M., Young, S.K., Johnson, G.L. and Worthen, G.S. (1996). Activation of MEKK by formyl-methionyl-leucyl-phenylalanine in human neutrophils. Mapping pathways for mitogen-activated protein kinase activation. *J. Biol. Chem.* **271**, 33 598–33 606.

20. Babior, B.M., Lambeth, J.D. and Nauseef, W. (2002). The neutrophil NADPH oxidase. *Arch. Biochem. Biophys.* **397**, 342–344.

21. Baggiolini, M., Dewald, B., and Moser, B. (1994). Interleukin-8 and related chemotactic cytokines. CXC and CC chemokines. *Adv. Immunol.* **55**, 97–179.

22. Baggiolini, M. (2001). Chemokines in pathology and medicine. *J. Internal Med.* **250**, 91–104.

23. Bauldry, S.A., Elsey, K.L. and Bass, D.A. (1992). Activation of NADPH oxidase and phospholipase D in permeabilized human neutrophils. Correlation between oxidase activation and phosphatidic acid production. *J. Biol. Chem.* **267**, 25 141–25 152.

24. Benard, V., Bohl, B.P. and Bokoch, G.M. (1999). Characterization of Rac and Cdc42 activation in chemoattractant-stimulated human neutrophils using a novel assay for active GTPases. *J. Biol. Chem.* **274**, 13198–13204.

25. Bengtsson, T., Rundquist, I., Stendahl, O., Wymann, M.P. and Andersson, T. (1988). Increased breakdown of phosphatidylinositol 4,5-bisphosphate is not an initiating factor for actin assembly in human neutrophils. *J. Biol. Chem.* **263**, 17385–17389.

26. Bengtsson, T. (1990). Correlation between chemotactic peptide-induced changes in chlorotetracycline fluorescence and F-actin content in human neutrophils: a role for membrane-associated calcium in the regulation of actin polymerization? *Exp. Cell Res.* **191**, 57–63.

27. Bennett, T.A., Foutz, T.D., Gurevich, V.V., Sklar, L.A. and Prossnitz, E.R. (2001). Partial phosphorylation of the N-formyl peptide receptor inhibits G-protein association independent of arrestin binding. *J. Biol. Chem.* **276**, 49195–49203.

28. Bennett, T.A., Key, T.A., Gurevich, V.V., Neubig, R., Prossnitz, E.R. and Sklar, L.A. (2001). Real-time analysis of G-protein-coupled receptor reconstitution in a solubilized system. *J. Biol. Chem.* **276**, 22453–22460.

29. Bhalla, U.S. and Iyengar, R. (1999). Emergent properties of networks of biological signaling pathways. *Science* **283**, 381–387.

30. Bhalla, U.S. and Iyengar, R. (2001). Robustness of the bistable behavior of a biological signaling feedback loop. *Chaos* **11**, 221–226.

31. Bhalla, U.S., Ram, P.T. and Iyengar, R. (2002). MAP kinase phosphatase as a locus of flexibility in a mitogen-activated protein kinase signaling network. *Science* **297**, 1018–1023.

32. Billah, M.M., Eckel, S., Mullmann, T.J., Egan, R.W. and Siegel, M.I. (1989). Phosphatidylcholine hydrolysis by phospholipase D determines phosphatidate and diacylglycerol levels in chemotactic peptide-stimulated human neutrophils. Involvement of phosphatidate phosphohydrolase in signal transduction. *J. Biol. Chem.* **264**, 17069–17077.

33. Bishop, A.L. and Hall, A. (2000). Rho GTPases and their effector proteins. *Biochem. J.* **348**, 241–255.

34. Blanchoin, L., Amann, K.J., Higgs, H.N., Machand, J.-B., Kaiser, D.A. and Pollard, T.D. (2000). Direct observation of dendritic actin filament networks nucleated by Arp2/3 complex and WASP/Scar proteins. *Nature* **404**, 1007–1011.

35. Bleul, C.C., Farzan, M., Choe, H., Parolin, C., Clark-Lewis, I., Sodroski, J. and Springer, T.A. (1996). The lymphocyte chemoattractant SDF-1 is a ligand for LESTR/fusin and blocks HIV-1 entry. *Nature* **382**, 829–833.

36. Bokoch, G.M., Katada, T., Northrup, J.K., Hewlett, E.L. and Gilman, A.G. (1983). Identification of the predominant substrate for ADP-ribosylation by islet activating protein. *J. Biol. Chem.* **258**, 2072–2075.

37. Bokoch, G.M. and Gilman, A.G. (1984). Inhibition of receptor-mediated release of arachidonic acid by pertussis toxin. *Cell* **39**, 301–308.

38. Bokoch, G.M., Bohl, B.P. and Chang, T.-H. (1994). Guanine nucleotide exchange regulates membrane translocation of Rac/Rho GTP-binding proteins. *J. Biol. Chem.* **269**, 31674–31679.

39. Bondeva, T., Pirola, L., Bulgarelli-Leva, G., Rubio, I., Wetzker, R. and Wymannn, M.P. (1998). Bifurcation of lipid and protein kinase signals of PI3Kγ to the protein kinases PKB and MAPK. *Science* **282**, 293–296.

40. Bonser, R.W., Thompson, N.T., Randall, R.W. and Garland, L.G. (1989). Phospholipase D activation is functionally linked to superoxide generation in the human neutrophil. *Biochem. J.* **264**, 617–620.

41. Bonser, R.W., Thompson, N.T., Randall, R.W., Tateson, J.E., Spacey, G.D., Hodson, H.F. and Garland, L.G. (1991). Demethoxyviridin and wortmannin block phospholipase C and D activation in the human neutrophil. *Br. J. Pharmacol.* **103**, 1237–1241.

42. Boulay, F., Naik, N., Giannini, E., Tardif, M. and Brouchon, L. (1997). Phagocytic chemoattractant receptors. *Annals New York Acad. Sci.* **832**, 69–84.

43. Boyden, S. (1962). The chemotactic effect of mixtures of antibody and antigen on polymorphonuclear leucocytes. *J. Exp. Med.* **115**, 453–466.

44. Brandes, M.E., Mai, U.E.H., Ohura, K. and Wahl, S.M. (1991). Type I transforming growth factor-β receptors on human neutrophils mediate chemotaxis to transforming growth factor-β. *J. Immunol.* **147**, 1600–1606.

45. Bresnick, A.R. (1999). Molecular mechanisms of nonmuscle myosin-II regulation. *Curr. Opin. Cell Biol.* **11**, 26–33.

46. Bretscher, A. (1999). Regulation of cortical structure by the ezrin-radixin-moesin protein family. *Curr. Opin. Cell Biol.* **11**, 109–116.

47. Brown, A.F. (1982). Neutrophil granulocytes: Adhesion and locomotion on collagen substrata and in collagen matrices. *J. Cell Sci.* **58**, 455–467.

48. Brown, D.A. and London, E. (1998). Structure and origin of ordered lipid domains in biological membranes. *J. Membrane Biol.* **164**, 103–114.

49. Buettner, H.M., Lauffenburger, D.A., and Zigmond, S.H. (1989). Measurement of leukocyte motility and chemotaxis parameters with the Millipore filter assay. *J. Immunol. Meth.* **123**, 25–37.

50. Buhl, A.M., Avdi, N. and Worthen, G.S. (1994). Mapping of the C5a receptor signal transduction network in human neutrophils. *Proc. Natl. Acad. Sci. U.S.A.* **91**, 9190–9194.

51. Bunting, M., Harris, E.S., McIntyre, T.M., Prescott, S.M. and Zimmerman, G.A. (2002). Leukocyte adhesion deficiency syndromes: adhesion and tethering defects involving β2 integrins and selectin ligands. *Curr. Opin. Hematol.* **9**, 30–35.

52. Burg, N.D. and Pillinger, M.H. (2001). The neutrophil: Function and regulation in innate and humoral immunity. *Clinical Immunol.* **99**, 7–17.

53. Campbell, J.J., Foxman, E.F. and Butcher, E.C. (1997). Chemoattractant receptor crosstalk as a regulatory mechanism in leukocyte adhesion and migration. *Eur. J. Immunol.* **27**, 2571–2578.

54. Carlier, M.-F. (1991). Actin: Protein structure and filament dynamics. *J. Biol. Chem.* **266**, 1–4.

55. Carlier, M.-F, Jean, C., Rieger, K.J., Lenfant, M. and Pantaloni, D. (1993). Modulation of the interaction between G-actin and thymosin β4 by the ATP/ADP ratio: Possible implication in the regulation of actin dynamics. *Proc. Natl. Acad. Sci. U.S.A.* **90**, 5034–5038.

56. Carlson, J.M. and Doyle, J. (2002). Complexity and Robustness. *Proc. Natl. Acad. Sci. U.S.A.* **99** (suppl. 1), 2538–2545.

57. Carp, H. (1982). Mitochondrial N-formylmethionyl proteins as chemoattractants for neutrophils. *J. Exp. Med.* **155**, 264–275.

58. Cassimeris, L., NcNeill, H. and Zigmond, S.H. (1990). Chemoattractant-stimulated polymorphonuclear leukocytes contain two populations of actin filaments that differ in their spatial distribution and relative stabilities. *J. Cell Biol.* **110**, 1067–1075.

59. Chang, L.-C. and Wang, J.-P. (2000). Activation of p38 mitogen-activated kinase by formyl-methionyl-leucyl-phenylalanine in rat neutrophils. *Eur. J. Pharmacol.* **390**, 61–66.

60. Christophe, T., Karlsson, A., Dugave, C., Rabiet, M.-J., Boulay, F. and Dahlgren, C. (2001). The synthetic peptide (Trp-Lys-Tyr-Met-Val-Met-NH_2) specifically activated neutrophils through FPRL1/lipoxin A_4 receptors and is an agonist for the orphan monocyte-expressed chemoattractant receptor FPRL2. *J. Biol. Chem.* **276**, 21585–21593.

61. Chuang, T.T., Sallese, M., Ambrosini, G., Parruti, G. and De Blasi, A. (1992). High expression of β-adrenergic receptor kinase in human peripheral blood leukocytes. *J. Biol. Chem.* **267**, 6886–6892.

62. Chuang, T.-H., Bohl, B.P. and Bokoch, G.M. (1993). Biologically active lipids are regulators of Rac-GDI complexation. *J. Biol. Chem.* **268**, 26206–26211.

63. Clark, E.R., Clark, E.L. and Rex, R.O. (1936). Observations on polymorphonuclear leukocytes in the living animal. *Am. J. Anat.* **59**, 123–173.

64. Coates, T.D., Watts, R.G., Hartman, R. and Howard, T.H. (1992). Relationship of F-actin distrubution to development of polar shape in human polymorphonuclear neutrophils. *J. Cell Biol.* **117**, 765–774.

65. Coffer, P.J., Geijsen, N., M'Rabet, L., Schweizer, R.C., Maikoe, T., Raaijmaker, J.A.M., Lammers, J.-W. J. and Koenderman, L. (1998). Comparison of the roles of mitogen activated protein kinase kinase and phosphatidylinositol 3-kinase signal transduction in neutrophil effector function. *Biochem. J.* **329**, 121–130.

66. Comandon, J. (1917). Phagocytose in vitro des hematozaires du calfat (enregistrement cinematographique). *Compt. Rend. Soc. Biol.* (Paris) **80**, 314–316

67. Condeelis, J., Bresnick, A., Demma, M., Dharmawardhane, S., Eddy, R., Hall, A.L., Sauterer, R. and Warren, V. (1990). Mechanisms of amoeboid chemotaxis: An evaluation of the cortical expansion model. *Devel. Genet.* **11**, 333–340.

68. Condeelis, J. (1992). Are all pseudopods created equal? *Cell Motil. Cytoskeleton* **22**, 1–6.

69. Condliffe, A.M., Kitchen, K. and Chilvers, E.R. (1998). Neutrophil priming: pathophysiological consequences and underlying mechanisms. *Clinical Sci.* **94**, 461–471.

70. Cox, E.A. and Huttenlocher, A. (1998). Regulation of integrin-mediated adhesion during cell migration. *Microsc. Res. Tech.* **43**, 412–419.

71. Craddock, P.R., Hammerschmitd, D., White, J.G., Dalmasso, A.P. and Jacob, H.S. (1977). Complement (C5a)-induced granulocyte aggregation *in vitro. J. Clin. Invest.* **60**, 260–264.

72. Cramer, L.P. and Mitchison, T.J. (1995). Myosin is involved in postmitotic cell spreading. *J. Cell Biol.* **131**, 179–189.

73. Cutler, J. (1974). A simple *in vitro* method for studies on chemotaxis. *Proc. Soc. Exp. Biol. Med.* **147**, 471–474.

74. Cyster, J.G. (2000). Leukocyte migration: Scent of the T zone. *Cur. Biol.* **10**, R30-R33.

75. Davis, B.H., Walter, R.J., Pearson, C.B., Becker, E.L. and Oliver, J.M. (1982). Membrane activity and topography of F-Met-Leu-Phe-treated polymorphonuclear leukocytes. *Am. J. Pathol.* **108**, 206–216.

76. Dennler, S., Goumans, M.-J. and ten Dijke, P. (2002). Transforming growth factor β signal transduction. *J. Leukoc. Biol.* **71**, 731–740.

77. Dewald, B., Thelen, M. and Baggiolini, M. (1988). Two transduction sequences are necessary for neutrophil activation by receptor agonists. *J. Biol. Chem.* **263**, 16179–16184.

78. Dharmawardhane, S., Brownson, D., Lennarts, M. and Bokoch, G.M. (1999). Localization of p21-activated kinase 1 (PAK1) to pseudopodia, membrane ruffles, and phagocytic cups in activated human neutrophils. *J. Leuk. Biol.* **66**, 521–527.

79. Dickinson, R.B., McCarthy, J.B. and Tranquillo, R.T. (1993). Quantitative characterization of cell invasion *in vitro*: Formulation and validation of a mathematical model of the collagen gel invasion assay. *Annals Biomed. Engin.* **21**, 679–697.

80. Dickinson, R.B. and Tranquillo, R.T. (1993). Optimal estimation of cell movement indices from the statistical analysis of cell tracking data. *AIChE J.* **39**, 1995–2010.

81. Didsbury, J.R., Uhing, R.J., Tomhave, E., Gerard, C., Gerard, N. and Snyderman, R. (1991). Receptor class desensitization of leukocyte chemoattractant receptors. *Proc. Natl. Acad. Sci. U.S.A.* **88**, 11564–11568.

82. Ding, J., Knaus, U.G., Lian, J.P., Bokoch, G.M. and Badwey, J.A. (1996). The renaturable 69 and 63 kDa protein kinases that undergo rapid activation in chemoattractant-stimulated guinea pig neutrophils are p21-activated kinases (PAKs). *J. Biol. Chem.* **271**, 24869–24873.

83. DiNubile, M.J., Cassimeris, L., Joyce, M. and Zigmond, S.H. (1995). Actin filament barbed-end capping activity in neutrophil lysates: The role of capping protein-β_2. *Mol. Biol. Cell* **6**, 1659–1671.

84. Dixon, H.M. and McCutcheon, M. (1936). Chemotropism of leucocytes in relation to their rate of locomotion. *Proc. Soc. Exp. Biol. Med.* **34**, 173–176.

85. Djafarzadeh, S. and Niggli, V. (1997). Signaling pathways involved in dephosphorylation and localization of the actin-binding protein cofilin in stimulated human neutrophils. *Exp. Cell Res.* **236**, 427–435.

86. Doerschuk, C.M., Winn, R.K., Coxson, H.O. and Harlan, J.M. (1990). CD18-dependent and -independent mechanisms of neutrophil emigration in the pulmonary and systemic mircocirculation of rabbits. *J. Immunol.* **144**, 2327–2333.

87. Dow, J.A.T., Lackie, J.M. and Crocket, K.V. (1987). A simple microcomputer-based system for real-time analysis of cell behavior. *J. Cell Sci.* **87**, 171–182.

88. Downey, G.P., Butler, J.R., Tapper, H., Fialkow, L., Saltiel, A.R., Rubin, B.B. and Grinstein, S. (1998). Importance of MEK in neutrophil microbicidal responsiveness. *J. Immunol.* **160**, 434–443.

89. Dunn, G.A. (1983). Characterizating a kinesis response: Time averaged measure of cell speed and directional persistence. *Agents Actions Suppl.* **12**, 14–33.

90. Dusi, S., Donini, M., Wientjes, F. and Rossi, F. (1996). Tyrosine phosphorylation and subcellular redistribution of p125 ras guanosine triphosphatase-activating protein in human neutrophils stimulated with FMLP. *FEBS Lett.* **383**, 181–184.

91. Eddy, R.J., Pierini, L.M., Matsumura, F. and Maxfield, F.R. (2000). Ca^{2+}-dependent myosin II activation is required for uropod retraction during neutrophil migration. *J. Cell Science* **113**, 1287–1298.

92. Edwards, D.C., Sanders, L.C., Bokoch, G.M. and Gill, G.N. (1999). Activation of LIM-kinase by Pak1 couples Rac/Cdc42 GTPase signaling to actin cytoskeletal dynamics. *Nature Cell Biol.* **1**, 253–259.

93. Ehrengruber, M.U., Boquet, P., Coates, T.D. and Deranleau, D.A. (1995). ADP-ribosylation of rho enhances actin polymerization-coupled oscillations in human neutrophils. *FEBS Lett.* **373**, 161–164.

94. Ehrengruber, M.U., Coates, T.D. and Deranleau, D.A. (1995). Shape oscillations: a fundamental response of human neutrophils stimulated by chemotactic peptides? *FEBS Lett.* **359**, 229–232.

95. Ehrengruber, M.U., Deranleau, D.A. and Coates, T.D. (1996). Shape oscillations of human neutrophil leukocytes: Characterization and relationship to cell motility. *J Exp. Biol.* **199**, 741–747.

96. Ember, J.A. and Hugli, T.E. (1997). Complement factors and their receptors. *Immunopharmacol.* **38**, 3–15.

97. English, D., Cui, Y. and Siddiqui, R.A. (1996). Messenger functions of phosphatidic acid. *Chem. Phys. Lipids* **80**, 117–132.

98. English, J., Pearson, G., Wilsbacher, J., Swantek, J., Karandikar, M., Xu, S. and Cobb, M.H. (1999). New insights into the control of MAP kinase pathways. *Exp. Cell Res.* **253**, 255–270.

99. Escrig, V., Ubeda, A., Ferrandiz, M.L., Darias, J., Sanchez, J.M., Alcaraz, M.J. and Paya, M. (1997). Vairabilin: A dual inhibitor of human secretory and cytosolic phospholipase A2 with anti-inflammatory activity. *J. Pharmacol. Exp. Therapeutics* **282**, 123–131.

100. Fay, S.P., Posner, R.G., Swann, W.N. and Sklar, L.A. (1991). Real-time analysis of the assembly of ligand, receptor, and G protein by quantitative fluorescence flow cytometry. *Biochem.* **30**, 5066–5075.

101. Fechheimer, M. and Zigmond, S.H. (1983). Changes in cytoskeletal proteins of polymorphonuclear leukocytes induced by chemotactic peptides. *Cell Motility* **3**, 349–361.

102. Felder, S. and Kam, Z. (1994). Human neutrophil motility: Time dependent three-dimensional shape and granule diffusion. *Cell Motil. Cytoskeleton* **28**, 285–302.

103. Fiore, S., Maddox, J.F., Perez, H.D. and Serhan, C.N. (1994). Identification of a human cDNA encoding a functional high affinity lipoxin A4 receptor. *J. Exp. Med.* **180**, 253–260.

104. Fisher, P.R. (1990). Pseudopodium activation and inhibition signals in chemotaxis. *Sem. Cell Biol.* **1**, 87–97.

105. Foxman, E.F., Campbell, J.J. and Butcher, E.C. (1997). Multistep navigation and the combinatorial control of leukocyte chemotaxis. *J. Cell Biol.* **139**, 1349–1360.

106. Foxman, E.F., Kunkel, E.J. and Butcher, E.C. (1999). Integrating conflicting chemotactic signals: The role of memory in leukocyte navigation. *J. Cell Biol.* **147**, 577–587.

107. Freer, R.J., Day, A.R., Muthukumaraswamy, N., Pinon, D., Wu, A., Showell, H.J. and Becker, E.L. (1982). Formyl peptide chemoattractants: a model of the receptor on rabbit neutrophils. *Biochem.* **21**, 257–263.

108. Fujita, K.-I., Murakami, M., Yamashita, F., Amemiya, K. and Kudo, I. (1996). Phospholipase D is involved in cytosolic phospholipase A_2-dependent selective release of acrachidonic acid by fMLP-stimulated rat neutrophils. *FEBS Lett.* **395**, 293–298.

109. Funk, C.D. (2001). Prostaglandins and leukotrienes: Advances in eicosanoid biology. *Science* **294**, 1871–1875

110. Gerisch, G. and Keller, H.U. (1981). Chemotactic reorientation of granulocytes stimulated with micropipettes containing fMet-Leu-Phe. *J. Cell Sci.* **52**, 1–10.

111. Gerthoffer, W.T. and Gunst, S.J. (2001). Focal adhesion and small heat shock proteins in the regulation of actin remodeling and contractility in smooth muscle. *J. Appl. Physiol.* **91**, 963–972.

112. Gierschik, P., Sidiropoulos, D. and Jakobs, K.H. (1989). Two distinct Gi-proteins mediate formyl peptide receptor signal transduction in human leukemia (HL-60) cells. *J. Biol. Chem.* **264**, 21 470–21 473.

113. Gilbert, T.L., Bennett, T.A., Maestas, D.C., Cimino, D.F. and Prossnitz, E.R. (2001). Internalization of the human N-formyl peptide and C5a chemoattractant receptors occurs via clathrin-independent mechanisms. *Biochem.* **40**, 3467–3475.

114. Glogauer, M., Hartwig, J. and Stossel, T. (2000). Two pathways through Cdc42 couple the N-formyl receptor to actin nucleation in permeabilized human neutrophils. *J. Cell Biol.* **150**, 785–796.

115. Guse, A.H. (1999). Cyclic ADP-ribose: A novel Ca^{2+}-mobilizing second messenger. *Cell. Signal.* **11**, 309–316.

116. Gutkind, J.S. (1998). The pathways connecting G-protein-coupled receptors to the nucleus through divergent mitogen-activated protein kinases cascades. *J. Biol. Chem.* **273**, 1839–1842.

117. Haddox, J.L., Pfister, R.R. and Sommers, C.I. (1991). A visual assay for quantitating neutrophil chemotaxis in a collagen gel matrix. A novel chemotactic chamber. *J. Immunol. Meth.* **141**, 41–52.

118. Haines, K.A., Kolasinski, S.L., Cronstein, B.N., Reibman, J., Gold, L.I. and Weissmann, G. (1993). Chemoattraction of neutrophils by substance P and transforming growth factor-β1 is inadequately explained by current models of lipid remodeling. *J. Immunol.* **151**, 1491–1499.

119. Harakawa, N., Sasada, M., Maeda, A., Asagoe, K., Nohgawa, M., Takano, K., Matsuda, Y., Yamamoto, K. and Okuma, M. (1997). Random migration of polymorphonuclear leukocytes induced by GM-CSF involving a signal transduction pathway different from that of fMLP. *J. Leukoc. Biol.* **61**, 500–506.

120. Haribabu, R., Richardson, R.M., Verghese, M.W., Barr, A.J., Zhelev, D.V., Snyderman, R. (2000). Function and regulation of chemoattractant receptors. *Immunol. Res.* **22**, 271–279.

121. Harris, E.S., McIntyre, T.M., Prescott, S.M. and Zimmerman, G.A. (2000). The leukocyte integrins. *J. Biol. Chem.* **275**, 23 409–23 412.

122. Hart, M.J., Jiang, X., Kozasa, T., Roscoe, W., Singer, W.D., Gilman, A.G., Sternweis, P.C. and Bollag, G. (1998). Direct stimulation of the guanine nucleotide exchange activity of p115RhoGEF by $G\alpha_{13}$. *Science* **280**, 2112–2114.

123. Hartman, R.S., Lau, K., Chou, W. and Coates, T.D. (1994). The fundamental motor of the human neutrophil is not random: Evidence for local non-Markov movement in neutrophils. *Biophys. J.* **67**, 2535–2545.

124. Hartwig, J.H., Bokoch, G.M., Carpenter, C.L., Janmey, P.A., Taylor, L.A., Toker, A. and Stossel, T.P. (1995). Thrombin receptor ligation and activated Rac uncap actin filament barbed ends through phosphoinositide synthesis in permeabilized human platelets. *Cell* **82**, 643–653.

125. Haston, W.S. and Shields, J.M. (1985). Neutrophil leucocyte chemotaxis: a simplified assay for measuring polarizing response to chemotactic factors. *J. Immunol. Meth.* **81**, 229–237.

126. Haston, W.S. and Wilkinson, P.C. (1987). Gradient perception by neutrophil leucocytes. *J. Cell. Sci.* **87**, 373–374.

127. Hellewell, P.G., Young, S.K., Henson, P.M. and Worthen, G.S. (1994). Disparate role of the β2-integrin CD18 in the local accumulation of neutrophils in pulmonary and cutaneous inflammation in the rabbit. *Am. J. Respir. Cell Mol. Biol.* **10**, 391–398.

128. Heiss, S.G. and Cooper, J.A. (1991). Regulation of capZ, an actin capping protein of chicken muscle, by anionic phospholipids. *Biochem.* **30**, 8753–8758.

129. Hess, B. and Boiteux, A. (1971). Oscillatory phenomena in biochemistry. *Ann. Rev. Biochem.* **40**, 237–258.

130. Heyworth, P.G., Bohl, B.P., Bokoch, G.M. and Curnutte, J.T. (1994). Rac translocates independently of the neutrophil NADPH oxidase components p47^{phox} and p67^{phox}. Evidence for its interaction with flavocytochrome b_{558}. *J. Biol. Chem.* **269**, 30749–30752.

131. Heyworth, P.G., Robinson, J.M., Ding, J., Ellis, B.A. and Badwey, J.A. (1997). Cofilin undergoes rapid dephosphorylation in stimulated neutrophils and translocates to ruffled membranes enriched in products of the NADPH oxidate complex. Evidence for a novel cycle of phosphorylation and dephosphorylation. *Histochem. Cell Biol.* **108**, 221–233.

132. Higgs, H.N. and Pollard, T.D. (1999). Regulation of actin polymerization by Arp2/3 complex and WASp/Scar proteins. *J. Biol. Chem.* **274**, 32531–32534.

133. Hinchliffe, K. (2000). Intracellular signaling: Is PIP$_2$ a messenger too? *Curr. Biol.* **10**, R104-R105.

134. Hinder, M. (1998). Investigation on the effect of experimental phospholipase A$_2$ inhibitors on the formyl-methionyl-leucyl-phenylalanine-stimulated chemotaxis of human leukocytes *in vitro*. *Arzneim.-Forsch./Drug Res.* **48**, 77–81.

135. Hirsch, E., Katanaev, V.L., Garlanda, C., Azzolino, O., Pirola, L., Silengo, L., Sozzani, S., Mantovani, A., Altruda, F. and Wymann, M.P. (2000). Central role for G-protein-coupled phosphoinositide 3-kinase γ in inflammation. *Science* **287**, 1049–1053.

136. Hoffman, J.F., Linderman, J.J., Omann, G.M. (1996). Receptor upregulation, internalization, and interconverting receptor states: critical components of a quantitative description of N-formyl peptide–receptor dynamics in the neutrophil. *J. Biol. Chem.* **271**, 18394–18404.

137. Hoffman, J.F., Omann, G.M. and Linderman, J.J. (1996). Interconverting receptor states at 4°C for the neutrophil N-formyl peptide receptor. *Biochem.* **35**, 13047–13055.

138. Hoffstein, S.T., Friedman, R.S. and Weissmann, G. (1982). Degranulation, membrane addition, and shape change during chemotactic factor-induced aggregation of human neutrophils. *J. Cell Biol.* **95**, 234–241.

139. Horn, J.M., Lehman, J.A., Alter, G., Horwitz, J. and Gomez-Cambronero, J. (2001). Presence of a phospholipase D (PLD) distinct from PLD1 or PLD2 in human neutrophils: immunobiochemical characterization and initial purification. *Biochim. Biophys. Acta* **1530**, 97–110.

140. Horwitz, D.A, and Garret, M.A. (1971). Use of leukocyte chemotaxis *in vitro* to assay mediators generated by immune reactions. I. Quantification of mononuclear and polymorphonuclear leukocyte chemotaxis with polycarbonate (Nucleopore) filters. *J. Immunol.* **106**, 649–655.

141. Houle, M.G. and Bourgoin, S. (1996). Small GTPase-regulated phospholipase D in granulocytes. *Biochem. Cell Biol.* **74**, 459–467.

142. Howard, T.H. and Meyer, W.H. (1984). Chemotactic peptide modulation of actin assembly and locomotion in neutrophils. *J. Cell Biol.* **98**, 1265–1271.

143. Hsu, M.H., Chiang, S.C., Ye, R.D. and Prossnitz, E.R. (1997). Phosphorylation of the N-formyl peptide receptor is required for receptor internalization but not chemotaxis. *J. Biol. Chem.* **272**, 29 426–29 429.

144. Huang, R., Lian, J.P., Robinson, D. and Badwey, J.A. (1998). Neutrophils stimulated with a variety of chemoattractants exhibit rapid activation of p21-activated kinases (Paks): Separate signals are required for activation and inactivation of Paks. *Mol. Cell. Biol.* **18**, 7130–7138.

145. Hyslop, P.A., Oades, Z.G., Jesaitis, A.J., Painter, R.G., Cochrane, C.G. and Sklar, L.A. (1984). Evidence for N-formyl chemotactic peptide–stimulated GTPase activity in human neutrophil homogenates. *FEBS lett.* **166**, 165–169.

146. Ibarrondo, F.J., Torres, M. and Coates, T.D. (1999). Periodic formation of nascent lamellae is driven by changes in the stable F-actin pool of polymorphonuclear neutrophils after stimulation with chemotactic peptide and cross-linking of CD18 or CD61. *Cell Motil. Cytoskeleton* **44**, 234–247.

147. Imhof, B.A. (1997). Basic mechanism of leukocyte migration. *Hormone and Metab. Res.* **29**, 614–621.

148. Islam, L.N., McKay, I.C. and Wilkinson, P.C. (1985). The use of collagen or fibrin gels for the assay of human neutrophil chemotaxis. *J. Immunol. Meth.* **85**, 137–151.

149. Janmey, P.A., Iida, K., Yin, H. and Stossel, T. (1987). Polyphosphoinositide micelles and polyphosphoinositide-containing vesicles dissociate endogenous gelsolin-actin complexes and promote actin assembly from the fast-growing end of actin filaments blocked by gelsolin. *J. Biol. Chem.* **262**, 12 228–12 236.

150. Janmey, P.A. and Stossel, T.P. (1987). Modulation of gelsolin function by phosphatidylinositol 4,5-bisphosphate. *Nature* **325**, 362–364.

151. Jeon, N.L., Baskaran, H., Dertinger, S.K.W., Whitesides, G.M., Van De Water, L. and Toner, M. (2002). Neutrophil chemotaxis in linear and complex gradients of interleukin-8 formed in a microfabricated device. *Nature Biotech.* **20**, 826–830.

152. Jeong, H., Tombor, B., Albert, R., Oltval, Z.N. and Barabasi, A.-L. (2000). The large-scale organization of metabolic networks. *Nature* **407**, 651–654.

153. Jesaitis, A.J., Naemura, J.R., Painter, R.G., Sklar, L.A. and Cochrane, C.G. (1983). The fate of an N-formylated chemotactic peptide in stimulated human granulocytes. Subcellular fractionation studies. *J. Biol. Chem.* **258**, 1968–1977.

154. Jesaitis, A.J., Naemura, J.R., Sklar, L.A., Cochrane, C.G. and Painter, R.G. (1984). Rapid modulation of N-formyl chemotactic peptide receptors on the surface of

human granulocytes: Formation of high-affinity ligand-receptor complexes in transient association with cytoskeleton. *J. Cell Biol.* **98**, 1378–1387.

155. Jesaitis, A.J., Tolley, J.O. and Allen, R.A. (1986). Receptor-cytoskeleton interactions and membrane traffic may regulate chemoattractant-induced superoxide production in human granulocytes. *J. Biol. Chem.* **261**, 13662–13669.

156. Jesaitis, A.J., Bokoch, G.M., Tolley, J.O. and Allen, R.A. (1988). Lateral segregation of neutrophil chemoattractant receptors into actin- and fodrin-rich plasma membrane microdomains depleted in guanyl nucleotide regulatory proteins. *J. Cell. Biol.* **107**, 921–928.

157. Jesaitis, A.J., Tolley, J.O., Bokoch, G.M. and Allen, R.A. (1989). Regulation of chemoattractant receptor interaction with transducing proteins by organizational control in the plasma membrane of human neutrophils. *J. Cell Biol.* **109**, 2783–2790.

158. Jesaitis, A.J., Erickson, R.W., Klotz, K.-N., Bommakanti, R.K. and Siemsen, D.W. (1993). Functional molecular complexes of human N-formyl chemoattractant receptors and actin. *J. Immunol.* **151**, 5653–5665.

159. Jones, G.E. 2000. Cellular signaling in macrophage migration and chemotaxis. *J. Leuk. Biol.* **68**, 593–602.

160. Kahler, C.M., Pischel, A., Kaufmann, G., and Wiedermann, C.J. (2001). Influence of neuropeptides on neutrophil adhesion and transmigration through a lung fibroblast barrier *in vitro*. *Exp. Lung Res.* **27**, 25–46.

161. Kaldi, K., Szeberenyi, J., Rada, B.K., Kovacs, P., Geiszt, M., Mocsai, A. and Ligeti, E. (2002). Contribution of phospholipase D and a brefeldin A-sensitive ARF to chemoattractant-induced superoxide production and secretion of human neutrophils. *J. Leuk. Biol.* **71**, 695–700.

162. Kanaho, Y., Kanoh, H., Saitoh, K. and Nozawa, Y. (1991). Phospholipase D activation by platelet activating factor, leukotriene B_4, and formyl-methionyl-leucyl-phenylalanine in rabbit neutrophils. Phospholipase D activation is involved in enzyme release. *J. Immunol.* **146**, 3536–3541.

163. Kandel, E.S. and Hay, N. (1999). The regulation and activities of the multifunctional serine/threonine kinase akt/PKB. *Exp. Cell Res.* **253**, 210–229.

164. Katanaev, V.L. and Wymann, M.P. (1998). GTPγS-induced actin polymerization *in vitro*: ATP- and phosphoinositide-independent signaling via Rho-family proteins and a plasma membrane-associated guanine nucleotide exchange factor. *J. Cell Sci.* **111**, 1583–1594.

165. Kawaguchi, A., Ohmori, M., Harada, K., Tsuruoka, S., Sugimoto K. and Fujimura, A. (2000). The effect of a Rho kinase inhibitor Y-27632 on superoxide production, aggregation and adhesion in human polymorphonuclear leukocytes. *Eur. J. Pharmacol.* **403**, 203–208.

166. Keil, M.L., Solomon, N.L., Lodhi, I.J., Stone, K.C., Jesaitis, A.J., Chang, P.S., Linderman, J.J. and Omann, G.M. (2003). Priming-induced localization of $G_{i\alpha2}$ in high density membrane microdomains. *Biochem. Biophys. Res. Commu.* **301**, 862–872.

167. Keller, H.U., Borel, J.F., Wilkinson, P.C., Hess, M.W. and H. Cottier. (1972). Re-assessment of Boyden's technique for measuring chemotaxis. *J. Immunol. Meth.* **1**, 165–168.

168. Keller, H.U. and Bessis, M. (1975). Migration and chemotaxis of anucleate cytoplasmic leukocyte fragments. *Nature* **258**, 723–724.

169. Keller, H.U. and Cottier, H. (1981). Crawling-like movements and polarization in non-adherent leucocytes. *Cell Biol. Intl. Reports* **5**, 3–7.

170. Keller, H.U. (1983). Motility, cell shape, and locomotion of neutrophil granulocytes. *Cell Motility* **3**, 47–60.

171. Keller, H. and Zimmerman, A. (1987). Shape, movement and function of neutrophil granulocytes. *Biomed. Pharmacotherapy* **41**, 285–289.

172. Keller, H., Niggli, V. and Zimmermann, A. (1991). Diversity in motile responses of human neutrophil granulocytes: functional meaning of cytoskeletal bias. *Adv. Exp. Biol. Med.* **297**, 23–37.

173. Kennelly, P.J. (2001). Protein phosphatases—a phylogenetic perspective. *Chem. Rev.* **101**, 2291–2312.

174. Kent, J.D., Sergeant, S., Burns, D.J. and McPhail, L.C. (1996). Identification and regulation of protein kinase C-δ in human neutrophils. *J. Immunol.* **157**, 4641–4647.

175. Key, T.A., Bennett, T.A., Foutz, T.D., Gurevich, V.V., Sklar, L.A. and Prossnitz, E.R. (2001). Regulation of formyl peptide receptor agonist affinity by reconstitution with arrestins and heterotrimeric G-proteins. *J. Biol. Chem.* **276**, 49204–49212.

176. Kim, C. and Dinauer, M.C. (2001). Rac2 is an essential regulator of neutrophil nicotinamide adenine dinucleotide phosphate oxidase activation in response to specific signaling pathways. *J. Immunol.* **166**, 1223–1232.

177. Kimura, K., Ito, M., Amano, M., Chihara, K., Fukata, Y., Nakafuku, M., Yamamori, B., Feng. J., Nakano, T., Okawa, K., Iwamatsu, A. and Kaibuchi, K. (1996). Regulation of myosin phosphatase by Rho and Rho-associated kinase (Rho-kinase). *Science* **273**, 245–248.

178. Kindzelskii, A.L., Eszes, M.M., Todd III, R.F. and Petty, H.R. (1997). Proximity oscillations of complement type 4 ($\alpha_x\beta_2$) and urokinase receptors on migrating neutrophils. *Biophys. J.* **73**, 1777–1784.

179. Kishimoto, T.K., Baldwin, E.T. and Anderson, D.C. (1999). The role of β2 integrins in inflammation, in *Inflammation: Basic Principles and Clinical Correlates*, 3rd ed. (J.I. Gallin and R. Snyderman, eds.), 537–569. Lippincott Williams and Wilkins, Philadelphia.

180. Kitayama, J., Carr, M.W., Roth, S.J., Buccola, J. and Springer, T.A. (1997). Contrasting responses to multiple chemotactic stimuli in transendothelial migration. Heterologous desensitization in neutrophils and augmentation of migration in eosinophils. *J. Immunol.* **158**, 2340–2349.

181. Klotz, K.-N. and Jesaitis, A.J. (1994). Physical coupling of N-formyl peptide chemoattractant receptors to G-protein is unaffected by desensitization. *Biochem. Pharmacol.* **48**, 1297–1300.

182. Klotz, K.-N., Krotec, K.L., Gripentrog, J. and Jesaitis, A.J. (1994). Regulatory interaction of N-formyl peptide chemoattractant receptors with the membrane skeleton in human neutrophils. *J. Immunol.* **152**, 801–810.

183. Knall, C., Young, S., Nick, J.A., Buhl, A.M., Worthen, G.S. and Johnson, G.L. (1996). Interleukin-8 regulation of the Ras/Raf/mitogen-activated protein kinase pathway in human neutrophils. *J. Biol. Chem.* **271**, 2832–2838.

184. Knall, C., Worthen, G.S. and Johnson, G.L. (1997). Interleukin 8–stimulated phosphatidylinositol-3-kinase activity regulates the migration of human

neutrophils independent of extracellular signal-regulated kinase and p38 mitogen-activated protein kinases. *Proc. Natl. Acad. Sci. U.S.A.* **94**, 3052–3057.

185. Knaus, U.G., Morris, S., Dong, H.-J., Chernoff, J. and Bokoch, G.M. (1995). Regulation of human leukocyte p21-activated kinases through G-protein coupled receptors. *Science* **269**, 221–223.

186. Knaus, U.G. and Bokoch, G.M. (1998). The p21$^{Rac/Cdc42}$-activated kinases (PAKs). *Intl. J. Biochem. Cell Biol.* **30**, 857–862.

187. Koo, C., Lefkowitz, R.J. and Snyderman, R. (1983). Guanine nucleotides modulate the binding affinity of the oligopeptide chemoattractant receptor on human polymorphonuclear leukocytes. *J. Clin. Invest.* **72**, 748–753.

188. Korchak, J.M., Vienne, K., Wilkenfeld, C., Roberts, C.S., Rutherford, L.E., Haines, K.A. and Weissmann, G. (1983). The role of calcium in neutrophil activation—mobilization of multiple calcium pools. *J. Cell Biol.* **97**, 160a.

189. Kraynov, V.S., Chamberlain, C., Bokoch, G.M., Schwartz, M.A., Slabaugh, S. and Hahn, K.M. (2000). Localized Rac activation dynamics visualized in living cells. *Science* **290**, 333–337.

190. Krump, E., Sanghera, J.S., Pelech, S.L., Furuya, W. and Grinstein, S. (1997). Chemotactic peptide N-formyl-Met-Leu-Phe activation of p38 mitogen-activated protein kinase (MAPK) and MAPK-activated protein kinase-2 in human neutrophils. *J. Biol. Chem.* **272**, 937–944.

191. Kuntz, R.M. and Saltzman, W.M. (1997). Neutrophil motility in extracellular matrix gels: Mesh size and adhesion affect speed of migration. *Biophys. J.* **72**, 1472–1480.

192. Kuroki, M. and O'Flaherty, J.T. (1997). Differential effects of a mitogen-activated protein kinase kinase inhibitor on human neutrophil responses to chemotactic factors. *Biochem. Biophys. Res. Commun.* **232**, 474–477.

193. Kvietys, P.R. and Sandig, M. (2001). Neutrophil diapedesis: paracellular or transcellular? *News Physiol. Sci.* **16**, 15–19.

194. Lachance, G., Levasseur, S. and Naccache, P.H. (2002). Chemotactic factor-induced recruitment and activation of Tec family kinases in human neutrophils. *J. Biol. Chem.* **277**, 21 537–21 541.

195. Lackie, J.M. and Wilkinson, P.C. (1984). Adhesion and locomotion of neutrophil leucocytes on 2-D and 3-D matrices. *Kroc Found. Ser.* **16**, 237–254.

196. Laffafian, I. and Hallett, M.B. (1995). Does cytosolic free Ca^{2+} signal neutrophil chemotaxis in response to formylated chemotactic peptide? *J. Cell Sci.* **108**, 3199–3205.

197. Landry, J., and Huot, J. (1995). Modulation of actin dynamics during stress and physiological stimulation by a signaling pathway involving p38 MAP kinase and heat-shock protein 27. *Biochem. Cell Biol.* **73**, 703–707.

198. Landry, J., and Huot, J. (1999). Regulation of actin dynamics by stress-activated protein kinase 2 (SAPK2)-dependent phosphorylation of heat-shock protein of 27 kDa (Hsp27). *Biochem. Soc. Symp.* **64**, 79–89.

199. Laudanna, C., Campbell, J.J. and Butcher, E.C. (1996). Role of Rho in chemoattractant-activated leukocyte adhesion through integrins. *Science* **271**, 981–983.

200. Laudanna, C., Mochly-Rosen, D., Liron, T., Constantin, G. and Butcher, E.C. (1998). Evidence of ζ protein kinase C involvement in polymorphonuclear

neutrophil integrin-dependent adhesion and chemotaxis. *J. Biol. Chem.* **273**, 30 306–30 315.

201. Lauffenburger, D. and Aris, R. (1979). Measurement of leukocyte motility and chemotaxis parameters using a quantitative analysis of the under-agarose migration assay. *Mathematical Biosciences* **44**, 121–138.

202. Lauffenburger, D.A. and Zigmond, S.H. (1981). Chemotactic factor concentration gradients in chemotaxis assay systems. *J. Immunol. Meth.* **40**, 45–60.

203. Lauffenburger, D.A. (1983). Measurement of phenomenological parameters for leukocyte motility and chemotaxis. *Agents Action Suppl.* **12**, 34–53.

204. Lauffenburger, D., Rothman, C. and Zigmond, S.H. (1983). Measurement of leukocyte motility and chemotaxis parameters with a linear under-agarose migration assay. *J. Immunol.* **131**, 940–947.

205. Lauffenburger, D.A., Tranquillo, R.T. and Zigmond, S.H. (1988). Concentration gradients of chemotactic factors in chemotaxis assays. *Meth. Enzymol.* **162**, 85–101.

206. Lauffenburger, D.A. and Linderman, J.J. (1993). *Receptors: Models for Binding, Trafficking, and Signaling.* Oxford University Press, New York.

207. Lauffenburger, D.A. and Horwitz, A.F. (1996). Cell migration: a physically integrated molecular process. *Cell* **84**, 359–369.

208. Lawson, M.A. and Maxfield, F.R. (1995). Ca^{2+} and calcineurin-dependent recycling of an integrin to the front of migrating neutrophils. *Nature* **377**, 75–79.

209. Lepidi, H., Benoliel, A.-M., Mege, J.L., Bongrand, P. and Capo, C. (1992). Double localization of F-actin in chemoattractant-stimulated polymorphonuclear leukocytes. *J. Cell Sci.* **103**, 145–156.

210. Leslie, C.C. (1997). Properties and regulation of cytosolic phospholipase A_2. *J. Biol. Chem.* **272**, 16 709–16 712.

211. LevchenkoA. and Iglesias, P.A. (2002). Models of eukaryotic gradient sensing: Application to chemotaxis of amoebae and neutrophils. *Biophys. J.* **82**, 50–63.

212. Lewis, W.H. (1934). On the locomotion of the polymorphonuclear neutrophils of the rat in autoplasma cultures. *Bull. Johns Hopkins Hospital* **55**, 273–279

213. Li, Z., Jiang, H., Xie, W., Zhang, Z., Smrcka, A.V. and Wu, D. (2000). Roles of PLC-β2 and -β3 and PI3Kγ in chemoattractant-mediated signal transduction. *Science* **287**, 1046–1049.

214. Lian, J.P., Marks, P.G., Wang, J.Y., Falls, D.L. and Badwey, J.A. (2000). A protein kinase from neutrophils that specifically recognizes Ser-3 in cofilin. *J. Biol. Chem.* **275**, 2869–2876.

215. Liscovitch, M., Czarny, M., Fiucci, G. and Tang, X. (2000). Phospholipase D: molecular and cell biology of a novel gene family. *Biochem. J.* **345**, 401–415.

216. Loitto, V.-M., Rasmusson, B. and Magnusson, K.-E. (2001). Assessment of neutrophil N-formyl peptide receptors by using antibodies and fluorescent peptides. *J. Leuk. Biol.* **69**, 762–771.

217. Lowell, C.A. and Berton, G. (1999). Integrin signal transduction in myeloid leukocytes. *J. Leukocyte Biol.* **65**, 313–320.

218. Lund, F.E., Cockayne, D.A., Randall, T.D., Solvason, N., Schuber, F. and Howard, M.C. (1998). CD38: A new paradigm in lymphocyte activation and signal transduction. *Immunol. Rev.* **161**, 79–93.

219. Machesky, L.M. and Insall, R.H. (1998). Scar1 and the related Wiskott-Aldrich syndrome protein, WASP, regulate the actin cytoskeleton through the Arp2/3 complex. *Current Biol.* **8**, 1347–1356.

220. Machesky, L.M. and Insall, R.H. (1999). Signaling to actin dynamics. *J. Cell Biol.* **146**, 267–272.

221. Machesky, L.M., Mullins, R.D., Higgs, H.N., Kaiser, D.A., Blanchoin, L., May, R.C., Hall, M.E. and Pollard, T.D. (1999). Scar, a WASp-related protein, activates nucleation of actin filaments by the Arp2/3 complex. *Proc. Natl. Acad. Sci. U.S.A.* **96**, 3739–3744.

222. Maciver, S.K. and Hussey, P.J. (2002). The ADF/cofilin family: actin-remodeling proteins. Genome Biology **3**, reviews3007.1–3007.12.

223. Maekawa, M., Ishizaki, T., Boku, S., Watanabe, N., Fujita, A., Iwamatsu, A., Obinata, T., Ohashi, K., Mizuno, K. and Narumiya, S. (1999). Signaling from Rho to the actin cytoskeleton through protein kinases ROCK and LIM-kinase. *Science* **285**, 895–898.

224. Maheshwari, G. and Lauffenburger, D.A. (1998). Deconstructing (and reconstructing) cell migration. *Microscopy Res. Tech.* **43**, 358–368.

225. Malawista, S.E and de Boisfleury Chevance, A. (1982). The cytokineplast: purified, stable, and functional motile machinery from human blood polymorphonuclear leukocytes. *J. Cell Biol.* **95**, 960–973.

226. Malawista, S.E. and de Boisfleury Chevance, A. (1997). Random locomotion and chemotaxis of human blood polymorphonuclear leukocytes (PMN) in the presence of EDTA: PMN in close quarters require neither leukocyte integrins nor external divalent cations. *Proc. Natl. Acad. Sci. U.S.A.* **94**, 11577–11582.

227. Malawista, S.E., de Boisfleury Chevance, A. and Boxer, L.A. (2000). Random locomotion and chemotaxis of human blood polymorphonuclear leukocytes from a patient with leukocyte adhesion deficiency-1. Normal displacement in close quarters via chimneying. *Cell Motil. Cytoskeleton* **46**, 183–189.

228. Mandeville, J.T., Lawson, M.A. and Maxfield, F.R. (1997). Dynamic imaging of neutrophil migration in three dimensions: mechanical interactions between cells and matrix. *J. Leukoc. Biol.* **61**, 188–200.

229. Mansfield, P.J., Hinkovska-Galcheva, V., Carey, S.S., Shayman, J.A. and Boxer, L.A. (2002). Regulation of polymorphonuclear leukocyte degranulation and oxidant production by ceramide through inhibition of phospholipase D. *Blood* **99**, 1434–1441.

230. Marasco, W.A., Showell, H.J. and Becker, E.L. (1981). Substance P binds to the formyl peptide chemotaxis receptor on the rabbit neutrophil. *Biochem. Biophys. Res. Commun.* **99**, 1065–1072.

231. Marasco, W.A., Phan, S.H., Krutzsch, H., Showell, H.J., Feltner, D.E., Nairn, R., Becker, E.L. and Ward, P.A. (1984). Purification and identification of formyl-methionyl-leucyl-phenylalanine as a major peptide neutrophil chemotactic factor produced by *Escherichia coli*. *J. Biol. Chem.* **259**, 5430–5439.

232. Martin, T.F.P. (1998). Phosphoinositide lipids as signaling molecules: common themes for signal transduction, cytoskeletal regulation, and membrane trafficking. *Annu. Rev. Cell Dev. Biol.* **14**, 231–264.

233. Martin, U., Bock, D., Arseniev, L., Tornetta, M.A., Ames, R.S., Bautsch, W., Kohl, J., Ganser, A. and Klos, A. (1997). The human C3a receptor is expressed on

neutrophils and monocytes, but not on B or T lymphocytes. *J. Exp. Med.* **186**, 199–207.

234. Matsui, T., Yonemura, S., Tsukita, S. and Tsukita, S. (1999). Activation of ERM proteins *in vivo* by Rho involves phosphatidylinositol-4-phosphate 5-kinase and not ROCK kinases. *Curr. Biol.* **9**, 1259–1262.

235. Maun, N.A., Speicher, D.W., DiNubile, M.J. and Southwick, F.S. (1996). Purification and properties of a Ca^{2+}-independent barbed-end actin filament capping protein, CapZ, from human polymorphonuclear leukocytes. *Biochem.* **35**, 3518–3524.

236. McCluskey, A., Sim, A.T.R. and Sakoff, J.A. (2002). Serine-threonine protein phosphatase inhibitors: Development of potential therapeutic strategies. *J. Medicinal Chem.* **45**, 1151–1175.

237. McCutcheon, M. (1946). Chemotaxis in leukocytes. *Physiol. Rev.* **26**, 319–336.

238. McKay, D.A., Kusel, J.R. and Wilkinson, P.C. (1991). Studies of chemotactic factor-induced polarity in human neutrophils. Lipid mobility, receptor distribution and the time-sequence of polarization. *J. Cell Sci.* **100**, 473–479.

239. McLeish, K.R., Gierschik, P. and Jakobs, K.H. (1989). Desensitization uncouples the formyl peptide receptor-guanine nucleotide-binding protein interaction in HL60 cells. *Molec. Pharmacol.* **36**, 384–390.

240. Merry, C., Puri, P. and Reen, D.J. (1998). Phosphorylation and the actin cytoskeleton in defective newborn neutrophil chemotaxis. *Pediatric Res.* **44**, 259–264.

241. Miller, W.E. and Lefkowitz, R.J. (2001). Expanding roles for β-arrestins as scaffolds and adaptors in GPCR signaling and trafficking. *Curr. Opin. Cell Biol.* **13**, 139–145.

242. Mitchison, T.J. and Cramer, L.P. (1996). Actin-based cell motility and cell locomotion. *Cell* **84**, 371–379.

243. Mocsai, A., Jakus, Z., Vantus, T., Berton, G., Lowell, C.L. and Ligeti, E. (2000). Kinase pathways in chemoattractant-induced degranulation of neutrophils: The role of p38 mitogen-activated protein kinase activated by Src family kinases. *J. Immunol.* **164**, 4321–4331.

244. Model M.A. and Omann, G.M. (1995). Desensitization of the actin polymerization response in human neutrophils at low cell density. *J. Leuk. Biol.* **58**, 331–341.

245. Model, M.A. and Omann, G.M. (1996). Cell polarization as a possible mechanism of response termination. *Biochem. Biophys. Res. Commun.* **224**, 516–521. (ERRATA, 226:943).

246. Model, M.A. and Omann, G.M. (1998). Experimental approaches for observing homologous desensitization and their pitfalls. *Pharmacol. Res.* **38**, 41–44.

247. Moghe, P.V., Nelson, R.D and Tranquillo, R.T. (1995). Cytokine-stimulated chemotaxis of human neutrophils in a 3-D conjoined fibrin gel assay. *J. Immunol. Meth.* **180**, 193–211.

248. Mogilner, A. and Oster, G. (1996). Cell motility driven by actin polymerization. *Biophys. J.* **71**, 3030–3045.

249. Mogilner, A. and Edelstein-Keshet, L. (2002). Regulation of actin dynamics in rapidly moving cells: A quantitative analysis. *Biophys. J.* **83**, 1237–1258.

250. Morland, C.M., Morland, B.J., Darbyshire, P.H. and Stockley, R.A. (2000). Migration of CD18-deficient neutrophils *in vitro*: evidence for a CD18-independent pathway induced by IL-8. *Biochim. Biophys. Acta* **1500**, 70–76.

251. Morohashi, M., Winn, A.E., Borisuk, M.T., Bolouri, H., Doyle, J. and Kitano, H. (2002). Robustness as a measure of plausibility in models of biochemical networks. *J. Theor. Biol.* **216**, 19–30.

252. Mousli, M., Bronner, C., Landry, Y., Bockaert, J. and Rouot, B. (1990). Direct activation of GTP-binding regulatory proteins (G-proteins) by substance P and compound 48/80. *FEBS Lett.* **259**, 260–262.

253. M'Rabet, L., Coffer, P.J., Wolthuis, R.M.G., Zwartkruis, F., Koenderman, L. and Bos, J.L. (1999). Differential fMet-Leu-Phe- and platelet-activating factor-induced signaling toward Ral activation in primary human neutrophils. *J. Biol. Chem.* **274**, 21 847–21 852.

254. Muller, W.A. and Randolph, G.J. (1999). Migration of leukocytes across the endothelium and beyond: molecules involved in the transmigration and fate of monocytes. *J. Leuk. Biol.* **66**, 698–704.

255. Mullmann, T.J., Siegel, M.I., Egan, R.W. and Billah, M.M. (1990). Complement C5a activation of phospholipase D in human neutrophils. A major route to the production of phosphatidates and diglycerides. *J. Immunol.* **144**, 1901–1908.

256. Murphy, P.M., Eide, B., Goldsmith, P., Brann, M., Gierschik, P., Spiegel, A. and Malech, H.L. (1987). Detection of multiple forms of $G_{i\alpha}$ in HL60 cells. *FEBS Lett.* **221**, 81–86.

257. Murphy, P.M. (1994). The molecular biology of the leukocyte chemoattractant receptors. *Annu. Rev. Immunol.* **12**, 593–633.

258. Murphy, P.M. (1996). Chemokine receptors: Structure, function, and role in microbial pathogenesis. *Cytokine Growth Factor Rev.* **7**, 47–64.

259. Murray, J., Vawter-Hugart, H., Voss, E. and Soll, D.R. (1992). Three-dimensional motility cycle in leukocytes. *Cell Motil. Cytoskeleton* **22**, 211–223.

260. Naccache, P.H., Volpi, M., Showell, H.J., Becker, E.L. and Sha'afi, R.I. (1979). Chemotactic factor-induced release of membrane calcium in rabbit neutrophils. *Science* **203**, 461–463.

261. Naccache, P.H., Levasseur, S., Lachance, G., Chakravarti, S., Bourgoin, S.G. and McColl, SR. (2000). Stimulation of human neutrophils by chemotactic factors is associated with the activation of phosphatidylinositol 3-kinase γ. *J. Biol. Chem.* **275**, 23 636–23 641.

262. Nagase, H., Miyamasu, M., Yamaguchi, M., Imanishi, M., Tsuno, N.H., Matsushima, K., Yamamoto, K., Morita, Y. and Hirai, K. (2002). Cytokine-mediated regulation of CXCR4 expression in human neutrophils. *J. Leukoc. Biol.* **71**, 711–717.

263. Narumiya, S., Ishizaki, T. and Watanabe, N. (1997). Rho effectors and reoganization of actin cytoskeleton. *FEBS Lett.* **410**, 68–72.

264. Nelson, R.D., Quie, P.G. and Simmons, R.L. (1975). Chemotaxis under agarose: a new and simple method for measuring chemotaxis and spontaneous migration of human polymorphonuclear leukocytes and monocytes. *J. Immunol.* **115**, 1650–1656.

265. Nelson, R.D. and Herron, M.J. (1988). Agarose method for human neutrophil chemotaxis. *Meth. Enzymol.* **162**, 50–59.

266. Neptune, E.R. and Bourne, H.R. (1997). Receptors induce chemotaxis by releasing the βγ subunit of G_i, not by activating G_q or G_s. *Proc. Natl. Acad. Sci. U.S.A.* **94**, 14 489–14 494.

267. Neptune, E.R., Iiri, T. and Bourne, H.R. (1999). $G_{\alpha i}$ is not required for chemotaxis mediated by G_i-coupled receptors. *J. Biol. Chem.* **274**, 2824–2828.

268. Neubig, R.R., and Sklar, L.A. (1993). Subsecond modulation of formyl peptide-linked guanine nucleotide-binding proteins by guanosine 5′-O-(3-thio)triphosphate in permeabilized neutrophils. *Mol. Pharm.* **43**, 734–740.

269. Neubig, R.R., and Siderovski, D.P. (2002). Regulators of G-protein signaling as new central nervous system drug targets. *Nature Rev. Drug Discovery* **1**, 187–195.

270. Nick, J.A., Avdi, N.J., Young, S.K., Knall, C., Gerwins, P., Johnson, G.L. and Worthen, G.S. (1997). Common and distinct intracellular signaling pathways in human neutrophils utilized by platelet-activating factor and FMLP. *J. Clin. Invest.* **99**, 975–986.

271. Nick, J.A., Avdi, N.J., Young, S.K., Lehman, L.A., McDonald, P.P., Frasch, S.C., Billstrom, M.A., Henson, P.M., Johnson, G.L. and Worthen, G.S. (1999). Selective activation and functional significance of p38 alpha mitogen–activated protein kinase in lipopolysaccharide-stimulated neutrophils. *J. Clin. Invest.* **103**, 851–858.

272. Niggli, V. and Keller, H. (1997). The phosphotidylinositol 3-kinase inhibitor wortmannin markedly reduces chemotactic peptide-induced locomotion and increases in cytoskeletal actin in human neutrophils. *Eu. J. Pharmacol.* **335**, 43–52.

273. Niggli, V. (1999). Rho-kinase in human neutrophils: a role in signaling for myosin light chain phosphorylation and cell migration. *FEBS Lett.* **445**, 69–72.

274. Niggli, V. (2000). A membrane-permeant ester of phosphatidylinositol 3,4,5-trisphosphate (PIP_3) is an activator of human neutrophil migration. *FEBS Lett.* **473**, 217–221.

275. Nijhuis, E., Lammers, J.-W.J., Koenderman, L. and Coffer, P.J. (2002). Src kinases regulate PKB activation and modulate cytokine and chemoattractant-controlled neutrophil functioning. *J. Leukoc. Biol.* **71**, 115–124.

276. Norgauer, J., Krutmann, J., Dobos, G.J., Traynor-Kaplan, A.E., Oades, Z.G. and Schraufstatter, I.U. (1994). Actin polymerization, calcium-transients, and phospholipid metabolism in human neutrophils after stimulation with interleukin-8 and N-formyl peptide. *J. Invest. Dermatol.* **102**, 310–314.

277. Nossal, R. and Zigmond, S.H. (1976). Chemotropism indices for polymorphonuclear leukocytes. *Biophys. J.* **16**, 1171–1182.

278. O'Flaherty, J.T., Kreutzer, D.L., Showell, H.S., Becker, E.L. and Ward, P.A. (1978). Desensitization of the neutrophil aggregation response to chemotactic factors. *Am. J. Pathol.* **93**, 693–706.

279. Okada, K., Takano-Ohmuro, H., Obinata, T. and Abe, H. (1996). Dephosphorylation of cofilin in polymorphonuclear leukocytes derived from peripheral blood. *Exp. Cell Res.* **227**, 116–122.

280. Oliver, J.M. (1978). Cell biology of leukocyte abnormalities—membrane and cytoskeletal function in normal and defective cells. A review. *Am. J. Pathol.* **93**, 221–270.

281. Oliver, J.M., Krawiec, J.A. and Becker, E.L. (1978). The distribution of actin during chemotaxis in rabbit neutrophils. *J. Reticuloendothel. Soc.* **24**, 697–704.

282. Olson, S.C. and Lambeth, J.D. (1996). Biochemistry and cell biology of phopholipase D in human neutrophils. *Chem. Phys. Lipids* **80**, 3–19.

283. Omann, G.M., Swann, W.N., Oades, Z.G., Parkos, C.A., Jesaitis, A.J. and Sklar, L.A. (1987). N-formyl peptide receptor dynamics, cytoskeletal activation,

and intracellular calcium response in human neutrophil cytoplasts. *J. Immunol.* **139**, 3447–3455.

284. Omann, G.M., Traynor, A.E., Harris, A. and Sklar, L.A. (1987). LTB$_4$-induced activation signals and responses in neutrophils are short-lived compared to formyl peptide. *J. Immunol.* **138**, 2626–2632.

285. Omann G.M., Porasik, M.M. and Sklar, L.A. (1989). Oscillating actin polymerization/depolymerization responses in human polymorphonuclear leukocytes. *J. Biol. Chem.* **264**, 16355–16358.

286. Omann G.M., Rengan, R., Hoffman, J.F. and Linderman, J.J. (1995). Rapid oscillations of actin polymerization/depolymerization in polymorphonuclear leukocytes stimulated by leukotriene B$_4$ and platelet-activating factor. *J. Immunol.* **155**, 5375–5381.

287. Omann, G.M., and Axelrod, D. (1996). Membrane-proximal calcium transients in stimulated neutrophils detected by total internal reflection fluorescence. *Biophys. J.* **71**, 2885–2891.

288. Omann, G.M. and Hinshaw, D.B. (1997). Inflammation, in *Surgery: Scientific Principles and Practice*, 2nd ed. (L.J. Greenfield, M.W. Mulholland, K.T. Oldham, G.B. Zelenock, and K.D. Lillemoe, eds.), 130–159. Lippincott-Raven Publishers, Philadelphia, PA.

289. Omann, G.M. (1998). Actin polymerization in human neutrophils, in *Phagocytic Function: A guide for Research and Clinical Evaluation* (J.P. Robinson and G.F. Babcock, eds.), 155–185. Wiley-Liss, Inc.

290. Painter, R.G., Dukes, R., Sullivan, J., Carter, R., Erdos, E.G. and Johnson, A.R. (1988). Function of neutral endopeptidase on the cell membrane of human neutrophils. *J. Biol. Chem.* **263**, 9456–9461.

291. Painter, R.G. and Aiken, M.L. (1995). Regulation of the N-formyl-methionyl-leucyl-phenylalanine receptor recycling by surface membrane neutral endopeptidase-mediated degradation of ligand. *J. Leukoc. Biol.* **58**, 468–476.

292. Palecek, S.P., Loftus, J.C., Ginsberg, M.H., Lauffenburger, D.A. and Horwitz, A.F. (1997). Integrin-ligand binding properties govern cell migration speed through cell-substratum adhesiveness. *Nature* **385**, 537–540.

293. Pantaloni, D., Hill, T.L., Carlier, M.-F. and Korn, E.D. (1985). A model for actin polymerization and the kinetic effects of ATP hydrolysis. *Proc. Natl. Acad. Sci. U.S.A.* **82**, 7207–7211.

294. Parekh, T., Saxena, B., Reibman, J., Cronstein, B.N. and Gold, L.I. (1994). Neutrophil chemotaxis in response to TGF-β isoforms (TGF-β1, TGF-β2, TGF-β3) is mediated by fibronectin. *J. Immunol.* **152**, 2456–2466.

295. Parent, C.A. and Devreotes, P.N. (1999). A cell's sense of direction. *Science* **284**, 765–770.

296. Parkhurst, M.R. and Saltzman, W.M. (1992). Quantification of human neutrophil motility in three-dimensional collagen gels. Effect of collagen concentration. *Biophys. J.* **61**, 306–315.

297. Partida-Sanchez, S., Cockayne, D.A., Monard, S., Jacobson, E.L., Oppenheimer, N., Garvy, B., Kusser, K., Goodrich, S., Howard, M., Harmsen, A., Randall, T.D. and Lund, F.E. (2001). Cyclic ADP-ribose production by CD38 regulates intracellular calcium release, extracellular calcium influx and chemotaxis in neutrophils and is required for bacterial clearance *in vivo*. *Nature Med.* **7**, 1209–1216.

298. Perez, H.D., Elfman, F., Lobo, E., Sklar, L., Chenoweth, D. and Hooper, C. (1986). A derivative of wheat germ agglutinin specifically inhibits formyl peptide-induced polymorphonuclear leukocyte chemotaxis by blocking reexpression (or recycling) of receptors. *J. Immunol.* **136**, 1803–1812.

299. Perez, H.D., Elfman, F., Marder, S., Lobo, E. and Ives, H.E. (1989). Formyl peptide-induced chemotaxis of human polymorphonuclear leukocytes does not require either marked changes in cytosolic calcium or specific granule discharge. Role of formyl peptide receptor reexpression (or recycling). *J. Clin. Invest.* **83**, 1963–1970.

300. Peskin, C., Odell, G. and Oster, G. (1993). Cellular motions and thermal fluctuations: the Brownian ratchet. *Biophys. J.* **65**, 316–324.

301. Pestonjamasp, K., Amieva, M.R., Strassel, C.P., Nauseef, W.M., Furthmsyr, H. and Luna, E.J. (1995). Moesin, ezrin, and p205 are actin-binding proteins associated with neutrophil plasma membranes. *Mol. Biol. Cell* **6**, 247–259.

302. Petty, H.R. (2000). Oscillatory signals in migrating neutrophils: effects of time-varying chemical and electrical fields, in *Self-Organized Biological Dynamics and Nonlinear Control by External Stimuli* (J. Walleczek, ed.), 173–192. Cambridge University Press, Cambridge, UK.

303. Phillips, M.R., Feoktistov, A., Pillinger, M.H. and Abramson, S.B. (1995). Translocation of p21^{rac2} from cytosol to plasma membrane is neither necessary nor sufficient for neutrophil NADPH oxidase activity. *J. Biol. Chem.* **270**, 11514–11521.

304. Pierce, K.L. and Lefkowitz, R.J. (2001). Classical and new roles of β-arrestins in the regulation of G-protein-coupled receptors. *Nature Rev.* **2**, 727–733.

305. Pierce, K.L., Premont, R.T. and Lefkowitz, R.J. (2002). Seven-transmembrane receptors. *Nature Rev. Molec. Cell Biol.* **3**, 639–650.

306. Pierini, L.M., Furoles, M., Seveau, S., Casulo, C. and Maxfield, F.R. (2000). Membrane cholesterol is critical for asymmetric Rac targeting during human neutrophil migration, in *American Society for Cell Biology 40th annual meeting program*; Dec. 9–13, San Francisco, CA Abstract # 2921.

307. Pierini, L.M., Lawson, M.A., Eddy, R.J., Hendey, B. and Maxfield, F.R. (2000). Oriented endocytic recycling of α5β1 in motile neutrophils. *Blood* **15**, 2471–2481.

308. Pollard, T.D., Blanchoin, L. and Mullins, R.D. (2000). Molecular mechanisms controlling actin filament dynamics in nonmuscle cells. *Annu. Rev. Biophys. Biomol. Struct.* **29**, 545–576.

309. Pontremoli, S., Melloni, E., Michetti, M., Sparatore, B., Salamino, F., Sacco, O. and Horecher, B.L. (1987). Phosphorylation by protein kinase C of a 20-kDa cytoskeletal polypeptide enhances its susceptibility to digestion by calpain. *Proc. Natl. Acad. Sci. U.S.A.* **84**, 398–401.

310. Posner, R.G., Fay, S.P., Domalewski, M.D. and Sklar, L.A. (1994). Continuous spectrofluorometric analysis of formylpeptide receptor ternary complex interactions. *Mol. Pharm.* **45**, 65–73.

311. Prescott, S.M., Zimmerman, G.A., Stafforini, D.M. and McIntyre, T.M. (2000). Platelet-activating factor and related lipid mediators. *Annu. Rev. Biochem.* **69**, 419–445.

312. Prossnitz, E.R., Kim, C.M., Benovic, J.L. and Ye, R.D. (1995a). Phosphorylation of the N-formyl peptide receptor carboxyl terminus by the G-protein coupled receptor kinase, GRK2. *J. Biol. Chem.* **270**, 1130–1137.

313. Prossnitz, E.R., Schreiber, R.E., Bokoch, G.M. and Ye, R.D. (1995b). Binding of low-affinity N-formyl peptide receptors to G-protein. Characterization of a novel inactive receptor intermediate. *J. Biol. Chem.* **270**, 10 686–10 694.

314. Prossnitz, E.R. (1997). Desensitization of N-formyl peptide receptor–mediated activation is dependent upon receptor phosphorylation. *J. Biol. Chem.* **272**, 15 213–15 219.

315. Prossnitz, E.R., Gilbert, T.L., Chiang, S., Campbell, J.J., Qin, S., Newman, W., Sklar, L.A. and Ye, R.D. (1999). Multiple activation steps of the N-formyl peptide receptor. *Biochem.* **38**, 2240–2247.

316. Pryzwansky, K.B., Wyatt, T.A., Nichols, H. and Lincoln, T.M. (1990). Compartmentalization of cyclic GMP-dependent protein kinase in formyl-peptide stimulated neutrophils. *Blood* **76**, 612–618.

317. Ptasznik, A., Traynor-Kaplan, A.E. and Bokoch, G.M. (1995). G protein-coupled chemoattractant receptors regulate Lyn tyrosine kinase-Shc adaptor protein signaling complexes. *J. Biol. Chem.* **270**, 19 969–19 973.

318. Ptasznik, A., Prossnitz, E.R., Yoshikawa, D., Smrcka, A., Traynor-Kaplan, A.E. and Bokoch, G.M. (1996). A tyrosine kinase signaling pathway accounts for the majority of phosphatidylinositol 3,4,5-trisphosphate formation in chemoattractant-stimulated human neutrophils. *J. Biol. Chem.* **271**, 25 204–25 207.

319. Quinn, M.T., Evans, T., Loetterle, L.R., Jesaitis, A.J. and Bokoch, G.M. (1993). Translocation of Rac correlates with NADPH oxidase activation. Evidence for equimolar translocation of oxidase components. *J. Biol. Chem.* **268**, 20 983–20 987.

320. Quinn, M.T. (1996). Role of small GTP-binding proteins in leukocyte signal transduction, in *Signal Transduction in Leukocytes. G Protein-Related and Other Pathways* (PM Lad, JS Kaptein, and C-K E Lin, eds.), 75–141. CRC Press, Boca Raton.

321. Ramsey, W.S. (1972). Analysis of individual leucocyte behavior during chemotaxis. *Exp. Cell Res.* **70**, 129–139.

322. Ramsey, W.S. (1972). Locomotion of human polymorphonuclear leucocytes. *Exp. Cell Res.* **72**, 489–501.

323. Rane, M.J., Coxon, P.Y., Powell, D.W., Webster, R., Klein, J.B., Pierce, W., Ping, P. and McLeish, K.R. (2001). p38 kinase-dependent MAPKAPK-2 activation functions as 3-phosphoinositide-dependent kinase-2 for akt in human neutrophils. *J. Biol. Chem.* **276**, 3517–3523.

324. Rao, K.M.K. and Varani, J. (1982). Actin polymerization induced by chemotactic peptide and concanavalin A in rat neutrophils. *J. Immunol.* **129**, 1605–1607.

325. Rebuck, J.W. and Crowley, J.H. (1955). A method of studying leukocyte functions *in vivo. Annu. NY Acad. Sci.* **59**, 757–805.

326. Redmond, T. and Zigmond, S.H. (1993). Distribution of F-actin elongation sites in lysed polymorphonuclear leukocytes parallels the distribution of endogenous F-actin. *Cell Motil. Cytoskeleton* **26**, 7–18.

327. Regier, D.S., Greene, D.G., Sergeant, S., Jesaitis, A.J. and McPhail, L.C. (2000). Phosphorylation of p22^{phox} is mediated by phospholipase D-dependent and -independent mechanisms. Correlation of NADPH oxidase activity and p22^{phox} phosphorylation. *J. Biol. Chem.* **275**, 28 406–28 412.

328. Reibman, J., Meixler, S., Lee, T.C., Gold, L.I., Cronstein, B.N., Haines, K.A., Kolasinski, S.L., and Weissmann, G. (1991). Transforming growth factor b1, a

potent chemoattractant for human neutrophils, bypasses classic signal-transduction pathways. *Proc. Natl. Acad. Sci. U.S.A.* **88**, 6805–6809.

329. Rengan, R. (1998). Signaling actin polymerization responses in hematopoietic cells. Ph.D. Thesis, University of Michigan.

330. Rengan, R. and Omann, G.M. (1999). Regulation of oscillations in filamentous actin content in polymorphonuclear leukocytes stimulated with leukotriene B_4 and platelet-activating factor. *Biochem. Biophys. Res. Commun.* **262**, 479–486.

331. Rengan, R. and Ochs, H.D. (2000). Molecular biology of the Wiskott-Aldrich syndrome. *Rev. Immunogenetics* **2**, 243–255.

332. Rengan, R., Ochs, H.D., Sweet, L.I., Keil, M., Gunning, W.T., Lachant, N.A., Boxer, L.A. and Omann, G.M. (2000). Actin cytoskeletal function is spared, but apoptosis is increased in WAS patient hematopoietic cells. *Blood* **95**, 1283–1292.

333. Ressad, F., Didry, D., Egile, C., Pantaloni, D. and Carlier, M.-F. (1999). Control of actin filament length and turnover by actin depolymerizing factor (ADF/cofilin) in the presence of capping proteins and Arp2/3 complex. *J. Biol. Chem.* **274**, 20 970–20 976.

334. Rhee, S.G. and Bae, Y.S. (1997). Regulation of phosphoinositide-specific phospholipase C isozymes. *J. Biol. Chem.* **272**, 15 045–15 048.

335. Richardson, R.M., Ali, H., Tomhave, E.D., Haribabu, B. and Snyderman, R. (1995). Cross-desensitization of chemoattractant receptors occurs at multiple levels. Evidence for a role for inhibition of phospholipase C activity. *J. Biol. Chem.* **270**, 27 829–27 833.

336. Rickert, P., Weiner, O.D., Wang, F., Bourne, H.R. and Servant, G. (2000). Leukocytes navigate by compass: role of PI3Kg and its lipid products. *Trends Cell Biol.* **10**, 466–473.

337. Roberts, A.W., Kim, C., Zhen, L., Lowe, J.B., Kapur, R., Petryniak, B., Spaetti, A., Pollock, J.D., Borneo, J.B., Bradford, G.B., Atkinson, S.J., Dinauer, M.C. and Williams, D.A. (1999). Deficiency of the hematopoietic cell-specific Rho family GTPase Rac2 is characterized by abnormalities in neutrophil function and host defense. *Immunity* **10**, 183–196.

338. Rosengren, S., Henson, P.M. and Worthen, G.S. (1994). Migration-associated volume changes in neutrophils facilitate the migratory process *in vitro*. *Am. J. Physiol.* **267** (Cell Physiol. 36), C1623-C1632.

339. Ryder, M.I., Weinreb, R.N. and Niederman, R. (1984). The organization of actin filaments in human polymorphonuclear leukocytes. *Anat. Rec.* **209**, 7–20.

340. Sanders, L.C., Matsumura, F., Bokoch, G.M. and de Lanerolle, P. (1999). Inhibition of myosin light chain kinase by p21-activated kinase. *Science* **283**, 2083–2085.

341. Sasaki, T., Irie-Sasaki, J., Jones, R.G., Oliveira-dos-Santos, A.J., Stanford, W.L., Bolon, B., Wakeham, A., Itie, A., Bouchard, D., Kozieradzki, I., Joza, N., Mak, T.W., Ohashi, P.S., Suzuki, A. and Penninger, J.M. (2000). Function of PI3Kγ in thymocyte development, T cell activation, and neutrophil migration. *Science* **287**, 1040–1046.

342. Schmalstieg, F.C., Rudloff, H.E., Hillman, G.R. and Anderson, D.C. (1986). Two-dimensional and three-dimensional movement of human polymorphonuclear leukocytes: two fundamentally different mechanisms of locomotion. *J. Leukoc. Biol.* **40**, 677–691.

343. Schmitt, M. and Keller, H. (1987). Video intensification microscopy of binding and processing of the fluoresceinated N-formyl chemotactic peptide TRITC-NFLPNTL by human neutrophils. *Adv. Biosci.* **66**, 285–294.

344. Schmitt, M. and Bultmann, B. (1990). Fluorescent chemotactic peptides as tools to identify the f-Met-Leu-Phe receptor on human granulocytes. *Biochem. Soc. Trans.* **18**, 219–222.

345. Sechi, A.S. and Wehland, J. (2000). The actin cytoskeleton and plasma membrane connection: PtdIns(4,5)*P2* influences cytoskeletal protein activity at the plasma membrane. *J. Cell Sci.* **113**, 3685–3695.

346. Senda, N., Tamura, H., Shibata, N., Yoshitake, J., Kondo, K. and Tanaka, K. (1975). The mechanism of the movement of leucocytes. *Exp. Cell Res.* **91**, 393–407.

347. Servant, G., Weiner, O.D., Neptune, E.R., Sedat, J.W. and Bourne, H.R. (1999). Dynamics of a chemoattractant receptor in living neutrophils during chemotaxis. *Mol. Biol. Cell* **10**, 1163–1178.

348. Servant, G., Weiner, O.D., Herzmark, P., Balla, T., Sedat, J.W. and Bourne, H.R. (2000). Polarization of chemoattractant receptor signaling during neutrophil chemotaxis. *Science* **287**, 1037–1040.

349. Seveau, S., Lopez, S., Lesavre, P., Guichard, J., Cramer, E.M., and Halbwachs-Mecarelli, L. (1997). Leukosialin (CD43, sialophorin) redistribution in uropods of polarized neutrophils is induced by CD43 cross-linking by antibodies, by colchicines or by chemotactic peptides. *J. Cell Sci.* **110**, 1465–1475.

350. Seveau, S., Keller, H., Maxfield, F.R., Piller, F. and Halbwachs-Mecarelli, L. (2000). Neutrophil polarity and locomotion are associated with surface redistribution of leukocialin (CD43), an antiadhesive membrane molecule. *Blood* **95**, 2462–2470.

351. Seveau, S., Eddy, R.J., Maxfield, F.R. and Pierini, L.M. (2001). Cytoskeleton-dependent membrane domain segregation during neutrophil polarization. *Mol. Biol. Cell* **12**, 3550–3562.

352. Shea, L.D. and Linderman, J.J. (1997). Mechanistic model of G-protein signal transduction: Determinants of efficacy and the effect of precoupled receptors. *Biochem. Pharm.* **53**, 519–530.

353. Shea, L.D., Neubig, R.R. and Linderman, J.J. (2000). Timing is everything: The role of kinetics in G-protein activation. *Life Sci.* **68**, 647–658.

354. Shields, J.M. and Haston, W.S. (1985). Behavior of neutrophil leucocytes in uniform concentrations of chemotactic factors: contraction waves, cell polarity, and persistence. *J. Cell Sci.* **74**, 75–93.

355. Simchowitz, L., Atkinson, J.P. and Spilberg, I. (1980). Stimulus-specific deactivation of chemotactic factor-induced cyclic AMP response and superoxide generation by human neutrophils. *J. Clin. Invest.* **66**, 736–747.

356. Sklar, L.A., Jesaitis, A.J., Painter, R.G. and Cochrane, C.G. (1982). Ligand/receptor internalization: Spectroscopic analysis and comparison of ligand-binding, cellular response and internalization by human neutrophils. *J. Cellular Biochem.* **20**, 193–202.

357. Sklar, L.A., Finney, D.A., Oades, Z.G., Jesaitis, A.J., Painter, R.G. and Cochrane, C.G. (1984). The dynamics of ligand-receptor interactions: Real-time analysis of association, dissociation, and internalization of an N-formyl peptide and its receptors on the human neutrophil. *J. Biol. Chem.* **259**, 5661–5669.

358. Sklar, L.A., Hyslop, P.A., Oades, Z.G., Omann, G.M., Jesaitis, A.J., Painter, R.G. and Cochrane, C.G. (1985). Signal transduction and ligand-receptor dynamics in the human neutrophil. Transient responses and occupancy-response relations at the formyl peptide receptor. *J. Biol. Chem.* **260**, 11 461–11 467.

359. Sklar, L.A., Omann, G.M. and Painter, R.G. (1985). Relationship of actin polymerization and depolymerization to light scattering in human neutrophils: dependency on receptor occupancy and intracellular calcium. *J. Cell Biol.* **101**, 1161–1166.

360. Sklar, L.A. (1987). Real-time spectroscopic analysis of ligand-receptor dynamics. *Annu. Rev. Biophys. Biophys. Chem.* **16**, 479–506.

361. Sklar, L.A., Mueller, H., Omann, G.M. and Oades, Z.G. (1989). Three states for the formyl peptide receptor on intact cells. *J. Biol. Chem.* **264**, 8483–8486.

362. Sklar, L.A., Muller, H., Swann, W.N., Comstock, C., Omann, G.M. and Bokoch, G.M. (1989). Dynamics of interaction among ligand, receptor and G-protein, in Luminescence Applications in *Biological, Chemical, Environmental, and Hydrological Sciences* (MC Goldberg, ed.), 52–69. ACS Publications.

363. Sklar, L.A. and Omann, G.M. (1990). Kinetics and amplification in neutrophil activation and adaptation. *Sem. Cell Biol.* **1**, 115–123.

364. Sklar, L.A., Swann, W.N., Fay, S.P. and Oades, Z.G. (1990). Real-time analysis of marcromolecular assembly during cell activation, in *Biology of Cellular Transducing signals* (JY Vanderhoek, ed.) 1–10. Plenum Publishing Corporation.

365. Smith, C.W., Hollers, J.C., Patrick, R.A. and Hassett, C. (1979). Motility and adhesiveness in human neutrophils. Effects of chemotactic factors. *J. Clin. Invest.* **63**, 221–229.

366. Smith, C.W. (2000). Possible steps involved in the transition to stationary adhesion of rolling neutrophils: a brief review. *Microcirculation* **7**, 385–394.

367. Smolen, J.E., Eisenstat, B.A. and Weissmann, G. (1982). The fluorescence response of chlorotetracycline-loaded human neutrophils. Correlation with lysosomal enzyme release and evidence for a "trigger pool" of calcium. *Biochim. Biophys. Acta* **717**, 422–431.

368. Snyderman R., Altman, L.C., Hausman, M.S. and Mergenhagen, S.E. (1972). Human mononuclear leukocyte chemotaxis: a quantitative assay for humoral and cellular chemotactic factors. *J. Immunol.* **108**, 857–860.

369. Stasia, M.-J., Jouan, A., Bourmeyster, N., Boquet, P. and Vignais, P.V. (1991). ADP-ribosylation of a small size GTP-binding protein in bovine neutrophils by the C3 exoenzyme of *Closteridium botulinum* and effect on cell motility. *Biochem. Biophys. Res. Commun.* **180**, 615–622.

370. Stephens, L., Hughes, K.T. and Irvine, R.F. (1991). Pathway of phosphatidylinositol(3,4,5)-trisphosphate synthesis in activated neutrophils. *Nature* **351**, 33–39.

371. Stephens, L., Eguinoa, A., Corey, S., Jackson, T. and Hawkins, P.T. (1993). Receptor stimulated accumulation of phosphatidylinositol (3,4,5)-trisphosphate by G-protein mediated pathways in human myeloid derived cells. *EMBO J.* **12**, 2265–2273.

372. Stephens, L., Smrcka, A., Cooke, F.T., Jackson, T.R., Sternweis, P.C. and Hawkins, P.T. (1994). A novel phosphoinositide 3 kinase activity in myeloid-derived cells is activated by G protein bg subunits. *Cell* **77**, 83–93.

373. Stokes, C.L., Rupnick, M.A., Williams, S.K. and Lauffenburger, D.A. (1990). Chemotaxis of human microvessel endothelial cells in response to acidic fibroblast growth factor. *Lab. Invest.* **63**, 657–668.

374. Stokes, C.L., Lauffenburger, D.A. and Williams, S.K. (1991). Migration of individual microvessel endothelial cells: stochastic model and parameter measurement. *J. Cell Sci.* **99**, 419–430.

375. Suchard, S.J., Nakamura, T., Abe, A., Shayman, J.A. and Boxer, L.A. (1994). Phospholipase D-mediated dyradylglycerol formation coincides with H_2O_2 and lactoferrin release in adherent human neutrophils. *J. Biol. Chem.* **269**, 8063–8068.

376. Sullivan, S.J. and Zigmond, S.H. (1980). Chemotactic peptide receptor modulation in polymorphonuclear leukocytes. *J. Cell Biol.* **88**, 644–647.

377. Sullivan, S.J., Daukas, G. and Zigmond, S.H. (1984). Asymmetric distribution of the chemotactic peptide receptor on polymorphonuclear leukocytes. *J. Cell Biol.* **99**, 1461–1467.

378. Suzuki, K., Yamaguchi, T., Tanaka, T., Kawanishi, T., Nishimaki-Mogami, T., Yamamoto, K., Tsuji, T., Irimura, T., Hayakawa, T. and Takahashi, A. (1995). Activation induces dephosphorylation of cofilin and its translocation to plasma membranes in neutrophil-like differentiated HL-60 cells. *J. Biol. Chem.* **270**, 19551–19556.

379. Symons, M., Derry, J.M.J., Karlak, B., Jiang, S., Lemahieu, B., McCormick, F., Franke, U. and Abo, A. (1996). Wiskott-Aldrich syndrome protein, a novel effector for the GTPase Cdc42Hs, is implicated in actin polymerization. *Cell* **84**, 723–734.

380. Tan, P., Luscinskas, F.W. and Homer-Vanniasinkam, S. (1999). Cellular and molecular mechanisms of inflammation and thrombosis. *Eur. J. Vasc. Endovasc. Surg.* **17**, 373–389.

381. Tan, J. and Saltzman, W.M. (2002). Topographical control of human neutrophil motility on micropatterned materials with various surface chemistry. *Biomaterials* **23**, 3215–3225.

382. Tardiff, M., Mery, L., Brouchon, L. and Boulay, F. (1993). Agonist-dependent phosphorylation of N-formyl peptide and activation peptide from the fifth component of C (C5a) chemoattractant receptors in differentiated HL60 cells. *J. Immunol.* **150**, 3534–3545.

383. Thelen, M., Wymann, M.P. and Langen, H. (1994). Wortmannin binds specifically to 1-phosphatidylinositol 3-kinase while inhibiting guanine nucleotide-binding protein-coupled receptor signaling in neutrophil leukocytes. *Proc. Natl. Acad. Sci U.S.A.* **91**, 4960–4964.

384. Thelen, M., Uguccioni, M. and Bosiger, J. (1995). PI 3-kinase-dependent and independent chemotaxis of human neutrophil leukocytes. *Biochem. Biophys. Res. Commun.* **217**, 1255–1262.

385. Thelen, M. and Didichenko, S.A. (1997). G-protein-coupled receptor–mediated activation of PI-3-kinase in neutrophils. *Ann. New York Acad. Sci.* **832**, 368–382.

386. Thompson, H.L., Shiroo, M. and Saklatvala, J. (1993). The chemotactic factor N-formylmethionyl-leucyl-phenylalanine activates microtubule-associated protein 2 (MAP) kinase and MAP kinase kinase in polymorphonuclear leucocytes. *Biochem. J.* **290**, 483–488.

387. Tilney, L.G., Bonder, E.M. and De Rosier, D.J. (1991). Actin filaments elongate from their membrane-associated ends. *J. Cell. Biol.* **90**, 485–494.

388. Tilton, B., Andjelkovic, M., Didichenko, S.A., Hemmings, B.A. and Thelen, M. (1997). G-protein-coupled receptors and Fcγ-receptors mediate activation of akt/protein kinase B in human phagocytes. *J. Biol. Chem.* **272**, 28 096–28 101.
389. Toker, A. and Cantley, L.C. (1997). Signaling through the lipid products of phosphoinositide-3-OH kinase. *Nature* **387**, 673–676.
390. Tolias, K.F., Hartwig, J.H., Ishihara, H., Shibasaki, Y., Cantley, L.C. and Carpenter, C.L. (2000). Type Iα phosphatidylinositol-4-phosphate 5-kinase mediates Rac-dependent actin assembly. *Curr. Biol.* **10**, 153–156.
391. Torres, M. and Coates, T.D. (1999). Function of the cytoskeleton in human neutrophils and method for evaluation. *J. Immunol. Meth.* **232**, 89–109.
392. Tranquillo, R.T., Lauffenburger, D.A. and Zigmond, S.H. (1988). A stochastic model for leukocyte random motility and chemotaxis based on receptor binding fluctuations. *J. Cell Biol.* **106**, 303–309.
393. Tranquillo, R.T., Zigmond, S.H. and Lauffenburger, D.A. (1988). Measurement of the chemotaxis coefficient for human neutrophils in the under-agarose migration assay. *Cell Motil. Cytoskeleton* **11**, 1–15.
394. Traynor-Kaplan, A.E., Harris, A., Thompson, B., Taylor, P. and Sklar, L.A. (1988). An inositol tetrakisphosphate-containing phospholipid in activated neutrophils. *Nature* **334**, 353–356.
395. Traynor-Kaplan, A.E., Thompson, B.L., Harris, A.L., Taylor, P., Omann, G.M. and Sklar, L.A. (1989). Transient increase in phosphatidylinositol 3,4-bisphosphate and phosphatidylinositol trisphosphate during activation of human neutrophils. *J. Biol. Chem.* **264**, 15 668–15 673.
396. Tsukita, S. and Yonemura, S. (1999). Cortical actin organization: Lessons from ERM (ezrin/radixin/moesin) proteins. *J. Biol. Chem.* **274**, 34 507–34 510.
397. Ueda, M., Sako, Y., Tanaka, T., Devreotes, P. and Yanagida, T. (2001). Single-molecule analysis of chemotactic signaling in *Dictyostelium* cells. *Science* **294**, 864–867.
398. Urwyler, N., Eggli, P. and Keller, H.U. (2000). Effects of the myosin inhibitor 2,3-butanedione monoxime (BDM) on cell shape, locomotion and fluid pinocytosis in human polymorphonuclear leukocytes. *Cell Biol. Intl.* **24**, 863–870.
399. van Biesen, T., Luttrell, L.M., Hawes, B.E. and Lefkowitz, R.J. (1996). Mitogenic signaling via G-protein-coupled receptors. *Endocrine Rev.* **17**, 698–714.
400. van Hagen, P.M., Hofland, L.J., ten Bokum, M.C., Lichtenauer-Kaligis, E.G.R., Kweekkeboom, D.J., Ferone, D. and Lamberts, S.W.J. (1999). Neuropeptides and their receptors in the immune system. *Ann. Med.* **31** Suppl. 2, 15–22.
401. Valerius, N.H., Stendahl, O., Hartwig, J.H. and Stossel, T.P. (1981). Distribution of actin-binding protein and myosin in polymorphonuclear leukocytes during locomotion and phagocytosis. *Cell* **24**, 195–202.
402. Vanhaesebroeck, B. and Waterfield, M.D. (1999). Signaling by distinct classes of phosphoinositide 3-kinases. *Exp. Cell Res.* **253**, 239–254.
403. Verghese, M.W., Charles, L., Jakoi, L., Dillon, S.B. and Snyderman, R. (1987). Role of a guanine nucleotide regulatory protein in the activation of phospholipase C by different chemoattractants. *J. Immunol.* **138**, 4374–4380.
404. Vlahos, C.J. and Matter, W.F. (1992). Signal transduction in neutrophil activation. Phosphatidylinositol 3-kinase is stimulated without tyrosine phosphorylation. *FEBS Lett.* **309**, 242–248.

405. Vlahos, C.J., Matter, W.F., Brown, R.F., Traynor-Kaplan, A.E., Heyworth, P.G., Prossnitz, E.R., Ye, R.D., Marder, P., Schelm, J.A. and Rothfuss, K.J. (1995). Investigation of neutrophil signal transduction using a specific inhibitor of phosphatidylinositol 3-kinase. *J. Immunol.* **154**, 2413–2422.

406. Wagner, J.G. and Roth, R.A. (2000). Neutrophil migration mechanisms, with an emphasis on the pulmonary vasculature. *Pharmacol. Reviews* **52**, 349–374.

407. Wallace, P.J., Wersto, R.P., Packman, C.H. and Lichtman, M.A. (1984). Chemotactic peptide-induced changes in neutrophil actin conformation. *J. Cell Biol.* **99**, 1060–1065.

408. Waller, A., Pipkorn, D., Sutton, K., Linderman, J.J. and Omann, G.M. (2001). Validation of flow cytometric competitive binding protocols and characterization of fluorescently labeled ligands. *Cytometry* **45**, 102–114.

409. Walter, R.J., Berlin, R.D. and Oliver, J.M. (1980). Asymmetric Fc receptor distribution on human PMN oriented in a chemotactic gradient. *Nature* **286**, 724–725.

410. Walter, R.L. and Marasco, W.A. (1987). Direct visualization of formylpeptide receptor binding on rounded and polarized human neutrophils: cellular and receptor heterogeneity. *J. Leukoc. Biol.* **41**, 377–391.

411. Wang, F., Herzmark, P., Weiner, O.D., Srinivasan, S., Servant, G. and Bourne, H.R. (2002). Lipid products of PI(3)Ks maintain persistent cell polarity and directed motility in neutrophils. *Nature Cell Biol.* **4**, 513–518.

412. Ward, P.A. and Becker, E.L. (1968). The deactivation of rabbit neutrophils by chemotactic factor and the nature of the activatable esterase. *J. Exp. Med.* **127**, 693–709.

413. Watts, R.G., Crispens, M.A. and Howard, T.H. (1991). A quantitative study of the role of F-actin in producing neutrophil shape. *Cell Motil. Cytoskeleton* **19**, 159–168.

414. Webb, B.J.L., Hirst, S.J. and Giembycz, M.A. (2000). Protein kinase C isoenzymes: a review of there structure, regulation and role in regulating airways smooth muscle tone and mitogensis. *Br. J. Pharmacol.* **130**, 1433–1452.

415. Weinbaum, D.L., Sullivan, J.A. and Mandell, G.L. (1980). Receptors for concanavalin. A cluster at the front of polarized neutrophils. *Nature* **286**, 725–727.

416. Weiner, O.D., Servant, G., Welch, M.D., Mitchison, T.J., Sedat, J.W. and Bourne, H.R. (1999). Spatial control of actin polymerization during neutrophil chemotaxis. *Nature Cell Biol.* **1**, 75–81.

417. Weiner, O.D., Neilsen, P.O., Prestwich, G.D., Kirschner, M.W., Cantley, L.C. and Bourne, H.R. (2002). A PtdInsP$_3$- and Rho GTPase-mediated positive feedback loop regulates neutrophil polarity. *Nature Cell Biol.* **4**, 509–512.

418. White, J.R., Naccache, P.H. and Sha'afi, R.I. (1983). Stimulation by chemotactic factor of actin association with the cytskeleton in rabbit neutrophils. *J. Biol. Chem.* **258**, 14041–14047.

419. Wicki, A. and Niggli, V. (2001). The Rho/Rho-kinase and the phosphatidylinositol 3-kinase pathways are essential for spontaneous locomotion of Walker 256 carcinosarcoma cells. *Int. J. Cancer* **91**, 763–771.

420. Wilde, M.W., Carlson, K.E., Manning, D.R. and Zigmond, S.H. (1989). Chemoattractant-stimulated GTPase activity is decreased on membranes from polymorphonuclear leukocytes incubated in chemoattractant. *J. Biol. Chem.* **264**, 190–196.

421. Wilkinson, P.C., Michl, J. and Silverstein, S.C. (1980). Receptor distribution in locomoting neutrophils. *Cell Biol. Intl. Rep.* **4**, 736.
422. Wilkinson, P.C., Shields, J.M. and Haston, W.S. (1982). Contact guidance of human neutrophil leukocytes. *Exp. Cell Res.* **140**, 55–62.
423. Wilkinson, P.C. and Lackie, J.M. (1983). The influence of contact guidance on chemotaxis of human neutrophil leukocytes. *Exp. Cell Res.* **145**, 255–264.
424. Wilkinson, P.C. (1988). Micropore filter methods for leukocyte locomotion. *Meth. Enzym.* **162**, 38–49.
425. Wilkinson, P.C., Komai-Koma, M. and Newman, I. (1997). Locomotion and chemotaxis in lymphocytes. *Autoimmunity* **26**, 55–72.
426. Wilkinson, P.C. (1998). Assays of leukocyte locomotion and chemotaxis. *J. Immunol. Meth.* **216**, 139–153.
427. Williams, D.A., Tao, W., Yang, F., Kim, C., Gu, Y., Mansfield, P., Levine, J.E., Petryniak, B., Derrow, C.W., Harris, C., Jia, B., Ambruso, D.R., Lowe, J.B., Atkinson, S.J., Dinauer, M.C. and Boxer, L. (2000). Dominant negative mutation of the hematopoietic-specific Rho GTPase, Rac2, is associated with a human phagocyte immunodeficiency. *Blood* **96**, 1646–1654.
428. Williams, M.A. and Solomkin, J.S. (1999). Integrin-mediated signaling in human neutrophil functioning. *J. Leukoc. Biol.* **65**, 725–736.
429. Witke, W., Sharpe, A.H., Hartwig, J.H., Azuma, T., Stossel, T.P. and Kwiatkowski, D.J. (1995). Hemostatic, inflammatory, and fibroblast responses are blunted in mice lacking gelsolin. *Cell* **81**, 41–51.
430. Worthen G.S., Avdi, N., Buhl, A.M., Suzuki, N. and Johnson, G.L. (1994). FMLP activates Ras and Raf in human neutrophils. Potential role in activation of MAP kinase. *J. Clin. Invest.* **94**, 815–823.
431. Worthen, G.S., Henson, P.M., Rosengren, S., Downey, G.P. and Hyde, D.M. (1994). Neutrophils increase volume during migration *in vivo* and *in vitro*. *Am. J. Respir. Cell Mol. Biol.* **10**, 1–7.
432. Wyatt, T.A., Lincoln, T.M. and Pryzwansky, K.B. (1991). Vimentin is transiently co-localized with and phosphorylated by cyclic GMP-dependent protein kinase in formyl-peptide-stimulated neutrophils. *J. Biol. Chem.* **266**, 21274–21280.
433. Wymann, M.P., Kernen, P., Deranleau, D.A, Dewald, B., vonTscharner, V. and Baggiolini, M. (1987). Oscillatory motion in human neutrophils responding to chemotactic stimuli. *Biochem. Biophys. Res. Commun.* **147**, 361–368.
434. Wymann, M.P., Kernen, P., Bengtsson, T., Andersson, T., Baggiolini, M. and Deranleau, D.A. (1990). Corresponding oscillations in neutrophil shape and filamentous actin content. *J. Biol. Chem.* **265**, 619–622.
435. Wymann, M.P. and Pirola, L. (1998). Structure and function of phosphoinositide 3-kinases. *Biochim. Biophys. Acta* **1436**, 127–150.
436. Yang, N., Higuchi, O., Ohashi, K., Nagata, K., Wada, A., Kangawa, K., Nishida, E. and Mizuno, K. (1998). Cofilin phosphorylation by LIM-kinase 1 and its role in Rac-mediated actin reorganization. *Nature* **393**, 809–812.
437. Yang, W.-C., Collette, Y., Nunes, J.A. and Olive, D. (2000). Tec kinases: A family with multiple roles in immunity. *Immunity* **12**, 373–382.
438. Yasui, K., Yamazaki, M., Miyabayashi, M., Tsuno, T. and Komiyama, A. (1994). Signal transduction pathway in human polymorphonuclear leukocytes for chemotaxis

induced by a chemotactic factor. Distinct from the pathway for superoxide anion production. *J. Immunol.* **152**, 5922–5929.

439. Yasui, K. and Komiyama, A. (2001). Roles of phosphatidylinositol 3-kinase and phospholipase D in temporal activation of superoxide production in FMLP-stimulated human neutrophils. *Cell Biochem. Funct.* **19**, 43–50.

440. Ye, R.D. and Boulay, F. (1997). Structure and function of leukocyte chemoattractant receptors. *Adv. Pharmacol.* **39**, 221–289.

441. Yokomizo, T., Izumi, T. and Shimizu, T. (2001). Leukotriene B_4: Metabolism and signal transduction. *Arch. Biochem. Biophys.* **385**, 231–241.

442. Yonemura, S., Matsui, T., Tsukita, S. and Tsukita, S. (2002). Rho-dependent and -independent activation mechanisms of ezrin/radixin/moesin proteins: an essential role for polyphosphoinositides *in vivo*. *J. Cell Sci.* **115**, 2569–2580.

443. Yoshinaga-Ohara, N., Takahashi, A., Uchiyama, T. and Sasada, M. (2002). Spatiotemporal regulation of moesin phosphorylation and rear relase by Rho and serine/threonine phsophatase during neutrophil migration. *Exp. Cell Res.* **278**, 112–122.

444. Zenke, F.T., King, C.C., Bohl, B.P. and Bokoch, G.M. (1999). Identification of a central phosphorylation site in p21-activated kinase regulating autophosphorylation and kinase activity. *J. Biol. Chem.* **274**, 32 565–32 573.

445. Zheng, L., Eckerdal, J., Dimitrijevic, I. and Andersson, T. (1997). Chemotactic peptide-induced activation of Ras in human neutrophils is associated with inhibition of p120-GAP activity. *J. Biol. Chem.* **272**, 23 448–23 454.

446. Zhu, C. and Skalak, R. (1988). A continuum model of protrusion of pseudopod in leukocytes. *Biophys. J.* **54**, 1115–1137.

447. Zigmond, S.H. and Hirsch, J.G. (1972). Effects of cytochalasin B on polymorphonuclear leucocyte locomotion, phagocytosis and glycolysis. *Exp. Cell Res.* **73**, 383–393.

448. Zigmond, S.H. and Hirsch, J.G. (1973). Leukocyte locomotion and chemotaxis. New methods for evaluation, and demonstration of a cell-derived chemotactic factor. *J. Exp. Med.* **137**, 387–410.

449. Zigmond, S.H. (1974). Mechanisms of sensing chemical gradients by polymorphonuclear leukocytes. *Nature* **249**, 450–452.

450. Zigmond, S.H. (1977). Ability of polymorphonuclear leukocytes to orient in gradients of chemotactic factors. *J. Cell Biol.* **75**, 606–616.

451. Zigmond, S.H. and Sullivan, S.J. (1979). Sensory adaptation of leukocytes to chemotactic peptides. *J. Cell Biol.* **82**, 517–527.

452. Zigmond, S.H., Levitsky, H.I. and Kreel, B.J. (1981). Cell Polarity: An examination of its behavioral expression and its consequences for polymorphonuclear leukocyte chemotaxis. *J. Cell Biol.* **89**, 585–592.

453. Zigmond, S.H., Sullivan, S.J. and Lauffenburger, D.A. (1982). Kinetic analysis of chemotactic peptide receptor modulation. *J. Cell Biol.* **92**, 34–43.

454. Zigmond, S.H. and Tranquillo, A.W. (1986). Chemotactic peptide binding by rabbit polymorphonuclear leukocytes. Presence of two compartments having similar affinities but different kinetics. *J. Biol. Chem.* **261**, 5283–5288.

455. Zigmond, S.H. (1988). Orientation chamber in chemotaxis. *Meth. Enzymol.* **162**, 65–72.

456. Zigmond, S.H., Slonczewski, J.L., Wilde, M.W. and Carson, M. (1988). Polymorphonuclear leukocyte locomotion is insensitive to lowered cytoplasmic calcium levels. *Cell Motil. Cytoskeleton* **9**, 184–189.

457. Zigmond, S.H. (1993). Recent quantitative studies of actin filament turnover during cell locomotion. *Cell Motil. Cytoskeleton* **25**, 309–316.

458. Zigmond, S.H., Joyce, M., Borleis, J., Bokoch, G.M. and Devreotes, P.N. (1997). Regulation of Actin polymerization in cell-free systems by GTPγS and Cdc42. *J. Cell Biol.* **138**, 363–374.

459. Zigmond, S.H., Joyce, M., Yang, C., Brown, K., Huang, M. and Pring, M. (1998). Mechanism of Cdc42-induced actin polymerization in neutrophil extracts. *J. Cell Biol.* **142**, 1001–1012.

460. Zlotnik, A. and Yoshie, O. (2000). Chemokines: a new classification system and their role in immunity. *Immunity* **12**, 121–127.

461. Zu, Y.-L., Qi, J., Gilchrist, A., Fernandez, G.A., Vazquez-Abad, D., Kreutzer, D.L., Huang, C.-K. and Sha'afi, R.I. (1998). p38 mitogen-activated protein kinase activation is required for human neutrophil function triggered by TNF-α or fMLP stimulation. *J. Immunol.* **160**, 1982–1989.

Sperm Chemotaxis

1. Introduction

Sperm chemotaxis is a widespread phenomenon. It was discovered in marine species in the mid-1960s, and in the last decade it was also demonstrated in amphibians and mammals. In some of these species— marine species [35, 102], frogs [5], mice [56, 114], rabbits [52], and humans [49, 128]—the evidence for sperm chemotaxis is conclusive. In others, the evidence (if it exists) is still indirect and open to doubt (for a review, see [48]).

Sperm chemotaxis is an example of a phenomenon in which the very same process serves different purposes in different species. In marine species, most if not all of the cells within a sperm population appear to be chemotactically responsive. The role of chemotaxis there is apparently to recruit as many sperm cells (spermatozoa) as possible to the egg. In humans, rabbits, and, perhaps, in mammals in general, the role of chemotaxis seemingly also involves sperm selection: only a fraction of the sperm population, consisting of ripe spermatozoa that have the capacity to fertilize the egg, is chemotactically responsive and apparently recruited to the egg [28, 29, 52].

2. Sperm Motility

Spermatozoa are fast-moving cells driven by their tail, which is a single large flagellum. In many species (e.g., mammals), the spermatozoa are immotile when produced in the testis. They acquire the ability to swim progressively (forward) prior to ejaculation, after they are driven out of the testis into the epididymis[a] of the male genital tract [66]. In contrast, in other species—primarily marine species, freshwater species, and amphibians—the ejaculated spermatozoa are immotile. They acquire their motility in response to environmental or female-originated signals (for a review, see [105]). Environmental signals include reduction of the extracellular K^+ concentration (e.g., in salmonid fishes) or osmolality changes (e.g., in teleosts such as carp and goldfish and in amphibians such as toads and frogs). Female-originated signals include substances secreted from the egg or from the female reproductive organs (e.g., in herring and the ascidian *Ciona*).

Sperm motility is essential for fertilization. For example, mammalian sperm samples with too few motile cells or with too sluggish cells cannot fertilize the female egg [1, 64, 113].

2.1. *The flagellum*

The sperm flagellum (Figure 1), like the flagella of other eukaryotic cells (see [45, 90, 133] for recent reviews on eukaryotic flagella in general), is an organ completely unrelated to bacterial flagellum, both structurally and functionally (*cf.* Table 2 of Chapter 3). The sperm flagellum or sperm tail consists of a contractile axoneme which, in cross-section electron microscopy, is seen to be composed of 9 double microtubules at the periphery and 2 microtubules (singlets) at the center (Figure 2). The doublets are linked to their neighboring doublets by dynein arms and nexin bridges. The peripheral microtubules are connected by radial spokes to the sheath surrounding the central microtubules. Although the axoneme is built from about 250 different polypeptides, most of its protein mass is contributed by tubulin ($\sim$70%), which polymerizes linearly to form the microtubules, and dynein ($\sim$15%). The dynein molecules are large multisubunit complexes that are responsible for force production. They are ATPase molecules, each hydrolyzing $\sim$50 ATP molecules/s. (A more thorough review of the structure of sperm flagellum can be found, for example, in [32, 53, 112].)

---

[a] The epididymis is a long, tightly convoluted duct connected to the testis. Newly formed (immotile) spermatozoa are transported into it from the testis and acquire there their ability to swim.

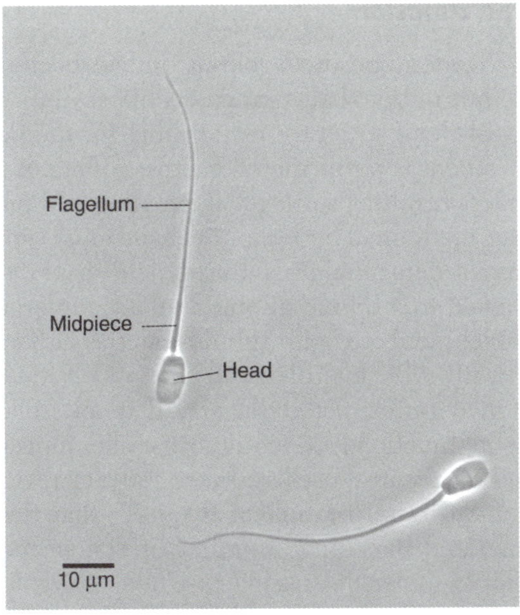

Figure 1. Electron micrograph of rabbit flagella. (Photographed by A. Gakamsky in the author's group.)

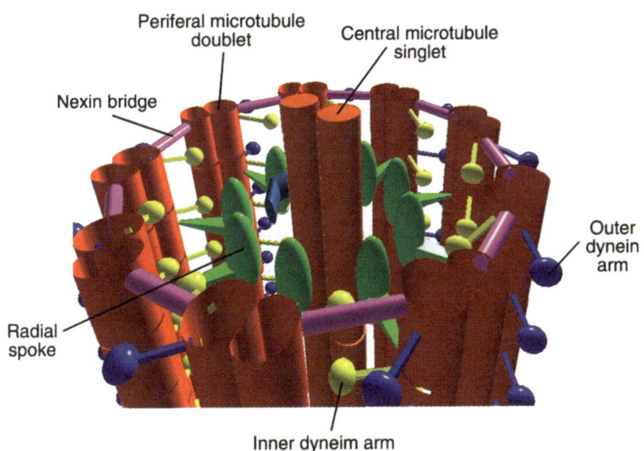

Figure 2. A three-dimensional reconstitution of the flagellar structure. Red, peripheral microtubule doublets; orange, central microtubule singlets; purple, nexin bridges; blue, outer dynein arms; yellow, inner dynein arms; green, radial spokes. (Kindly provided by Dr Jacky Cosson, CNRS, France.)

2.2. *Flagellar function*

To remind the reader, sperm flagellum, unlike bacterial flagellum, does not rotate but rather "beats" and acts like a whip. The axoneme is the basic cytoskeletal structure responsible for the flagellar movement. The movement is performed by active sliding of the nine doublet microtubules, resulting in flagellar bending and pushing against the surrounding medium. The basic mechanism of sliding is carried out by attachment/detachment of the dynein arms: a dynein arm, permanently linked to a tubule of one doublet, undergoes a cycle of attachment/detachment to the tubule of the adjacent doublet, concomitantly with ATP hydrolysis. It was nicely demonstrated in detergent- or glycerin-extracted flagella that an isolated axoneme fully maintains its function as a motile organelle, indicating that it is a self-contained mechanical oscillator (see [90] for review). Shingyoji *et al.* [142] demonstrated by optical tweezers that the direction of movement generated by a small number of dynein molecules (perhaps even a single molecule) oscillates when the dynein molecules work against an elastic load. The oscillation frequency is similar to the beat of an intact flagellum, with a maximum of $\sim 70\,\mathrm{Hz}$. These findings suggest that the oscillation of the dynein underlies the mechanism of flagellar beating [69, 142]. The mechanism for the dynein power stroke was recently revealed by electron microscopy and image processing of isolated dynein molecules at the start and end of their power stroke [20]. In addition to Mg^{2+}-ATP requirement, the flagellar function appears to be regulated by a number of factors and processes, including Ca^{2+}, cAMP-dependent phosphorylation, phosphatases, and a yet-undefined protease(s) (for reviews, see [32, 41, 53, 90, 105, 167, 168]). Recently, Ren *et al.* [129] nicely demonstrated that CatSper, a putative Ca^{2+} channel specific for the principal piece of the sperm tail, is likely involved in flagellar function. Spermatozoa of homozygous male mice lacking CatSper were poorly motile, nonfertile, and missing cAMP-induced Ca^{2+} influx. These findings suggest that CatSper is essential for cAMP-mediated Ca^{2+} influx in sperm and, hence, for sperm motility.

2.3. *Techniques for measuring sperm motility*

Over the years a number of approaches have been developed for the measurement of sperm motility. The main ones are as follows.

2.3.1. *Determination of the sperm head position as a function of time*

This is the simplest and fastest approach for determination of sperm motility. This approach yields the trajectories made by the swimming spermatozoa, the fraction of motile cells in the sperm population, and the swimming velocity. From these data, a number of conventional parameters can be calculated [40, 109], including the following: The straight-line velocity (VSL, also termed "progressive velocity") is defined as the time-average velocity of the sperm head along a straight line from its first position to its last position (light blue path in Figure 3). The curvilinear velocity (VCL; red path in Figure 3) is defined as the time-average velocity of the sperm head along its actual trajectory. It is calculated by summing incremental distances made by the sperm head along the path and dividing by the total time of the track. The average path velocity (VAP; dark blue path in Figure 3) is calculated by deriving a smoothed path and dividing by the time for the track. The linearity of the movement (LIN) is defined as the ratio VSL/VCL. The linear index is defined as the ratio VSL/VAP. The amplitude of lateral head displacement (ALH; black double arrow in Figure 3) is defined as the amplitude of the variations of the actual sperm-head trajectory about its average trajectory. Another parameter, which is sometimes used to identify a motility type termed "hyperactivated motility," to be discussed below, is the fractal dimension (FD). This parameter represents the degree to

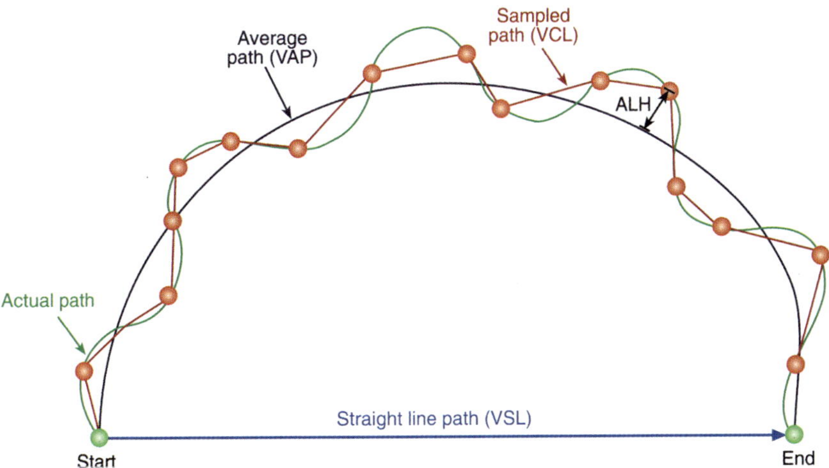

Figure 3. A trajectory made by a sperm head and some of the parameters derived from it. See text for definitions of terms and abbreviations.

which the sperm trajectory fills a plane, and it is calculated on the basis of the planar extent of the curve (i.e., the maximum distance between the origin and any plotted point) and the curvilinear path [108]. Nowadays, all these measurements are mostly carried out by commercially available computerized motion analysis systems [39, 107, 138].

2.3.2. Analysis of the sperm tail movement

This analysis, which nowadays is computerized as well, provides information on the pattern and strength of the tail movement [18, 157].

2.3.3. Distribution of swimming distances

This approach involves the measurement of the distribution of swimming distances made by the spermatozoa in various chambers, such as curved microchannels in microchips [87] or cervical mucus-filled capillaries [3, 82, 106]. The swimming distances are a measure of the progressive velocity and the percent motile cells in the sperm population.

2.3.4. Migration through a membrane

This approach involves determination of the extent to which spermatozoa can migrate through a membrane (e.g., a Nucleopore membrane with 5 μm pore size). This extent correlates primarily with the fraction of fast and straight-swimming cells in the sperm sample and with the sperm progressive velocity [68].

2.3.5. Intrinsic motility forces

Measurement of the intrinsic motility forces of individual spermatozoa is carried out by optical trapping. Individual spermatozoa are optically trapped by means of laser tweezers, and the minimal trapping power required to hold the sperm cell in the optical trap (that is, the power below which the spermatozoon escapes from the trap) is measured ([85] and references cited therein).

2.4. Types of sperm motility in different species

There are large variations in sperm motility both between species and within a given species—depending on the surroundings (see [74] for a comparison between mammalian species; apparently, such a comparison between nonmammalian species has not been published). The different motility types reflect different patterns of flagellar movement [74]. The most active spermatozoa in mammals are those of ram, beating—with almost symmetrical waves—at a frequency of 20.3–21.1 Hz at 37°C and having a progressive velocity of 105–127 μm/s. The beat frequency of golden hamster spermatozoa is significantly lower (12.4 Hz at 37°C), but their progressive velocity appears to be the highest (319–471 μm/s). In this case the beating waveform is asymmetrical. Human spermatozoa mostly swim in rather straight lines, resulting from the almost symmetrical waveform of their flagella. As they swim, their heads oscillate with a frequency similar to that of their flagellar beating. The spermatozoa rotate around their longitudinal axes synchronously with the beating [74]. With a beat frequency of 8.9–11.8 Hz, their progressive velocity is 34–53 μm/s at 37°C.

The rotation of spermatozoa around their long axis is not restricted to human spermatozoa; the spermatozoa of all species appear to do so [75]. In each species the spermatozoa rotate both clockwise and counterclockwise, but the proportion of the clockwise-rotating and counterclockwise-rotating populations varies between species. Interestingly, the majority of sea urchin and starfish spermatozoa rotate clockwise (viewed from the anterior end), and their flagella form right-handed waves. In contrast, the majority of mammalian spermatozoa rotate counterclockwise and their flagella form left-handed waves which, in the case of human and bull spermatozoa, look like a three-dimensional left-handed helicoid [75].

A given spermatozoon can change its direction of rotation, indicating that the three-dimensional geometry of flagellar movement can change [75]. This ability to change the swimming mode is a must for chemotaxis.

The motility of spermatozoa of nematodes such as *Caenorhabditis elegans* is exceptionally different. These spermatozoa have an amoeboid movement and they crawl to the site of fertilization by means of polymerization and depolymerization of the major sperm protein [130]. Spermatozoa with this type of motility will not be dealt with in this chapter.

2.5. *Types of sperm motility within the female genital tract*

Perhaps the most striking changes in sperm motility occur as mammalian spermatozoa move through the female genital tract (Figure 4). In species

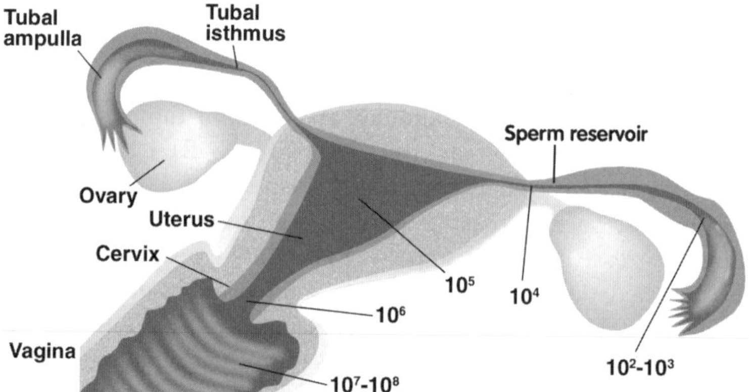

Figure 4. Sperm distribution in woman's genital tract subsequent to insemination. (Based on data reviewed in [11, 66, 184]. Taken with permission from Eisenbach and Tur-Kaspa [49].)

whose spermatozoa are deposited in the vagina at coitus (e.g., rabbit, cattle, sheep, and human), the spermatozoa first move into the cervix, where they must swim through the cervical mucus. In humans, only about one-tenth of the total motile cells inseminated reach the cervix. Of these, again only about one-tenth succeed in passing through into the uterus, and only about one-tenth of those in the uterus make it into the Fallopian tube (oviduct) [49] (Figure 4). In humans, spermatozoa passing through mid-cycle cervical mucus move in rather straight lines, their progressive velocity is higher than in physiological saline with a similar viscosity, the beat frequency of their flagella is also higher, the beating is restricted to the distal part of the flagellum, and the amplitude of the beating is small [74]. It appears that the ability of human spermatozoa to penetrate through cervical mucus depends mainly on the degree of lateral sperm head displacement: the higher the width of lateral displacement, the larger is the chance that the collision between a spermatozoon and the mucus will result in penetration [4]. (Note, however, that a significant proportion of the spermatozoa are transported to the uterus passively by muscular movements of the female tract and by the activity of cilia that are lining the lumina along the first sections of the female tract [61].)

As the spermatozoa move through the uterus they continue to exhibit vigorous motility (see [83] for a review). A few of them may exhibit hyperactivated motility (see below). For traversing the uterus-oviduct junction, the spermatozoa should apparently have progressive motility [113]. In humans, only about 0.004% of the total motile cells inseminated are found in both Fallopian tubes together [49] (Figure 4).

Once the spermatozoa enter the oviductal isthmus, things become more complicated. The mammalian oviductal isthmus was convincingly demonstrated to serve as a sperm reservoir or storage site after ejaculation (Figure 4) [11, 67, 71, 117, 159, 160]. (Defective, nonmotile spermatozoa and cells with disrupted membranes are apparently not stored; they appear to be passively and rapidly transported away towards the peritoneal cavity—see [116] for a review.) Spermatozoa that are not yet capacitated, that is, immature spermatozoa that have not yet acquired fertilizing potential [80, 186], attach to the oviductal epithelium and are released only when they become capacitated, i.e., only when they acquire a state of readiness to fertilize the egg.

A motility pattern—termed "hyperactivation" and characterized by increased velocity, decreased linearity, increased amplitude of lateral head displacement, and flagellar whiplash movement [21, 108, 110, 158, 186]—is thought to assist capacitated spermatozoa to detach from the oviductal epithelium [44]. During hyperactivation, the flagellar beats become more planar, and this, together with increasing asymmetry of the flagellar beat, causes free-swimming spermatozoa to move in nonlinear trajectories. The asymmetry may become so extreme that the spermatozoa are completely nonprogressive and move in circles, or even figure-of-eight–shaped trajectories [83]. Only capacitated spermatozoa can apparently acquire hyperactivated motility. Because of the efficiency of hyperactivated spermatozoa in penetrating viscous and viscoelastic substances, it is reasonable that hyperactivation assists the capacitated spermatozoa, just released from the sperm reservoir, to penetrate mucoid oviductal secretions [158].

Next the spermatozoa have to penetrate the cumulus oophorus (dense layers of cells surrounding the egg), to bind to the zona pellucida (the egg coat), to undergo the acrosome reaction (a release of proteolytic enzymes enabling sperm penetration through the egg coat [86, 182, 186]), and to penetrate the zona pellucida (for a review on the cumulus oophorus and the zona pellucida, see [22]). Penetration of the cumulus oophorus does not depend on the degree of progressiveness of movement, but it does require reasonable velocity [113]. Hyperactivation does not seem to be essential either [113, 166], but it may well assist the penetration [156, 158]. The zona pellucida is a more formidable barrier than the cumulus oophorus, for which reason penetration requires vigorous sperm motility [81, 113]. Here, too, it is believed that hyperactivation is important, if not essential, for penetration [4, 156, 158]. Figure 5 schematically summarizes the steps for which the spermatozoa must be capacitated and hyperactivated.

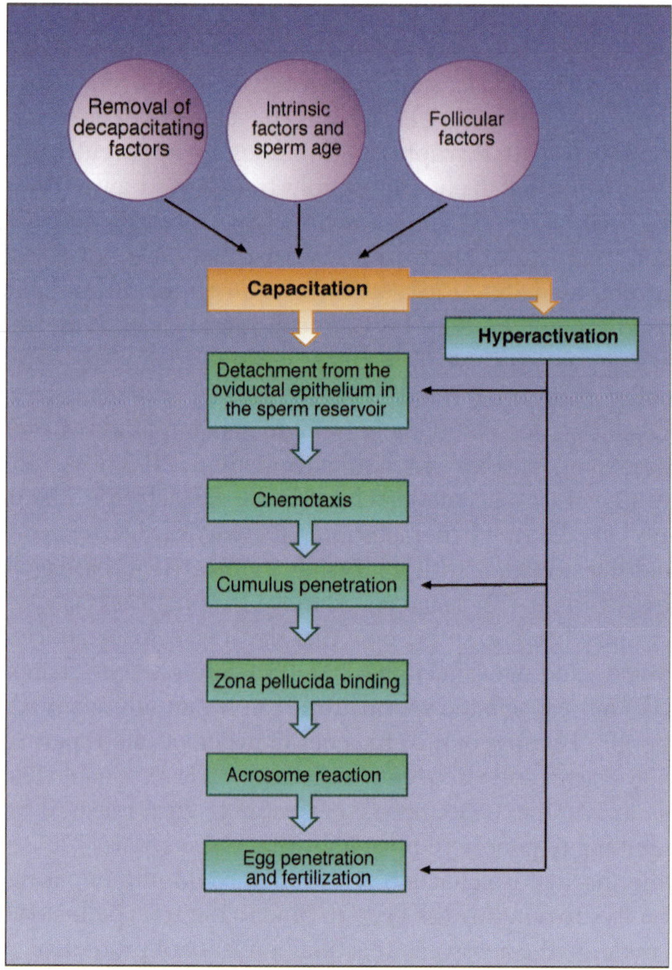

Figure 5. The steps for which mammalian spermatozoa must be capacitated and hyperactivated. The simplified scheme shows the events that affect, and the processes in the female genital tract that are regulated by, sperm capacitation. Bold colored arrows represent sequence of events. Thin black arrows represent modulation of processes. (Taken with permission from Jaiswal and Eisenbach [80].)

3. Criteria and Assays for Sperm Chemotaxis

One of the characteristics of sperm chemotaxis is sperm accumulation at the optimal chemoattractant concentration. However, sperm accumulation can also be caused by speed enhancement (chemokinesis) and trapping. The latter may result from a negative effect of a stimulus on motility, from a gradient-independent change in swimming behavior at a particular stimulus concentration, from mechanical effects such as adsorption to glass or capillary, or from any combination thereof. Therefore, it is essential to apply clear-cut criteria for distinguishing between chemotaxis and the other processes that may cause sperm accumulation. Such a criterion is the directional change of movement of spermatozoa towards the source of the chemoattractant—a unique feature of chemotaxis.

Two major types of assays are used: assays based on sperm accumulation and assays based on the directionality of sperm swimming. Below I discuss each of these assay types in more detail.

3.1. *Accumulation assays*

The accumulation-based assays for sperm chemotaxis can be divided into three main categories.

3.1.1. *Sperm accumulation in an ascending chemoattractant gradient*

The most commonly used category involves an accumulation assay in which spermatozoa sense an ascending gradient of the chemoattractant and accumulate near or at its source. The principle is that spermatozoa from one reservoir accumulate in another reservoir that contains the chemoattractant. The two reservoirs are connected and the chemoattractant gradient is established by diffusion. Variations on this assay include an apparatus in which the spermatozoa- and chemoattractant-containing wells are separated from each other by a thin polycarbonate membrane (Figure 6A) or are in direct contact with each other [174], an apparatus in which the spermatozoa- and chemoattractant-containing wells are connected via a tube (Figure 6B) or a channel [139], and a chemoattractant-containing capillary immersed in a drop or a well that contains a sperm suspension (Figure 6C) [48]. A modification of the latter is microinjection of the chemoattractant from a micropipette into a drop of sperm suspension [181]. The main disadvantage of this assay with all its variations is that it cannot distinguish between chemotaxis and other causes of sperm accumulation.

3.1.2. *Sperm accumulation in a descending chemoattractant gradient*

A similar technique that does distinguish between chemotaxis and other means of sperm accumulation is the inverted capillary assay (Figure 7).

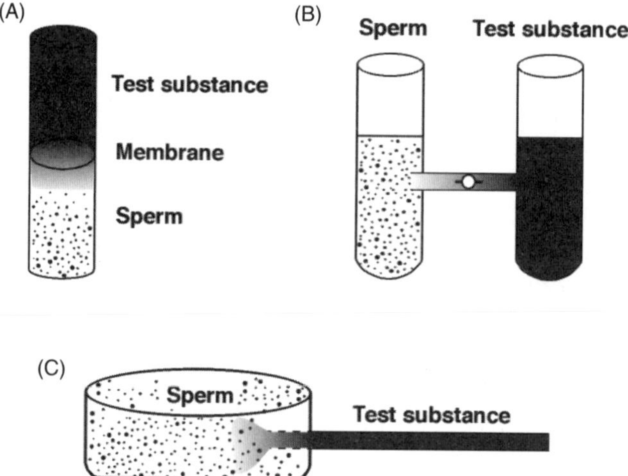

Figure 6. Various designs used to study sperm accumulation in an ascending gradient of a chemoattractant. (A) An accumulation assay in an apparatus consisting of two wells separated by a thin polycarbonate membrane [59]. (B) An accumulation assay in an apparatus consisting of two wells connected via a tube [28]. (C) Sperm accumulation in a capillary assay [128]. (Adapted with permission from Eisenbach [48].)

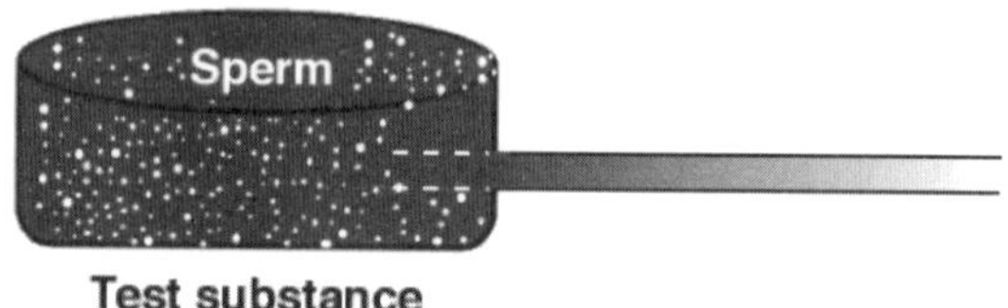

Figure 7. An inverted capillary assay designed to study sperm accumulation in an ascending gradient of a chemoattractant [128]. (Adapted with permission from Eisenbach [48].)

In this assay, the spermatozoa in the well are suspended in a solution containing the chemoattractant. The capillary, containing either a control buffer or the chemoattractant, is immersed in the sperm suspension. When the capillary contains buffer only, the spermatozoa sense a descending gradient of the chemoattractant as they move from the well to the capillary (Figure 7). When the chemoattractant is in both the capillary and the well, they sense no gradient at all. A comparison is made between these two conditions. This assay thus measures the sperm tendency to leave the chemoattractant rather than to accumulate in it. In the case of sperm chemotaxis, sperm accumulation in the capillary is expected to be relatively low when the well alone contains the chemoattractant; when there is no gradient and the chemoattractant (or buffer) is everywhere, sperm accumulation in the capillary is expected to be relatively high. By counting the spermatozoa accumulated in the capillaries in these settings, it is possible to distinguish between chemotaxis, chemokinesis, and trapping [128]. This is because only chemotaxis, unlike the chemokinetic and trapping effects, is dependent on the presence of a chemical gradient.

3.1.3. *Choice assays*

Another commonly used technique is a "choice" assay in which spermatozoa choose between two wells (or two chambers), one containing the chemoattractant and the other containing buffer as a control. A number of experimental designs have been published: a sealed chamber for microscopic measurements (Figure 8A), and—for macroscopic measurements—apparatuses with two (Figure 8B), three (Figure 8C), or five (Figure 8D) wells or chambers, connected by a tube or a groove [48]. Such assays can distinguish between chemotaxis and chemokinesis, but they cannot distinguish between chemotaxis and trapping.

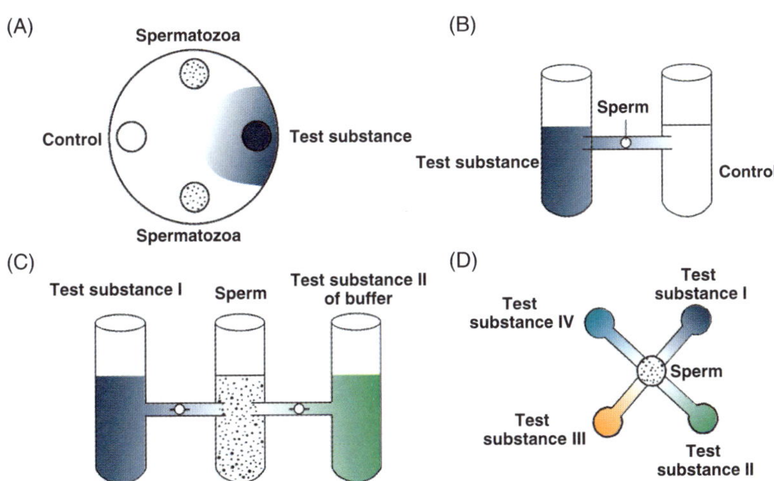

Figure 8. Choice assays. (A) A microscopic choice assay in sealed chamber (top view) [93]. (B) A choice assay in an apparatus consisting of two wells [176]. (C) A choice assay in an apparatus consisting of three wells [79]. (D) A choice assay in an apparatus consisting of five wells (top view) [146, 175]. (Adapted with permission from Eisenbach [48].)

3.2. *Directionality assays*

Directionality assays test the criterion for chemotaxis directly. They analyze (manually or by a computerized motion analysis system) video-recorded tracks, made by spermatozoa in a gradient of a chemo-attractant [49]. Two major assay types have been used. In one assay type, an attempt is made to distinguish between spermatozoa that reach a chemoattractant-containing well by changing their swimming direction according to the chemoattractant gradient (e.g., the green and yellow tracks in Figure 9) and spermatozoa that reach the well coincidentally (e.g., the red track in Figure 9). When this analysis is performed for both chemoattractant-containing and chemoattractant-free wells, it is possible to estimate the fraction of chemotactically responsive cells in a sperm population [79, 128]. The disadvantage of this assay is that the distinction between intentional and coincidental arrival to the well is subjective and not always clear-cut. This disadvantage is partly due to the fact that the chemoattractant gradient is two-dimensional.

The other assay therefore employs a one-dimensional chemoattractant gradient, established in a Zigmond chamber, which was originally designed for the evaluation of leukocyte chemotaxis [194]. In this chamber, which is built from two parallel, rectangular wells separated by a wall (Figure 10), a one-dimensional concentration gradient is

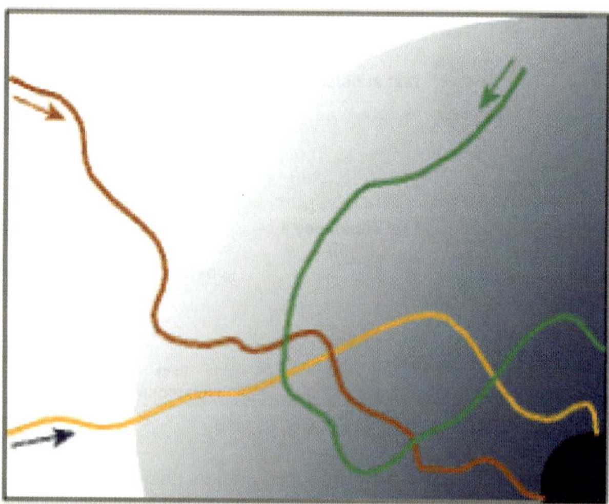

Figure 9. Schematic tracks of spermatozoa in a chemoattractant gradient. (Based on results obtained with human spermatozoa by Ralt *et al.* [128]. Adapted with permission from Eisenbach [48].)

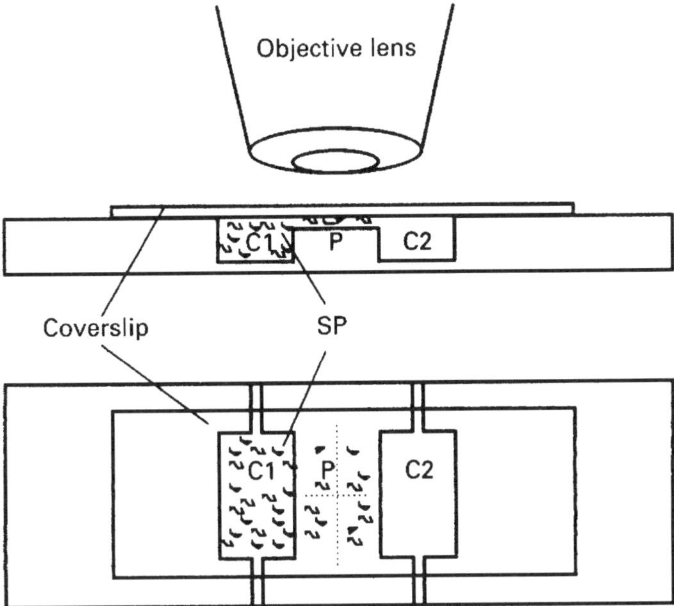

Figure 10. Chemotaxis-detection system in a Zigmond chamber. *Top*: a transverse section. *Bottom*: a plan view. A concentration gradient is formed between the coverslip and the partition wall (p) separating two wells (C1 and C2). See text for details. (Taken with permission from Giojalas *et al.* [57].)

established by diffusion. When this chamber was first used for demonstrating chemotaxis of mouse spermatozoa, the parameter measured was the net distance traveled by spermatozoa in the direction of the gradient [56, 114]. Thus, if the X-axis is defined as the direction of the chemoattractant gradient, the mean net distance traveled along the chemoattractant gradient ($\overline{\Delta X}$) was compared to $\overline{\Delta X}$ in the case of a no-gradient control [56, 114]. However, the velocity and pattern of swimming of the spermatozoa may bias the net distance that they travel. Therefore, nowadays, two additional directionality-based parameters are measured in parallel: the percentage of cells whose net distance of swimming was towards the chemoattractant well (cells with $\Delta X > 0$), and the percentage of cells traveling a longer distance in the direction of the chemoattractant gradient than in a gradient-less direction, perpendicular to the former (cells with $\Delta X / |\Delta Y| > 1$) [52]. In the case of random movement, the expected values are ~ 0 μm for $\overline{\Delta X}$, $\sim 50\%$ for the percentage of cells with $\Delta X > 0$, and $\sim 25\%$ for the percentage of cells with $\Delta X / |\Delta Y| > 1$. A significant deviation from these values is an indication of chemotaxis [52].

Table 1. Comparison between assays for mammalian sperm chemotaxis.

Assay	Distinction between chemotaxis and chemokinesis	Distinction between chemotaxis and trapping	References
Sperm accumulation in an ascending gradient	–	–	[6, 28, 59, 73, 88, 111, 127, 128, 139]
Sperm accumulation in a descending gradient	+	+	[128, 192]
Choice assays	+	–	[28, 79, 93, 94, 128, 146–149, 175–177, 191]
Directionality assays	+	+	[52, 56, 79, 99, 114, 128, 161, 189]

A comparison between the assays reviewed above is made in Table 1.

4. Chemotaxis of Nonmammalian Spermatozoa

Since the discovery of sperm attraction to the female gametes in ferns over a century ago [121–123], the process has been established in a large variety of species, from plants (e.g., bracken fern and fucus) and algae to marine invertebrates (for extensive reviews, see references [35, 102]), fish [105] and amphibians [5]. Of these, the species that has been most investigated is the sea urchins *Arbacia punctulata* and *Strongylocentrotus purpuratus*. A typical demonstration of sperm chemotaxis of sea urchin is shown in Figure 11, where resact, a peptide isolated from the jelly layer of *A. punctulata* eggs, is microinjected into a sperm suspension [181]. Transient sperm accumulation at the injection site is clearly observed. As discussed above, sperm accumulation is only one, insufficient, parameter of sperm chemotaxis. The other parameter, a directional change of swimming towards the source of the chemoattractant, was also demonstrated in many, though not all, of the investigated nonmammalian species [102].

Generally speaking, there is no single rule with respect to sperm chemotaxis. In some species (for example, in hydroids such as *Campanularia* or tunicate such as *Ciona*), the swimming direction of the spermatozoa changes abruptly towards the chemoattractant source (Figure 12A). In others (e.g., in hydromedusa, fern, or fish such as Japanese bitterlings), the approach to the chemoattractant source is

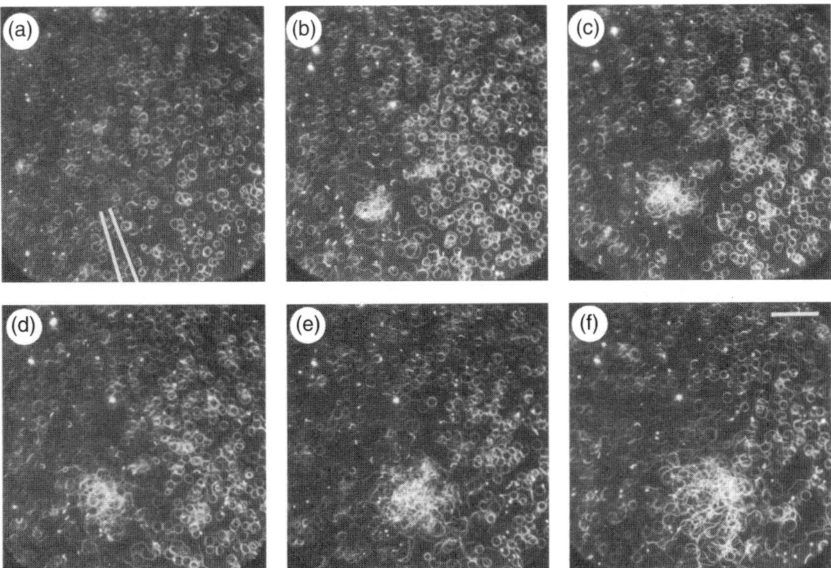

Figure 11. Effect of the chemoattractant resact on *A. punctulata* spermatozoa. A micropipette was inserted into a 20 μl drop of sperm suspension [position of the micropipette is indicated in (a)]. 1 nl of 10 nM resact was injected into the drop and 5 s later the micropipette was removed. Photographs (1 s exposures) were taken (a) 5 s before insertion of the micropipette, and (b) 20 s, (c) 40 s, (d) 50 s, (e) 70 s, (f) 90 s postinjection. Bar, 200 μm. (Taken with permission from Ward *et al.* [181].)

indirect and the movement is by repetitive loops of small radii (Figure 12, panel B vs. control panels C and D). Still in others (e.g., in Arthropoda, where fertilization is internal), sperm chemotaxis does not appear to occur at all. In some species (e.g., herring or the ascidian *Ciona*) activation of motility precedes chemotaxis [35, 102, 105].

The sperm chemoattractants in nonmammalian species vary to a large extent. Some examples are shown in Table 2. So far, most sperm chemoattractants that have been identified in animals are peptides or low-molecular-weight proteins (1–20 kDa), which are heat stable and sensitive to proteases [35, 102]. Exceptions to the rule are the sperm chemoattractants of corals—lipid-like substances of 140–250 Da [30], and the chemoattractants of the ascidians *Ciona*—a sulfated steroid [187, 189]. In plants such as ferns, the partially ionized form of malic acid and a large variety of unsaturated four-carbon *cis*-dicarboxylic acids are sperm chemoattractants [35]. In algae, low-molecular-weight

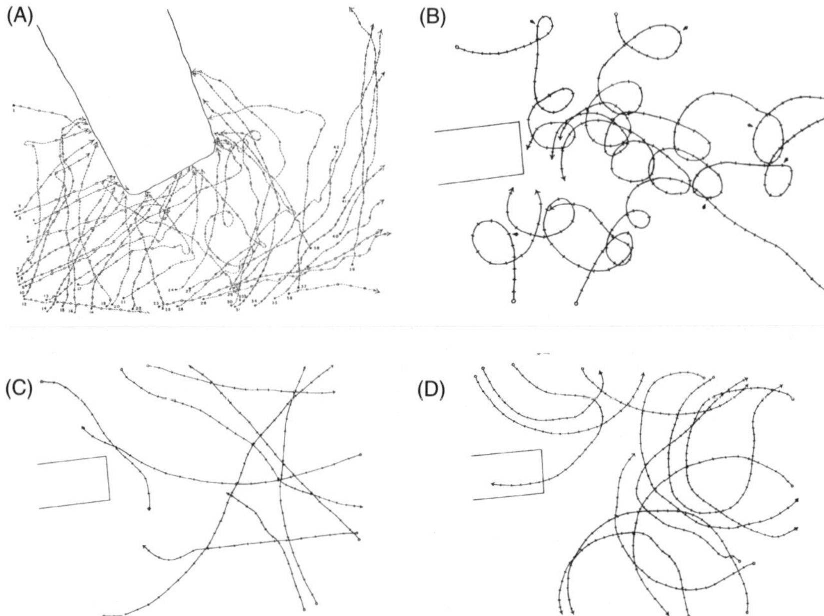

Figure 12. Sperm trails in nonmammalian species. (A) Spermatozoa of the hydroide *Campanularia flexuosa* approaching the female gonangium—an example of spermatozoa whose swimming direction changes abruptly towards the chemoattractant source. The solid circles indicate the start of each trail. The open circles are 0.45 s apart. Three types of trails can be seen: trails directed straight to the gonangium (16 trails), trails that go past the gonangium in straight line (10 trails), and trails that turn to enter or strike the gonangium (16 trails). (Taken with permission from Miller [98].) (B) Spermatozoa of the small pelagic chordate *Oikopleura dioica* approaching a pipette injecting *O. dioica* egg extract—an example of spermatozoa that approach the chemoattractant source by indirect movement with repetitive loops. (C, D) Trajectories of *O. dioica* spermatozoa near a pipette injecting, as a control, sea water. Panel C shows mainly straight or curved trails, whereas panel D shows mainly curved trails. Pipette diameter is 30 μm. Each interval on the trail represents 0.08 s. (Panels B–D were taken with permission from Miller and King [101].)

pheromones[b] are sperm chemoattractants [35, 92]. Perhaps the most investigated metazoan sperm chemoattractants are those of sea urchins. Resact—a 14-residue peptide isolated from the egg jelly layer—was demonstrated to be not only a specific sperm chemoattractant for the sea urchin A. *punctulata* [181], but also a stimulator of sperm motility and respiration. It thus belongs to a large family of sperm-activating

---

[b] Pheromones are substances secreted by an individual and sensed by a second individual of the same species, in which they trigger a specific action.

Table 2. Some sperm chemoattractants in nonmammalian species.

Species	Chemoattractant	References
Algae	Low-molecular-weight unsaturated pheromones of cyclic or linear structure (for example 532 Da pentosylated hydroquinone in the case of *Chlamydomonas allensworthii*)	[35, 92, 155]
Amphibians	Allurin—a 21 kDa protein (for *Xenopus*)	[5, 115]
Ascidians	SAAF—a sulfated steroid: 3, 4, 7, 26-tetrahydroxycholestane-3, 26-disulfate (for *Ciona savignyi* and *intestinalis*)	[187–189]
Corals	A lipid-like long chain fatty alcohol CH_3–$(CH_2)_8$–CH=CH–CH=CH–CH2OH (for *Montipora digitata*)	[30]
Ferns	Dicarboxylic acids, for example malic acid in its partially ionized form (for *Pteridium aquilinum*)	[17]
Mollusks	SepSAP—a 6-residue peptide-amide with the sequence PIDPGV-$CONH_2$ (for *Sepia officinalis*)	[193]
Sea urchins	Resact—a 14-residue peptide with the sequence CVTGAPGCVGGGRL-NH_2 (for *Arbacia punctulata*)	[181]
Starfish	Startrak—a 13 kDa heat-stable protein (for *Pycnopodia helianthoides*)	[97]

peptides [164]. Most other peptides of this family have not been demonstrated to be chemoattractants, although there is some indirect evidence which suggests that the peptide speract may be a sperm chemoattractant for the sea urchin *S. purpuratus* [31].

5. Chemotaxis of Mammalian Spermatozoa

The occurrence of sperm chemotaxis in mammals was established *in vitro* only at the last decade. As a matter of fact, until then there had been resistance to the concept of mammalian sperm chemotaxis for two main reasons. First, in mammals in which fertilization is internal and very large numbers of spermatozoa are ejaculated directly into the female reproductive tract (4–40×10^7 in humans [185]; 10^7–10^9 in mammals in general [66]), chemotaxis was believed to be unnecessary because a sufficient number of spermatozoa would reach the egg coincidentally. Second, technical difficulties in studying mammalian sperm chemotaxis prevented the acquisition of conclusive evidence. The primary technical difficulty was a very low signal-to-noise ratio in the measurements, resulting from the fact that only a small fraction of the sperm population is chemotactically responsive at a given moment [28]. (As I will discuss below, there is a good physiological reason for the small fraction of chemotactically responsive spermatozoa.) This low signal-to-noise ratio taken together with the large variability between sperm samples, the fact that only about one half of the follicular fluids (commonly used as a source for female-derived chemoattractants) are chemotactically active, the fact that many studies used too high chemoattractant concentrations (for example, insufficiently-diluted follicular fluid) that yielded false results, and the fact that many studies did not examine whether the criteria for chemotaxis were fulfilled and did not distinguish between chemotaxis and other processes that may cause sperm accumulation, all resulted in inconsistent results and ambiguity [49]. Only when mammalian sperm behavior was analyzed according to parameters that distinguish chemotaxis from chemokinesis and trapping (Table 1) was the occurrence of sperm chemotaxis established.

5.1. *Early observations interpreted as sperm chemotaxis*

Observations interpreted as sperm chemotaxis in mammals were made in the 1950s and 1960s. Moricard and Bossu [104] noticed the presence of rat spermatozoa near dissociated follicular cells and interpreted this selective localization of spermatozoa as the result of chemotactic attraction by the oocyte. Schwartz *et al.* [137] found that human ovarian cyst fluids, the outer liquid of egg white, and follicular fluid (only a single follicular fluid was tested) caused, among other effects, sperm accumulation. This accumulation was also interpreted as sperm chemotaxis [137]. Dickman [46] found that when rat and rabbit eggs were transferred into oviducts of previously mated rabbits, a larger number of spermatozoa collected within the rabbit eggs than on the zonae of the rat eggs. These findings were interpreted as chemotaxis of rabbit spermatozoa to the egg. However, lack of controls for other processes that might cause sperm accumulation (e.g., sperm trapping, adhesion, or swimming speed modulation), renders these early observations more suggestive than definitive [47]. Conversely, Bronson and Hamada [19] found *in vitro* that the cumulus oophorus secretes a substance that alters the pattern of mouse sperm movement. They noted that spermatozoa traversing microcapillary tubes in the environment of unfertilized eggs moved in erratic paths, owing to repeated adherence of the sperm head to the wall of the microcapillary tube. However, when the cumulus oophorus had been removed, the spermatozoa moved in linear trajectories as they do in the absence of eggs. Although Bronson and Hamada [19] suggested that there may be a selective trapping of mature spermatozoa by the cumulus oophorus, their observations could also be an example of sperm chemotaxis. Other examples of this sort are reviewed in Eisenbach and Ralt [47].

5.2. *Recent studies of sperm chemotaxis in mammals*

5.2.1. *Chemotaxis to follicular fluid*

Because it is very difficult, if not impossible, to carry out direct measurements of mammalian sperm chemotaxis *in vivo*, all published studies have been carried out *in vitro*. Most of the studies have investigated sperm chemotaxis to follicular fluid, primarily because follicular fluid contains secretions from the egg and its surrounding cells (prior to ovulation) in addition to being available. Follicular fluid *per se* may have no physiological role after ovulation because only small quantities of it are transported into the oviduct [63]. Sperm chemotaxis to follicular fluid was first established in humans [127, 128, 175, 176], then it was demonstrated in mice [114] and rabbits [51], and was implied in horses [111] and pigs [139]. The results of the studies of chemotaxis to follicular fluid are summarized in Table 3. Although not all the studies employed assays that can distinguish between chemotaxis and other sperm-accumulating processes, the results of all the studies, except one, were consistent and demonstrated the occurrence of sperm chemotaxis to follicular fluid. The establishment of sperm chemotaxis to follicular fluid *in vitro* strongly suggests the occurrence of sperm chemotaxis to female-derived factors *in vivo* as well.

5.2.2. *Chemotaxis to other fluids*

Conditioned media
Sperm chemotaxis can occur only after the release of an egg ready to be fertilized (ovulation). However, all published studies were carried out with follicular fluids containing substances secreted within the follicle *prior to* ovulation. For this reason it was relieving when Sun *et al.* [162] demonstrated human sperm chemotaxis to conditioned media of human cumulus-oocyte cells, i.e., media pre-incubated with mature eggs, their surrounding cumulus cells, or both. These media, therefore, contain secretions of cumulus cells and mature, ready-to-be-fertilized eggs.

Table 3. Mammalian sperm chemotaxis to follicular fluid.

Study[a]	Species	Apparent chemotaxis observed	Assay type[b]	Distinction between chemotaxis and all other accumulation-causing processes
Villanueva-Díaz et al. [175]	Human	+	Choice	−
Ralt et al. [127]	Human	+	Accumulation (ascending)	−
Makler et al. [93, 94]	Human	−[c]	Choice	−
Villanueva-Díaz et al. [176]	Human	+	Accumulation (ascending)	−
Ralt et al. [128]	Human	+	Accumulation (ascending & descending), choice, directionality	+
Cohen-Dayag et al. [28]	Human	+	Accumulation (ascending)	−
Navarro et al. [111]	Stallion	−	Choice	−
Giojalas and Rovasio [56]	Mouse	+	Directionality	+
Oliveira et al. [114]	Mouse	+	Directionality	+
Jaiswal et al. [79]	Human	+	Choice, directionality	+
Tacconis et al. [165]	Human	+	Accumulation (ascending)	−
Serrano et al. [139]	Pig	+	Accumulation (ascending)	−
Fabro et al. [52]	Rabbit	+	Directionality	+

[a]Chronological order.
[b]According to Table 1.
[c]The reasons for the negative results of Makler et al. [93, 94]—chiefly insufficient dilution of follicular fluid—were discussed by Eisenbach and Tur-Kaspa [49].

Oviductal fluid

An intriguing observation was made a few years ago by Oliveira et al. [114] with mouse spermatozoa. They found that not only mouse follicular fluid, but also mouse oviductal fluid attracts spermatozoa by chemotaxis. This finding raised the possibility that sperm chemotaxis within the female genital tract occurs in steps [114]. This possibility of sequential chemotaxis awaits further evidence.

Repulsive egg secretions
In the mouse, rat, and pig, the initial distribution of spermatozoa at the site of fertilization tends to be approximately one spermatozoon per egg across most of the eggs in the cell mass of the cumulus ([71] and references cited therein). (However, the number of spermatozoa may increase slowly and progressively with time after ovulation.) The fact that multiple spermatozoa do not arrive together at a single egg, nor are some eggs free from spermatozoa, suggests that each spermatozoon is guided to a single egg and that there is a mechanism (perhaps negative chemotaxis), which reports that an egg has just been activated by a spermatozoon and prevents the arrival of other spermatozoa for, at least, a short time (for an excellent review, see Hunter [71].) The possibility of negative chemotaxis in the vicinity of the egg immediately after sperm penetration into the egg membrane, suggests that the egg may secrete chemorepellents at the appropriate time. This possibility has not yet been examined experimentally.

5.3. *Mammalian sperm chemoattractants*

The identity of the chemoattractant(s) in follicular fluid is not known. Fractionation of human follicular fluid by high-pressure liquid chromatography (HPLC) and thin-layer chromatography revealed two substances, < 1.3 and $13 \pm 1\,\mathrm{kDa}$ in size, which retained the chemotactic activity of follicular fluid [95]. Both substances, probably peptides because of their sensitivity to proteases, were heat- and acid-stable. The identity of these substances is not yet known. Another unidentified protein, 8.6 kDa in size and suspected as chemoattractant, was isolated by preparative sodium dodecyl sulfate (SDS)-polyacrylamide gel electrophoresis from pig's follicular fluid [139]. In other studies, additional constituents of follicular fluid, including heparin, the 8-kDa chemokine RANTES (Regulated on Activation Normal T Expressed and Secreted Chemokine), and the hormones progesterone, relaxin, atrial natriuretic peptide (ANP), adrenaline, oxytocin, calcitonin, and acetylcholine, have been claimed to be the chemoattractants (Table 4). The evidence for these claims is summarized below. However, in view of the complete lack of species specificity with respect to chemotaxis between humans, cows, and rabbits (Section 6) and the resulting conclusion that the sperm chemoattractants of these mammals are common or very similar, it cannot be stated that the chemoattractants have been identified. Their conclusive identification awaits further experimental work.

Table 4. Substances reported to cause sperm accumulation in mammals.[a]

Substance	Species	Assay type[b]	Reference	Chemotaxis criterion fulfilled
Acetylcholine	Mouse	III	[149]	−
Adrenaline	Mouse	III	[148]	−
Antithrombin III	Pig	I	[88]	−
Atrial natriuretic peptide	Human	I II III	[6, 192]	+/−[c]
Calcitonin	Mouse	III	[149]	−
β-endorphin	Mouse	III	[151]	−
Heparin	Human	III	[146]	−
	Mouse	III	[147]	−
Hyaluronic acid	Human	III	[150]	−
Oxytocin	Mouse	III	[148]	−
Peptides (<1.3 and 13 ± 1 kDa)	Human	I II	[95]	+/−[c]
Peptide (8.6 kDa)	Pig	I	[139]	−
Progesterone	Human	I III	[174, 177, 180]	−
RANTES	Human	I II	[76]	+/−[c]
Relaxin	Human	I	[174]	−
Substance P	Mouse	III	[151]	−
Synthetic N-formylated peptides	Human, bull	I	[59, 73]	−[d]

[a]Adapted with permission from Eisenbach [48].
[b]Assay types: (I) Sperm accumulation in an ascending gradient; (II) sperm accumulation in a descending gradient; (III) a choice assay.
[c]A distinction between chemotaxis and other accumulation-causing processes has been made. However, a sperm track analysis has not been carried out.
[d]Chemotaxis was apparently ruled out as a cause of sperm accumulation [93, 99].

5.3.1. *Progesterone*

Villanueva-Díaz *et al.* [177] demonstrated in a choice assay that progesterone causes human sperm accumulation, that preincubation of spermatozoa with a progesterone receptor antagonist eliminates the accumulation, that dialysis of follicular fluid causes loss of this activity, that a lipid extract of follicular fluid causes sperm accumulation as does crude follicular fluid, and that heat or trypsin treatment does not affect the accumulation in follicular fluid. On the basis of these observations they suggested that progesterone is the chemoattractant in follicular fluid. Wang *et al.* [180] made a similar suggestion. However, this suggestion appears to be in conflict with earlier results that demonstrated absence of correlation between sperm accumulation in follicular fluid and the concentration of progesterone in the fluid [127], as well as absence of correlation between the characteristics of the active fractions of follicular fluid and those of progesterone [95]. Jaiswal *et al.* [79] demonstrated that progesterone (1–100 µg/ml) does bring about sperm accumulation, but that this accumulation is primarily due to physiological trapping rather than to chemotaxis, thus resolving the apparent contradiction. This conclusion was reached by using track analysis, which demonstrated that most of the spermatozoa present near the progesterone-containing well apparently arrived there by coincidence, not by changing the direction of the swimming path. Physiological trapping was apparently caused by an acquisition of motility patterns resembling hyperactivation (Section 2.5 above) [79]. Progesterone is known to cause sperm hyperactivation [171]. Jaiswal *et al.* found that, upon approaching a progesterone-containing well, a significant portion of the spermatozoa present in the accumulation zone acquired hyperactivation-like motility, resulting in very small progressive motility in spite of the vigorous motion. Consequently, the spermatozoa remained in the vicinity of the well. In this manner, some of the spermatozoa that happened to reach the neighborhood of the progesterone-containing well by chance were essentially trapped there. Further evidence that progesterone is, at least, not the main chemoattractant in follicular fluid was provided by the demonstration that removal of progesterone from follicular fluid does not eliminate the chemotactic activity of the fluid but does eliminate its ability to cause hyperactivation [79].

5.3.2. *Atrial natriuretic peptide (ANP)*

ANP is a polypeptide hormone that is secreted in large quantities by the atrial portion of the heart and from a variety of other mammalian cell types. It exerts many of its actions via activation of particulate guanylate cyclase [16, 134]. ANP is present in human follicular fluids [163] and specific ANP receptors have been identified in human spermatozoa [144]. Sperm chemotaxis to ANP was demonstrated by sperm accumulation in capillaries with ascending [6] and descending [192] gradients and by choice assays [192] (Table 4). It is not yet known whether ANP is involved in sperm chemotaxis *in vivo* and whether the physiological chemoattractant for human spermatozoa is an ANP-like substance. Since chemotaxis to ANP at physiological concentrations can be observed only in the presence of a neutral endopeptidase inhibitor such as phosphoramidon, which is probably absent in follicular fluid [192], there seem to be two alternatives: either that follicular fluid contains a neutral endopeptidase inhibitor, or that ANP is not the chemoattractant in follicular fluid. According to the latter alternative, ANP may directly affect guanylate cyclase *in vitro* in a manner similar to that caused by the physiological chemoattractant *in vivo* [192]. In line with this possibility, no correlation was found between the chemotactic activities of follicular fluids and their ANP content [6].

5.3.3. *Other hormones*

In a series of studies, Sliwa [146–149, 151] demonstrated mouse sperm accumulation in acetylcholine, adrenaline, calcitonin, β-endorphin, oxytocin, and substance P. Negative mouse sperm accumulation (i.e., apparent repulsion) was demonstrated with glucagon and vasopressin. However, since only a single assay was used in these studies (a choice assay that did not distinguish between chemotaxis and trapping), the significance of these observations with respect to chemotaxis is not clear.

5.3.4. *Other substances in follicular fluid*

A number of other constituents of follicular fluid have been shown to cause sperm accumulation (Table 4): RANTES for human spermatozoa [76], heparin for human [146] and mouse [147] spermatozoa, purified antithrombin III for boar spermatozoa [88], and hyaluronic acid for human spermatozoa [150]. RANTES, which appears to be a potent chemoattractant for eosinophils, monocytes, and T lymphocytes, was demonstrated to cause sperm accumulation in ascending and descending gradients [76] (but a sperm track analysis for direct demonstration of chemotaxis has not been carried out). It was further demonstrated that the mRNA for RANTES receptors is present in human spermatozoa [76]. The physiological significance of these observations is not at all clear, because RANTES is produced, prior to ovulation, by granulosa cells and is upregulated in some diseases associated with infertility [76]. In the cases of all other constituents mentioned above, a distinction between chemotaxis and the other processes that may cause sperm accumulation has not been made. Furthermore, heparin has been shown to induce capacitation of bovine spermatozoa [120] and an acrosome reaction of human spermatozoa [43], and it is possible that, like progesterone, it causes hyperactivation and, therefore, trapping.

Similarly, antithrombin III [88] and hyaluronic acid [62] have been shown to enhance sperm motility. Sperm accumulation in them could, therefore, be the consequence of chemokinesis.

5.3.5. *Odorants*

Just before the submission of this book to the publisher, Spehr *et al.* [154] reported the identification of the odorant receptor hOR17-4 on human spermatozoa. They further found that the ligands for this receptor, for example, the odorant bourgeonal, act as sperm chemoattractants. It is not known whether bourgeonal (or a similar compound) is, at all, secreted within the female genital tract. An intriguing observation made in this study was that most of the spermatozoa responded to bourgeonal [154]. This suggests that the response to bourgeonal is not restricted to capacitated spermatozoa only, unlike the case of sperm chemotaxis to follicular fluid (Section 5.4.2). This observation may have far-reaching significance (see below).

5.3.6. *Other potential chemoattractants and chemorepellents*

Small synthetic N-formylated peptides, such as N-formyl-Met-Leu-Phe (fMLP), are chemoattractants for neutrophils and macrophages [136]. Such peptides were reported to bind to specific sites on human spermatozoa [10, 60] and to cause accumulation of both human [59] and bull [73] spermatozoa. However, studies using the choice assay for human spermatozoa [93] and track analysis of bull spermatozoa [99] ruled out chemotaxis as the cause of accumulation. At least in the case of bull spermatozoa, the accumulation was demonstrated to result from sperm adhesion to the glass surface inside the peptide-containing capillaries [99]. These complications again demonstrate the importance of carrying out assays that distinguish chemotaxis from other accumulation-causing processes.

Not only potential sperm chemoattractants have been found. Tso *et al.* [170] demonstrated, by accumulation assays with ascending and descending chemical gradients, that p-nitro-phenyl-glycerol is a chemorepellent for rat spermatozoa, in addition to being an inhibitor of sperm motility. Makler *et al.* [94] studied hydrochloride acid, sodium hydroxide, ethanol, and glutaraldehyde as potential chemorepellents for human spermatozoa, but found no evidence for sperm repulsion. No other studies of potential sperm chemorepellents have been reported.

In conclusion, follicular fluid, bourgeonal, ANP, and RANTES are the only substances that have been demonstrated to act as chemoattractants for spermatozoa by assays that differentiate between chemotaxis and other processes causing accumulation (Table 1). ANP may not be a direct physiological chemoattractant but rather a guanylate cyclase activator. The significance of sperm chemotaxis to bourgeonal and RANTES is an open question. The identity of the chemoattractant(s) in follicular fluid has yet to be revealed.

5.4. *Physiological significance of mammalian sperm chemotaxis*

5.4.1. *Involvement in fertilization*

In contrast to the older view, discussed above, that sperm chemotaxis might be unnecessary in mammals because a sufficient number of spermatozoa would reach the egg coincidentally, the number of spermatozoa that actually reach the oviduct is extremely low, about six orders of magnitude less than the number of spermatozoa deposited in the vagina [11, 66, 184] (in humans, for example, only about 250 (range 80–1400) spermatozoa are found in both Fallopian tubes [184]). This low number reduces the chance of a coincidentally successful collision between a spermatozoon and the egg within the tubal ampulla to unrealistic values. Furthermore, although there is no significant difference between the number of spermatozoa found within the ovulatory tube compared with the non-ovulatory tube, the difference between the sperm distribution within the tubes is significant: the ovulatory tubal ampulla, where fertilization occurs, contains a significantly larger percentage of spermatozoa than are found in the contralateral ampulla [184]. This, again, cannot be explained by the older dogma. Chemotaxis may be one of the mechanisms responsible for this sperm distribution. Furthermore, the finding, discussed above, that the initial distribution of spermatozoa at the site of fertilization tends to be approximately one spermatozoon per egg across most of the eggs in the cell mass of the cumulus [71] also suggests the involvement of chemotaxis in fertilization (Section 5.2.2).

It is hard to provide more direct evidence for the involvement of sperm chemotaxis in fertilization, given the lack of *in vivo* data. Even when *in vivo* data are available, they are often difficult to interpret. For example, a number of *in vivo* studies ([65, 77] and references therein) have provided evidence that the products of ovulation are essential for sperm transport in the oviduct. These studies compared, for instance, superovulated with non-ovulated animals, or untreated animals with animals in which the products of ovulation were blocked from entering the oviduct (for example, by ligation). There were much higher numbers of spermatozoa in nonblocked oviducts compared with blocked ones, and in oviducts of superovulated animals compared with those of nonovulated animals. (The finding, mentioned above, that the ovulatory tubal ampulla contains a significantly larger percentage of spermatozoa than the ampulla of the contralateral tube [184] is consistent with this notion.) These observations could potentially be interpreted as an indication of the involvement of chemotaxis in fertilization. However, they do not distinguish between the spermatozoa and the oviduct as the effectors on which the ovulatory products act. In view of the evidence suggesting that the contractions of the oviduct move liquid from the isthmus to the ampulla [12] and that the Fallopian tube may act as a peristaltic pump [183], a distinction between passive and active sperm movement in the oviduct is essential and awaits experimental data.

5.4.2. *Role in fertilization*

The first *in vitro* indication that sperm chemotaxis may indeed be involved in fertilization came from the observation that not all follicular fluids are active in causing sperm accumulation and that all the active fluids are only from follicles containing eggs that can be fertilized [127]. Likewise, pathological semen specimens are chemotactically non-responsive [165]. The first hint of the potential physiological role of sperm chemotaxis *in vivo* came from the findings that, in a given sperm population, there are chemotactically responsive and chemotactically non-responsive spermatozoa [28, 52, 114, 165], and that the fraction of responsive cells to follicular fluid in the total sperm population is small (2–12% in humans [28], ∼10% in mice [56, 114] and up to 23% in rabbits [52]). In addition, the chemotactic responsiveness was demonstrated to be temporary, and the responsive spermatozoa were found to change with time, that is, there is a continuous replacement of chemotactic spermatozoa within a sperm population [28]. Finally it was demonstrated that only capacitated spermatozoa are chemotactic, that they acquire their chemotactic responsiveness as part of the capacitation process, and that they lose this responsiveness when the capacitated state is terminated [29, 52]. This conclusion relied on the similar percentages of chemotactic and capacitated spermatozoa in a sperm population, on the occurrence of continuous replacement, with similar kinetics, of capacitated and chemotactically responsive spermatozoa, and on the fact that deliberate depletion of capacitated spermatozoa results in total loss of chemotaxis and, *vice versa*, depletion of chemotactically responsive spermatozoa results in depletion of capacitated spermatozoa [29].

The association of chemotactic responsiveness with the capacitated state raised the possibility that, *in vivo*, the role of human sperm chemotaxis is not to direct many spermatozoa to the egg, but rather to recruit a selective population of spermatozoa, that is, capacitated spermatozoa, for fertilizing the egg [29, 47]. The results of a number of much earlier *in vivo* studies were nonexclusively in line with the notion of sperm selection by the female, e.g., the orders of magnitude higher fertilizing potential of spermatozoa recovered from the oviduct of rabbits, hamsters and rats relative to ejaculated spermatozoa or spermatozoa recovered from the uterus [25, 26, 36, 104, 140] (for reviews, see [27, 47]). A possible role of the continuous replacement of capacitated/chemotactic spermatozoa may be to ensure availability of capacitated spermatozoa for an extended period of time despite the short life span of the capacitated state in any one spermatozoon [29, 47]. This possibility was recently put to test by examining the timing of sperm capacitation and chemotactic responsiveness between rabbits (where the eggs are ovulated roughly ~10 h subsequent to mating and there is no obvious need for extending the availability of capacitated/chemotactic cells) and humans (where ovulation is periodic and unlinked to mating) [58]. Human spermatozoa became capacitated/chemotactic within less than an hour and maintained a steady-state level of capacitated/chemotactic cells for at least 30 h, whereas rabbit spermatozoa became capacitated/chemotactic only after a long delay and remained at this stage for a relatively narrow time window, which fitted well with the known time window of egg survival in the rabbit oviduct following ovulation. These observations suggested that mammalian spermatozoa become capacitated/chemotactic when, and for as long as, they have a chance to find a fertilizable egg in the oviduct: extended in periodic ovulators and synchronized with ovulation in induced ovulators [58].

What does happen to spermatozoa that reach the post-capacitated/chemotactic stage? Two facts are known about such spermatozoa. First, they have intact acrosomes, as evident from the observation that the level of acrosome-less spermatozoa does not increase during the continuous replacement of capacitated spermatozoa [29]. Second, once a cell becomes post-capacitated/chemotactic, it is a dead end; the cell cannot undergo the acrosome reaction when the appropriate stimulus appears and it cannot become chemotactically responsive again:

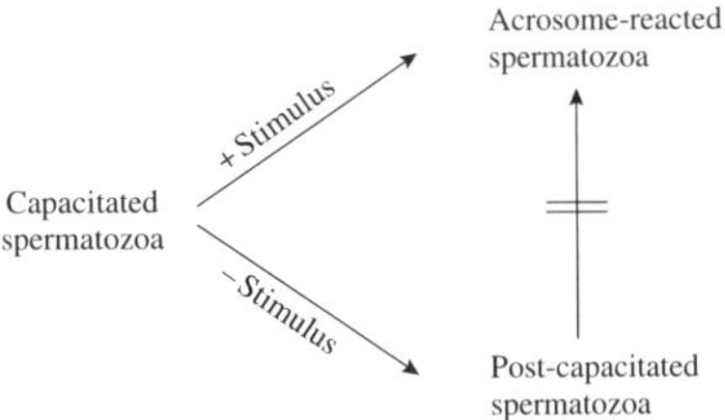

This is evident from the observation that a sperm subpopulation rich in post-capacitated/chemotactic spermatozoa does not acquire with time the ability to undergo the acrosome reaction upon stimulation [29]. These facts taken together with the findings that leukocytes (primarily neutrophils) are attracted to spermatozoa in the female genital tract and phagocyte some of the spermatozoa [8, 13, 118, 124] (as do vaginal epithelial cells [125] or, after fertilization, isthmic epithelial cells [23]) and that this phagocytosis does not interfere with fecundity [169], raise the possibility that spermatozoa in the female genital tract are phagocytosed when they reach the post-capacitated stage [50]. These cells, which lost their fertilizing ability and became functionless, apparently recruit leukocytes and then undergo apoptosis and phagocytosis and, thereby, are removed from the female genital tract. This fast removal probably prevents severe inflammation that could have been caused by necrotic products of sperm cells that remain functionless in the tract [50].

5.4.3. *Potential locations of sperm chemotaxis* in vivo

As already mentioned, a considerable fraction of the spermatozoa ejaculated into the female reproductive tract is retained with reduced motility in storage sites (usually the oviductal isthmus; Figure 4). Apparently, as the spermatozoa move up the oviductal isthmus, they encounter high mucus-containing narrow lumen, which impedes their forward progression. They frequently come into contact with the oviductal epithelium, where they can bind strongly to carbohydrate moieties on glycoproteins or glycolipids on the surface of the oviductal epithelium, and consequently become stored there [159, 160]. When ovulation occurs, some of the spermatozoa in this sperm reservoir (also termed "storage site") resume high motility and travel the distance between the storage site and the fertilization site at the oviductal ampulla within minutes [11, 71, 117].

(In mammals such as humans, where the egg transport is slow, the fertilization site is in the ampulla; in mammals like rabbits, where the egg transport is rapid, the site is the isthmic-ampullary junction; Figure 4.) Only capacitated spermatozoa are detached from the epithelium and released from the storage site [89, 153]. In view of the relatively small number of spermatozoa released from the storage site and their small dimensions, a navigation mechanism appears essential [71]. One possibility is that chemotaxis is involved in directing the released capacitated spermatozoa toward the egg. However, because of the oviductal contractions discussed above, a gradient of chemoattractant probably cannot be established over long distances in the oviduct, that is, the range of sperm chemotaxis in the oviduct may be relatively short. It is, therefore, more likely that chemotaxis only occurs at close proximity to the cumulus-egg complex. The observation that the cumulus oophorus secretes a substance that alters the pattern of sperm movement [19] is consistent with this possibility. Sperm chemotaxis may also occur within the cumulus. The following observations are compatible with this alternative: (a) In mammals, the few first spermatozoa that enter the cumulus find the egg very effectively [14]. (b) *In vitro*, bull spermatozoa preferentially penetrate intact cumulus-oocyte complexes rather than cumulus-free oocytes [24]. (c) Only capacitated spermatozoa can penetrate the cumulus oophorus (for a review, see [80]). The possibility of negative chemotaxis, discussed above, is also expected to occur within the cumulus in view of the observation that the initial distribution of spermatozoa at the site of fertilization usually tends to be approximately one spermatozoon per egg across most eggs in the cell mass of the cumulus [71].

If sperm chemotaxis in the oviduct is short ranged, how do the capacitated spermatozoa, released from the sperm reservoir, reach so effectively the fertilization site? One possibility is that the oviductal contractions passively drive the spermatozoa from the isthmus to the fertilization site at the ampulla or isthmic-ampullary junction. Another possibility is that the spermatozoa direct themselves from the isthmus by a multistep chemotaxis process, each step sequentially directing the capacitated spermatozoa over a relatively short range. The finding that both follicular fluid and oviductal fluid contain chemoattractants for mouse spermatozoa is in line with this alternative [114]. A third possibility is that spermatozoa employ the ovulation-dependent temperature difference within the female genital tract for guidance [38, 70, 72]. In the rabbit, this difference amounts to $\sim 2°$C between the cooler sperm reservoir's site and the fertilization site [9, 38]. Bahat *et al.* [9] recently

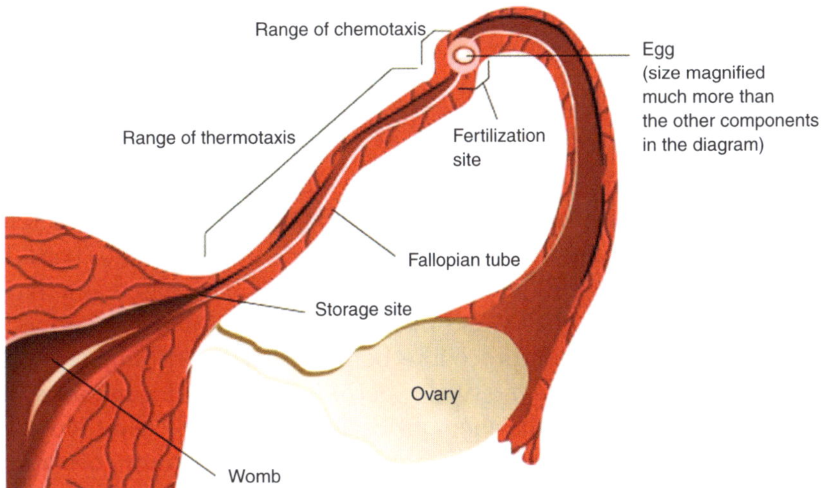

Figure 13. A scheme indicating the potential location of sperm chemotaxis and thermotaxis in the female genital tract. The pink halo surrounding the egg represents the cumulus oophorus.

examined whether rabbit spermatozoa can sense and respond to such a temperature difference by thermotaxis. Employing *in vitro* directionality assays, they found that spermatozoa, capacitated only, can indeed respond to a temperature difference as small as 0.5°C (perhaps even lower) by swimming toward the warmer temperature. They obtained similar results with human spermatozoa [9]. Since a temperature gradient is insensitive to the oviductal contractions, a reasonable possibility is that sperm thermotaxis and chemotaxis are long- and short-range mechanisms, respectively, which occur consecutively—each in a region where the other is not functional. First, capacitated spermatozoa, released from the isthmic sperm reservoir, may be guided by thermotaxis towards the warmer fertilization site. Then, at close proximity to the egg and within the cumulus mass, the guidance is likely carried out by chemotaxis [9] (Figure 13). It is also possible that all three alternatives function in harmony within the female genital tract.

Because noncapacitated cells constitute the majority of a human sperm population [29], the recent finding that almost all spermatozoa chemotactically respond to bourgeonal [154] suggests that the responsive cells are, predominantly or exclusively, noncapacitated ones. If so, it is reasonable that bourgeonal or its *in vivo* analog would be involved in directing non-capacitated spermatozoa, e.g., from the uterus, through the small opening of the Fallopian tube, to the storage site in the isthmus. Future experiments will tell whether or not this is so and what the role of bourgeonal and similar compounds is.

5.4.4. *Potential applications to fertility and contraception*

In view of the notion that chemotaxis is required for the selection of capacitated spermatozoa, faulty precontact sperm–egg communication may be one of the causes of male infertility, female infertility, or both. It is reasonable that, in the future, chemotaxis may be exploited as a diagnostic tool for sperm quality and male infertility and may be used as a biological sperm-selection procedure before *in vitro* fertilization, especially before intracytoplasmic single sperm insertion or for intrauterine insemination. Moreover, the chemotactic activity of a follicular fluid may be an effective way to monitor the possible effects of different ovulation induction protocols on the maturational status of oocytes aspirated for *in vitro* fertilization [2]. On the other hand, interference with sperm chemotaxis may represent an exciting new approach to contraception.

6. Species Specificity of Sperm Chemotaxis

Is sperm chemotaxis specific, i.e., are the chemoattractants specific for each species or are there chemoattractants common to several species? There is no single answer to this question. With some exceptions, species specificity appears to be the rule in marine species [102, 103]. There, the gametes are released in to seawater, and gametes of different species may be in close proximity. Therefore, in these cases, chemotaxis may be needed as one of the means to avoid interspecies fertilization. Indeed, in some marine groups (e.g., sea urchins, hydromedusae and certain ophiuroids), the specificity of sperm chemotaxis is very high. In other groups (e.g., starfish), the specificity is at the family level and, within the family, there is no specificity. In contrast, in mollusks, there appears to be no specificity at all ([35, 102, 103] and references therein).

In plants, a unique simple compound [for example, fucoserratene—a linear, unsaturated alkene (1,3-trans 5-cis-octatriene)] might be a chemoattractant for various species [92]. These differences in specificity between species may reflect the different physiological tasks that sperm chemotaxis fulfills in different species.

What is the situation in mammals? The need for species specificity in mammalian sperm chemotaxis is not obvious. On the one hand, specificity might be desired as a mean to avoid cross-species fertilization.

On the other hand, there are many other means for this purpose—behavioral barriers that prevent matings between individuals of different species [61] as well as effective physiological barriers at a number of levels, from the level of sperm transport within the tract to the level of the sperm's ability to penetrate ova vestments [61, 91, 131, 132, 143, 152]. Therefore, chemotaxis might not be needed as a safeguard for species-specific reproduction. The answer was recently provided by Sun *et al.* [161], who demonstrated that human spermatozoa can respond chemotactically not only to human's egg-related factors, but also to rabbit's and bovine's egg-related factors, and that, likewise, rabbit spermatozoa can respond chemotactically to human's and bovine's egg-related factors. This complete lack of species specificity between these three remote mammals suggests lack of chemotactic species specificity between most, if not all mammals. It further suggests that the egg-related sperm chemoattractants in mammals are common or very similar [161].

7. Molecular Mechanism of Sperm Chemotaxis

Very little is known about the molecular mechanism of sperm chemotaxis, but recent discoveries made in the sea urchin A. *punctulata* may change this situation. The receptor for the chemoattractant of this species, resact, was identified long ago as a membrane guanylyl cyclase, localized to the entire length of the flagellum [126, 141, 145]. (Guanylyl cyclases appear to represent a family of odorant/pheromone receptors, in addition to the family of G-protein-coupled receptors that activate specific adenylyl cyclases and, consequently, activate cyclic nucleotide-gated channels [55].) Recently, a great increase in the understanding of sperm chemotaxis of A. *punctulata* was made by Kaupp *et al.* [84]. Using rapid mixing techniques and novel caged compounds of cyclic nucleotides and of the chemoattractant resact, they demonstrated that the first event following resact stimulation is rapid and transient rise in the cGMP concentration, followed by a transient increase in the Ca^{2+} concentration. Interestingly, resact triggers two distinct Ca^{2+} responses—an early and a late response. The cGMP response and the early Ca^{2+} response are very sensitive: the binding of a single resact molecule can elicit a Ca^{2+} response, and 50–100 bound molecules saturate the response [84]. Similar chemoattractant responses were recently found in starfish as well [96]. The requirement of sperm chemotaxis for Ca^{2+} is well established in sea urchins, hydroids, and in the ascidian *Ciona savignyi* [31, 33, 34, 100, 181, 188,

190]. A number of models for signal transduction during chemotaxis of sea urchin spermatozoa have been proposed [31, 37, 84]. On the basis of the more recent experimental results, a reasonable sequence of events is that binding of resact to its guanylate cyclase receptor triggers rapid rise in intracellular cGMP, and this rise activates a Ca^{2+}-permeable channel with a consequent rapid elevation of intracellular Ca^{2+} (Ca^{2+}_{in}). This activation of the Ca^{2+} channel may be mediated by an unknown cGMP-mediated step [84]. The elevated Ca^{2+}_{in} level is thought to cause flagellar bending and, consequently, a change in swimming direction. Subsequently, a slower rise of cAMP occurs, which may mediate the second phase of Ca^{2+} entry. This second rise of Ca^{2+} may be involved in adaptation [84]. Cation-selective channels, regulated by cAMP, were recently identified in the sea urchin [135], and an increase in the cAMP level in response to a chemoattractant-induced hyperpolarization was demonstrated in the ascidian *Ciona intestinalis* [78].

The molecular mechanism of sperm chemotaxis in mammals is even more obscure. On the one hand, assuming universality, it may be similar to the guanylate cyclase-mediated mechanism of sperm chemotaxis in sea urchin, discussed above. The finding that ANP is chemotactically active (Table 4) and the suggestion that it may directly affect guanylate cyclase in a manner similar to that caused by the physiological attractant [192] are in line with this possibility. On the other hand, the finding of G protein-coupled olfactory receptors in sperm [7, 15, 42, 54, 119, 172, 173, 178, 179] raised the possibility that some of these proteins may be the chemotaxis receptors [119, 173]. The observation that only a limited population of spermatozoa is stained by an antibody against a testis olfactory receptor in dogs [178] is consistent with the fact that only a small fraction of the sperm population is chemotactically responsive at a given moment (Section 5.4.2 above). In dogs, the olfactory receptors were found to be localized to the midpiece of the tail of mature spermatozoa (a region rich in mitochondria), consistent with a role for these receptors in transducing chemotactic signals [178]. The strongest support for this possibility came from the recent identification of the odorant receptor hOR17-4 on human spermatozoa and demonstration of sperm chemotaxis to its agonist bourgeonal [154]. Bourgeonal induced transient Ca^{2+} influx in ~36% of the cells, a response that was inhibited by an adenylate cyclase inhibitor [154]. Binding of a chemoattractant to its receptor on the sperm may thus trigger a signal transduction pathway similar to that in the olfactory system. This may seem a valid possibility in view of the finding that male germ cells appear to contain all the elements of the signaling cascade present in olfactory cells [42, 54]. However, the apparent lack of consistency between the fraction of spermatozoa that chemotactically responded to bourgeonal ($\geq$90%) and the fraction of spermatozoa in which bourgeonal-induced Ca^{2+} influx was observed ($\sim$36%) [154], awaits clarification.

8. Open Questions

As could be concluded from reading this chapter, many major questions related to sperm chemotaxis, in every level, are still open. Some of the very basic questions are: How does a spermatozoon direct itself in a chemoattractant gradient? What gradient does it sense—temporal or spatial? What are the intracellular molecular events that change the flagellar beating and bring about a change in the direction of swimming? What are the identities of the chemoattractants secreted in the female genital tract and their respective receptors? What are the cellular origins of the chemoattractants? Where exactly does chemotaxis occur *in vivo*? Is it a single event in the female genital tract or are there several sequential steps of sperm chemotaxis? Does negative chemotaxis (repulsion) play a role in bringing spermatozoa to the egg as well? Future experiments should provide answers to all these intriguing questions.

References

1. Afzelius, B.A. (1985). The immotile-cilia syndrome: a microtubule-associated defect. *CRC Crit. Rev. Biochem.* **19**, 63–87.
2. Aitken, J. (1991). Do sperm find eggs attractive? *Nature* **351**, 19–20.
3. Aitken, R.J., Sutton, M., Warner, P. and Richardson, D.W. (1985). Relationship between the movement characteristics of human spermatozoa and their ability to penetrate cervical mucus and zona-free hamster oocytes. *J. Reprod. Fertil.* **73**, 441–449.
4. Aitken, R.J. (1990). Motility parameters and fertility, in *Controls of Sperm Motility: Biological and Clinical Aspects* (C. Gagnon, ed.), pp. 285–302. CRC Press, Boca Raton, FL.
5. Al-Anzi, B. and Chandler, D.E. (1998). A sperm chemoattractant is released from Xenopus egg jelly during spawning. *Dev. Biol.* **198**, 366–375.
6. Anderson, R.A., Feathergill, K.A., Rawlins, R.G., Mack, S.R. and Zaneveld, L.J.D. (1995). Atrial natriuretic peptide: a chemoattractant of human spermatozoa by a guanylate cyclase-dependent pathway. *Mol. Reprod. Dev.* **40**, 371–378.
7. Asai, H., Kasai, H., Matsuda, Y., Yamazaki, N., Nagawa, F., Sakano, H. and Tsuboi, A. (1996). Genomic structure and transcription of a murine odorant receptor gene: differential initiation of transcription in the olfactory and testicular cells. *Biochem. Biophys. Res. Commun.* **221**, 240–247.

8. Austin, C.R. (1957). Fate of spermatozoa in the uterus of the mouse and rat. *J. Endocrinol.* **14**, 335–342.

9. Bahat, A., Tur-Kaspa, I., Gakamsky, A., Giojalas, L.C., Breitbart, H. and Eisenbach, M. (2003). Thermotaxis of mammalian sperm cells: A potential navigation mechanism in the female genital tract. *Nat. Med.* **9**, 149–50.

10. Ballesteros, L.M., Delgado, N.M., Rosado, A., Correa, C. and Hernandez-Perez, O. (1988). Binding of chemotactic peptide to the outer surface and to whole human spermatozoa with different affinity states. *Gamete Res.* **20**, 233–239.

11. Barratt, C.L.R. and Cooke, I.D. (1991). Sperm transport in the human female reproductive tract—a dynamic interaction. *Int. J. Androl.* **14**, 394–411.

12. Battalia, D.E. and Yanagimachi, R. (1979). Enhanced and coordinated movement of the hamster oviduct during the periovulatory period. *J. Reprod. Fert.* **56**, 515–520.

13. Bedford, J.M. (1965). Effect of environment on phagocytosis of rabbit spermatozoa. *J. Reprod. Fertil.* **9**, 249–256.

14. Bedford, J.M. and Kim, H.H. (1993). Cumulus oophorus as a sperm sequestering device, *in vivo*. *J. Exp. Zool.* **265**, 321–328.

15. Branscomb, A., Seger, J. and White, R.L. (2000). Evolution of odorant receptors expressed in mammalian testes. *Genetics* **156**, 785–797.

16. Brenner, B.M., Ballerman, B.J., Gunning, M.E. and Zeidel, M.L. (1990). Diverse biological actions of atrial natriuretic peptide. *Physiol. Rev.* **70**, 665–699.

17. Brokaw, C.J. (1958). Chemotaxis of bracken spermatozoids. The role of bimalate ions. *J. Exp. Zool.* **35**, 192–196.

18. Brokaw, C.J. (1990). Computerized analysis of flagellar motility by digitization and fitting of film images with straight segments of equal length. *Cell Mot. Cytoskeleton* **17**, 309–316.

19. Bronson, R. and Hamada, Y. (1977). Gamete interactions in vitro. *Fertil. Steril.* **28**, 570–576.

20. Burgess, S.A., Walker, M.L., Sakakibara, H., Knight, P.J. and Oiwa, K. (2003). Dynein structure and power stroke. *Nature* **421**, 715–718.

21. Burkman, L.J. (1990). Hyperactivated motility of human spermatozoa during *in vitro* capacitation and implications for fertility, in *Controls of Sperm Motility: Biological and Clinical Aspects* (C. Gagnon, ed.), pp. 303–329. CRC Press, Boca Raton.

22. Cardullo, R.A. and Thaler, C.D. (2002). Function of the egg's extracellular matrix. *In*: Fertilization (D.M. Hardy, ed.) pp. 119–152. Academic Press, San Diego.

23. Chakraborty, J. and Nelson, L. (1975). Fate of surplus sperm in the fallopian tube of the white mouse. *Biol. Reprod.* **12**, 455–463.

24. Chian, R.C., Park, C.K. and Sirard, M.A. (1996). Cumulus cells act as a sperm trap during *in vitro* fertilization of bovine oocytes. *Theriogenology* **45**, 258.

25. Cohen, J. and McNaughton, D.C. (1974). Spermatozoa: The probable selection of a small population by the genital tract of the female rabbit. *J. Reprod. Fert.* **39**, 297–310.

26. Cohen, J. and Tyler, K.R. (1980). Sperm populations in the female genital tract of the rabbit. *J. Reprod. Fert.* **60**, 213–218.

27. Cohen, J. and Adeghe, A.J.-H. (1987). The other spermatozoa: fate and functions, in *New Horizons in Sperm Cell Research* (H. Mohri, ed.), pp. 125–134. Japan Sci. Soc. Press Gordon and Breach Sci. Publ., Tokyo and New York.

28. Cohen-Dayag, A., Ralt, D., Tur-Kaspa, I., Manor, M., Makler, A., Dor, J., Mashiach, S. and Eisenbach, M. (1994). Sequential acquisition of chemotactic responsiveness by human spermatozoa. *Biol. Reprod.* **50**, 786–790.

29. Cohen-Dayag, A., Tur-Kaspa, I., Dor, J., Mashiach, S. and Eisenbach, M. (1995). Sperm capacitation in humans is transient and correlates with chemotactic responsiveness to follicular factors. *Proc. Natl. Acad. Sci. U.S.A.* **92**, 11 039–11 043.

30. Coll, J.C. and Miller, R.L. (1992). The nature of sperm chemoattractants in coral and starfish, in *Comparative Spermatology: 20 Years After* (B. Baccetti, ed.), pp. 129–134. Raven Press, New York.

31. Cook, S.P., Brokaw, C.J., Muller, C.H. and Babcock, D.F. (1994). Sperm chemotaxis: egg peptides control cytosolic calcium to regulate flagellar responses. *Dev. Biol.* **165**, 10–19.

32. Cosson, J. (1996). A moving image of flagella: news and views on the mechanisms involved in axonemal beating. *Cell Biol. Int.* **20**, 83–94.

33. Cosson, M.P., Carré, D., Cosson, J. and Sardet, C. (1983). Calcium mediates sperm chemotaxis in siphonophores. *J. Submicrosc. Cytol.* **15**, 89–93.

34. Cosson, M.P., Carre, D. and Cosson, J. (1984). Sperm chemotaxis in siphonophores. II. Calcium-dependent asymmetrical movement of spermatozoa induced by the attractant. *J. Cell Sci.* **68**, 163–181.

35. Cosson, M.P. (1990). Sperm chemotaxis, in *Controls of Sperm Motility: Biological and Clinical Aspects* (C. Gagnon, ed.), pp. 103–135. CRC Press, Boca Raton, FL.

36. Cummins, J.M. and Yanagimachi, R. (1982). Sperm-egg ratios and the site of the acrosome reaction during *in vivo* fertilization in the hamster. *Gamete Res.* **5**, 239–256.

37. Darszon, A., Beltran, C., Felix, R., Nishigaki, T. and Trevino, C.L. (2001). Ion transport in sperm signaling. *Develop. Biol.* **240**, 1–14.

38. David, A., Vilensky, A. and Nathan, H. (1972). Temperature changes in the different parts of the rabbit's oviduct. *Int. J. Gynaec. Obstet.* **10**, 52–56.

39. Davis, R.O. and Katz, D.F. (1993). Operational standards for CASA instruments. *J. Androl.* **14**, 385–394.

40. Davis, R.O. and Siemers, R.J. (1995). Derivation and reliability of kinematic measures of sperm motion. *Reprod. Fertil. Dev.* **7**, 857–869.

41. de Lamirande, E., Cosson, M.-P. and Gagnon, C. (1990). Protease involvement in sperm motility, in *Controls of Sperm Motility: Biological and Clinical Aspects* (C. Gagnon, ed.), pp. 241–250. CRC Press, Boca Raton, FL.

42. Defer, N., Marinx, O., Poyard, M., Lienard, M.O., Jegou, B. and Hanoune, J. (1998). The olfactory adenylyl cyclase type 3 is expressed in male germ cells. *FEBS Lett.* **424**, 216–220.

43. Delgado, N.M., Reyes, R., Hora-Galindo, J. and Rosado, A. (1988). Size-uniform heparin fragments as nuclear decondensation and acrosome reaction inducers in human spermatozoa. *Life Sci.* **42**, 2177–2183.

44. DeMott, R.P. and Suarez, S.S. (1992). Hyperactivated sperm progress in the mouse oviduct. *Biol. Reprod.* **46**, 779–785.

45. DiBella, L.M. and King, S.M. (2001). Dynein motors of the *Chlamydomonas* flagellum. *Int. Rev. Cytol.* **210**, 227–268.

46. Dickman, Z. (1963). Chemotaxis of rabbit spermatozoa. *J. Exp. Biol.* **40**, 1–5.

47. Eisenbach, M. and Ralt, D. (1992). Precontact mammalian sperm-egg communication and role in fertilization. *Am. J. Physiol.* **262** (Cell Physiol. 31), C1095–C1101.

48. Eisenbach, M. (1999). Sperm chemotaxis. *Rev. Reprod.* **4**, 56–66.

49. Eisenbach, M. and Tur-Kaspa, I. (1999). Do human eggs attract spermatozoa? *BioEssays* **21**, 203–210.

50. Eisenbach, M. (2003). Why are sperm cells phagocytosed by leukocytes in the female genital tract? *Med. Hypoth.* **60**, 590–592.

51. Fabro, G., Eisenbach, M., Rovasio, R.A. and Giojalas, L.C. (2001). Capacitated rabbit spermatozoa respond by chemotaxis to follicular fluid. Submitted.

52. Fabro, G., Rovasio, R.A., Civalero, S., Frenkel, A., Caplan, S.R., Eisenbach, M. and Giojalas, L.C. (2002). Chemotaxis of capacitated rabbit spermatozoa to follicular fluid revealed by a novel directionality-based assay. *Biol. Reprod.* **67**, 1565–1571.

53. Gagnon, C. (1995). Regulation of sperm motility at the axonemal level. *Reprod. Fertil. Dev.* **7**, 847–855.

54. Gautier-Courteille, C., Salanova, M. and Conti, M. (1998). The olfactory adenylyl cyclase III is expressed in rat germ cells during spermiogenesis. *Endocrinology* **139**, 2588–2599.

55. Gibson, A.D. and Garbers, D.L. (2000). Guanylyl cyclases as a family of putative odorant receptors. *Annu. Rev. Neurosci.* **23**, 417–439.

56. Giojalas, L.C. and Rovasio, R.A. (1998). Mouse spermatozoa modify their dynamic parameters and chemotactic response to factors from the oocyte microenvironment. *Int. J. Androl.* **21**, 201–206.

57. Giojalas, L.C., Eisenbach, M., Fabro, G., Frenkel, A., Civalero, S., H., B., Caplan, S.R. and Rovasio, R.A. (2002). Directionality-based assay for sperm chemotaxis, independent of chemokinesis and the locomotion pattern. Submitted.

58. Giojalas, L.C., Rovasio, R.A., Fabro, G., Gakamsky, A. and Eisenbach, M. (2003). Timing of sperm capacitation appears to be programmed according to egg availability in the female genital tract. Submitted.

59. Gnessi, L., Ruff, M.R., Fraioli, F. and Pert, C.B. (1985). Demonstration of receptor-mediated chemotaxis by human spermatozoa. A novel quantitative bioassay. *Exp. Cell Res.* **161**, 219–230.

60. Gnessi, L., Fabbri, A., Silvestroni, L., Moretti, C., Fraioli, F., Pert, C.B. and Isidori, A. (1986). Evidence for the presence of specific receptors for N-formyl chemotactic peptides on human spermatozoa. *J. Clin. Endocrinol. Metab.* **63**, 841–846.

61. Gomendio, M., Harcourt, A.H. and Roldán, E.R.S. (1998). Sperm competition in mammals, in *Sperm Competition and Sexual Selection* (T.R. Birkhead and A.P. Moller, eds.), pp. 667–751. Academic Press, London.

62. Hamamah, S., Wittemer, C., Barthelemy, C., Richet, C., Zerimech, F., Royere, D. and Degand, P. (1996). Identification of hyaluronic acid and chondroitin sulfates in human follicular fluid and their effects on human sperm motility and the outcome of *in vitro* fertilization. *Reprod. Nutr. Dev.* **36**, 43–52.

63. Hansen, C., Srikandakumar, A. and Downey, B.R. (1991). Presence of follicular fluid in the porcine oviduct and its contribution to the acrosome reaction. *Mol. Reprod. Dev.* **30**, 148–153.

64. Hargreave, T.B. (1990). Clinical management of men with disorders of sperm motility, in *Controls of Sperm Motility: Biological and Clinical Aspects* (C. Gagnon, ed.), pp. 341–348. CRC Press, Boca Raton, FL.

65. Harper, M.J.K. (1973). Stimulation of sperm movement from the isthmus to the site of fertilization in the rabbit oviduct. *Biol. Reprod.* **8**, 369–377.

66. Harper, M.J.K. (1982). Sperm and egg transport, in *Germ Cells and Fertilization* (C.R. Austin and R.V. Short, eds.), pp. 102–127. Cambridge University Press, Cambridge, England.

67. Harper, M.J.K. (1994). Gamete and zygote transport, in *The physiology of Reproduction* (E. Knobil and J.D. Neill, eds.), pp. 123–187. Raven Press, New York.

68. Hong, C.Y., Chao, H.T., Tsai, K.L. and Ng, H.T. (1991). Evaluation of human sperm motility by means of transmembrane migration method and computer-assisted semen analysis—a comparison study. *Andrologia* **23**, 7–10.

69. Hunt, A.J. (1998). Keeping the beat. *Nature* **393**, 624–625.

70. Hunter, R.H.F. and Nichol, R. (1986). A preovulatory temperature gradient between the isthmus and the ampulla of pig oviducts during the phase of sperm storage. *J. Reprod. Fert.* **77**, 599–606.

71. Hunter, R.H.F. (1993). Sperm : egg ratios and putative molecular signals to modulate gamete interactions in polytocous mammals. *Mol. Reprod. Dev.* **35**, 324–327.

72. Hunter, R.H.F. (1998). Sperm-epithelial interactions in the isthmus and ampulla of the Fallopian tubes and their ovarian control, in *Gamettes: Develoment and Function* (A. Lauria, F. Gandolfi, G. Enne and L. Gianaroli, eds.), pp. 355–367. Serono Symposia, Rome.

73. Iqbal, M., Shivaji, S., Vijayasarathy, S. and Balaram, P. (1980). Synthetic peptides as chemoattractants for bull spermatozoa: structure activity correlations. *Biochem. Biophys. Res. Commun.* **96**, 235–242.

74. Ishijima, S. and Mohri, H. (1990). Beating patterns of mammalian spermatozoa, in *Controls of Sperm Motility: Biological and Clinical Aspects* (C. Gagnon, ed.), pp. 29–42. CRC Press, Boca Raton, FL.

75. Ishijima, S., Hamaguchi, M.S., Naruse, M., Ishijima, S.A. and Hamaguchi, Y. (1992). Rotational movement of a spermatozoon around its long axis. *J. Exp. Biol.* **163**, 15–31.

76. Isobe, T., Minoura, H., Tanaka, K., Shibahara, T., Hayashi, N. and Toyoda, N. (2002). The effect of RANTES on human sperm chemotaxis. *Hum. Reprod.* **17**, 1441–1446.

77. Ito, M., Smith, T.T. and Yanagimachi, R. (1991). Effect of ovulation on sperm transport in the hamster oviduct. *J. Reprod. Fert.* **93**, 157–163.

78. Izumi, H., Márián, T., Inaba, K., Oka, Y. and Morisawa, M. (1999). Membrane hyperpolarization by sperm-activating and -attracting factor increases cAMP level and activates sperm motility in the Ascidian *Ciona intestinalis*. *Develop. Biol.* **213**, 246–256.

79. Jaiswal, B.S., Tur-Kaspa, I., Dor, J., Mashiach, S. and Eisenbach, M. (1999). Human sperm chemotaxis: Is progesterone a chemoattractant? *Biol. Reprod.* **60**, 1314–1319.

80. Jaiswal, B.S. and Eisenbach, M. (2002). Capacitation, in *Fertilization* (D.M. Hardy, ed.), pp. 57–117. Academic Press, San Diego.

81. Johnson, L.R., Pilder, S.H., Bailey, J.L. and Olds-Clarke, P. (1995). Sperm from mice carrying one or two *t* haplotypes are deficient in investment and oocyte penetration. *Dev. Biol.* **168**, 138–149.

82. Katz, D.F., Overstreet, J.W. and Hanson, F.W. (1980). A new quantitative test for sperm penetration into cervical mucus. *Fertil. Steril.* **33**, 179–186.

83. Katz, D.F. (1991). Characteristics of sperm motility. *Ann. N.Y. Acad. Sci.* **637**, 409–423.

84. Kaupp, U.B., Solzin, J., Hildebrand, E., Brown, J.E., Helbig, A., Hagen, V., Beyermann, M., Pampaloni, F. and Weyand, I. (2003). The signal flow and

motor response controlling chemotaxis of sea urchin sperm. *Nature Cell Biol.* **5**, 109–117.

85. König, K., Svaasand, L., Liu, Y.G., Sonek, G., Patrizio, P., Tadir, Y., Berns, M.W. and Tromberg, B.J. (1996). Determination of motility forces of human spermatozoa using an 800 nm optical trap. *Cell. Mol. Biol.* **42**, 501–509.

86. Kopf, G.S. and Gerton, G.L. (1991). Mammalian sperm acrosome and the acrosome reaction, *Elements of Mammalian Fertilization* (P.M. Wassarman, ed.), pp. 153–203. CRC Press, Boca Raton, FL.

87. Kricka, L.J., Faro, I., Heyner, S., Garside, W.T., Fitzpatrick, G., McKinnon, G., Ho, J. and Wilding, P. (1997). Micromachined analytical devices: microchips for semen testing. *J. Pharm. Biomed. Anal.* **15**, 1443–1447.

88. Lee, S.-L., Kao, C.-C. and Wei, Y.-H. (1994). Antithrombin III enhances the motility and chemotaxis of boar sperm. *Comp. Biochem. Physiol.* **107A**, 277–282.

89. Lefebvre, R. and Suarez, S.S. (1996). Effect of capacitation on bull sperm binding to homologous oviductal epithelium. *Biol. Reprod.* **54**, 575–582.

90. Lindemann, C.B. and Kanous, K.S. (1997). A model for flagellar motility. *Int. Rev. Cytol.* **173**, 1–72.

91. Maddock, M.B. and Dawson, W.D. (1974). Artificial insemination of deermice (*Peromyscus maniculatus*) with sperm from other rodent species. *J. Embryol. Exp. Morphol.* **31**, 621–634.

92. Maier, I. and Müller, D.G. (1986). Sexual pheromones in algae. *Biol. Bull.* **170**, 145–175.

93. Makler, A., Reichler, A., Stoller, J. and Feigin, P.D. (1992). A new model for investigating in real time the existence of chemotaxis in human spermatozoa. *Fertil. Steril.* **57**, 1066–1074.

94. Makler, A., Stoller, J., Reichler, A., Blumenfeld, Z. and Yoffe, N. (1995). Inability of human sperm to change their orientation in response to external chemical stimuli. *Fertil. Steril.* **63**, 1077–1082.

95. Manor, M. (1994). Identification and purification of female-originated chemotactic factors. Ph.D. Thesis, The Weizmann Institute of Science.

96. Matsumoto, M., Solzin, J., Helbig, A., Hagen, V., Ueno, S.-I., Kawase, O., Maruyama, Y., Ogiso, M., Godde, M., Kaupp, U.B., Hoshi, M. and Weyand, I. (2003). A sperm-activating peptide controls a cGMP-signaling pathway in starfish sperm. *Dev. Biol.* In press.

97. Miller, R.D. and Vogt, R. (1996). An N-terminal partial sequence of the 13kDa *Pycnopodia helianthoides* sperm chemoattractant "startrak" possesses sperm-attracting activity. *J. Exp. Biol.* **199**, 311–318.

98. Miller, R.L. (1966). Chemotaxis during fertilization in the hydroid *Campanularia*. *J. Exp. Zool.* **162**, 23–44.

99. Miller, R.L. (1982). Synthetic peptides are not chemoattractants for bull sperm. *Gamete Res.* **5**, 395–401.

100. Miller, R.L. (1982). Sperm chemotaxis in ascidians. *Amer. Zool.* **22**, 827–840.

101. Miller, R.L. and King, K.R. (1983). Sperm chemotaxis in *Oikopleura dioica* FOL, 1872 (Urochordata: Larvacea). *Biol. Bull.* **165**, 419–428.

102. Miller, R.L. (1985). Sperm chemo-orientation in the metazoa, in *Biology of Fertilization* (C.B. Metz and A. Monroy, eds.), pp. 275–337. Academic Press, New York.

103. Miller, R.L. (1997). Specificity of sperm chemotaxis among great barrier reef shallow-water holothurians and ophiuroids. *J. Exp. Zool.* **279**, 189–200.

104. Moricard, R. and Bossu, J. (1951). Arrival of fertilizing sperm at the follicular cell of the secondary oocyte. *Fertil. Steril.* **2**, 260–266.

105. Morisawa, M. (1994). Cell signaling mechanisms for sperm motility. *Zool. Sci.* **11**, 647–662.

106. Mortimer, D., Pandya, I.J. and Sawers, R.S. (1986). Relationship between human sperm motility characteristics and sperm penetration into cervical mucus *in vitro*. *J. Reprod. Fertil.* **78**, 93–102.

107. Mortimer, D. and Fraser, L. (1996). Consensus workshop on advanced diagnostic andrology techniques. *Hum. Reprod.* **11**, 1463–1479.

108. Mortimer, S.T., Swan, M.A. and Mortimer, D. (1996). Fractal analysis of capacitating human spermatozoa. *Hum. Reprod.* **11**, 1049–1054.

109. Mortimer, S.T. (1997). A critical review of the physiological importance and analysis of sperm movement in mammals. *Hum. Reprod.* Update **3**, 403–439.

110. Mortimer, S.T., Schoevaert, D., Swan, M.A. and Mortimer, D. (1997). Quantitative observations of flagellar motility of capacitating human spermatozoa. *Hum. Reprod.* **12**, 1006–1012.

111. Navarro, M.C., Valencia, J., Vazquez, C., Cozar, E. and Villanueva, C. (1998). Crude mare follicular fluid exerts chemotactic effects on stallion spermatozoa. *Reprod. Domest. Anim.* **33**, 321–324.

112. Oko, R. and Clermont, Y. (1990). Mammalian spermatozoa: structure and assembly of the tail, in *Controls of Sperm Motility: Biological and Clinical Aspects* (C. Gagnon, ed.), pp. 3–27. CRC Press, Boca Raton, FL.

113. Olds-Clarke, P. (1996). How does poor motility alter sperm fertilizing ability? *J. Androl.* **17**, 183–186.

114. Oliveira, R.G., Tomasi, L., Rovasio, R.A. and Giojalas, L.C. (1999). Increased velocity and induction of chemotactic response in mouse spermatozoa by follicular and oviductal fluids. *J. Reprod. Fertil.* **115**, 23–27.

115. Olson, J.H., Xiang, X.Y., Ziegert, T., Kittelson, A., Rawls, A., Bieber, A.L. and Chandler, D.E. (2001). Allurin, a 21-kDa sperm chemoattractant from *Xenopus* egg jelly, is related to mammalian sperm-binding proteins. *Proc. Natl. Acad. Sci. U.S.A.* **98**, 11 205–11 210.

116. Overstreet, J.W. and Katz, D.F. (1990). Interaction between the female reproductive tract and spermatozoa, in *Controls of Sperm Motility: Biological and Clinical Aspects* (C. Gagnon, ed.), 63–75. CRC Press, Boca Raton, FL.

117. Overstreet, J.W. and Drobnis, E.Z. (1991). Sperm transport in the female tract, in *Advances in Donor Insemination* (C.L.R. Barratt and I.D. Cooke, eds.), pp. 33–49. Cambridge University Press, Cambridge, England.

118. Pandya, I.J. and Cohen, J. (1985). The leukocytic reaction of the human uterine cervix to spermatozoa. *Fertil. Steril.* **41**, 417–421.

119. Parmentier, M., Libert, F., Schurmans, S., Schiffmann, S., Lefort, A., Eggerickx, D., Ledent, C., Mollereau, C., Gerard, C., Perret, J., Grootegoed, A. and Vassart, G. (1992). Expression of members of the putative olfactory receptor gene family in mammalian germ cells. *Nature* **355**, 453–455.

120. Parrish, J.J., Suako-Parrish, J., Winer, M.A. and First, N.L. (1988). Capacitation of bovine sperm by heparin. *Biol. Reprod.* **38**, 1171–1180.

121. Pfeffer, W. (1883). Lokomotorische Richtungsbewegungen durch chemische Reize. *Ber. Deutsche Botan. Gesellschaft* **1**, 524–533.

122. Pfeffer, W. (1884). Lokomotorische richtungsbewegungen durch chemische reize. *Untersuch. aus d. Botan. Inst. Tübingen* **1**, 363–482.

123. Pfeffer, W. (1888). Ueber chemotaktische bewegungen von bakterien, flagellaten und volvocineen. *Unters. Botan. Inst. Tubingen* **2**, 582–661.

124. Phillips, D.M. and Mahler, S. (1977). Leukocyte emigration and migration in the vagina following mating in the rabbit. *Anat. Rec.* **189**, 45–59.

125. Phillips, D.M. and Mahler, S. (1977). Phagocytosis of spermatozoa by the rabbit vagina. *Anat. Rec.* **189**, 61–71.

126. Quill, T.A. and Garbers, D.L. (1998). Fertilization: Common molecular signaling pathways across the species, in *Hormones and Signaling* (B.W. O'Malley, ed.), pp. 167–207. Academic Press, San Diego.

127. Ralt, D., Goldenberg, M., Fetterolf, P., Thompson, D., Dor, J., Mashiach, S., Garbers, D.L. and Eisenbach, M. (1991). Sperm attraction of follicular factor(s) correlates with human egg fertilizability. *Proc. Natl. Acad. Sci. U.S.A.* **88**, 2840–2844.

128. Ralt, D., Manor, M., Cohen-Dayag, A., Tur-Kaspa, I., Makler, A., Yuli, I., Dor, J., Blumberg, S., Mashiach, S. and Eisenbach, M. (1994). Chemotaxis and chemokinesis of human spermatozoa to follicular factors. *Biol. Reprod.* **50**, 774–785.

129. Ren, D.J., Navarro, B., Perez, G., Jackson, A.C., Hsu, S.F., Shi, Q., Tilly, J.L. and Clapham, D.E. (2001). A sperm ion channel required for sperm motility and male fertility. *Nature* **413**, 603–609.

130. Roberts, T.M. and Stewart, M. (1995). Nematode sperm locomotion. *Curr. Opin. Cell Biol.* **7**, 13–17.

131. Roldan, E.R., Vitullo, A.D., Merani, M.S. and von Lawzewitsch, I. (1985). Cross fertilization *in vivo* and *in vitro* between three species of vesper mice, *Calomys* (Rodentia, Cricetidae). *J. Exp. Zool.* **233**, 433–442.

132. Roldan, E.R.S. and Yanagimachi, R. (1989). Cross-fertilization between Syrian and Chinese hamsters. *J. Exp. Zool.* **250**, 321–328.

133. Rosenbaum, J.L. and Witman, G.B. (2002). Intraflagellar transport. *Nature Rev. Mol. Cell Biol.* **3**, 813–825.

134. Ruskoaho, H. (1992). Atrial natriuretic peptide: synthesis, release, and metabolism. *Pharmacol. Rev.* **44**, 481–602.

135. Sánchez, D., Labarca, P. and Darszon, A. (2001). Sea urchin sperm cation-selective channels directly modulated by cAMP. *FEBS Lett.* **503**, 111–115.

136. Schiffmann, E., Corcoran, B.A. and Wahl, S.M. (1975). N-formylmethionyl peptides as chemoattractants for leucocytes. *Proc. Natl. Acad. Sci. U.S.A.* **72**, 1059–1062.

137. Schwartz, R., Brooks, W. and Zinsser, H.H. (1957). Evidence of chemotaxis as a factor in sperm motility. *Fertil. Steril.* **9**, 300–308.

138. Seed, J., Chapin, R.E., Clegg, E.D., Dostal, L.A., Foote, R.H., Hurtt, M.E., Klinefelter, G.R., Makris, S.L., Perreault, S.D., Schrader, S., Seyler, D., Sprando, R., Treinen, K.A., Veeramachaneni, D.N.R. and Wise, L.D. (1996). Methods for assessing sperm motility, morphology, and counts in the rat, rabbit, and dog: A consensus report. *Reprod. Toxicol.* **10**, 237–244.

139. Serrano, H., Canchola, E. and García-Suárez, M.D. (2001). Sperm-attracting activity in follicular fluid associated to an 8.6 kDa protein. *Biochem. Biophys. Res. Commun.* **283**, 782–784.

140. Shalgi, R. and Kraicer, P.F. (1978). Time of sperm transport, sperm penetration, and cleavage in the rat. *J. Exp. Zool.* **204**, 353–360.

141. Shimomura, H., Dangott, L.J. and Garbers, D.L. (1986). Covalent coupling of a resact analogue to guanylate cyclase. *J. Biol. Chem.* **261**, 15778–15782.

142. Shingyoji, C., Higuchi, H., Yoshimura, M., Katayama, E. and Yanagida, T. (1998). Dynein arms are oscillating force generators. *Nature* **393**, 711–714.

143. Short, R.V. (1976). The origin of species, in *The Evolution of Reproduction* (C.R. Austin and R.V. Short, eds.), pp. 110–131. Cambridge University Press, Cambridge.

144. Silvestroni, L., Palleschi, S., Guglielmi, R. and Croce, C.T. (1992). Identification and localization of atrial natriuretic factor receptors in human spermatozoa. *Arch. Androl.* **28**, 75–82.

145. Singh, S., Lowe, D.G., Thorpe, D.S., Rodriguez, H., Kuang, W.J., Dangott, L.J., Chinkers, M., Goeddel, D.V. and Garbers, D.L. (1988). Membrane guanylate cyclase is a cell-surface receptor with homology to protein kinases. *Nature* **334**, 708–712.

146. Sliwa, L. (1993). Effect of heparin on human spermatozoa migration *in vitro*. *Arch. Androl.* **30**, 177–181.

147. Sliwa, L. (1993). Heparin as a chemoattractant for mouse spermatozoa. *Arch. Androl.* **31**, 149–152.

148. Sliwa, L. (1994). Chemotactic effect of hormones in mouse spermatozoa. *Arch. Androl.* **32**, 83–88.

149. Sliwa, L. (1995). Chemotaction of mouse spermatozoa induced by certain hormones. *Arch. Androl.* **35**, 105–110.

150. Sliwa, L. (1999). Hyaluronic acid and chemoattractant substance from follicular fluid: *in vitro* effect of human sperm migration. *Arch. Androl.* **43**, 73–76.

151. Sliwa, L. (2001). Substance P and beta-endorphin act as possible chemoattractants of mouse sperm. *Arch. Androl.* **46**, 135–140.

152. Smith, T.T., Koyanagi, F. and Yanagimachi, R. (1988). Quantitative comparison of the passage of homologous and heterologous spermatozoa through the uterotubal junction of the golden hamster. *Gamete Res.* **19**, 227–234.

153. Smith, T.T. and Yanagimachi, R. (1991). Attachment and release of spermatozoa from the caudal isthmus of the hamster oviduct. *J. Reprod. Fertil.* **91**, 567–573.

154. Spehr, M., Gisselmann, G., Poplawski, A., Riffell, J.A., Wetzel, C.H., Zimmer, R.K. and Hatt, H. (2003). Identification of a testicular odorant receptor mediating human sperm chemotaxis. *Science* **299**, 2054–2058.

155. Starr, R.C., Marner, F.J. and Jaenicke, L. (1995). Chemoattraction of male gametes by a pheromone produced by female gametes of chlamydomonas. *Proc. Natl. Acad. Sci. U.S.A.* **92**, 641–645.

156. Stauss, C.R., Votta, T.J. and Suarez, S.S. (1995). Sperm motility hyperactivation facilitates penetration of the hamster zona pellucida. *Biol. Reprod.* **53**, 1280–1285.

157. Stephens, D.T. and Hoskins, D.D. (1990). Computerized quantitation of flagellar motion in mammalian sperm, in *Controls of Sperm Motility: Biological and Clinical Aspects* (C. Gagnon, ed.), pp. 251–260. CRC Press, Boca Raton, FL.

158. Suarez, S.S. (1996). Hyperactivated motility in sperm. *J. Androl.* **17**, 331–335.

159. Suarez, S.S. (1998). The oviductal sperm reservoir in mammals: mechanisms of formation. *Biol. Reprod.* **58**, 1105–1107.

160. Suarez, S.S. (2002). Gamete transport, in *Fertilization* (D.M. Hardy, ed.), pp. 3–28. Academic Press, San Diego.

161. Sun, F., Giojalas, L.C., Rovasio, R.A., Tur-Kaspa, I., Sanchez, R. and Eisenbach, M. (2003). Lack of species-specificity in mammalian sperm chemotaxis. *Dev. Biol.* **255**, 423–427.

162. Sun, F., Tur-Kaspa, I. and Eisenbach, M. (2003). Chemotaxis of human spermatozoa to conditioned media of human cumulus-oocyte cells. In preparation.

163. Sundfjord, J.A., Forsdahl, F. and Thibault, G. (1989). Physiological levels of immunoreactive ANH-like peptides in human follicular fluid. *Acta Endocrinol.* **121**, 578–580.

164. Suzuki, N. (1995). Structure, function, and biosynthesis of sperm-activating peptides and fucose sulfate glycoconjugate in the extracellular coat of sea urchin eggs. *Zool. Sci.* **12**, 13–27.

165. Tacconis, P., Revelli, A., Massobrio, M., Battista La Sala, G. and Tesarik, J. (2001). Chemotactic responsiveness of human spermatozoa to follicular fluid is enhanced by capacitation but is impaired in dyspermic semen. *J. Assist. Reprod. Genet.* **18**, 36–44.

166. Talbot, P., DiCarlantonio, G., Zao, P., Penkala, J. and Haimo, L.T. (1985). Motile cells lacking hyaluronidase can penetrate the hamster oocyte cumulus complex. *Dev. Biol.* **108**, 387–398.

167. Tash, J.S. (1990). Role of cAMP, calcium, and protein phosphorylation in sperm motility, in *Controls of Sperm Motility: Biological and Clinical Aspects* (C. Gagnon, ed.), pp. 229–240. CRC Press, Boca Raton, FL.

168. Tash, J.S. and Brancho, G.E. (1994). Regulation of sperm motility: emerging evidence for a major role for protein phosphatases. *J. Androl.* **15**, 505–509.

169. Taylor, N.J. (1982). Investigation of sperm-induced cervical leucocytosis by a double mating study in rabbits. *J. Reprod. Fertil.* **66**, 157–160.

170. Tso, W.-W., Lee, W.M. and Wong, M.-Y. (1979). Spermatozoa repellent as a contraceptive. *Contraception* **19**, 207–211.

171. Uhler, M.L., Leung, A., Chan, S.Y.W. and Wang, C. (1992). Direct effects of progesterone and antiprogesterone on human sperm hyperactivated motility and acrosome reaction. *Fertil. Steril.* **58**, 1191–1198.

172. Vanderhaeghen, P., Schurmans, S., Vassart, G. and Parmentier, M. (1993). Olfactory receptors are displayed on dog mature sperm cells. *J. Cell Biol.* **123**, 1441–1452.

173. Vanderhaeghen, P., Schurmans, S., Vassart, G. and Parmentier, M. (1997). Specific repertoire of olfactory receptor genes in the male germ cells of several mammalian species. *Genomics* **39**, 239–246.

174. Vawda, A.I., Wiley, R.E., Younglai, E.V. and Lobb, D.K. (1999). A simple apparatus for studying sperm attraction: chemotaxis induced by relaxin and progesterone. *Med. Sci. Res.* **27**, 693–696.

175. Villanueva-Díaz, C., Vadillo-Ortega, F., Kably-Ambe, A., Diaz-Perez, M.A. and Krivitzky, S.K. (1990). Evidence that human follicular fluid contains a chemoattractant for spermatozoa. *Fertil. Steril.* **54**, 1180–1182.

176. Villanueva-Díaz, C., Arias-Martínez, J., Bustos-López, H. and Vadillo-Ortega, F. (1992). Novel model for study of human sperm chemotaxis. *Fertil. Steril.* **58**, 392–395.

177. Villanueva-Díaz, C., Arias-Martínez, J., Bermejo-Martínez, L. and Vadillo-Ortega, F. (1995). Progesterone induces human sperm chemotaxis. *Fertil. Steril.* **64**, 1183–1188.

178. Walensky, L.D., Roskams, A.J., Lefkowitz, R.J., Snyder, S.H. and Ronnett, G.V. (1995). Odorant receptors and desensitization proteins colocalize in mammalian sperm. *Mol. Med.* **1**, 130–141.

179. Walensky, L.D., Ruat, M., Bakin, R.E., Blackshaw, S., Ronnett, G.V. and Snyder, S.H. (1998). Two novel odorant receptor families expressed in spermatids undergo 5′-splicing. *J. Biol. Chem.* **273**, 9378–9387.

180. Wang, Y., Storeng, R., Dale, P.O., Åbyholm, T. and Tanbo, T. (2001). Effects of follicular fluid and steroid hormones on chemotaxis and, motility of human spermatozoa *in vitro*. *Gynecol. Endocrinol.* **15**, 286–292.

181. Ward, G.E., Brokaw, C.J., Garbers, D.L. and Vacquier, V.D. (1985). Chemotaxis of *Arbacia punctulata* spermatozoa to resact, a peptide from the egg jelly layer. *J. Cell Biol.* **101**, 2324–2329.

182. Wassarman, P.M. (1987). The biology and chemistry of fertilization. *Science* **235**, 553–560.

183. Wildt, L., Kissler, S., Licht, P. and Becker, W. (1998). Sperm transport in the human female genital tract and its modulation by oxytocin as assessed by hysterosalpingoscintigraphy, hysterotonography, electrohysterography and Doppler sonography. *Hum. Reprod.* Update **4**, 655–666.

184. Williams, M., Hill, C.J., Scudamore, I., Dunphy, B., Cooke, I.D. and Barratt, C.L.R. (1993). Sperm numbers and distribution within the human Fallopian tube around ovulation. *Hum. Reprod.* **8**, 2019–2026.

185. World Health Organization (1993). *Laboratory Manual for the Examination of Human Semen and Semen–Cervical Mucus Interaction*. Cambridge University Press, New York.

186. Yanagimachi, R. (1994). Mammalian fertilization, in *The Physiology of Reproduction* (E. Knobil and J. Neill, eds.), pp. 189–317. Raven Press, New York.

187. Yoshida, M., Inaba, K. and Morisawa, M. (1993). Sperm chemotaxis during the process of fertilization in the ascidians *Ciona-Savignyi* and *Ciona-Intestinalis*. *Dev. Biol.* **157**, 497–506.

188. Yoshida, M., Inaba, K., Ishida, K. and Morisawa, M. (1994). Calcium and cyclic AMP mediate sperm activation, but Ca^{2+} alone contributes sperm chemotaxis in the ascidian, *Ciona savignyi*. *Dev. Growth Dif.* **36**, 589–595.

189. Yoshida, M., Murata, M., Inaba, K. and Morisawa, M. (2002). A chemoattractant for ascidian spermatozoa is a sulfated steroid. *Proc. Natl. Acad. Sci. U.S.A.* **99**, 14831–14836.

190. Yoshida, M., Ishikawa, M., Izumi, H., De Santis, R. and Morisawa, M. (2003). Store-operated calcium channel regulates the chemotactic behavior of ascidian sperm. *Proc. Natl. Acad. Sci. U.S.A.* **100**, 149–154.

191. Zamir, N., Skofitsch, G., Eskay, R. and Jacobowitz, D.M. (1986). Distribution of immunoreactive atrial natriuretic peptide in the central nervous system of the rat. *Brain Res.* **365**, 105–111.

192. Zamir, N., Riven-Kreitman, R., Manor, M., Makler, A., Blumberg, S., Ralt, D. and Eisenbach, M. (1993). Atrial natriuretic peptide attracts human spermatozoa *in vitro*. *Biochem. Biophys. Res. Commun.* **197**, 116–122.

193. Zatylny, C., Marvin, L., Gagnon, J. and Henry, J.L. (2002). Fertilization in *Sepia officinalis*: the first mollusk sperm-attracting peptide. *Biochem. Biophys. Res. Commun.* **296**, 1186–1193.

194. Zigmond, S.H. (1977). Ability of polymorphonuclear leukocytes to orient in gradients of chemotactic factors. *J. Cell Biol.* **75**, 606–616.

Chemotropic Guidance of Axons in the Nervous System *

1. Growthcone Guidance by Short- and Long-Range Cues

Neurons are highly polarized cells composed of three different functional domains: the cell body or soma, containing the nucleus and major cytoplasmic organelles; multiple dendrites, tapered processes extending from the soma; and a single axon, a smooth process extending from the soma over a long distance (Figure 1). The terminals of axons are connected to the soma or dendrites of other neurons via chemical junctions known as synapses. One of the central issues of neural development is how the axons grow through their stereotyped pathways, find their target neurons, and form specific synaptic connections with them. During development, differentiated neurons start to elongate processes from the soma: one of the processes differentiates into an axon while the others become dendrites. At the growing tip of the axon and the dendrites, there is a spiky enlargement known as the growthcone (Figure 1). The growthcone is composed of lamellipodia, which are fan-shaped sheets with ruffled membranes, and filopodia, which are multiple microspikes protruding from the lamellipodia. The

* Written by Atusuhi Tamada and Fujio Murakami, Laboratory of Neuroscience, Graduate School of Frontier Bioscience, Osaka University, Machikaneyama 1–3, Toyonaka, Osaka 560-8531, Japan.

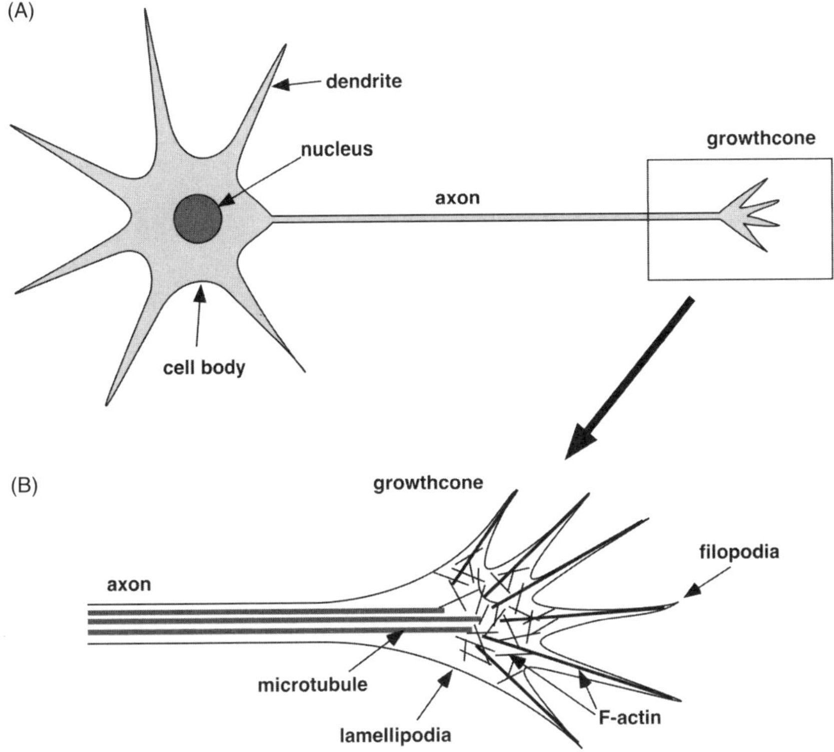

Figure 1. Structure of a neuron (A) and a growthcone (B) at growing tip of an axon.

growthcone comprises an engine that produces movement and a steering apparatus that directs itself along the proper pathway by sensing environmental cues. The cues that guide growthcones have been classified into short-range and long-range types depending on the effective distance at which they act [58]. Short-range cues, mediated by cell surface molecules or extracellular matrix molecules, contact-dependently affect the behavior of the growthcones. Long-range cues, mediated by secreted molecules, affect the growthcones at a distance from the source of the cues. These cues are also divided into attractive and repulsive types depending on their effects on the axons. Among long-range cues, chemotropic cues can control the direction of growthcone advance. Chemoattractive cues, mediated by chemoattractant molecules, attract the growthcones toward the source of chemoattractants at the final or intermediate target of the axons. In contrast, chemorepulsive cues, mediated by chemorepellents, direct the growth-cones away from the source at the nontarget tissues.

2. Chemoattraction and Chemorepulsion in the Nervous System

2.1. Chemoattraction

2.1.1. Nerve growth factor

Ramón y Cajal observed growthcones at the tip of growing axons for the first time in the late nineteenth century. The finding that leucocytes can migrate upon detecting a gradient of a diffusible chemoattractant led him to propose that nerve growthcones might also be attracted by gradients of target-derived molecules. The first demonstration that growthcones can be attracted by a diffusible molecule was not obtained until nearly a century later, by Gundersen and Barrett [15]. They observed the behavior of growthcones of sympathetic and sensory axons exposed to a point source of nerve growth factor (NGF) [32], a diffusible molecule that supports the survival of embryonic sympathetic and sensory neurons. Although there is no evidence to suggest that NGF guides these axons *in vivo*, regenerating sympathetic and sensory axons are reoriented toward NGF ejected from a tip of a glass micropipette, indicating that growthcones can respond to a gradient of chemoattractant.

2.1.2. Maxillary arch

The first evidence that a target-derived chemoattractant guides growthcones was obtained in an *in vitro* experiment in which explants of trigeminal ganglia were cocultured with explants of maxillary arch, the target of trigeminal axons, placed 300–500 μm away in collagen gel matrices [34, 35]. The collagen gel matrix appears to stabilize gradients of target-derived molecules extending over a long distance (Figure 2). While no axonal outgrowth occurred from trigeminal ganglia cultured alone, they demonstrated extensive outgrowth toward explants of maxillary arch. This directed growth was taken to reflect the detection of a concentration gradient of a target-derived chemoat-tractant, because (1) some of the axons turned toward the target and (2) when two trigeminal ganglia were cultured in tandem next to a maxillary arch, more axons extended toward the target from the closer ganglion.

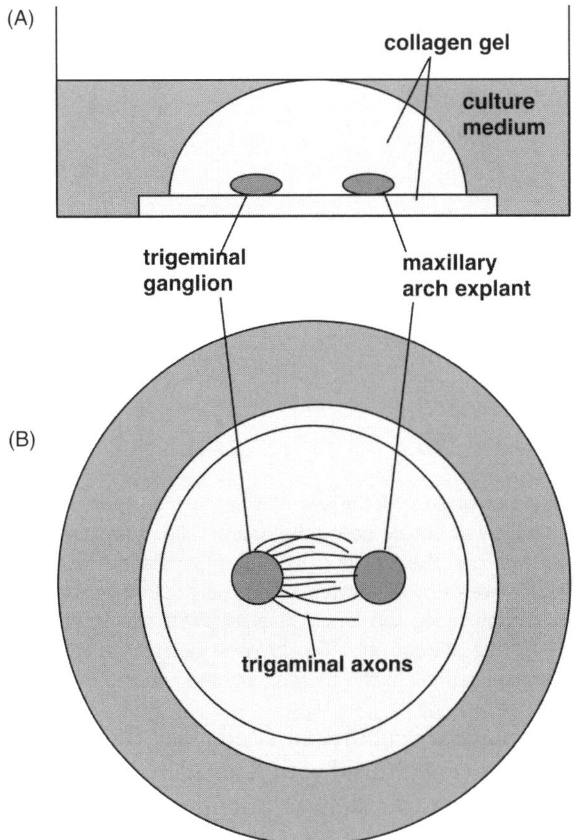

Figure 2. Schematic view of collagen gel coculture of a trigeminal ganglion with a maxillary arch explant. Side view (A) and top view (B) of the culture dish.

2.1.3. *Floor plate*

A fundamental characteristic of the vertebrate central nervous system (CNS) is bilateral symmetry along the midline (Figure 3). This symmetrical structure divides neuronal projections into two types, uncrossed projections to ipsilateral targets and crossed projections to contralateral targets. Over the past decade, the mechanism involved in the formation of crossed projections has been extensively studied in spinal cord commissural projections. Spinal commissural axons, which derive from the dorsal part of the neural tube (alar plate), initially project ventrally along the circumferential axis and then cross the ventral midline [10]. These axons have been shown to be guided by a diffusible molecule released

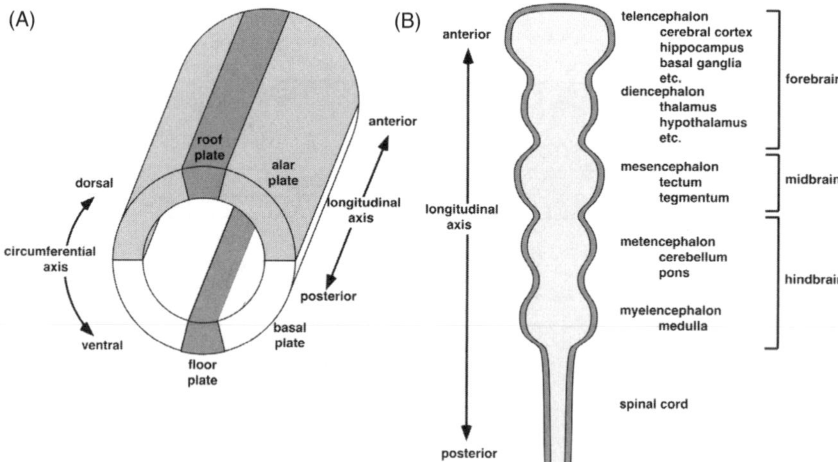

Figure 3. Basic structure of the vertebrate central nervous system. The CNS is derived from a hollow structure called the neural tube. It has two orthogonal axes, the circumferential axis and the longidudinal axis. Four distinct domains called the roof, alar, basal, floor plates are generated along the circumferential axis (A). The neural tube can also divided along the longitudinal axis into the spinal cord and the brain, which can be further divided into five brain vesicles, the telencephalon, the diencephalon, the mesencephalon, the metencephalon and the myelencephalon (B).

from the floor plate, a structure situated along the ventral midline of the neural tube. When cocultured with floor plate explants in collagen gels, spinal commissural axons show reoriented growth toward floor plate explants [43, 44, 59]. Consistent with these *in vitro* findings are the *in vivo* observations that spinal commissural axons in chick embryo grow toward ectopically-transplanted floor plate [44, 61], and that these axons show abnormal trajectories in the floor plate–deficient zebrafish mutant, cyclops [3, 4], floor plate–deleted zebrafish [3]; and mouse mutant, Danforth's short tail [5].

The fact that the basic structures of the neural tube such as floor plate, basal plate, and alar plate extend up to the caudal diencephalon (Figure 3) and that the brain contains a number of ventrally decussating axons coursing through the floor plate raised the question of whether mechanisms of axon guidance for spinal commissural axons also apply to commissural axons in the more rostral parts of the CNS. Indeed this turned out to be the case [52, 53, 56]. Axons from the deep cerebellar nuclei located in the alar plate of rostral hindbrain (metencephalon) grow ventrally along the circumferential axis of the neural tube, cross the floor plate, and project to contralateral targets [53]. Alar plate

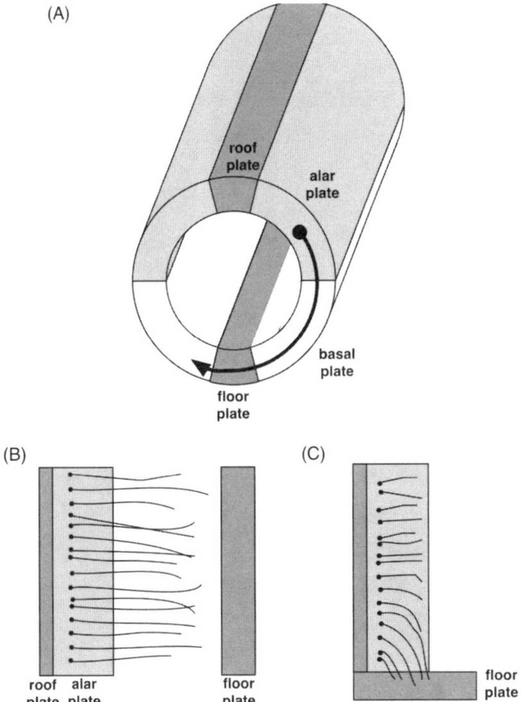

Figure 4. Chemoattraction of spinal cord commissural axons by the floor plate. (A) Axon growth pattern of commissural neurons. (B, C) Collagen gel culture of an alar plate explant with a floor plate explant. Extensive outgrowth of commissural axons occurred toward the floor plate explant (B). Commissural axons were reoriented toward the floor plate explant (C).

explants taken from the metencephalon extend axons toward floor plate explants in the same fashion as spinal commissural axons [53], when cocultured with floor plate explants in collagen gels (Figure 5). These findings suggest that cerebellofugal axons are guided toward the ventral midline by a floor plate–derived diffusible chemoattractant. Similarly, axons originating from the alar plate of the caudal hindbrain (myelencephalon) (Figure 5) and the midbrain (mesencephalon) are attracted by the floor plate from corresponding levels along the rostrocaudal axis [52, 56]. Moreover, floor plate at different axial level can attract alar plate neurons. Together these findings suggest that alar plate–derived ventrally decussating axons from the spinal cord up to the level of the mesencephalon are guided toward the ventral midline by a common cue: a diffusible chemoattractant(s) released from the floor plate.

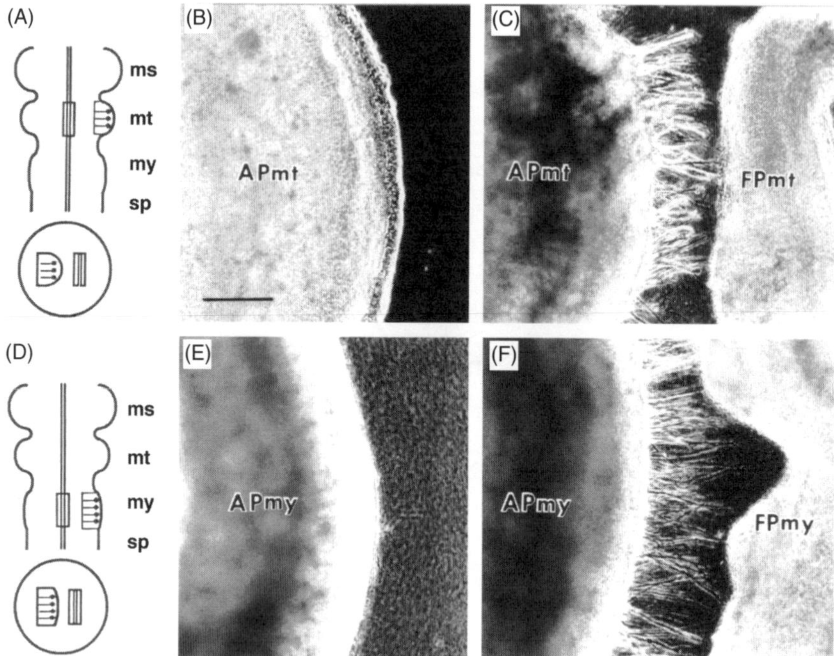

Figure 5. Chemoattraction of alar plate commissural axons in the brain by the floor plate. (A, D) Flattened view of the brain from the mesencephalon to the spinal cord, showing the position of the explants used in collagen gel culture (B, C, E, F). When metencephalic alar plate explants were cultured alone, no axon outgrowth occurred into collagen gels (B). In contrast, when they were cocultured with floor plate explants, extensive axon outgrowth occurred toward floor plate explants (C). Similar axon outgrowth patterns were observed in cultures of myelencephalic alar plate explants (E, F). Scale bar (shown in B) for panels B, C, E, and F = 200 μm.

2.2. *Chemorepulsion*

2.2.1. *Septum*

Pini was the first to demonstrate chemorepulsion of axons [42]. Olfactory axons grow along a pathway called the lateral olfactory tract, which lies on the lateral surface of the telencephalon (Figure 6). He presumed that repellent activity of the septum at the midline contributes to the establishment of this trajectory and therefore tested this hypothesis by coculturing an explant of the olfactory bulb of the rat embryo, in collagen gels, with tissue from the septum. He found that axons emanating from the tissue of the olfactory bulb turned and grew away from the septal explant, indicating that chemorepulsion by the septum contributes to the guidance of olfactory axons (Figure 6).

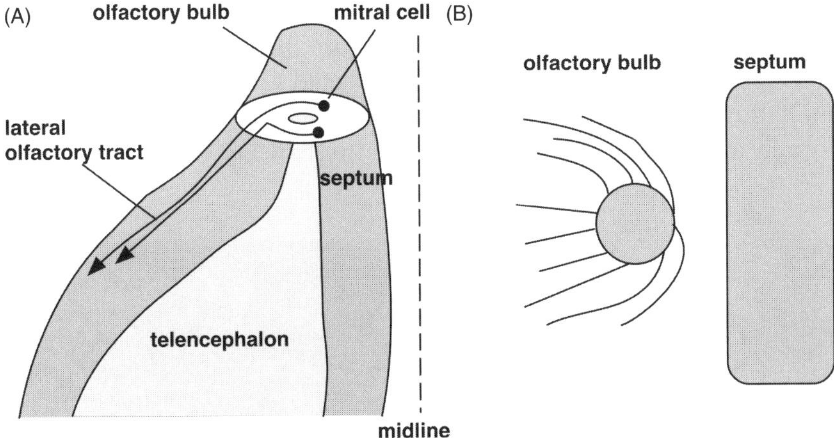

Figure 6. Chemorepulsion of mitral cell axons by septum. (A) Axon growth pattern of mitral cells in the olfactory bulb. Mitral cell axons grow laterally away from the septum and form the lateral olfactory tract in the lateral telencephalon. (B) Collagen gel coculture of olfactory bulb explants with septal explants.

2.2.2. *Ventral spinal cord*

The spinal cord receives somatosensory information through peripheral sensory afferents from dorsal root ganglia (DRG). Small diameter myelinated and unmyelinated afferents, many of which mediate noxious mechanical, thermal, and chemical cutaneous sensory information, are NGF-dependent and terminate predominantly in the most superficial region of the dorsal horn. In contrast, large diameter low-threshold cutaneous afferents, which are NGF-independent, terminate in deeper laminae, while group I muscle afferents terminate in deeper ventral horn. When an explant of DRG is cocultured with ventral spinal cord from E14–15 rat embryos in collagen gels in the presence of NGF, DRG axons extend less on the side opposite the spinal cord explant [14]. These findings suggest that the ventral spinal cord secretes a chemorepellent toward NGF-dependent DRG axons.

2.2.3. *Floor plate*

In the zebrafish cyclops mutant which lacks floor plate, normally uncrossed axons in the spinal cord [3] and the hindbrain [18] often cross the ventral midline. An intriguing possibility that arises from this observation is that the floor plate releases a diffusible chemorepellent(s) that prevents axons from crossing the midline. Indeed, the floor plate repels uncrossed axons originating from the alar plate of the mesencephalon, near the dorsal midline *in vitro* [56] (Figure 8). Axons extending from this region in rat embryo are initially directed toward the ventral midline but course ipsilaterally without crossing it. Axonal subsets growing dorsally away from the floor plate were also shown to be repelled by the

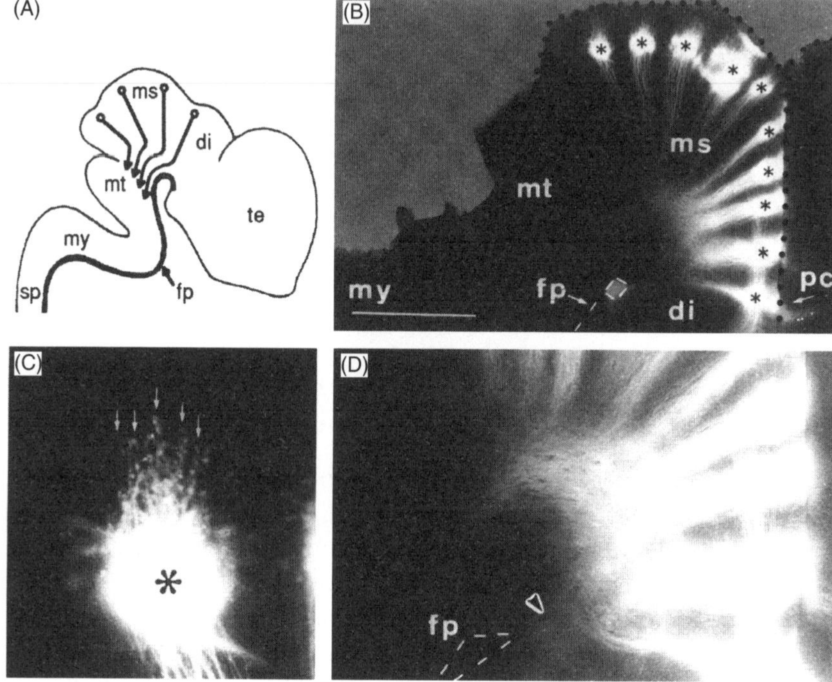

Figure 7. Growth of mesencephalic dorsal alar plate (DAPms) axons *in vivo*. (A) Schematic lateral view of rat embryonic CNS, including trajectory of DAPms axons. Open circles and curved arrows show DAPms cells and their axons, respectively. (B–D) Fluorescence photomicrographs of flat-mounted E15 rat brain cut along the dorsal (black dotted line) and ventral midlines and implanted with lipophilic tracer, DiI into the DAP of the mesencephalon and the caudal diencephalon. DiI injections (asterisks) retrogradely labeled cells (shown at high magnification and indicated by arrows in C) near the dorsal midline and anterogradely axons that initially run ventrally along the circumferential axis toward the invaginated ventral midline floor plate (fp) (white broken line) and then gradually turn caudally, constantly maintaining a distance from the fp (B). The axons turned at varying distances from the FP. The labeled axons are shown at high magnification in D. The arrowhead in D indicates FP-crossing axons. te, telencephalon; di, diencephalon; ms, mesencephalon; mt, metencephalon; my, myelencephalon; sp, spinal cord; pc, posterior commissure. Scale bar (shown in B) μ 1 mm for B, 200 μm for C, and 500 μm for D.

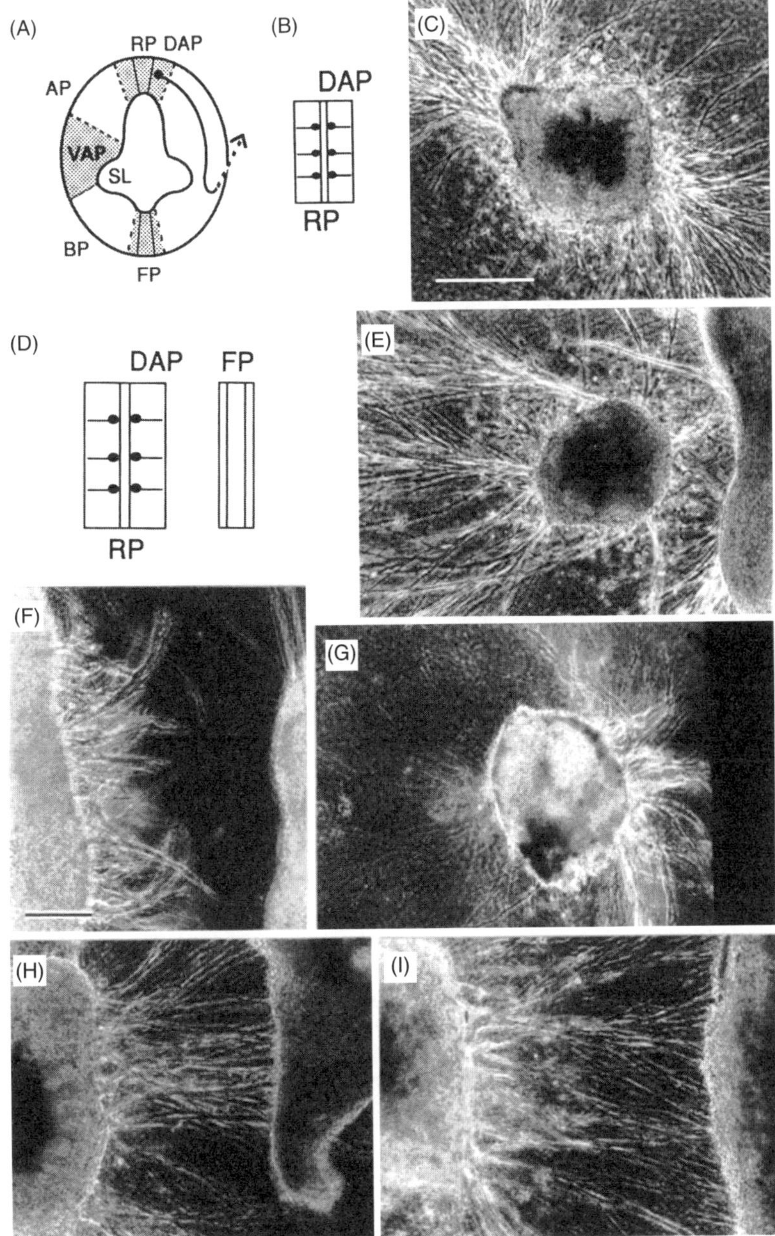

Figure 8. Turning of DAPms axons away from floor explants in collagen gel culture. (A) Diagrammatic representation of coronal view of the mesencephalon showing explant origins (hatched areas). RP, roof plate; AP, alar plate; VAP, ventral alar plate; BP, basal plate; SL, sulcus limitans, the AP/BP boundary. (B, D) Diagrams showing

floor plate. This includes axons from the basal plate of the mesencephalon [56], caudal diencephalon [52], trochlear motor neuron [11], and visceral motor and branchiomotor neuron subclasses [16]. Thus, the floor plate repels at least two types of uncrossed axons: dorsally migrating axons and initially ventrally migrating axons (Figure 9).

2.2.4. *Roof plate*

Spinal commissural axons are shown to be guided toward the ventral midline by a chemoattractant from the floor plate. However, this can only explain the guidance mechanism near the ventral midline, raising the possibility that there may be another guidance mechanism.

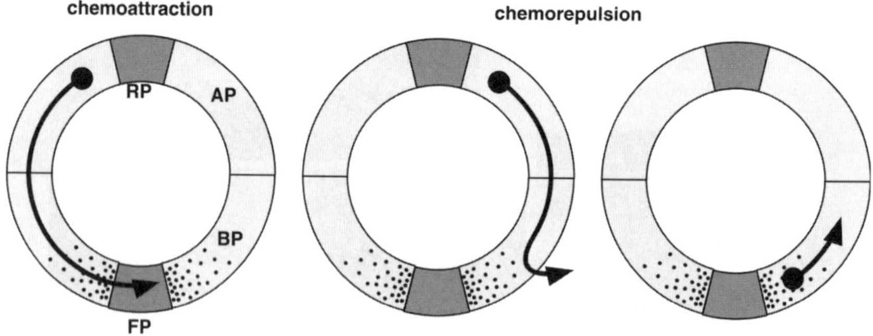

Figure 9. Role of floor plate chemoattraction and chemorepulsion in axon guidance along the circumferential axis. Coronal view of the neural tube showing proposed axonal guidance of crossed alar plate (AP) axons by floor plate (FP) chemoattraction (left panel), and of uncrossed AP (center panel) and basal plate (BP) (right panel) axons by chemorepulsion. Presumable release of diffusible factors from the FP is represented by dots. FP chemorepulsion may act in two ways: to prevent ventrally directed axons from crossing the ventral midline, allowing them to grow along the anteroposterior axis (center), and to direct BP axons away from the ventral midline (right).

---

orientation and arrangement of explants. B corresponds to C, D to E–H. DAPms neurons and the direction of axonal growth *in vivo* are shown by filled circles with short bars. (C) E–I, Phase-contrast photomicrographs of DAPms explants cultured alone (C), or cocultured with live FPms (E–G), heat-treated FPms (H) or VAPms (I) explants. In G, DAPms and FPms explants were separated by a permeable membrane filter (a white arrow shows the edge of the filter). Filters alone did not affect the growth of the axons (n = 3). Note that DAPms axons turn away from FPms explants (E–G), without contacting them or their processes (F–G). Scale bar (shown in C) = 500 μm for C and E. Scale bar (shown in F) = 200 μm for F–I.

Recently, Augsburger and her colleagues found that the roof plate, which lies adjacent to the spinal commissural neurons, has chemorepellent activity to this set of axons [2]. They further found that bone morphogenic protein-7 (BMP-7) mimics the repellent activity of the roof plate *in vitro*. Spinal commissural axons juxtaposed to roof plate explants or BMP-7 expressing cell aggregates are deflected away from the explants or cell aggregates. These results suggest the chemorepellent activity of the roof plate, which is mediated by BMPs, contributes to the early phase of commissural axon extension in the spinal cord by steering the axons ventrally away from the dorsal midline.

3. Molecular Mechanism of Chemoattraction and Chemorepulsion

3.1. *Chemoattractants and chemorepellents*

3.1.1. *Netrins*

After the discovery of chemoattraction by the floor plate, Tessier-Lavigne and his colleagues tried to identify the chemoattractant secreted by the floor plate. They eventually identified two laminin-related secreted proteins that mimic the diffusible neurite outgrowth-promoting activity of the floor plate on commissural neurons, and named them Netrin-1 and Netrin-2 [26, 50]. Netrin-1 is expressed at the floor plate, and exhibits a chemoattractive activity to commissural axons. These findings suggest that Netrin-1 is a floor plate–derived chemoattractant to spinal commissural axons. In support of this notion, in netrin-1 deficient mice most spinal commissural axons fail to reach the ventral midline, despite extending halfway toward the midline [49]. This indicates that Netrin-1 plays a crucial role in the guidance of ventrally growing spinal commissural axons *in vivo*, although it is not required for initial ventral growth of the axons.

Netrin-1 is expressed in the floor plate at all axial levels from the spinal cord through to the caudal diencephalon [26], suggesting that Netrin-1 is also involved in the guidance of commissural axons at higher axial levels such as those from the metencephalon. In support of this view, Netrin-1, *in vitro*, can mimic the activity of floor plate explants toward cerebellofugal axons as well as those from the mesencephalic and the myelencephalic alar plate [52, 53]. It is therefore likely that Netrin-1 plays a global role in the guidance of ventrally growing commissural axons in the CNS, at least up to the level of the mesencephalon.

Netrin-1 is a vertebrate homologue of nematode *unc-6* gene, the mutation of which disrupts circumferential dorsal and ventral growth of axons in the body wall of *C. elegans* [21, 23]. In the nematode, *unc-6* function requires other genes, *unc-40* and *unc-5* (see below), and the involvement of these genes in receptor function of *unc-6* has been implicated [7]. *dcc* (deleted in colorectal cancer; [13]), a human tumor suppresser gene, is a vertebrate homologue of *unc-40* and DCC has turned out to be a Netrin-1 receptor molecule, or a component of the receptor [25]. Consistent with this idea are the findings that *dcc* mRNA is expressed in the region containing commissural neurons in the spinal cord and that DCC protein is expressed on commissural axons of the rat embryo [25]. However, anti-DCC antibody does not block chemotropic activity of the floor plate *in vitro*, although it blocks neurite outgrowth-promoting activity of Netrin-1 [25] suggesting that a floor plate–derived chemoattractant other than Netrin-1 might also be present.

In *unc-6* mutant, circumferential dorsal and ventral growth of axons in the body wall of *C. elegans* [21, 23] are disrupted. This led to the idea that Netrin-1 may also be involved in the guidance of dorsally migrating axons [26]. This hypothesis was directly tested by an *in vitro* experiment with collagen gels, in which it was demonstrated that floor plate explants repel trochlear motor axons that normally grow dorsally away from the floor plate *in vivo* [11]. The same study showed that aggregates of Netrin-1-expressing cells mimicked these activities of the floor plate. Together, these findings suggest that Netrin-1 functions as a chemorepellent as well as a chemoattractant. Similar *in vitro* results have been obtained for visceral motor and branchiomotor axons [60], raising the possibility that Netrin-1 plays a global role in the guidance of dorsally migrating axons by repelling them.

In the floor plate–deficient zebrafish mutant, cyclops, the formation of the posterior commissure (PC), a dorsal commissure near the rostral end of the mesencephalon, is severely disrupted [19, 41]; this suggests that the floor plate is also involved in the formation of the posterior commissure, through secretion of a diffusible chemorepellent(s) that drives dorsally decussating axons away from the floor plate. In contrast to the case of trochlear and other motor axons described above, heterologous cells secreting recombinant Netrin-1 do not repel PC axons, while the floor plate shows a clear chemorepellent effect [52]. These findings raise the possibility that the repellent effect of the floor plate on PC axons is mediated by a factor distinct from Netrin-1. The observation that the posterior commissure and trochlear motor axons *in vivo* develop apparently normally in mice deficient in Netrin-1 or DCC function [12, 49] is consistent with this notion and further supports the likelihood of a Netrin-1- independent mechanism in the guidance of dorsally extending neurons.

Studies in *C. elegans* have suggested that the bifunctional activities of Netrin-1 appear to be mediated by different receptors. UNC-5, a transmembrane protein that appears to be involved in dorsally oriented guidance of axons in the nematode [31], has been proposed to act as a receptor of UNC-6 repellent activity on the basis of the effects of mutations in *unc-5* gene on dorsal axon growth [17]. To date, three vertebrate homologues of *unc-5* have been cloned [1, 30]. These are all expressed in the developing brain and two of them have been shown to bind to Netrin-1 [30], implicating that they may be involved in the receptor function of Netrin-1-mediated repulsion.

3.1.2. Class III semaphorins

The semaphorins are a large family of secreted and transmembrane proteins defined by characteristic semaphorin domains in the extracellular region. The semaphorins have been categorized by the arrangement of additional domains into eight classes [48]. Among them, class III semaphorins are secreted proteins found in vertebrates. The first member of class III semaphorins (Collapsin-1 or Sema-3A) was identified as a protein that causes the collapse of sensory growthcones in culture [36]. Messersmith *et al.* [37] later demonstrated that recombinant Sema-3A protein functions as a chemorepellent to axons of dorsal root ganglia, and that Sema-3A mimicked the repellent activity of ventral spinal cord reported by Fitzgerald *et al.* [14]. Taken together with the observation that Sema-3A mRNA is expressed in the ventral spinal cord at the time of sensory axon ingrowth, Sema-3A was suggested to pattern small diameter myelinated and unmyelinated afferents by repelling these axons.

To date, six members of class III semaphorins (Sema-3A to Sema-3F) have been identified, and these members have been reported by a number of studies to function as chemorepellents to distinct sets of axons (reviewed in [45]). Furthermore, there are studies demonstrating that some members of class III semaphorins can act as chemoattractants as well as chemorepellents.

The receptor for class III semaphorins was identified independently by two groups [20, 29]. They screened a plasmid library by probing library-transfected COS cells with Sema-3A/alkaline-phosphatase chimeric proteins, and found that a known molecule, Neuropilin-1 is a receptor for Sema-3A. Subsequently, a second neuropilin, Neuropilin-2, was identified and found to bind different members of class III semaphorins [8]. Neuropilins form dimers (homodimers or heterodimers) and the dimerization appears to determine the response specificity to a variety of class III semaphorins [9, 55]. Neuropilins are likely to require transmembrane receptor proteins, plexins, for class III semaphorin signals [55, 57]. There is evidence that plexins form complexes with neuropilins, that such complexes increase the binding affinity with semaphorins, and that plexins are required for collapsing signals by semaphorins.

3.1.3. *Slits*

Another important family of chemorepellents was recently identified through genetic studies in Drosophila. In the roundabout (*robo*) mutant, axons fail to respect the midline boundary [47]. Robo protein was later identified as a transmembrane protein of the immunoglobulin superfamily with five immunoglobulin domains and three fibronectin type III repeats, and it was shown to be abundantly expressed on uncrossing axons [28]. These findings led to the hypothesis that Robo functions as a receptor for a repellent of the midline.

A ligand for Robo was subsequently identified in 1999 by genetic and biochemical analyses, and it turned out to be a known molecule, Slit [6, 27]. Slit was first identified through Drosophila slit mutant, which exhibits a collapsed midline. Slit is a large secreted protein containing four leucine-rich repeats and seven EGF-like repeats and is expressed on midline glial cells [46]. Slit protein has been shown to bind Robo protein [6]. To date, it is known that there is another member of Robo in Drosophila, Robo2, as well as three mammalian homologues of Drosophila Robo, Robo1, Robo2, and Rig1, and three mammalian homologues of Drosophila Slit, Slit1, Slit2, and Slit3 [6, 22, 24, 28, 33, 39, 62, 63]. Direct demonstration of the chemorepellent activity of Slit protein has been shown by *in vitro* collagen gel assay in vertebrates, with the protein acting as a chemorepellent to a variety of axons including olfactory bulb axons, spinal motor axons, and hippocampal axons [6, 33, 40]. Taken together with the observation that Slit2 is expressed in the septum, Slit2 is suggested to mediate the chemorepulsion of olfactory bulb axons by the septum that was originally demonstrated by Pini (1993).

4. Switching Mechanism of Chemoattraction and Chemorepulsion

Developing axons reach their final targets as a result of a series of axonal projections to successive intermediate targets. Long-range chemoattraction by intermediate targets is thought to play a key role in this process. Growing axons, however, do not stall at intermediate targets, where the chemoattractant concentration is expected to be maximal. Commissural axons in the metencephalon, initially attracted by a chemoattractant released from the floor plate [53], lose responsiveness to the chemoattractant when they cross the floor plate *in vitro* [51]. Such changes in axon responsiveness to chemoattractants may enable developing axons to continue to navigate toward their final destinations.

More recently, evidence has suggested that commissural axons acquire responsiveness to a chemorepellent after crossing the midline. Zou *et al.* [64] reported that floor plate explants suppress the outgrowth of spinal cord commissural axons after crossing the midline but not the outgrowth of those before crossing the midline, and that floor plate suppressing activity can be mimicked by Slit2, Sema-3B, and Sema-3F. These findings suggest that both loss and gain of responsiveness to chemotropic molecules at the intermediate targets play a key role in navigation toward the final targets.

Moo-Ming Poo and his colleagues revealed that the same molecule elicits two different types of growthcone responses, either attractive or repulsive, by micropipette application of molecules, the method first used successfully by Gundersen and Barrett to detect chemotropic activity of NGF [15]. They used dissociated spinal neurons from Xenopus and exposed them to neurotrophins, neurotransmitters, or Netrin-1 that were ejected from the tip of a micropipette. When Netrin-1 is applied to those neurons, it induces an attractive response toward the micropipette tip. Netrin-1, however, induces a repulsive turning of the same growthcone, when Rp-cAMPS, a nonhydrolyzable analog competitor of cAMP, or KT5720, a specific inhibitor of protein kinase A [38, 54] is applied. Thus the responsiveness of growthcones to diffusible molecules may be dependent on the internal state of the growthcones.

5. Future Challenges

Chemotropic theory of axon guidance assumes that there is a concentration gradient of molecules and that growthcones detect this concentration gradient and approach or depart from the source of chemoattractant or chemorepellent, respectively. Growthcones then should somehow be able to detect the concentration gradient, thereby deciding on the direction of growth. However, it remains totally unknown how they are able to achieve this. A more fundamental issue is that there has been no demonstration of a concentration gradient of any candidate chemoattractant or chemorepellent molecule that acts at a distance. Although a variety of growthcones can be reoriented in collagen gels thought to form a gradient of diffusible molecules, it remains unknown whether a gradient of diffusible molecules is actually present *in vivo*.

References

1. Ackerman, S.L., Kozak, L.P., Przyborski, S.A., Rund, L.A., Boyer, B.B. and Knowles, B.B. (1997). The mouse rostral cerebellar malformation gene encodes an UNC-5-like protein. *Nature* **386**, 838–842.
2. Augsburger, A., Schuchardt, A., Hoskins, S., Dodd, J. and Butler, S. (1999). BMPs as mediators of roof plate repulsion of commissural neurons. *Neuron* **24**, 127–141.
3. Bernhardt, R.R., Nguyen, N. and Kuwada, J.Y. (1992). Growthcone guidance by floor plate cells in the spinal cord of zebrafish embryos. *Neuron* **8**, 869–882.
4. Bernhardt, R.R., Patel, C.K., Wilson, S.W. and Kuwada, J.Y. (1992). Axonal trajectories and distribution of GABAergic spinal neurons in wild-type and mutant zebrafish lacking floor plate cells. *J. Comp. Neurol.* **326**, 263–272.
5. Bovolenta, P. and Dodd, J. (1991). Perturbation of neuronal differentiation and axon guidance in the spinal cord of mouse embryos lacking a floor plate: analysis of Danforth's short-tail mutation. *Development* **113**, 625–639.
6. Brose, K., Bland, K.S., Wang, K.H., Arnott, D., Henzel, W., Goodman, C.S., Tessier-Lavignek, M. and Kidd, T. (1999). Slit proteins bind robo receptors and have an evolutionarily conserved role in repulsive axon guidance. *Cell* **96**, 795–806.
7. Chan, S.S. , Zheng, H., Su, M.W., Wilk, R., Killeen, M.T., Hedgecock, E.M. and Culotti, J.G. (1996). UNC-40, a C. elegans homologues of DCC (Deleted in Colorectal Cancer), is required in motile cells responding to UNC-6 netrin cues. *Cell* **87**, 187–195.
8. Chen, H., Chédotal, A., He, Z., Goodman, C.S. and Tessier-Lavigne, M. (1997). Neuropilin-2, a novel member of the neuropilin family, is a high-affinity receptor for the semaphorins sema E and sema IV but not sema III. *Neuron* **19**, 547–559.
9. Chen, H., He, Z., Bagri, A. and Tessier-Lavigne, M. (1998). Semaphorin-neuropilin interactions underlying sympathetic axon responses to class III semaphorins. *Neuron* **21**, 1283–1290.
10. Colamarino, S.A. and Tessier-Lavigne, M. (1995a). The role of the floor plate in axon guidance. *Ann. Rev. Neurosci.* **18**, 497–529.
11. Colamarino, S.A. and Tessier-Lavigne, M. (1995b). The axonal chemoattractant netrin-1 is also a chemorepellent for trochlear motor axons. *Cell* **81**, 621–629.
12. Fazeli, A., Dickinson, S.L., Hermiston, M.L., Tighe, R.V., Steen, R.G., Small, C.G., Stoeckli, E.T, Keino-Masu, K., Masu, M., Rayburn, H., Simons, J., Bronson, R.T.,

Gordon, J.I., Tessier-Lavigne, M. and Weinberg, R.A. (1997). Phenotype of mice lacking functional Deleted in colorectal cancer (Dcc) gene. *Nature* **386**, 796–804.

13. Fearon, E.R., Cho, K.R., Nigro, J.M., Kern, S.E., Simons, J.W., Ruppert, J.M., Hamilton, S.R., Preisinger, A.C., Thomas, G., Kinzler, K.W. and Vogelstein, B. (1990). Identification of a chromosome 18q gene that is altered in colorectal cancers. *Science* **247**, 49–56.

14. Fitzgerald, M., Kwiat, G.C., Middleton, J. and Pini, A. (1993). Ventral spinal cord inhibition of neurite outgrowth from embryonic rat dorsal root ganglia. *Development* **117**, 1377–1384.

15. Gundersen, R.W. and Barrett, J.N. (1979). Neuronal chemotaxis: chick dorsal-root axons turn toward high concentrations of nerve growth factor. *Science* **206**, 1079–1080.

16. Guthrie, S. and Pini, A. (1995). Chemorepulsion of developing motor axons by the floor plate. *Neuron* **14**, 1117–1130.

17. Hamelin, M., Zhou, Y., Su, M.W., Scott, I.M. and Culotti, J.G. (1993). Expression of the UNC-5 guidance receptor in the touch neurons of C. elegans steers their axons dorsally. *Nature* **364**, 327–330.

18. Hatta, K. (1992). Role of the floor plate in axonal patterning in the zebrafish CNS. *Neuron* **9**, 629–642.

19. Hatta, K., Puschel A.W. and Kimmel, C.B. (1994). Midline signaling in the primordium of the zebrafish anterior central nervous system. *PNAS U.S.A.* **91**, 2061–2065.

20. He, Z. and Tessier-Lavigne, M. (1997). Neuropilin is a receptor for the axonal chemorepellent semaphorin III. *Cell* **90**, 739–751.

21. Hedgecock, E.M., Culotti, J.G. and Hall, D.H. (1990). The unc-5, unc-6, and unc-40 genes guide circumferential migrations of pioneer axons and mesodermal cells on the epidermis in C. elegans. *Neuron* **4**, 61–85.

22. Holmes, G.P. , Negus, K., Burridge, L., Raman, S., Algar, E., Yamada T. and Little, M.H. (1998). Distinct but overlapping expression patterns of two vertebrate slit homologs implies functional roles in CNS development and organogenesis. *Mech. Dev.* **79**, 57–72.

23. Ishii, N., Wadsworth, W.G., Stern, B.D., Culotti, J.G. and Hedgecock, E.M. (1992). UNC-6, a laminin-related protein, guides cell and pioneer axon migrations in C. elegans. *Neuron* **9**, 873–881.

24. Itoh, A., Miyabayashi, T., Ohno, M. and Sakano, S. (1998). Cloning and expressions of three mammalian homologs of Drosophila slit suggest possible roles for Slit in the formation and maintenance of the nervous system. *Mol. Brain Res.* **62**, 175–185.

25. Keino-Masu, K., Masu, M., Hinck, L., Leonardo, E.D., Chan, S.S.-Y., Culotti, J.G. and Tessier-Lavigne, M. (1996). Deleted in colorectal cancer (DCC) encodes a netrin receptor. *Cell* **87**, 175–185.

26. Kennedy, T.E., Serafini, V., de la Torre, J.R. and Tessier-Lavigne, M. (1994). Netrins are diffusible chemotropic factors for commissural axons in the embryonic spinal cord. *Cell* **78**, 425–435.

27. Kidd, T., Bland K.S. and Goodman, C.S. (1999). Slit is the midline repellent for the robo receptor in drosophila. *Cell* **96**, 785–794.

28. Kidd, T. Brose, K., Mitchell, K.J., Fetter, R.D., Tessier-Lavigne, M., Goodman, C.S. and Tear, G. (1998). Roundabout controls axon crossing of the CNS midline and

defines a novel subfamily of evolutionarily conserved guidance receptors. *Cell* **92**, 205–215.

29. Kolodkin, A.L., Levengood, D.V., Rowe, E.G., Tai, Y.T.,. Giger R.J and Ginty, D.D. (1997). Neuropilin is a semaphorin III receptor. *Cell* **90**, 753–762.

30. Leonardo, E.D., Hinck, L., Masu, M., Keino-Masu, K., Ackerman, S.L. and Tessier-Lavigne, M. (1997). Vertebrate homologues of C. elegans UNC-5 are candidate netrin receptors. *Nature* **386**, 833–838.

31. Leung-Hagesteijn, C., Spence, A.M., Stern, B.D., Zhou, Y., Su, M.W., Hedgecock, E.M. and Culotti, J.G. (1992). UNC-5, a transmembrane protein with immunoglobulin and thrombospondin type 1 domains, guides cell and pioneer axon migrations in C. elegans. *Cell* **71**, 289–299.

32. Levi-Montalcini, R. (1982). Developmental neurobiology and the natural history of nerve growth factor. *Ann. Rev. Neurosci.* **5**, 341–362.

33. Li, H.-S., Chen, J.-H., Wu, W., Fagaly, T., Zhou, L., Yuan, W., Dupuis, S., Gick, C., Ornitz, D.M., Wu, J.Y. and Rao, Y. (1999). Vertebrate slit, a secreted ligand for the transmembrane protein roundabout, is a repellent for olfactory bulb axons. *Cell* **96**, 807–818.

34. Lumsden, A.G. and Davies, A.M. (1983). Earliest sensory nerve fibres are guided to peripheral targets by attractants other than nerve growth factor. *Nature* **306**, 786–788.

35. Lumsden, A.G. and Davies, A.M. (1986). Chemotropic effect of specific target epithelium in the developing mammalian nervous system. *Nature* **323**, 538–539.

36. Luo, Y., Raible, D. and Raper, J.A. (1993). Collapsin: a protein in brain that induces the collapse and paralysis of neuronal growth cones. *Cell* **75**, 217–227.

37. Messersmith, E.K., Leonardo, E.D., Shatz, C.J., Tessier-Lavigne, M., Goodman, C.S. and Kolodkin, A.L. (1995). Semaphorin III can function as a selective chemorepellent to pattern sensory projections in the spinal cord. *Neuron* **13**, 1001–1014.

38. Ming, G.-L., Song, H.-J., Benedikt, B., Holt, C.E., Tessier-Lavigne, M. and Poo, M.-M. (1997). cAMP-dependent growthcone guidance by Netrin-1. *Neuron* **19**, 1225–1235.

39. Nakayama, M., Nakajima, D., Nagase, T., Nomura, N., Seki, N. and Ohara, O. (1998). Identification of high-molecular-weight proteins with multiple EGF-like motifs by motif-trap screening. *Genomics* **51**, 27–34.

40. Nguyen Ba-Charvet, K.T., Brose, K., Marillat, V., Kidd, T., Goodman, C.S., Tessier-Lavigne, M., Sotelo, C. and Chédotal, A. (1999). Slit2-mediated chemorepulsion and collapse of developing forebrain axons. *Neuron* **22**, 463–473.

41. Patel, C.K., Rodriguez, L.C. and Kuwada, L.C. (1994). Axonal outgrowth within the abnormal scaffold of brain tracts in a zebrafish mutant. *J. Neurobiol.* **25**, 345–360.

42. Pini, A. (1993). Chemorepulsion of axons in the developing mammalian central nervous system. *Science* **261**, 95–98.

43. Placzek, M., Tessier-Lavigne, M., Jessell, T.M. and Dodd, T. (1990). Orientation of commissural axons *in vitro* in response to a floor plate-derived chemoattractant. *Development* **110**, 19–30.

44. Placzek, M., Tessier-Lavigne, M., Yamada, T., Dodd, J. and Jessell, T.M. (1990). Guidance of developing axons by diffusible chemoattractants. *Cold Spring Harbor Symposia On Quantitative Biology* **55**, 279–289.

45. Raper, J.A. (2000). Semaphorins and their receptors in vertebrates and invertebrates. *Curr. Opi. Neurobiol.* **10**, 88–94.
46. Rothberg, J.M., Jacobs, J.R., Goodman, C.S. and Artavanis-Tsakonas, S. (1990). Slit: an extracellular protein necessary for development of midline glia and commissural axon pathways contains both EGF and LRR domains. *Genes Dev.* **4**, 2169–2187.
47. Seeger, M., Tear, G., Ferres-Marco, D. and Goodman, C.S. (1993). Mutations affecting growth cone guidance in Drosophila: genes necessary for guidance toward or away from the midline. *Neuron* **10**, 409–426.
48. Semaphorin-Nomenclature-Committee (1999). Unified nomenclature for the semaphorins/collapsins. *Cell* **97**, 551–552.
49. Serafini, T., Colamarino, S.A., Leonardo, D., Wang, H., Beddington, R., Skarnes, W.C. and Tessier-Lavigne, M. (1996). Netrin-1 is required for commissural axon guidance in the developing vertebrate nervous system. *Cell* **87**, 1001–1014.
50. Serafini, T., Kennedy, T.E., Galko, M.J., Mirzayan, C., Jessell, T.M. and Tessier-Lavigne, M. (1994). The netrins define a family of axon outgrowth-promoting proteins homologous to C. elegans UNC-6. *Cell* **78**, 409–424.
51. Shirasaki, R., Katsumata. R. and Murakami, F. (1998). Change in chemoattractant responsiveness of developing axons at an intermediate target. *Science* **279**, 106–108.
52. Shirasaki, R., Mirzayan, C., Tessier-Lavigne, M. and Murakami, F. (1996). Guidance of circumferentially growing axons by netrin-dependent and -independent floor plate chemotropism in the vertebrate brain. *Neuron* **17**, 1079–1088.
53. Shirasaki, R., Tamada, A., Katsumata, R. and Murakami, F. (1995). Guidance of cerebellofugal axons in the rat embryo: directed growth toward the floor plate and subsequent elongation along the longitudinal axis. *Neuron* **14**, 961–972.
54. Song, H.-J., Ming, G.-L., He, Z., Lehmann, M., McKerracher, L., Tessier-Lavigne, M. and Poo, M.-M. (1998). Conversion of neuronal growthcone responses from repulsion to attraction by cyclic nucleotides. *Science* **281**, 1515–1518.
55. Takahashi, T., Fournier, A., Nakamura, F., Wang, L.-H., Murakami, Y., Kalb, R.G., Fujisawa, H. and Strittmatter, S.M. (1999). Plexin-neuropilin-1 complexes form functional semaphorin-3A receptors. *Cell* **99**, 59–69.
56. Tamada, A., Shirasaki, R. and Murakami, F. (1995). Floor plate chemoattracts crossed axons and chemorepels uncrossed axons in the vertebrate brain. *Neuron* **14**, 1083–1093.
57. Tamagnone, L., Artigiani, S., Chen, H., He, Z., Ming, G.-L., Song, H.-J., Chedotal, A., Winberg, M.L., Goodman, C.S., Poo, M.-M., Tessier-Lavigne, M. and Comoglio, P.M. (1999). Plexins are a large family of receptors for transmembrane, secreted, and GPI-anchored semaphorins in vertebrates. *Cell* **99**, 71–80.
58. Tessier-Lavigne, M. and Goodman, C.S. (1996). The molecular biology of axon guidance. *Science* **274**, 1123–1133.
59. Tessier-Lavigne, M., Placzek, M., Lumsden, A.G., Dodd, J. and Jessell, T.M. (1988). Chemotropic guidance of developing axons in the mammalian central nervous system. *Nature* **336**, 775–778.
60. Varela-Echavarria, A., Tucker, A., Puschel, A.W. and Guthrie, S. (1997). Motor axon subpopulations respond differentially to the chemorepellents netrin-1 and semaphorin D. *Neuron* **18**, 193–207.

61. Yaginuma, H. and Oppenheim, R.W. (1991). An experimental analysis of *in vivo* guidance cues used by axons of spinal interneurons in the chick embryo: evidence for chemotropism and related guidance mechanisms. *J. Neurosci.* **11**, 2598–2613.

62. Yuan, S.S., Cox, L.A., Dasika, G.K. and Lee, E.Y. (1999). Cloning and functional studies of a novel gene aberrantly expressed in RB-deficient embryos. *Dev. Biol.* **207**, 62–75.

63. Yuan, W., Zhou, L., Chen, J.H., Wu, J.Y., Rao, Y. and Ornitz, D.M. (1999). The mouse Slit family: secreted ligands for ROBO expressed in patterns that suggest a role in morphogenesis and axon guidance. *Dev. Biol.* **212**, 290–306.

64. Zou, Y., Stoeckli, E., Chen, H. and Tessier-Lavigne, M. (2000). Squeezing axons out of the gray matter: a role for slit and semaphorin proteins from midline and ventral spinal cord. *Cell* **102**, 363–375.

Commonality and Diversity of Chemotaxis

What can we learn from the variety of chemotaxis systems described in this book? Can one generalize about the mechanism of chemotaxis and about the functions that this process fulfills? Let us examine these questions at the behavioral and molecular levels, starting with the mechanisms of motility.

1. Motility Mechanisms

From Table 1, which summarizes the mechanisms of motility in the systems reviewed in this book, it appears that, for motility, commonality and diversity go hand in hand. Many species share similar motility systems: many bacteria move by rotating their flagella; sperm cells of many species—from marine species to mammals—swim by bending and pushing their flagella; and many eukaryotic cells move slowly by repeated cycles of cell extension and retraction (amoeboid movement). On the other hand, there are a large variety of motility systems. The largest variability, no doubt, exists in prokaryotic cells (bacteria and archaea). Due to their cell wall, these species are incapable of amoeboid movement, which requires cell shape changes. Instead, they have acquired other mechanisms of motility, which reflect the large diversity of habitats in which they reside: no less than six modes of motility are known in prokaryotes, some of which include at least two different mechanisms (Table 1). On the other hand, eukaryotic cells have adopted

Table 1. Mechanisms of motility.

Motility type	Energy source	Mechanism	Species	Book chapter
Swimming by rotating flagella	H^+-motive force	Flux of H^+ ions through H^+ channels rotates the flagellar motor	Some bacteria (e.g., *E. coli*), archaea	3
	Na^+-motive force	Flux of Na^+ ions through Na^+ channels rotates the flagellar motor	Some bacteria (e.g., alkalophilic *Bacilli*, *Vibrio*)	3
Swimming by a bending-and-pushing flagellum	ATP	Active sliding of microtubules, carried out by attachment/ detachment of the dynein arms	Spermatozoa of most species	7
Swimming without flagella	Not known	Not known	Some marine cyanobacteria (e.g., *Synechococcus*)	3
Swarming	H^+-motive force	Flux of H^+ ions through H^+ channels rotates the flagellar motor	Some bacteria (e.g., *E. coli*, *Vibrio*)	3
Gliding	Not known	Movement of cell surface parts?	Some bacteria (e.g. cyanobacteria, mycobacteria, myxobacteria, *Cytophaga*)	3
	ATP	Pilus adherence and retraction	Myxobacteria	3
	Not known	Not known	Some parasitic eukaryotic protozoa, some algae (e.g., *Desmidiaceae*, *Diatomeae*)	2
Twitching	ATP	Pilus adherence and retraction	Some bacteria (e.g., *Pseudomonas aeruginosa*)	3
Crawling	ATP	Movement of crawling devices at the ventral side of the cell	Some ciliates (e.g., *Stylonychia*)	2
Propulsion by filaments	ATP	Actin polymerization	Some bacteria (e.g., *Listeria*, *Shigella*)	3
Amoeboid movement (cell extension on one side and retraction on the other side)	ATP	Actin polymerization; interaction between cytoskeletal filaments (actin, myosin)	Amoebae (e.g., *Dictyostelium discoideum*), some mammalian cells (e.g., white blood cells, mesenchymal cells, tumor cells), spermatozoa of nematodes	5, 6, 7

only two mechanisms of motility, possibly reflecting their limited and specific habitats: either an amoeboid movement for relatively sluggish cells, or propagation by a bending-and-pushing flagellum for fast-moving cells (e.g., sperm cells of most animals).

The energy source for the motility is, again, more diverse in prokaryotes: proton-motive force, sodium-motive force, or ATP. In eukaryotes, the energy source appears to be solely ATP. This may simply be due to the fact that, in prokaryotes only, the proton- or sodium-motive force is readily available for the motility organelle. There, the ion gradient is maintained across the cytoplasmic membrane. In eukaryotic cells, the ion gradient is maintained across the mitochondrial membrane and is, therefore, not directly available for motility. The outcome is that the molecular mechanisms underlying motility of eukaryotic cells are less diverse than those of prokaryotic cells.

The diversity of prokaryotic motility is an example of the amazing power of prokaryotes to adapt to a large variety of environments. This is especially prominent in species that have two motility machineries, each being used in a different environment. For example, bacteria like *E. coli* and *Salmonella* can swim as individuals in aqueous solutions, driven by a few rotating flagella. However, when on a hard surface, they are able to differentiate into filamentous, hyperflagellated cells that translocate together as a colony on the surface (swarming). Similarly, when *Vibrio* cells are in non-viscous media, they swim by rotating their polar flagella, driven by a flux of Na^+ ions. However, when these bacteria are in viscous media, they use lateral flagella, driven by a flux of protons. These flagella are also used for swarming. And, finally, myxobacteria can glide on hard surfaces either as groups or as individuals. In each case they use a different, independent motility system.

2. Behavioral Mechanisms of Chemotaxis

Table 2 summarizes the behavioral mechanisms of chemotaxis in the systems reviewed in this book. If these systems faithfully represent the mechanisms of chemotaxis in nature, it appears that prokaryotes mainly employ trial-and-error chemotaxis mechanisms—the phobic response and klinokinesis, defined in Table 2 and, in more detail, in Chapter 1. In contrast, eukaryotes appear to primarily employ the more direct mechanism—modulation of the direction of movement according to the stimulant gradient. A number of non-mutually-exclusive reasons may account for this difference. First, the mechanisms employed by bacteria

Table 2. Behavioral mechanisms of chemotaxis.

Behavioral mechanism	Species	Book chapter
Gradient-stimulated modulation of the direction of movement (classical, narrow definition of chemotaxis)	Amoebae (e.g., *Dictyostelium discoideum*), cells with amoeboid movement (e.g., tumor cells, white blood cells), spermatozoa (of, e.g., hydroids and tunicate)	5, 6, 7
Gradient-stimulated modulation of the linear velocity followed by a directional change (phobic response)	Bacteria (e.g., *Rhodobacter sphaeroides*, *Sinorhizobium meliloti*), archaea (e.g., some cells of *Halobacterium salinarium*)	3
Gradient-stimulated modulation of the frequency of spontaneous directional changes (klinokinesis)	Bacteria (e.g., *E. coli*, *Pseudomonas* spp., *Spirillum volutans*), archaea (e.g., some cells of *Halobacterium salinarium*), spermatozoa (of, e.g., hydromedusa, fern, and Japanese bitterlings)	3, 7

and archaea may be more primitive mechanisms, whereas the mechanism employed by the evolutionarily more recent eukaryotic cells may be more advanced. Second, the apparent difference between prokaryotes and eukaryotes may reflect the difference in the dimensions of their cells. Thus, because prokaryotic cells are relatively small, the differences between the stimulant concentrations at opposing sides of the cell may be within the range of thermal fluctuations and, therefore, too small to be detected. Prokaryotes may have had, therefore, to adopt a mechanism that does not rely on a spatial comparison between different ends of the cells. The phobic response and klinokinesis are such mechanisms. In contrast, the dimensions of eukaryotic cells are sufficiently large to allow spatial comparison and their movement can, therefore, be modulated more directly by the chemical gradient. Third, the speed of movement may be a contributing factor as well. Fast-moving cells may have to employ the less direct mechanisms because, for them, spatial information may be too brief. This possibility can be tested when the behavioral mechanism of fast-moving eukaryotic cells (e.g., sperm cells) and slow-moving prokaryotic cells (e.g., gliding bacteria) is revealed.

3. Molecular Mechanisms of Chemotaxis

Some generalizations can be made about all molecular mechanisms of chemotaxis. The first component of the signal-transduction pathway is a detector—a receptor that interacts with the stimulant. Often, the receptor is membrane-bound; it senses changes in the concentration of specific stimulants in the exterior environment and transmits this information to the interior of the cell. Some receptors are intracellular and respond to stimulants that have to be transported inside and possibly metabolized (Chapter 3). Next, the receptor produces an excitation signal that involves amplification. The outcome is a behavioral response that usually results (directly or indirectly) in a change of the direction of movement. Once the response has occurred, a termination mechanism comes into action and ends the signal. Most, if not all, chemotaxis systems also operate a feedback mechanism, adaptation, which alters the sensitivity of the receptor and adjusts it to the basal level of the stimulant.

Beyond this universal, general framework, the molecular mechanisms of chemotaxis involve both commonalities and diversities. Generally speaking, the molecular mechanisms of prokaryotes and eukaryotes are different. Within the prokaryotes, however, commonality seems to govern. Within the eukaryotes, both commonality and diversity appear to dominate (Table 3).

Table 3. Molecular mechanisms of chemotaxis.

Chemotaxis system	Receptor[a]	Signal-transduction pathway	Book chapter
Bacteria and archaea	MCP or Enzyme II of the PTS	Two-component regulatory system: receptor-modulated autophosphorylation of a histidine kinase and phosphotransfer to a response regulator	3
Amoeba	GPCR	G-protein activation, PIP_3 regulation of rho family G proteins, cGMP regulation of myosin kinases, ERK activation	5
Mesenchymal cells	Tyrosine kinase	PIP_3 regulation of rho family G proteins, PLCγ, ERK activation	5
White blood cells	GPCR	G-protein activation, lipid remodeling, protein kinase activation, and Ca^{2+} elevation	6
Sperm cells	Guanylate cyclase[b]	Receptor-modulated elevation of cGMP, rise in Ca^{2+}_{in}, possibly as a consequence of Ca^{2+} channels opening, cAMP rise	7
	GPCR[c]	G-protein activation and Ca^{2+} elevation	

[a]Abbreviations: ERK, extracellular signal-regulated kinase; GPCR, G-protein-coupled receptor; MCP, methyl-accepting chemotaxis protein; PIP_3, phosphatidylinositol triphosphate; PLCγ, phospholipase Cγ; PTS, phospho*enol*pyruvate-dependent carbohydrate phosphotransferase system.
[b]At least in the sea urchin.
[c]At least in human spermatozoa, where the odorant receptor hOR17-4 was identified.

In prokaryotes, the receptors for most chemoattractants (excluding PTS carbohydrates) and chemorepellents are methyl-accepting chemotaxis proteins (MCPs) or MCP-like proteins (Chapter 3). Although some of them are membrane proteins and others are soluble proteins, and although they appear in a variety of forms, they all share a high degree of similarity. The downstream signaling pathway is also similar, because all known chemotaxis machineries in prokaryotes belong to the superfamily of two-component regulatory systems [3]. They involve a sensor—a histidine kinase that undergoes autophosphorylation on a conserved histidine residue. The sensor usually belongs to the receptor supramolecular complex and its extent of phosphorylation depends on the receptor occupancy. The sensor transfers the phosphoryl group to a conserved aspartate residue of the response regulator. The response regulator—CheY in most cases of prokaryotic chemotaxis—is activated by this phosphorylation. Activated CheY modulates the bacterial swimming mode. Even when the chemotaxis receptor is not an MCP, as is the case for PTS chemoattractants, the signal from the receptor channels into the regular pathway that involves the same histidine kinase and response regulator. Two-component regulatory systems, consisting of a histidine kinase and a response regulator, are highly abundant in prokaryotes and are also found in eukaryotic microorganisms [15] and in the mustard plant, *Arabidopsis* [6]. Thus, in prokaryotes, "the name of the game" is commonality.

Although the molecular mechanisms of chemotaxis in eukaryotes also involve phospho-transfers, they are different from those in prokaryotes. The eukaryotic receptors are either enzymes such as tyrosine kinase and guanylate cyclase or G-protein-coupled proteins. The signaling pathways are quite diverse, although similarity between some systems is evident (Table 3). In eukaryotic microorganisms such as the amoeba *Dictyostelium*, the prokaryotic and eukaryotic signaling pathways appear to merge. There, two-component systems seem to modulate downstream mitogen-activated protein (MAP) kinase cascades [15]. Thus, it appears that the molecular mechanisms of chemotaxis reflect both evolutionary development and adjustment to the specific needs of the individual.

4. Other Taxes

Limited information is available on other types of active taxis (listed in Chapter 1). Of them, relatively more information is available on thermotaxis and phototaxis. Generally speaking, the behavioral responses to a gradient of temperature or of light are indistinguishable from those to a gradient of a chemical stimulant. As briefly summarized below, these taxes also share a high degree of commonality at the molecular level.

4.1. *Thermotaxis*

The physiology and biological activities of all organisms are tightly dependent on temperature and are restricted by it. Therefore, most organisms have means to respond to temperature changes, and thermotaxis is quite prevalent. This process was found in a variety of species and cells—from bacteria [13, 18], through ciliates [33], amoebae [5, 10, 25] and nematodes [27, 34], to mammalian sperm cells [4], trophoblastic cells [9], and white blood cells [20]. From the systems in which thermotaxis was investigated in greater detail, it appears that it shares with chemotaxis a high degree of commonality, at both the behavioral and molecular levels.

Thus, at the behavioral level, the response of bacteria such as *E. coli* to a temperature change is indistinguishable from their response to a respective chemical stimulant. In both cases the response is a change in the direction of flagellar rotation, resulting in a change of the swimming behavior [13]. Likewise, in mammalian spermatozoa, the thermotactic response is very similar to the chemotactic response, and both responses are restricted to capacitated cells, i.e., cells in a state of readiness for fertilizing the egg [4].

Similarly, at the molecular level, the signal transduction steps that follow thermosensing appear to be shared with those of chemotaxis, at least in systems whose molecular mechanisms are partially known. However, in the absence of specific temperature-sensing organs that can be studied and the variety of nonspecific effects caused by temperature changes, the mechanism of the first step of thermotaxis, thermosensing, is poorly understood [24]. An exception is thermotaxis of *E. coli*, which is relatively well characterized [13]. In *E. coli*, four of the main chemoreceptors, termed MCPs (for methyl-accepting chemotaxis proteins, reviewed in section 6.1 of Chapter 3), are thermosensors. Three of them—Tsr, Tar, and Trg—are "warmth sensors," namely, they mediate attraction to higher temperatures and repulsion from lower temperatures [16, 19, 23]. The fourth thermosensor, Tap, is a "cold sensor": it mediates attraction to lower temperatures and repulsion from higher temperatures [23]. Tar has been of special interest for the investigators because its thermosensing function is inverted following adaptation to chemoattractants [19]. It appears that methylation of a single glutamyl residue of this protein during adaptation is sufficient to convert it from a "warmth sensor" into a "cold sensor" [24]. However, the molecular events that occur in the thermosensors upon temperature changes remain to be revealed. In eukaryotic cells, the mechanisms of thermosensing appear to involve temperature-sensitive ion channels. Thus, in the ciliate *Paramecium*, two types of Ca^{2+}-channels are responsible for thermosensing: heat-activated and cold-activated channels, each with different ion selectivity [12]. Peripheral thermosensors of mammals that measure changes of skin temperature as well as thermosensors of sponges appear to involve two-pore domain K^+ channels [17, 38]. In nematodes, specific sensory neurons and interneurons that mediate thermotaxis have been identified [21, 28, 34].

4.2. Phototaxis

Many organisms use light as their energy source. For them, it is essential to move to regions with optimal light intensity and wavelength and to avoid regions with hazardous light. The process by which this is done is phototaxis. An excellent example is the archaeon *Halobacterium salinarium*. Its photosensory receptors are two rhodopsins—sensory rhodopsins I and II (SRI and SRII, respectively). Like rhodopsins of other organisms—from eukaryotic microorganisms, such as the algae *Chlamydomonas* and other halophilic archaea, to the mammalian eye— these seven transmembrane helix, retinal-containing proteins are spectrally tuned throughout the visible spectrum to communicate information to the cell regarding the intensity and color of light in the environment [31]. SRI senses an attractant light (green-orange light) and a repellent light at near-ultraviolet wavelengths. SRII senses a repellent light at blue wavelengths. Both sensory rhodopsins are parts of the receptor supramolecular complex. They convey the sensed information to chemotaxis receptors, MCPs, termed Htr in *H. salinarium* [1, 2, 11, 30, 32, 35–37] (Chapter 3). The Htrs thus mediate both the chemotactic and phototactic response in this species. *H. salinarium* possesses 13 Htrs that can be divided into three subfamilies on the basis of their structure. Two of the subfamilies include membrane-bound proteins, and the other subfamily contains soluble Htrs [36]. So it seems that in this system, as well, the molecular mechanism of chemotaxis is shared with phototaxis: first, specific photoreceptors sense changes in light intensity and wavelength, and then they transduce the sensed information to the proper MCPs and, thereby, initiate the chemotactic signaling cascade. The recently determined three-dimensional structure of the transmembrane helices of HtrII bound to SRII provided evidence for a common mechanism of signal transduction in phototaxis and chemotaxis of the archaeon *Natronobacterium pharaonis* [8]. In other light-responsive organisms (e.g., cyanobacteria, algae, and the amoeba *Dictyostelium discoideum*), the mechanism of signal transduction during phototaxis has not yet been elucidated [7, 14, 22, 26, 29].

To conclude, from the little that is known, it appears that commonality is the rule when chemotaxis, thermotaxis, and phototaxis are compared. The sensing mechanism is, of course, specific for each stimulus type, employing chemoreceptors, photoreceptors, and thermoreceptors, respectively. However, the subsequent signal-transduction processes, culminating at a behavioral response, appear to be common for all these taxes. Both the signal transduction cascade and the behavioral response appear to be independent of whether the stimulus is a change in a chemical concentration, a temperature change, or a change in the light intensity at a specific wavelength range.

References

1. Alam, M., Lebert, M., Oesterhelt, D. and Hazelbauer, G.L. (1989). Methyl-accepting taxis proteins in *Halobacterium halobium*. *EMBO J.* **8**, 631–639.
2. Alam, M. and Hazelbauer, G.L. (1991). Structural features of methyl-accepting taxis proteins conserved between archaebacteria and eubacteria revealed by antigenic cross-reaction. *J. Bacteriol.* **173**, 5837–5842.
3. Appleby, J.L., Parkinson, J.S. and Bourret, R.B. (1996). Signal transduction via the multi-step phosphorelay: Not necessarily a road less traveled. *Cell* **86**, 845–848.
4. Bahat, A., Tur-Kaspa, I., Gakamsky, A., Giojalas, L.C., Breitbart, H. and Eisenbach, M. (2003). Thermotaxis of mammalian sperm cells: A potential navigation mechanism in the female genital tract. *Nature Med.* **9**, 149–150.
5. Bonner, J.T., Clarke, W.W.J., Neely, C.L.J. and Slifkin, M.K. (1950). The orientation to light and the extremely sensitive orientation to temperature gradients in the slime mold *Dictyostelium discoideum*. *J. Cell. Comp. Physiol.* **36**, 149–158.
6. Chang, C., Kwok, S.F., Bleecker, A.B. and Meyerowitz, E.M. (1993). Arabidopsis ethylene-response gene *ETR1* : similarity of product to two-component regulators. *Science* **262**, 539–544.
7. Darcy, P.K., Wilczynska, Z. and Fisher, P.R. (1994). The role of cGMP in photosensory and thermosensory transduction in *Dictyostelium discoideum*. *Microbiology* **140**, 1619–1632.
8. Gordeliy, V.I., Labahn, J., Moukhametzianov, R., Efremov, R., Granzin, J., Schlesinger, R., Büldt, G., Savopol, T., Scheidig, A.J., Klare, J.P. and Engelhard, M. (2002). Molecular basis of transmembrane signaling by sensory rhodopsin II-transducer complex. *Nature* **419**, 484–487.

9. Higazi, A.A., Kniss, D., Manuppello, J., Barnathan, E.S. and Cines, D.B. (1996). Thermotaxis of human trophoblastic cells. *Placenta* **17**, 683–687.

10. Hong, C.B., Fontana, D.R. and Poff, K.L. (1983). Thermotaxis of *Dictyostelium discoideum* amoebae and its possible role in pseudoplasmodial thermotaxis. *Proc. Natl. Acad. Sci. U.S.A.* **80**, 5646–5649.

11. Hou, S.B., Brooun, A., Yu, H.S., Freitas, T. and Alam, M. (1998). Sensory rhodopsin II transducer HtrII is also responsible for serine chemotaxis in the archaeon *Halobacterium salinarum*. *J. Bacteriol.* **180**, 1600–1602.

12. Imada, C. and Oosawa, Y. (1999). Thermoreception of *Paramecium*: different Ca^{2+} channels were activated by heating and cooling. *J. Membr. Biol.* **168**, 283–287.

13. Imae, Y. (1985). Molecular mechanism of thermosensing in bacteria, in *Sensing and Response in Microorganisms* (M. Eisenbach and M. Balaban, eds.), pp. 73–81. Elsevier, Amsterdam.

14. Kreimer, G. (1994). Cell biology of phototaxis in flagellate algae. *Int. Rev. Cytol.* **148**, 229–310.

15. Loomis, W.F., Kuspa, A. and Shaulsky, G. (1998). Two-component signal transduction systems in eukaryotic microorganisms. *Curr. Opin. Microbiol.* **1**, 643–648.

16. Maeda, K. and Imae, Y. (1979). Thermosensory transduction in *Escherichia coli*: inhibition of the thermoresponse by L-serine. *Proc. Natl. Acad. Sci. U.S.A.* **76**, 91–95.

17. Maingret, F., Lauritzen, I., Patel, A.J., Heurteaux, C., Reyes, R., Lesage, F., Lazdunski, M. and Honorè, E. (2000). TREK-1 is a heat-activated background K^+ channel. *Embo J.* **19**, 2483–2491.

18. Metzner, P. (1919). Die Bewegung und Reizbeantwortung der bipolar gegeisselten Spirillen. *Jahrb. Wiss. Botan.* **59**, 325–412.

19. Mizuno, T. and Imae, Y. (1984). Conditional inversion of the thermoresponse in *Escherichia coli*. *J. Bacteriol.* **159**, 360–367.

20. Mizuno, T., Kawasaki, K. and Miyamoto, H. (1992). Construction of a thermotaxis chamber providing spatial or temporal thermal gradients monitored by an infrared video camera system. *Analyt. Biochem.* **207**, 208–213.

21. Mori, I. and Ohshima, Y. (1995). Neural regulation of thermotaxis in *Caenorhabditis elegans*. *Nature* **376**, 344–348.

22. Mullineaux, C.W. (2001). How do cyanobacteria sense and respond to light? *Mol. Microbiol.* **41**, 965–971.

23. Nara, T., Lee, L. and Imae, Y. (1991). Thermosensing ability of Trg and Tap chemoreceptors in *Escherichia coli*. *J. Bacteriol.* **173**, 1120–1124.

24. Nishiyama, S., Umemura, T., Nara, T., Homma, M. and Kawagishi, I. (1999). Conversion of a bacterial warm sensor to a cold sensor by methylation of a single residue in the presence of an attractant. *Mol. Microbiol.* **32**, 357–365.

25. Raper, K.B. (1940). Pseudoplasmodium formation and organization in *Dictyostelium discoideum*. *J. Elisha Mitchell Sci. Soc.* **56**, 241–82.

26. Ridge, K.D. (2002). Algal rhodopsins: Phototaxis receptors found at last. *Curr. Biol.* **12**, R588–R590.

27. Ryu, W.S. and Samuel, A.D.T. (2002). Thermotaxis in *Caenorhabditis elegans* analyzed by measuring responses to defined thermal stimuli. *J. Neurosci.* **22**, 5727–5733.

28. Samuel, A.D.T., Silva, R.A. and Murthy, V.N. (2003). Synaptic activity of the AFD neuron in *Caenorhabditis elegans* correlates with thermotactic memory. *J. Neurosci.* **23**, 373–376.

29. Schlenkrich, T., Fleischmann, P. and Häder, D.-P. (1995). Biochemical and spectroscopic characterization of the putative photoreceptor for phototaxis in amoebae of the cellular slime mould *Dictyostelium discoideum. J. Photochem. Photobiol.* **B 30**, 139–143.

30. Spudich, E.N., Takahashi, T. and Spudich, J.L. (1989). Sensory rhodopsins I and II modulate a methylation/demethylation system in *Halobacterium halobium* phototaxis. *Proc. Natl. Acad. Sci. U.S.A.* **86**, 7746–7750.

31. Spudich, J.L. and Luecke, H. (2002). Sensory rhodopsin II: functional insights from structure. *Curr. Opin. Struct. Biol.* **12**, 540–546.

32. Sundberg, S.A., Alam, M., Lebert, M., Spudich, J.L., Oesterhelt, D. and Hazelbauer, G.L. (1990). Characterization of *Halobacterium halobium* mutants defective in taxis. *J. Bacteriol.* **172**, 2328–2335.

33. Tawada, K. and Miyamoto, H. (1973). Sensitivity of *Paramecium* thermotaxis to temperature change. *J. Protozool.* **20**, 209–292.

34. Thomas, J.H. (1995). Thermosensation. Some like it hot. *Curr. Biol.* **5**, 1222–1224.

35. Wegener, A.-A., Klare, J.P., Engelhard, M. and Steinhoff, H.-J. (2001). Structural insights into the early steps of receptor-transducer signal transfer in archaeal phototaxis. *EMBO J.* **20**, 5312–5319.

36. Zhang, W.S., Brooun, A., McCandless, J., Banda, P. and Alam, M. (1996). Signal transduction in the Archaeon *Halobacterium salinarium* is processed through three subfamilies of 13 soluble and membrane-bound transducer proteins. *Proc. Natl. Acad. Sci. U.S.A.* **93**, 4649–4654.

37. Zhang, X.N., Zhu, J.Y. and Spudich, J.L. (1999). The specificity of interaction of archaeal transducers with their cognate sensory rhodopsins is determined by their transmembrane helices. *Proc. Natl. Acad. Sci. U.S.A.* **96**, 857–862.

38. Zocchi, E., Carpaneto, A., Cerrano, C., Bavestrello, G., Giovine, M., Bruzzone, S., Guida, L., Franco, L. and Usai, C. (2001). The temperature-signaling cascade in sponges involves a heat-gated cation channel, abscisic acid, and cyclic ADP-ribose. *Proc. Natl. Acad. Sci. U.S.A.* **98**, 14859–14864.

Index

MotY 74–75
MS ring 64, 68, 73, 75, 79, 81–82
MTLn3 254–255, 260, 262
multicellular behavior 216–217, 223–224, 244
mutants of *P. dendritiformis* defective in
    pattern formation 227
mutants
  Che- 66, 70
  Fla- 66
  Mot- 66, 70, 73
mycoplasma 60
myosin 13, 43, 253, 256–258, 284–287, 293,
    331, 363–367, 369, 371, 478, 481
myosin heavy chain 261, 285
myosin heavy chain kinase 261, 265, 285, 293
myosin light chain 261, 269, 291, 363–365
myosin light chain kinase 266, 269, 285, 337,
    363–364
myxobacteria 59, 217, 222, 226, 230–232,
    234–237, 241–242, 478–479
Myxococcus xanthus 3, 59, 107, 216, 233, 239
myxospores 13, 230, 242–243

Na^+-motive force 73
Na^+ channel 73–75, 478
N-acyl homoserine lactones 216, 222
NADPH oxidase 310, 362, 373, 375
nebula 226
nerve growth factor 458
Netrin 467–469, 471
neural tube 459–460, 466
neurons 14, 35, 44, 46, 456, 458, 461,
    466–469, 471, 484
neuropilin 470
neurotoxins 168
neutrophils 253–254, 256, 279, 292, 308–318
nexin 410–411
N-formylated peptides 432, 435
NGF 458, 463, 471
nocodazole 263
nozzle 59, 232–233

odorant 103, 126, 435, 443
odorant receptor 435, 444, 481
olfactory axons 462
oocyte 428–429, 440
oogamy 30
optical trapping 414
optical tweezers 99, 412
orthokinesis 2, 15
oscillatory motor 337
osmotic forces 336

ovulation 429, 431, 434, 436–440, 442
oxygen as a chemotactic stimulus 97

P ring 64, 67, 82
P. aeruginosa, see Pseudomonas aeruginosa
P. alvei 227
P. dendritiformis, see Paenibacillus
    dendritiformis, see Paemibacillus
    dendritiformis
P. glucanolyticus 225
P. lautus 225
P. mirabilis 222–223, 230
P. thiaminolyticus 224, 228
P. vortex, see Paenibacillus vortex
p120-GAP 356, 367–368
p21-activated kinase 266, 287, 331, 362
p21 activated protein kinases 284
p21 regulated kinase 284
p38 MAPK 369
Paenibacillus dendritiformis 224
Paenibacillus, see pattern formation in the
    Paenibacilli
PAK 284, 363–364, 366–367
PAS domain 19, 22, 156
path velocity 413
pattern formation 216–217, 219–221,
    223–224, 226–228, 230
pattern formation by *E. coli* and
    Salmonella 217
pattern formation by *P. vortex* 226
pattern formation by *Salmonella* in complex
    medium 220
pattern formation in minimal medium *see*
    pattern formation by *E. coli* and
    Salmonella
pattern formation in *Paenibacilli* 224
patterning 219–220, 224–230
patterns 84–217–221, 224–225, 415, 433, 462
paxillin 269
PE 236–237, 240–242
PE-diC16:15c 236
peptides 16, 33–35, 44, 141, 216, 223, 242,
    309–310, 425, 427, 431–432, 435
periplasmic binding proteins 17, 21, 112,
    124–125
peritrichous bacteria 99
peritrichous flagella 221–222
permease 124
persistence time 317, 320, 324, 326–327
pertussis toxin 348, 352, 354, 367, 376
phallotoxin 318, 321
PhdA 284

random motility coefficient 317, 319, 324–325
random walk 14, 84, 86, 326–327, 329
random walk-like motion 225
range 16, 44, 59, 90, 125, 127, 148, 153–154,
 161, 168, 225, 253–254, 273, 285, 290,
 311, 322, 328, 436, 440, 480, 485
RANTES 431–432, 434, 436
Ras 147, 282–283, 288, 291–292, 351–352,
 356, 367–370
Ras binding domain 281–282
Ras p21 141
RasC 282
RasG 282
RasGTP 280, 282, 368
receptor desensitization 339–340, 347
receptor internalization 345, 348
receptor phosphorylation 343–344, 348,
 351, 369
receptor recycling 341, 346
receptor supramolecular complex 119–121,
 129, 145, 150–152, 154–155, 158–159,
 161, 482, 484
receptor tyrosine kinases 256, 259–270, 274,
 279, 287–291, 293
receptor upregulation 344
receptors
 aggregation 153, 279
 chemotaxis-specific 109, 113, 115, 124, 135
 clustering 119, 153
 dual-function 109, 123
 for chemorepellents 125–126
 periplasmic 124
 soluble 122
regnum 25–26
repellent 89, 102–105, 125–126, 157, 159,
 462, 467–470, 484
response regulator 22, 79, 111, 127, 130, 134,
 238, 240–241, 481–482
reversal frequency 237–238, 243
RflH 83
Rhizobium lupini 107
Rho 264, 289–291, 355, 356, 358, 364–367,
 372–373, 481
Rho GDIs 356–357
Rho kinase 266, 364–366
RhoA 266, 269, 366, 374
Rho-associated coiled-coil forming kinase
 (ROCK) 365–366
Rho-associated kinase 269
Rhodobacter sphaeroides 2, 55–56, 108, 480
Rhodospirillum rubrum 168
Rickettsia 61
ring-forming assay on semisolid agar 91, 95

ripples 234
rippling 242–243
robust 162, 378
role of the Tar receptor *see* pattern formation
 by *E. coli* and *Salmonella*
rolling 310, 313–314, 359
roof plate 465–467
rotor 73, 76, 78
ruffles 259, 329–330, 363, 365
ryanodine-sensitive calcium 377

S ring 64
S. liquefaciens, see Serratia liquefaciens
S. marcescens 223, 230
S-adenosylmethionine 117
Salmonella 54–57, 217, 219–223, 479
Sarcina ureae 108
second-site suppression analysis 135, 143, 148
semaphorin 469–470
sensory rhodopsin 122, 167
sensory transduction 43, 244
sensory transduction systems 217, 237
Serratia marcescens 56
severing proteins 264, 289
shc 288, 352–353
Shigella 61, 478
sialomucins 313–314
signal amplification 153–154
signal transduction
 in large bacterial species 147, 168
 multiplicity of pathways 155, 164, 168
 nonconventional pathway 163–164
 variations 155, 164
signaling pathways 254, 270, 273, 287–288,
 290, 293, 341, 347, 352, 356, 366,
 371–372, 378, 482
single cell tracking 317, 320, 325–326
Sinorhizobium meliloti 56, 480
skin chamber 317
Slit 470–471
small GTP-binding proteins 349, 351,
 356–358, 366
S-motility 232–233, 237–239
social (S) motility 59–60, 123, 233, 239
soma 456
son of sevenless 288–289
sos 288–289, 356, 367
spatial comparisons 270
spatial gradient 15, 88–90, 220, 259, 270–272,
 274, 276, 292, 325
spatial sensing 270, 272–273, 291–292
sperm chemoattractants 425–427, 431,
 435, 443